BIRKHÄUSER

Modern Birkhäuser Classics

Many of the original research and survey monographs in pure and applied mathematics published by Birkhäuser in recent decades have been groundbreaking and have come to be regarded as foundational to the subject. Through the MBC Series, a select number of these modern classics, entirely uncorrected, are being re-released in paperback (and as eBooks) to ensure that these treasures remain accessible to new generations of students, scholars, and researchers.

Hans Triebel

Theory of Function Spaces II

Reprint of the 1992 Edition

Birkhäuser

Prof. Dr. Hans Triebel
Mathematisches Institut
Friedrich-Schiller-Universität Jena
07743 Jena
Germany
hans.triebel@uni-jena.de

ISBN 978-3-0346-0418-5 e-ISBN 978-3-0346-0419-2
DOI 10.1007/978-3-0346-0419-2

Library of Congress Control Number: 2010924183

2000 Mathematics Subject Classification 46E35, 46-02, 46E15

Originally printed under the same title as volume 84 in the Monographs in Mathematics series by Birkhäuser Verlag, Switzerland, ISBN 978-3-7643-2639-5
Reprinted 2010 by Springer Basel AG

Printed on acid-free paper

Springer Basel AG is part of Springer Science+Business Media

www.birkhauser.ch

Preface

This book deals with the theory of function spaces of type B^s_{pq} and F^s_{pq} as it stands at the end of the eighties. These two scales of spaces cover many well-known spaces of functions and distributions such as Hölder-Zygmund spaces, Sobolev spaces, fractional Sobolev spaces (previously also often referred to as Bessel-potential spaces), Besov spaces, inhomogeneous Hardy spaces, spaces of BMO-type and local approximation spaces which are closely connected with Morrey–Campanato spaces.

The book is essentially self-contained, although it seems to be reasonable to call it part II of the author's book [Triß] with the same title, which dealt mainly with the same two scales of spaces B^s_{pq} and F^s_{pq} and which reflected the situation around 1980. But since that time many new discoveries changed the picture completely (this applies both to simpler proofs of known basic assertions and to new results and new applications). This may justify starting again from scratch and not simply continuing where we stopped in [Triß]. However those topics where we have nothing new to say will not again be treated in detail; we shall refer to [Triß] or other relevant sources. But in order to make the book self-contained we give in any case a detailed description of all needed results.

This book is the author's third monograph about spaces of this type (beside the two research reports [Tri4,5] and the joint monograph [ScT] with H.-J. Schmeisser which is related to the subject but not in a direct line). The two other directly related books are the above-mentioned [Triß] and [Triα]. The latter deals with some of these spaces from the standpoint of interpolation theory, whereas [Triß] is mainly based on Fourier-analytical techniques combined with maximal inequalities and applications to PDE's (partial differential equations). The present book is characterized by local means and local methods with applications to ψDE's (pseudodifferential equations). Although the book is mainly based on the author's results obtained in the last few years we try to give a sufficiently comprehensive picture of these spaces and of some of their applications, at least of those topics which are treated here in detail.

The book has seven chapters. Chapter 1 is a self-contained historically-orientated survey of those function spaces and their roots which are treated in the book, including the related devices for measuring smoothness (derivatives, differences of functions, boundary values of harmonic and thermic functions, local approximations, sharp maximal functions, interpolation methods, Fourier-analytical representations, atomic decompositions, etc.). This chapter surveys the rather different techniques developed in the last 50 years (some of them are quite recent) without proofs but with many references. The aim of this introductory chapter is twofold. First, it serves as an independent survey readable (so we hope) also for non-specialists who are not so much interested in technical details but who wish to learn recent trends in the theory of function spaces including some of their historical roots. Secondly, it prepares from a historical point of view what follows and it emphasizes the main goal of the book, that is, to show how at the end of the eighties all these apparently different devices come together, and to demonstrate that they are only different ways of characterizing the same function spaces.

In Chapters 2 and 3 we start again (now in earnest): We define the spaces B^s_{pq} and F^s_{pq} and prove equivalence assertions which cover the above-mentioned more or less classical spaces, including their related, seemingly different, devices for measuring smoothness. We develop the technical instruments which are the basis for the rest of the book. This includes atomic representations and local approximations. Morrey–Campanato spaces (including BMO-spaces) will be treated as a by-product of the developed techniques.

Based on the two preceding chapters we give in Chapter 4 new (simpler) proofs of some (more or less known) crucial theorems for the spaces B^s_{pq} and F^s_{pq}: invariance under diffeomorphic maps of $\mathbb{R}^n$ onto itself, pointwise multipliers, traces on hyperplanes, extensions of these spaces from $\mathbb{R}^n_+$ to $\mathbb{R}^n$.

The spaces $B^s_{pq}(\Omega)$, $F^s_{pq}(\Omega)$ and corresponding Morrey–Campanato spaces on bounded C^∞ domains Ω in $\mathbb{R}^n$ are treated in Chapter 5. We are mainly interested in intrinsic characterizations.

Mapping properties of pseudodifferential and Fourier integral operators of type $S^\mu_{\rho,\delta}$ will be studied in Chapter 6. Of special interest is the exotic class characterized by $\rho = \delta = 1$.

Chapter 7 deals with spaces of type F^s_{pq} and B^s_{pq} (including all the mentioned special cases) on Riemannian manifolds (bounded geometry, positive injectivity radius) and on Lie groups.

The reader is expected to have a working knowledge of functional analysis as presented in the classical textbooks (including the standard facts of the theory of distributions). A familiarity with the basic results of the spaces of differential functions, L_p-spaces and Sobolev spaces would be helpful.

The book is organized by the decimal system. "n.k.m" refers to subsection n.k.m, "Theorem n.k.m/l" means the theorem l in n.k.m etc. All unimportant positive numbers will be denoted by c (with additional indices if there are several c's in the same formula).

Jena, Summer 1990 Hans Triebel

Contents

Chapter 1

How To Measure Smoothness

1.1 Introduction

This book deals with spaces of functions and distributions of Hölder, Sobolev, Hardy, Campanato, Besov type and their descendants. So it is quite clear that we restrict our attention also in this introductory chapter to spaces of these types. There is hardly any doubt that these types of spaces were and are the backbone (in the widest sense of the word) of the theory of function spaces from the very beginning up to our time. On the other hand, in the last few decades many other types of function spaces have been treated extensively, some of them closely connected with the spaces considered here (anisotropic and weighted spaces, spaces with dominating mixed derivatives, spaces on non-smooth domains and on general structures), while others have been based on different principles (Lorentz and Orlicz spaces, etc.). The following books cover a large variety of different aspects of the theory of function spaces in the widest sense: [Sob4, Nik2, BIN, KJF, Pee6, Ste1, StW2, Ada, Maz, MaS, JoW, Triα, Triß, ScT]. Finally, we refer to the recent surveys [BKLN, KuN1] (1988) concentrating mainly, but not exclusively, on developments in the Soviet Union.

The aim of this chapter is at least twofold. First we wish to sketch some historical roots and how these spaces emerged. But we do not present a carefully balanced history of function spaces not even in the above-mentioned restricted sense. Secondly (and this is more important for us) we survey the different devices for measuring smoothness developed in the last few decades; we wish to emphasize that all these techniques, if put in the right frame, end up with the same classes of function spaces B^s_{pq} and F^s_{pq}. In other words, if smoothness is expressed via these two scales of function spaces, then one has an armada of different devices at hand. This gives these spaces a high degree of flexibility which was unexpected even a few years ago (for example, for the author at the time when [Triß] was written).

Our approach to function spaces is characterized by possible applications, for instance, to boundary-value problems for linear and nonlinear PDE's, to mapping properties connected with ψDE's, etc. In this book we shall not stress this point, but it explains why we strongly prefer spaces which are called, in technical terms, inhomogeneous (see 1.4.5). But Chapter 6 may be considered as an application of

H. Triebel, *Theory of Function Spaces II*, Modern Birkhäuser Classics,
DOI 10.1007/978-3-0346-0419-2_1, © Birkhäuser Verlag 1992

the core of the book (these are the Chapters 2–5 and 7): it deals with ψDE's. We shall not discuss the interesting applications to nonlinear problems and refer to the recent survey [Run2].

1.2 Derivatives and differences of functions

1.2.1 Basic spaces

We use standard notations: $\mathbb{R}^n$ denotes the real euclidean n-space with $\mathbb{R} = \mathbb{R}^1$. Furthermore $\mathbb{N}$ denotes the collection of all natural numbers and $\mathbb{N}_0 = \mathbb{N} \cup \{0\}$.

Let Ω be a domain in $\mathbb{R}^n$. Then, by definition, $C(\Omega)$ is the collection of all complex-valued bounded and uniformly continuous functions in Ω, equipped with the norm

$$\|f \mid C(\Omega)\| = \sup_{x\in\Omega} |f(x)|. \tag{1}$$

Obviously, $C(\Omega)$ is a Banach space. Let $k \in \mathbb{N}$, then

$$C^k(\Omega) = \{f \in C(\Omega)\colon D^\alpha f \in C(\Omega) \ \text{ if } \ |\alpha| \leqq k\} \tag{2}$$

are Banach spaces equipped with the norm

$$\|f \mid C^k(\Omega)\| = \sum_{|\alpha|\leqq k} \|D^\alpha f \mid C(\Omega)\|. \tag{3}$$

We use standard notations: $\alpha = (\alpha_1, \ldots, \alpha_n)$ with $\alpha_j \in \mathbb{N}_0$ is a multiindex, $|\alpha| = \sum_{j=1}^n \alpha_j$ and

$$D^\alpha f(x) = \frac{\partial^{|\alpha|} f}{\partial x_1^{\alpha_1} \cdots \partial x_n^{\alpha_n}}(x) \tag{4}$$

For convenience we write occasionally $C(\Omega) = C^0(\Omega)$.

Furthermore dx stands for the Lebesgue measure in $\mathbb{R}^n$ and

$$\|f \mid L_p(\Omega)\| = \left(\int_\Omega |f(x)|^p dx\right)^{1/p}, \qquad 0 < p < \infty, \tag{5}$$

with the usual modification

$$\|f \mid L_\infty(\Omega)\| = \operatorname*{ess\,sup}_{x\in\Omega} |f(x)| \tag{6}$$

have the usual meaning. By definition, $L_p(\Omega)$ with $0 < p \leqq \infty$, is the collection of all complex-valued Lebesgue-measurable functions in Ω such that $\|f \mid L_p(\Omega)\|$ is finite. If $1 \leqq p \leqq \infty$ then $L_p(\Omega)$ are Banach spaces, otherwise, i.e., if $0 < p < 1$, they are only quasi-Banach spaces.

We recall what is meant by a quasi-Banach space. Let A be a (real or complex) linear space, then $\|a \mid A\|$ is said to be a quasi-norm if $\|a \mid A\|$ satisfies the usual conditions of a norm with exception of the triangle inequality, which is replaced by

$$\|a_1 + a_2 \mid A\| \leqq c(\|a_1 \mid A\| + \|a_2 \mid A\|). \tag{7}$$

More precisely: There exists a constant c such that (7) holds for all $a_1 \in A$ and all $a_2 \in A$. Of course we have $c \geqq 1$ (if $c = 1$ is admissible then $\|a \mid A\|$ is a norm). A quasi-normed space is said to be a quasi-Banach space if it is complete with respect to the quasi-norm in question.

We are looking at function spaces from the point of view of possible applications to PDE's and ψDE's. Then $L_p(\Omega)$ with $1 < p < \infty$ are good spaces; whereas $L_1(\Omega)$, $L_\infty(\Omega)$ and $C^k(\Omega)$ with $k \in \mathbb{N}_0$ have well-known shortcomings; and the spaces $L_p(\Omega)$ with $0 < p < 1$ are really nasty. We shall be interested to find substitutes for these bad spaces (from our point of view) which on the one hand preserve some characteristic features of these spaces but which on the other hand are well adapted to PDE's and ψDE's. This task leads us to (inhomogeneous) Hardy spaces and Zygmund classes.

1.2.2 Hölder–Zygmund spaces

For sake of simplicity we restrict ourselves in the sequel to function spaces defined on $\mathbb{R}^n$ and omit "$\mathbb{R}^n$". In other words, C^k with $k \in \mathbb{N}_0$, and L_p with $0 < p \leqq \infty$ stand for $C^k(\mathbb{R}^n)$ and $L_p(\mathbb{R}^n)$, respectively. The first step beyond the spaces C^k with $k \in \mathbb{N}_0$ are the Hölder spaces C^s with $0 < s \neq$ integer (filling the gaps between the spaces C^k with $k \in \mathbb{N}_0$). Let $0 < \sigma < 1$, then we introduce the norm

$$\|f \mid C^\sigma\| = \|f \mid C\| + \sup \frac{|f(x) - f(y)|}{|x - y|^\sigma}, \tag{1}$$

where the supremum is taken over all $x \in \mathbb{R}^n$ and $y \in \mathbb{R}^n$ with $x \neq y$. Let $s \in \mathbb{R}$, then we put

$$s = [s] + \{s\} = [s]^- + \{s\}^+, \tag{2}$$

where $[s]$ and $[s]^-$ are integers, whereas $0 \leqq \{s\} < 1$ and $0 < \{s\}^+ \leqq 1$.

Definition 1 (Hölder spaces). Let $0 < s \neq$ integer, then

$$C^s = \Big\{ f \in C \colon \|f \mid C^s\| = \|f \mid C^{[s]}\| + \sum_{|\alpha|=[s]} \|D^\alpha f \mid C^{\{s\}}\| < \infty \Big\}.$$

Remark 1. We use (1.2.1/3,4) and (1). Hölder spaces have been employed by J. Schauder and other mathematicians since the mid-thirties in connection with boundary value problems for second order elliptic differential equations. The final step in this direction is due to C. Miranda, see his book [Mir] where one finds also many references. S. Agmon, A. Douglis and L. Nirenberg extended these considerations to boundary value problems for higher order elliptic differential equations, see [ADN].

Next we introduce the differences of functions

$$\Delta_h^k f(x) = \sum_{j=0}^{k} (-1)^{k-j} \binom{k}{j} f(x+hj), \quad k \in \mathbb{N},\ h \in \mathbb{R}^n,\ x \in \mathbb{R}^n, \tag{3}$$

where $\binom{m}{j}$ are binomial coefficients. Recall that Δ_h^k can also be defined iteratively via

$$\Delta_h^1 f(x) = f(x+h) - f(x) \quad \text{and} \quad \Delta_h^{k+1} = \Delta_h^1 \Delta_h^k, \quad k \in \mathbb{N}. \tag{4}$$

Definition 2 (Zygmund spaces). Let $s > 0$, then

$$\begin{aligned} \mathcal{C}^s = \Big\{ f \in C \colon \|f \mid \mathcal{C}^s\| = \|f \mid C^{[s]^-}\| \\ + \sum_{|\alpha|=[s]^-} \sup_{0 \neq h \in \mathbb{R}^n} |h|^{-\{s\}^+} \|\Delta_h^2 D^\alpha f \mid C\| < \infty \Big\}. \end{aligned} \tag{5}$$

Remark 2. Usually $\mathcal{C}^1$, and more generally $\mathcal{C}^k$ with $k \in \mathbb{N}$, are denoted as Zygmund classes. By the theorem below C^s and $\mathcal{C}^s$ coincide if s is a fractional number. In this sense we speak about Hölder–Zygmund spaces. They are Banach spaces.

Theorem. (i) Let $s > 0$; $k \in \mathbb{N}_0$ with $k < s$; and $m \in \mathbb{N}$ with $m > s - k$, then

$$\mathcal{C}^s = \Big\{ f \in C \colon \|f \mid \mathcal{C}^s\|_{k,m} = \|f \mid C^k\| \qquad\qquad (6)$$

$$+ \sum_{|\alpha|=k} \sup_{0 \neq h \in \mathbb{R}^n} |h|^{-(s-k)} \|\Delta_h^m D^\alpha f \mid C\| < \infty \Big\}$$

where all admissible norms $\|f \mid \mathcal{C}^s\|_{k,m}$ are mutually equivalent.

(ii) If $0 < s \neq$ integer then

$$C^s = \mathcal{C}^s. \qquad (7)$$

If $s = k \in \mathbb{N}$ then

$$C^k \subset \mathcal{C}^k \quad \text{and} \quad C^k \neq \mathcal{C}^k. \qquad (8)$$

Remark 3. It was A. Zygmund [Zyg1] who discovered in 1945 the remarkable fact that sometimes second differences $\Delta_h^2 f$ of functions f are much more useful than first differences $\Delta_h^1 f$. By the above theorem $\mathcal{C}^s$ extends C^s from fractional values of s to all $s > 0$, where $\mathcal{C}^k$ with $k \in \mathbb{N}$ looks more complicated than C^k but it is the better space seen from the standpoint of PDE's. Furthermore by (6) derivatives are not necessary in order to define $\mathcal{C}^s$ (we may assume $k = 0$ in any case) but differences are indispensable.

1.2.3 Sobolev spaces

Perhaps the Sobolev spaces are the best known function spaces beyond the basic spaces from 1.2.1 and the Hölder spaces. They were introduced by S.L. Sobolev in the mid-thirties, see [Sob1–3]. Since the publication of [Sob4] in 1950 they are widely accepted as one of the crucial instruments in functional analysis; in particular, in connection with PDE's. There are at least two books with the title "Sobolev spaces," [Ada, Maz], and almost all books mentioned in 1.1 deal with Sobolev spaces more or less extensively. Embeddings of the Sobolev spaces W_p^k in L_q-spaces and in Hölder spaces play a major role. It should be mentioned that in the case of $n = 1$ (spaces on $\mathbb{R}$) this part of the theory had been anticipated by G.H. Hardy and J.E. Littlewood, see [HaL2] and [HLP: 383 and 383[a]]. A more detailed discussion of this historical point may be found in [Tri21].

Now the spaces $L_p = L_p(\mathbb{R}^n)$ with $1 \leqq p \leqq \infty$ are interpreted as subspaces of $S' = S'(\mathbb{R}^n)$, where the latter stands for the collection of all complex-valued tempered distributions on $\mathbb{R}^n$. Derivatives must be understood in the sense of distributions.

Definition. Let $1 < p < \infty$ and $k \in \mathbb{N}_0$ then

$$W_p^k = \{f \in S' : D^\alpha f \in L_p \ \text{ if } \ |\alpha| \leqq k\}. \tag{1}$$

Remark. For sake of simplicity we write again W_p^k instead of $W_p^k(\mathbb{R}^n)$. It is well-known that W_p^k equipped with the norm

$$\|f \mid W_p^k\| = \left(\sum_{|\alpha| \leqq k} \|D^\alpha f \mid L_p\|^p \right)^{1/p} \tag{2}$$

is a Banach space (Hilbert space if $p = 2$). Furthermore, we have $W_p^0 = L_p$. Of course, (1) makes sense also if $p = 1$ or $p = \infty$, but the corresponding spaces are not useful for applications to PDE's, in sharp contrast to the spaces with $1 < p < \infty$.

1.2.4 How to fill the gaps

Around 1950 the situation was the following: On the one hand there was the continuous scale $\mathcal{C}^s$, with $s > 0$, of the Hölder–Zygmund spaces; on the other hand, the discrete scales of the Sobolev spaces W_p^k, with $1 < p < \infty$ fixed and $k \in \mathbb{N}_0$, started to make their way. It was quite natural (afterwards one is always very wise) to ask for spaces W_p^s with s fractional, filling the gaps between L_p, W_p^1, $W_p^2, \ldots$. Several proposals were made in the fifties and these attempts culminated around 1960. The spaces introduced by S.M. Nikol'skij in 1951, see [Nik1], can be obtained from (1.2.2/6) if one replaces there the C-norm by an L_p-norm: Let $0 < s \neq$ integer and $1 < p < \infty$, then $B^s_{p\infty}$ is the collection of all $f \in L_p$ such that

$$\|f \mid B^s_{p\infty}\| = \|f \mid W_p^{[s]}\| + \sum_{|\alpha| = [s]} \sup_{0 \neq h \in \mathbb{R}^n} |h|^{-\{s\}} \|\Delta_h D^\alpha f \mid L_p\| \tag{1}$$

is finite (we used more recent notations and $\Delta_h = \Delta_h^1$). It is quite clear that (1) combines with ideas of Hölder spaces, see (1.2.2/1), and Sobolev spaces, see (1.2.3/2). N. Aronszajn [Aro], L.N. Slobodeckij [Slo] and E. Gagliardo [Gag] suggested in 1955–1958 to replace the sup-norm in (1) with respect to h by an L_p-norm: Let $0 < s \neq$ integer and $1 < p < \infty$ then B^s_{pp} is the collection of all $f \in L_p$ such that

$$\|f \mid B^s_{pp}\| = \|f \mid W_p^{[s]}\| + \sum_{|\alpha| = [s]} \left(\int_{\mathbb{R}^n} |h|^{-\{s\}p} \|\Delta_h D^\alpha f \mid L_p\|^p \frac{dh}{|h|^n} \right)^{1/p} \tag{2}$$

is finite (again we used more recent notations and $\Delta_h = \Delta_h^1$). Let, for example, $0 < s < 1$, then we have

$$\|f \mid B^s_{pp}\| = \|f \mid L_p\| + \left(\int_{\mathbb{R}^n \times \mathbb{R}^n} \frac{|f(x) - f(y)|^p}{|x-y|^{sp+n}} dxdy \right)^{1/p} \tag{3}$$

which makes clear that we replaced the sup-norm in the Hölder spaces, see (1.2.2/1), by an L_p-norm.

At the same time it was noted that at least for $p = 2$ both the Sobolev spaces W^k_2 with $k \in \mathbb{N}_0$ and the Slobodeckij spaces B^s_{22} with $0 < s \neq$ integer can be described in Fourier analytical terms. Recall that S' stands for the collection of all complex-valued tempered distributions on $\mathbb{R}^n$. Then the Fourier transform $\hat{f}$ of $f \in S'$ is given by

$$\hat{f}(\xi) = (2\pi)^{-n/2} \int_{\mathbb{R}^n} e^{-ix\xi} f(x) dx, \quad \xi \in \mathbb{R}^n, \tag{4}$$

(usual interpretation), where $x\xi = \sum_{j=1}^n x_j \xi_j$ stands for the scalar product in $\mathbb{R}^n$. Furthermore, the inverse Fourier transform $\check{f}$ is given by (4) with i instead of $-i$. It is well-known that both $f \to \hat{f}$ and $f \to \check{f}$ map S' onto itself and L_2 isometrically onto itself. We have

$$\|f \mid W^k_2\| \sim \|(1+|\xi|^2)^{k/2} \hat{f}(\xi) \mid L_2\| = \|((1+|\xi|^2)^{k/2} \hat{f})^\vee \mid L_2\| \tag{5}$$

if $k \in \mathbb{N}_0$ and

$$\|f \mid B^s_{22}\| \sim \|(1+|\xi|^2)^{s/2} \hat{f}(\xi) \mid L_2\| = \|((1+|\xi|^2)^{s/2} \hat{f})^\vee \mid L_2\| \tag{6}$$

if $0 < s \neq$ integer, where "$\sim$" indicates equivalent norms. In other words the spaces B^s_{22} with fractional values of s fill in a natural way the gaps between the spaces W^k_2 with $k \in \mathbb{N}_0$.

1.2.5 Besov spaces

Around 1960 the situation was the following: On the one hand there was the continuous scale $\mathcal{C}^s$, with $s > 0$, of the Hölder–Zygmund spaces; on the other hand, we have the Sobolev spaces W^k_p, with $1 < p < \infty$ and $k \in \mathbb{N}_0$, and the proposals for the spaces $B^s_{p\infty}$ and B^s_{pp} described in the last subsection. Furthermore, Fourier analytical tools lurked behind the scene. In retrospective it seems to be natural to combine Zygmund's idea, i.e., to use second and higher differences of functions, with Sobolev's approach, and to extend (1.2.4/1,2) to values $s \in \mathbb{N}$. Furthermore the chosen notations (which are different from the original ones) suggest that $B^s_{p\infty}$ and B^s_{pp} may be special cases of a larger class B^s_{pq} equipped with a genuine third parameter. This results in the nowadays classical Besov spaces which we are going to describe. Another possibility is the attempt to replace L_2 on the right hand side of (1.2.4/5,6) by L_p. This leads to the fractional Sobolev spaces H^s_p which will be discussed in 1.3.2.

Definition (Besov spaces). Let $s > 0$, $1 < p < \infty$ and $1 \leqq q < \infty$, then

$$B^s_{pq} = \Big\{ f \in L_p \colon \| f \mid B^s_{pq} \| = \| f \mid W^{[s]^-}_p \| \tag{1}$$

$$+ \sum_{|\alpha|=[s]^-} \left(\int_{\mathbb{R}^n} |h|^{-\{s\}^+ q} \| \Delta^2_h D^\alpha f \mid L_p \|^q \frac{dh}{|h|^n} \right)^{1/q} < \infty \Big\}$$

and

$$B^s_{p\infty} = \Big\{ f \in L_p \colon \| f \mid B^s_{p\infty} \| = \| f \mid W^{[s]^-}_p \| \tag{2}$$

$$+ \sup_{0 \neq h \in \mathbb{R}^n} |h|^{-\{s\}^+} \| \Delta^2_h D^\alpha f \mid L_p \| < \infty \Big\}$$

with $s = [s]^- + \{s\}^+$ in the sense of (1.2.2/2).

Remark 1. These Banach spaces have been introduced by O.V. Besov in 1959/60, see [Bes1,2]. The analogy with the spaces $\mathcal{C}^s$ from (1.2.2/5) is obvious. If $s > 0$ is not an integer then Δ^2_h in (1) and (2) can be replaced by Δ_h ($= \Delta^1_h$) in the sense of equivalent norms. This justifies our notations in the above definition on the one hand and in (1.2.4/1,2) on the other hand. If there is no danger of confusion then we adopt occasionally the little sloppy convention not to distinguish between equivalent norms of a given Banach space.

Theorem. Let $s > 0$, $1 < p < \infty$ and $1 \leqq q \leqq \infty$. Let $k \in \mathbb{N}_0$ with $k < s$ and $m \in \mathbb{N}$ with $m > s - k$. Then

$$B^s_{pq} = \Big\{ f \in L_p \colon \| f \mid B^s_{pq} \|_{k,m} = \| f \mid W^k_p \| \tag{3}$$

$$+ \sum_{|\alpha|=k} \left(\int_{\mathbb{R}^n} |h|^{-(s-k)q} \| \Delta^m_h D^\alpha f \mid L_p \|^q \frac{dh}{|h|^n} \right)^{1/q} < \infty \Big\}$$

(with the usual modification if $q = \infty$) where all admissible norms $\| f \mid B^s_{pq} \|_{k,m}$ are mutually equivalent.

Remark 2. This is the direct counterpart of Theorem 1.2.2(i). Furthermore, it is now clear that the spaces from (1.2.4/1,2) are covered by (3).

Remark 3. What about assertions in the sense of Theorem 1.2.2(ii)? Here the situation is quite different in dependence whether $p = 2$ or $p \neq 2$. On the one hand we have

$$W_2^k = B_{22}^k \quad \text{if} \quad k \in \mathbb{N}_0 \tag{4}$$

in good agreement with our discussion at the end of 1.2.4. On the other hand, if $k \in \mathbb{N}_0$, $1 < p < \infty$ and $p \neq 2$ then W_p^k does not coincide with any space B_{rq}^σ where $\sigma > 0$, $1 < r < \infty$ and $1 \leqq q \leqq \infty$. Even worse, the Banach structure of W_p^k and B_{rq}^σ is quite different. W_p^k is isomorphic to L_p whereas B_{rq}^σ is isomorphic to $l_q(l_r)$, i.e., to the l_r-valued sequence spaces l_q. But L_p with $p \neq 2$ and $l_q(l_r)$ with $1 \leqq q \leqq \infty$ and $1 < r < \infty$ are not isomorphic to each other. In particular B_{pp}^s with $1 < p < \infty$ is isomorphic to l_p. Structure results for the spaces B_{pq}^s are due to J. Peetre, see [Pee6, p. 190] and the author, see [Tri1]. We refer to [Pee6, Chapter 9] and [Triα, 2.11, 2.12] for further results and details.

1.3 The spectral approach

1.3.1 The abstract background

We begin with some remarks about self-adjoint operators in Hilbert spaces. The aim is twofold. First we wish to shed new light on what has been said at the end of 1.2.4, formulas (1.2.4/5,6). Secondly, we pave the way for the fractional Sobolev spaces and, more generally, for the Fourier analytical approach to the spaces B_{pq}^s and F_{pq}^s.

Let H be a complex separable Hilbert space endowed with the norm $\|\cdot\|$. Let T be a positive-definite self-adjoint operator in H with the spectral decomposition

$$Tf = \int_0^\infty t \, dE(t) f \tag{1}$$

where $\{E(t)\}_{t>0}$ stands for the spectral measure of T. Let $s \geqq 0$, then

$$T^s f = \int_0^\infty t^s \, dE(t) f \tag{2}$$

are the fractional powers of T, and

$$D(T^s) = \left\{ f \in H \colon \|f\|_s = \left(\int_0^\infty t^{2s} \, d\|E(t)f\|^2 \right)^{1/2} < \infty \right\} \tag{3}$$

are the corresponding domains of definition endowed in a natural way with a norm. Then $\{D(T^s)\}_{s \geqq 0}$ is a scale of Hilbert spaces (of course, only the case of unbounded operators T is of any interest).

Let $H = L_2$, where we again adopt the previous convention $L_2 = L_2(\mathbb{R}^n)$ (with respect to the Lebesgue measure). Then

$$T = \mathrm{id} - \Delta \quad \text{with} \quad D(T) = W_2^2 \tag{4}$$

is a positive-definite self-adjoint operator, where id stands for the identity and $\Delta = \sum_{j=1}^n \frac{\partial^2}{\partial x_j^2}$ is the Laplacian. The fractional powers T^s of T are given by (2,3).

Proposition. Let $s > 0$, then $D(T^s) = B_{22}^{2s}$.

Remark. We refer to (1.2.4/5,6) and (1.2.5/4). In other words, the scale $\{B_{22}^s\}_{s \geqq 0}$ describes fractional smoothness with respect to the Laplacian. To provide a better understanding of what follows we add some remarks about the proof of the proposition. We use the notations introduced at the end of 1.2.4. By well-known properties of the Fourier transform we have

$$(\mathrm{id} - \Delta) f = \left((1 + |\xi|^2) \hat{f} \right)^\vee \tag{5}$$

with the domain of definition

$$D(\mathrm{id} - \Delta) = \{ f \in L_2 \colon (1 + |\xi|^2) \hat{f}(\xi) \in L_2 \}. \tag{6}$$

The fractional powers of the multiplication operator $g \to (1 + |\xi|^2) g$ are given by $g \to (1 + |\xi|^2)^s g$. By (5) this can be transferred to the fractional powers of $\mathrm{id} - \Delta$, because the Fourier transform is an isomorphic map of L_2 onto itself:

$$I_s f = (\mathrm{id} - \Delta)^{s/2} f = \left((1 + |\xi|^2)^{s/2} \hat{f} \right)^\vee, \tag{7}$$

where we introduce the operator I_s which will be useful later on. The proposition follows now from (1.2.4/5,6).

1.3.2 Fractional Sobolev spaces

We return to the situation around 1960. Following the way paved by S.L. Sobolev and A. Zygmund, and the proposals made by S.M. Nikol'skij, N. Aronszajn, L.N. Slobodeckij and E. Gagliardo the Besov spaces emerged finally. All these spaces can be described in terms of (generalized) derivatives and differences of functions. However by Remark 1.2.5/3 the situation is not completely satisfactory: If $1 < p < \infty$ and $p \neq 2$ then even the Banach space structure (isomorphy classes) of W_p^k and B_{pq}^s is different (although this was surely unknown at that time and it is a little bit curious to argue backwards in time). Looking for a cure of this shortcoming one may attempt to extend Proposition 1.3.1 from L_2 to L_p, what on a tentative basis means that we wish to replace L_2 on the extreme right hand sides of (1.2.4/5,6) by L_p. If one wishes that such a spectral approach in the sense of 1.3.1 covers the original Sobolev spaces W_p^k then assertions of the following type are desirable: Let $\alpha = (\alpha_1, \ldots, \alpha_n)$ be a multi-index, $\xi = (\xi_1, \ldots, \xi_n) \in \mathbb{R}^n$ and $\xi^\alpha = \xi_1^{\alpha_1} \cdots \xi_n^{\alpha_n}$ then

$$f \to (\xi^\alpha (1 + |\xi|^2)^{-|\alpha|/2} \hat{f})^\vee \tag{1}$$

is a linear and bounded map from L_p into itself. We have given a somewhat more careful version of this heuristical argument in [Triß, 2.2.3] where we tried to find a criterion under what conditions a function space should be denoted as a good one (from our point of view, i.e., possible applications to PDE's). However, the desired property is completely covered by the multiplier theorem for L_p-spaces with $1 < p < \infty$ proved by S.G. Michlin [Mic1,2] and L. Hörmander [Hör1]. Further references about Fourier multipliers may be found in [Triα, 2.2.4]. In other words, after sufficiently strong Fourier multiplier theorems had been established in 1960, it was reasonable to extend the spectral approach described in 1.3.1 from L_2 to L_p.

Definition (Fractional Sobolev spaces). Let $1 < p < \infty$ and $s \in \mathbb{R}$ then

$$H_p^s = \{f \in S' \colon ((1 + |\xi|^2)^{s/2} \hat{f})^\vee \in L_p\}. \tag{2}$$

Remark 1. We recall that we omit $\mathbb{R}^n$ in S', L_p and now also in H_p^s. It should be noted that we extended (1.2.4/5,6) not only to $1 < p < \infty$ but also to $s \leqq 0$. This makes sense in any case. H_p^s equipped with the norm $\|((1 + |\xi|^2)^{s/2} \hat{f})^\vee \mid L_p\|$ becomes a Banach space and it is quite obvious that

$$I_s \colon ((1 + |\xi|^2)^{s/2} \hat{f})^\vee \tag{3}$$

is a lift which maps H_p^σ isomorphically onto $H_p^{\sigma-s}$. In particular all spaces H_p^s are isomorphic to L_p.

Remark 2. The above definition goes back to N. Aronszajn, K.T. Smith [ArS] and A.P. Calderón [Cal1,2]. The notation "fractional Sobolev spaces" sounds a little bit suspicious, in particular for the spaces H_p^s with $s < 0$, however, it is apparently nowadays in common use and the theorem below may be considered as a justification. The spaces H_p^s are also called Liouville spaces (in particular in the Russian literature) or Bessel potential spaces. The latter notation comes from the fact that the operator I_s with $s < 0$ can be represented as a convolution operator where the kernel $[(1+|\xi|^2)^{s/2}]^\vee$ can be described explicitely via Bessel functions, see [Nik2, 8.1].

Theorem. (i) Let $1 < p < \infty$ and $k \in \mathbb{N}_0$, then

$$W_p^k = H_p^k. \tag{4}$$

(ii) Let $1 < p < \infty$, $s \in \mathbb{R}$ and $\sigma \in \mathbb{R}$, then

$$I_s\, H_p^\sigma = H_p^{\sigma-s}. \tag{5}$$

(iii) Let $1 < p < \infty$, $1 \leqq q \leqq \infty$, $\sigma > 0$, $s \in \mathbb{R}$ and $\sigma - s > 0$, then

$$I_s\, B_{pq}^\sigma = B_{pq}^{\sigma-s}, \qquad I_s\, \mathcal{C}^\sigma = \mathcal{C}^{\sigma-s}. \tag{6}$$

(iv) Let $s > 0$ then

$$H_2^s = B_{22}^s. \tag{7}$$

Remark 3. The proof of (4) is based on Fourier multiplier assertions for operators of type (1). The lifting property (5) is quite obvious by the above discussion. Furthermore, (7) is covered by (1.2.4/6) and (1.2.5/4). We complemented (5) by the lifting properties (6) which are not so clear at least at this stage in the game. But one can take (6) as the basis of some speculations. First of all, the restrictions $\sigma > 0$ and $\sigma - s > 0$ in (6) are disturbing compared with the elegant formula (5). What about using (6) with $\sigma > 0$, $s \in \mathbb{R}$ and $\sigma - s \leqq 0$ as a definition of spaces $B_{pq}^{\sigma-s}$ and $\mathcal{C}^{\sigma-s}$? It comes out that this makes sense, in particular, $B_{pq}^{\sigma-s}$ and $\mathcal{C}^{\sigma-s}$ depend only on $\sigma - s$ and not on the chosen numbers σ and s. This is one reasonable way to introduce spaces B_{pq}^s and $\mathcal{C}^s$ with $s \leqq 0$, although we prefer later on a slightly different method which yields the same result. Furthermore (6) and the above discussion suggest to search for a Fourier analytical characterization of the spaces B^s_{pq} and $\mathcal{C}^s$ with $s > 0$ which can be naturally extended to values $s \leqq 0$. Recall that all spaces defined on $\mathbb{R}^n$ are considered as subspaces of S'.

Remark 4. We restricted (2) to L_p with $1 < p < \infty$. What about L_1, L_∞, C or C^k with $k \in \mathbb{N}$ as basic space instead of L_p? Of course, the definition itself makes sense, but the outcome is rather poor (seen from the point of view of possible applications to PDE's). The reason is that the spaces L_1, L_∞ and C^k with $k \in \mathbb{N}_0$ have not enough Fourier multipliers, see [Triß, 2.2.3] where we discussed this problem and where we gave also references to papers dealing with non-multiplier theorems for these ugly spaces.

1.3.3 Littlewood–Paley theorems

We summarize what has been achieved up to now. On the one hand, we have the spaces W^k_p, $\mathcal{C}^s$ and B^s_{pq} with $k \in \mathbb{N}_0$, $s > 0$, $1 < p < \infty$, $1 \leqq q \leqq \infty$, defined in terms of derivatives and differences. On the other hand, there is the spectral theoretical approach, motivated by some simple observations in the framework of the theory of self-adjoint operators in Hilbert spaces, see 1.3.1, and rigorously based on more sophisticated Fourier multiplier theorems. As a result we defined the fractional Sobolev spaces H^s_p with $s \in \mathbb{R}$ and $1 < p < \infty$, and Theorem 1.3.2 strongly suggests to have a closer look on the spectral theoretical approach.

To provide a better understanding of what is going on we return shortly to 1.3.1. We assume that H, T and in particular $D(T^s)$ have the same meaning as in 1.3.1, see (1.3.1/1–3). Let $\varphi(t)$ be a continuous function on $\mathbb{R}$ or the characteristic function of an interval, then $\varphi(T)$,

$$\varphi(T)f = \int_0^\infty \varphi(t)\, dE(t) f, \tag{1}$$

$$D(\varphi(T)) = \left\{ f \in H \colon \|\varphi(T)f\| = \left(\int_0^\infty |\varphi(t)|^2 \, d\|E(t)f\|^2 \right)^{1/2} < \infty \right\}, \tag{2}$$

has the usual meaning. We call $\{\varphi_j(t)\}_{j=0}^\infty$ a dyadic resolution of unity if the φ_j's are either continuous functions or step functions on $\mathbb{R}$,

$$\begin{cases} \operatorname{supp} \varphi_0 \subset \{t \in \mathbb{R} \colon |t| \leqq 2\}, \\ \operatorname{supp} \varphi_k \subset \{t \in \mathbb{R} \colon 2^{k-1} \leqq t \leqq 2^{k+1}\}, \quad k \in \mathbb{N}, \end{cases} \tag{3}$$

$$\sup_{t\in\mathbb{R}, j\in\mathbb{N}_0} |\varphi_j(t)| < \infty, \tag{4}$$

and

$$\sum_{j=0}^{\infty} \varphi_j(t) = 1 \quad \text{if} \quad t \geqq 0. \tag{5}$$

Now it follows easily from (1.3.1/2,3)

$$T^s = \sum_{j=0}^{\infty} T^s \, \varphi_j(T) \tag{6}$$

and

$$\|T^s f\| = \|f\|_s \sim \left(\sum_{j=0}^{\infty} 2^{2js} \, \|\varphi_j(T) f\|^2 \right)^{1/2} \tag{7}$$

where "$\sim$" indicates equivalent norms. We obtained a dyadic decomposition both of the operator T^s and of its norm. We remark that $T^s \varphi_j(T)$ is a bounded operator; it is self-adjoint if φ_j is real and in this case the spectrum is contained in $\operatorname{supp} \varphi_j$. In this sense, (6,7) is a dyadic spectral decomposition.

As in 1.3.1 we apply these remarks to $H = L_2$ (i.e., $L_2(\mathbb{R}^n)$) and $T = \mathrm{id} - \Delta$. Then the concretization of (1) is given by

$$\varphi(\mathrm{id} - \Delta) f = (\varphi(1 + |\xi|^2) \hat{f})^{\vee}. \tag{8}$$

Now if we decompose T^s in the sense of (6) with the help of the above dyadic decomposition $\{\varphi_j(t)\}_{j=0}^{\infty}$, then we arrive at the following procedure: First decompose $\mathbb{R}^n$ by functions φ_j satisfying the $\mathbb{R}^n$-counterparts of (3–5), then construct operators of type (8), and use these operators in order to decompose smoothness (at this moment expressed by $\mathrm{id} - \Delta$ and its powers) in the sense of (6,7). The question is whether this scheme can be extended from L_2 to L_p with $1 < p < \infty$ and how such a generalization should look like. Such an extension requires the above dyadic structure which was up to this moment completely immaterial.

First we modify the above approach slightly. Let φ be a continuous function on $\mathbb{R}^n$ or the sum of finitely many characteristic functions of cubes in $\mathbb{R}^n$, then $\varphi(D)$ is defined by

$$\varphi(D) f = (\varphi \hat{f})^{\vee}. \tag{9}$$

If φ is a compactly supported C^∞ function in $\mathbb{R}^n$, then (9) makes sense for any $f \in S'$.

We call $\{\varphi_j\}_{j=0}^{\infty}$ a smooth dyadic resolution of unity in $\mathbb{R}^n$ if the φ_j's are C^∞ functions on $\mathbb{R}^n$,

$$\begin{cases} \operatorname{supp} \varphi_0 \subset \{\xi \in \mathbb{R}^n : |\xi| \leqq 2\} \\ \operatorname{supp} \varphi_k \subset \{\xi \in \mathbb{R}^n : 2^{k-1} \leqq |\xi| \leqq 2^{k+1}\}, \quad k \in \mathbb{N}, \end{cases} \tag{10}$$

for any multi-index α holds

$$\sup_{\xi \in \mathbb{R}^n, j \in \mathbb{N}_0} 2^{j|\alpha|} |D^\alpha \varphi_j(\xi)| < \infty \tag{11}$$

and

$$\sum_{j=0}^{\infty} \varphi_j(\xi) = 1 \quad \text{if} \quad \xi \in \mathbb{R}^n. \tag{12}$$

It is rather easy to construct such resolutions of unity: Let φ_0 be an arbitrary C^∞ function on $\mathbb{R}^n$ with $\operatorname{supp} \varphi_0 \subset \{\xi \in \mathbb{R}^n: |\xi| < 2\}$ and $\varphi_0(\xi) = 1$ if $|\xi| \leqq 1$. Then $\{\varphi_j\}_{j=0}^{\infty}$ with $\varphi_j(\xi) = \varphi_0(2^{-j}\xi) - \varphi_0(2^{-j+1}\xi)$ if $j \in \mathbb{N}$ is a system with the required properties.

Theorem (of Littlewood–Paley type). Let $\{\varphi_j\}_{j=0}^{\infty}$ be a smooth dyadic resolution of unity. Let $s \in \mathbb{R}$ and $1 < p < \infty$ then

$$H_p^s = \left\{ f \in S' : \left\| \left(\sum_{j=0}^{\infty} 2^{2js} |\varphi_j(D) f(\cdot)|^2 \right)^{1/2} \mid L_p \right\| < \infty \right\} \tag{13}$$

(equivalent norms).

Remark 1. Of course, $\varphi_j(D)f$ must be understood in the sense of (9). Because φ_j has a compact support $\varphi_j(D)f$ is well-defined for any $f \in S'$ and $\varphi_j(D)f(x)$ is an entire analytic function with respect to $x \in \mathbb{R}^n$ (Paley–Wiener–Schwartz theorem). In particular, (13) makes sense.

Remark 2. Let $p = 2$, then (13) coincides essentially with (7), however, there are some immaterial technical modifications because we prefer now (9) instead of (8). By Proposition 1.3.1 we arrive at (1.3.2/7).

Remark 3. Characterization (13) is a theorem of Littlewood–Paley type. A simple proof based on a l_2-version of Hörmander's Fourier multiplier theorem may be found in [Triα, 2.3.3, 2.2.4]. The first proof of a Fourier multiplier theorem for l_2 valued L_p-spaces with $1 < p < \infty$ is due to P.I. Lizorkin, see [Liz1]. Littlewood–Paley theorems play a crucial role in L_p harmonic analysis, they attracted much attention. A large variety of different aspects of Littlewood–Paley assertions may be found in the books [EdG, Ste1,2] including references to the original papers by J.E. Littlewood and R.E.A.C. Paley (1931–1937), see also [ScT, 3.4.4] for further references.

We preferred in (13) smooth functions φ_j for reasons which will be clear later on (and which enabled us to prove (13) on the basis of a l_2-version of Hörmander's Fourier multiplier theorem). However, mostly Littlewood–Paley theorems are formulated on the basis of characteristic functions of "dyadic" cubes instead of smooth functions. We give a brief description which will also be of interest later on. Let

$$K_0 = \{x \in \mathbb{R}^n : x = (x_1, \ldots, x_n),\ |x_k| < 1\} \tag{14}$$

and

$$K_m = \begin{cases} \{x \in \mathbb{R}^n : x = (x_1, \ldots, x_n),\ |x_k| < 2^m\} \backslash \\ \{x \in \mathbb{R}^n : x = (x_1, \ldots, x_n),\ |x_k| \leqq 2^{m-1}\}, \quad m \in \mathbb{N}. \end{cases} \tag{15}$$

Let χ_j be the characteristic function of K_j with $j \in \mathbb{N}_0$, then $\chi_j(D)$ is defined by (9).

Proposition. Let $s \in \mathbb{R}$ and $1 < p < \infty$, then

$$H^s_p = \Big\{ f \in S' : \Big\| \Big(\sum_{j=0}^{\infty} 2^{2js} |\chi_j(D) f(\cdot)|^2 \Big)^{1/2} \mid L_p \Big\| < \infty \Big\} \tag{16}$$

(equivalent norm).

Remark 4. One can prove (16) if one combines (13) with a vector-valued multiplier theorem due to P.I. Lizorkin, see [Liz2]. We refer also to [Triß, 2.5.4].

1.3.4 Spectral decompositions

Theorem 1.3.3 and also Proposition 1.3.3 suggest the following speculation: Let $\{\varphi_j\}_{j=0}^{\infty}$ be a smooth dyadic resolution of unity in $\mathbb{R}^n$ in the sense of (1.3.3/10–12). Then any $f \in S'$ can be represented as

$$f = \sum_{j=0}^{\infty} \varphi_j(D)f \qquad \text{(convergence in } S'\text{)} \tag{1}$$

where any $\varphi_j(D)f(x)$ is an entire analytic function (by the Paley–Wiener–Schwartz theorem). Furthermore, (1) can also be interpreted as a spectral decomposition of f based on the Laplacian. Now (1.3.3/13) suggests to measure smoothness of $f \in S'$ by examining the behaviour of $\{2^{js}\varphi_j(D)f(x)\}_{j=0}^{\infty}$ with respect to $L_p(l_q)$-norms (where we generalized l_2 in (1.3.3/13) by l_q). Furthermore one may change the order of L_p and l_q, i.e., one asks a corresponding question with respect to $l_q(L_p)$-norms. Technically this is even much simpler because any analytic function $\varphi_j(D)f(x)$ can be treated separately.

Theorem. (i) Let $s > 0$, $1 < p < \infty$ and $1 \leqq q \leqq \infty$, then

$$B^s_{pq} = \left\{ f \in S' \colon \left(\sum_{j=0}^{\infty} 2^{jsq} \| \varphi_j(D)f \mid L_p \|^q \right)^{1/q} < \infty \right\} \tag{2}$$

(usual modification if $q = \infty$) in the sense of equivalent norms.

(ii) Let $s > 0$, then

$$C^s = \left\{ f \in S' \colon \sup_{x \in \mathbb{R}^n, j \in \mathbb{N}_0} 2^{js} |\varphi_j(D)f(x)| < \infty \right\} \tag{3}$$

in the sense of equivalent norms.

Remark 1. This is the Fourier analytical characterization of the classical Besov spaces from Definition 1.2.5 we are looking for. It is due to J. Peetre [Pee1] in 1967 (with a forerunner in [Hör2] if $p = q = 2$ which is covered by Theorem 1.3.3). Further comments may also be found in the notes at the end of the chapters in [Pee6]. The characterization (3) of C^s has been proved in [Tri6] (the homogeneous case is due to B. Grevhold [Gre], see also [Pee2]).

Remark 2. The question is whether (2) can be taken as a starting point to define B^s_{pq} for reasonable values of s, p, q, i.e., so far for $s \in \mathbb{R}$, $1 \leqq p \leqq \infty$ and $1 \leqq q \leqq \infty$. This can be done and has been done in [Pee1]; however we postpone a detailed discussion to a later moment. Then (3) can be reformulated as $C^s = B^s_{\infty\infty}$ and (1.3.2/6) holds for all $s \in \mathbb{R}$ and $\sigma \in \mathbb{R}$.

Recall that we took the Littlewood–Paley theorem (1.3.3/13) as a motivation to measure smoothness via the decomposition (1). However if one compares (1.3.3/13) and (2) then we replaced by $L_p(l_2)$ by $l_q(L_p)$. It seems to be at least reasonable to replace $L_p(l_2)$ by $L_p(l_q)$.

Definition. Let $s \in \mathbb{R}$, $1 < p < \infty$ and $1 < q < \infty$, then

$$F^s_{pq} = \left\{ f \in S' \colon \left\| \left(\sum_{j=0}^{\infty} 2^{jsq} |\varphi_j(D) f(\cdot)|^q \right)^{1/q} \mid L_p \right\| < \infty \right\}. \tag{4}$$

Remark 3. We have $F^s_{p2} = H^s_p$, see (1.3.3/13), and $F^s_{pp} = B^s_{pp}$ if in addition $s > 0$, see (2). Hence F^s_{pq} links these two prominent scales of spaces. The question arises whether (4) makes sense, i.e., whether F^s_{pq} is independent of the chosen smooth dyadic resolution of unity in $\mathbb{R}^n$ in the sense of equivalent norms. An affirmative answer can be given on the basis of a $L_p(l_q)$-version of Hörmander's Fourier multiplier theorem, see [Triα: 2.2.4, 2.3.2], in particular, F^s_{pq} is a Banach space.

Beside (1.3.3/13) we have the Littlewood–Paley assertion (1.3.3/16) for H^s_p which is even nearer to what usually is done in Littlewood–Paley theories. So it is quite natural to ask whether corresponding assertions hold for B^s_{pq} and F^s_{pq} spaces. In the following proposition $\chi_j(D)$ has the same meaning as in (1.3.3/16).

Proposition. (i) Let $s > 0$, $1 < p < \infty$ and $1 \leqq q \leqq \infty$, then

$$B^s_{pq} = \left\{ f \in S' \colon \left(\sum_{j=0}^{\infty} 2^{jsq} \|\chi_j(D) f \mid L_p\|^q \right)^{1/q} < \infty \right\} \tag{5}$$

(usual modification if $q = \infty$) in the sense of equivalent norms.

(ii) Let $s \in \mathbb{R}$, $1 < p < \infty$ and $1 < q < \infty$, then

$$F^s_{pq} = \left\{ f \in S' \colon \left\| \left(\sum_{j=0}^{\infty} 2^{jsq} |\chi_j(D) f(\cdot)|^q \right)^{1/q} \mid L_p \right\| < \infty \right\} \tag{6}$$

in the sense of equivalent norms.

Remark 4. Assertion (6) can be proved on the basis of the l_q-valued L_p Fourier multiplier theorem due to P.I. Lizorkin, see [Liz2], whereas the l_q-valued version of Hörmander's theorem is not sufficient for this purpose. Formula (5) follows from the simpler scalar L_p Fourier multiplier theorem, where $s > 0$ can be replaced by $s \in \mathbb{R}$. As we shall see later on both (2) and (4) can be extended to $s \in \mathbb{R}$, $p = 1$, $1 \leqq q \leqq \infty$ (and even to $p < 1$ and/or $q < 1$). However, then one needs smooth resolutions of unity and the equivalence assertions (5) and (6) are no longer valid. Also (3) with χ_j instead of φ_j is wrong. In other words, for limiting cases such as (3) and for the extensions we have in mind, in particular beyond the realm of Banach spaces, we need smooth resolutions of unity in $\mathbb{R}^n$.

Remark 5. The spaces F^s_{pq} in the above form, including what has been said in Remark 3, are known to the author since 1970. But first it was unclear whether these spaces are of any use or whether a substantial theory can be expected. First results about F^s_{pq} spaces about interpolation, duality and embeddings, based on the above definition, may be found in [Tri2,3]. At that time we learned that spaces of type F^s_{pq} in the form of (6) had been introduced independently by P.I. Lizorkin, see [Liz3,5], with the l_q-valued L_p Fourier multiplier theorem of [Liz2] as basis.

1.3.5 An outlook

We pause for a moment looking backwards and forwards. What has been done up to now, including the characterization (1.3.4/2) (maybe extended to $s \in \mathbb{R}$ and $1 \leqq p \leqq \infty$) and the definition (1.3.4/4) describe the situation at the beginning of the seventies. The next task is to extend the definitions of B^s_{pq} and F^s_{pq} to $0 < p \leqq \infty$ and $0 < q \leqq \infty$. This is possible on the basis of (1.3.4/2,4) in a rather formal way, but the story behind is closely connected with the theory of Hardy spaces which will be sketched in 1.4. Afterwards we arrive in 1.5 at the spaces B^s_{pq} and F^s_{pq} in their full generality. Then we are in the middle of the seventies. We include in 1.5 a discussion about approximation procedures connected with entire analytic functions. This subject attracted much attention, especially in the Russian literature, mostly connected with classical Sobolev–Besov spaces. In 1.6–1.8 we go backwards in time and describe some other devices how to measure smoothness which began in the sixties and which are connected with the spaces under consideration. Interpolation theory (real and complex) was the first unifying method (earlier in time than the above-described spectral approach) which converted a plethora of diverse function spaces into a scheme governed by few simple abstract rules. This will be described in 1.6. In 1.7 we outline local approximation methods which go back to BMO-spaces and Campanato spaces and which are now, at the end of the eighties, closely connected with B^s_{pq} and F^s_{pq} spaces. In 1.8 we deal with the problem that B^s_{pq} and F^s_{pq} on $\mathbb{R}^n$ can be described as traces of harmonic and thermic functions on $\mathbb{R}^{n+1}_+$. In the remaining three sections 1.9–1.11 some more recent topics will be treated: Atoms in function spaces in 1.9, some latest news about B^s_{pq} and F^s_{pq} spaces on smooth domains in 1.10, blended with some historical remarks, and finally how to extend function spaces from $\mathbb{R}^n$ to other structures such as Riemannian manifolds and Lie groups in 1.11. In this chapter we lok at all these topics from a historical point of view. However as far as these developments are very recent ones, they will be treated again in the following chapters in detail.

1.4 Hardy spaces

1.4.1 Spaces on the unit disk

Hardy spaces have their beginning within complex function theory (of one variable). Let $0 < p < \infty$, then an analytic function f in the unit disk of the complex plane is said to be an element of the space $H_p(D)$ if

$$\mu_p(r, f) = \left(\frac{1}{2\pi}\int_0^{2\pi} |f(re^{i\theta})|^p d\theta\right)^{1/p} \tag{1}$$

is bounded for $0 \leqq r < 1$. Of special interest are the spaces $H_p(D)$ with $0 < p \leqq 1$. These spaces were introduced by G.H. Hardy in 1914, see [Har]. Afterwards F. Riesz [Rie] gave a simplified and systematic approach to these spaces. For details of this part of the theory we refer to A. Zygmund's treatise [Zyg2: VII,7]. We formulate a typical result due to G.H. Hardy and J.E. Littlewood some 60 years ago, [HaL1], which not only sheds light on what has been done at that time but which resembles in a striking way more recent assertions in the framework of the real n-dimensional theory. We follow [Zyg2: VII, (7.36)]. For any $0 \leqq r < 1$, $0 \leqq \theta \leqq 2\pi$, let $\Omega_r(\theta)$ denote the domain bounded by the two tangents from the point $e^{i\theta}$ to the circle $|z| = r$ and by the larger of the two arcs of that circle between the points of contact. (For $r = 0$, the domain reduces to the radius through $e^{i\theta}$). Let $0 < p < \infty$, $f \in H_p(D)$ and

$$N(\theta) = N_{r,f}(\theta) = \sup_{z \in \Omega_r(\theta)} |f(z)|. \tag{2}$$

Then $N(\theta) \in L_p(0, 2\pi)$ and

$$\left(\int_0^{2\pi} N^p(\theta)d\theta\right)^{1/r} \leqq c_p \lim_{\rho\to 1} \mu_p(\rho, f). \tag{3}$$

In recent terms this is an L_p- estimate of the non-tangential maximal function $N(\theta)$ by the boundary values of $f \in H_p(D)$. In our time estimates of such a type play a decisive role in the real n-dimensional theory of Hardy spaces. We shall not stress this point later on but it reminds us that the germs of recent L_p harmonic analysis are contained in the work of Hardy and Littlewood.

1.4.2 The real variable approach

The next aim is to extend the theory of Hardy spaces from one complex dimension (or two real dimensions) to n real dimensions. For this purpose it seems to be appropriate to get rid of complex function theory in the unit disk and to replace it by something like a L_p harmonic analysis in the upper half-plane. Whereas this substitution in 2 dimensions can be done in an obvious way, the extension to higher dimensions caused a lot of trouble and has been done by E.M. Stein and G. Weiss in 1960, see [StW1], we refer also to [StW2] and [FeS2]. We give a description following the just quoted references. Let

$$\mathbb{R}^{n+1}_+ = \{(x,t) \in \mathbb{R}^{n+1}: x \in \mathbb{R}^n, t > 0\} \tag{1}$$

and $1 < p < \infty$. Then, by definition, $H_p(\mathbb{R}^{n+1}_+)$ is the collection of all complex-valued harmonic functions $u(x,t)$ in $\mathbb{R}^{n+1}_+$ such that

$$\|u \mid H_p(\mathbb{R}^{n+1}_+)\| = \sup_{t>0} \left(\int_{\mathbb{R}^n} |u(x,t)|^p dx\right)^{1/p} < \infty. \tag{2}$$

It is well-known that any $u \in H_p(\mathbb{R}^{n+1}_+)$ can be represented uniquely as a Poisson integral

$$u(x,t) = c_n \int_{\mathbb{R}^n} \frac{t}{(|x-y|^2+t^2)^{\frac{n+1}{2}}} f(y)dy, \qquad f \in L_p(\mathbb{R}^n), \tag{3}$$

$(x,t) \in \mathbb{R}^{n+1}_+$ with

$$\|u \mid H_p(\mathbb{R}^{n+1}_+)\| \sim \|f \mid L_p(\mathbb{R}^n)\| \tag{4}$$

(equivalent norms) where we indicated for sake of clarity the domains of definition. In this sense we can identify $u \in H_p(\mathbb{R}^{n+1}_+)$ with its boundary values $u(\cdot,0) = f(\cdot) \in L_p(\mathbb{R}^n)$. Identifications of this type will be characteristic for the recent theory of Hardy spaces.

If $p \leqq 1$ then something like (2–4) cannot be expected, even not in the case $p = 1$. We follow E.M. Stein and G. Weiss, see the above references. Let $\frac{n-1}{n} < p < \infty$, then, by definition, $H_p(\mathbb{R}^{n+1}_+)$ is the collection of all complex-valued harmonic functions $u(x,t)$ on $\mathbb{R}^{n+1}_+$ for which there exists a $(n+1)$-tupel of harmonic functions $u = u_0, \ldots, u_n$ on $\mathbb{R}^{n+1}_+$ such that with $t = x_0$ this $(n+1)$-tupel satisfies

$$\begin{aligned} &\frac{\partial u_j}{\partial x_i} = \frac{\partial u_i}{\partial x_j} \quad \text{with } i = 0, \ldots, n \text{ and } j = 0, \ldots, n \\ &\sum_{j=0}^{n} \frac{\partial u_j}{\partial x_j} = 0 \end{aligned} \tag{5}$$

and

$$\|u \mid H_p(\mathbb{R}^{n+1}_+)\| = \sup_{t>0} \left(\int_{\mathbb{R}^n} \left(\sum_{j=0}^{n} |u_j(x,t)|^2 \right)^{p/2} dx \right)^{1/p} < \infty. \tag{6}$$

The function u_0 uniquely determines $u_1, \ldots, u_n$. If $1 < p < \infty$ then this definition is consistent with the previous one. Furthermore if $n = 1$, then (5) coincides with the Cauchy–Riemann equations and we have $0 < p < \infty$. In other words, in this case the just given definition is a reasonable real variable version of the original Hardy spaces from 1.4.1. Let $\frac{n-1}{n-1+k} < p < \infty$ with $k \in \mathbb{N}$, then one has to replace the above vector function $(u_0, \ldots, u_n)$ by a tensor function of rank k which satisfies appropriate generalizations of (5,6). We do not go into detail and refer to [FeS2]. But in any case $u \in H_p(\mathbb{R}^{n+1}_+)$ with $0 < p < \infty$ is a harmonic function in $\mathbb{R}^{n+1}_+$.

We formulate the counterpart of the classical assertion (1.4.1/3). Let $u(x,t)$ be a continuous function in $\mathbb{R}^{n+1}_+$, then

$$u^*(x) = \sup_{|x-y|<t} |u(y,t)|, \qquad x \in \mathbb{R}^n, \tag{7}$$

denotes the non-tangential maximal function, the counterpart of $N(\theta)$ from (1.4.1/2).

Proposition. Let $0 < p < \infty$, then

$$H_p(\mathbb{R}^{n+1}_+) = \{u\colon \text{harmonic in } \ \mathbb{R}^{n+1}_+, u^* \in L_p(\mathbb{R}^n)\}. \tag{8}$$

Remark. A proof may be found in [FeS2: Theorem 9]. We have

$$\|u \mid H_p(\mathbb{R}^{n+1}_+)\| \sim \|u^* \mid L_p(\mathbb{R}^n)\|, \tag{9}$$

(equivalent quasi-norms). One can take (8) with (9) as a definition. If $p \leqq 1$, then this definition is much simpler than the original one.

1.4.3 The Fefferman–Stein approach

Two aspects of the paper [FeS2] (1972) are of interest to us. First if $u(x,t) \in H_p(\mathbb{R}^{n+1}_+)$ then

$$f = \lim_{t \downarrow 0} u(x,t) \quad \text{in} \quad S'(\mathbb{R}^n) \tag{1}$$

exists and, by definition, $H_p(\mathbb{R}^n)$ is the collection of all these traces. By (1.4.2/3,4) we have $H_p(\mathbb{R}^n) = L_p(\mathbb{R}^n)$ if $1 < p < \infty$. Secondly the characterization (1.4.2/8) via the maximal function can be generalized essentially. Let $\varphi \in S(\mathbb{R}^n)$, where the latter space has the usual meaning, i.e., the collection of all complex-valued infinitely differentiable rapidly decreasing functions. Let $t > 0$, then

$$\varphi(tD)f = (\varphi(t\cdot)\hat{f})^\vee, \qquad f \in S'(\mathbb{R}^n), \tag{2}$$

see (1.3.3/9).

Theorem. Let $0 < p < \infty$ and $\varphi \in S(\mathbb{R}^n)$ with $\hat{\varphi}(0) = 1$. Then

$$H_p(\mathbb{R}^n) = \{f \in S'(\mathbb{R}^n) \colon \|f^* \mid L_p\| < \infty\} \tag{3}$$

with

$$f^*(x) = \sup_{|x-y|<t} |\varphi(tD)f(y)| \tag{4}$$

and

$$H_p(\mathbb{R}^n) = \left\{ f \in S'(\mathbb{R}^n) \colon \|\sup_{t>0} |\varphi(tD)f(\cdot)| \mid L_p\| < \infty \right\}. \tag{5}$$

Remark 1. A proof of this crucial theorem may be found in [FeS2: Theorem 11]. Instead of harmonic extensions $f \to u(x,t)$ and related maximal functions (1.4.2/7) we have the extension $f \to \varphi(tD)f(x)$ and (4). The harmonic extension can be recovered if one chooses $\varphi(\xi) = e^{-|\xi|}$ (which is not smooth at the origin, but this is unimportant). With $\varphi(\xi) = e^{-|\xi|^2}$ one obtains thermic extensions, i.e., $\varphi(tD)f(x)$ obeys the heat equation in $\mathbb{R}^{n+1}_+$. We return to this subject later on in connection with characterizations of function spaces in terms of Gauss–Weierstrass and Cauchy–Poisson semi-groups.

Remark 2. The characterizations (1.3.4/2–4) of the spaces B^s_{pq}, $\mathcal{C}^s$ and F^s_{pq} on the one hand and of $H_p(\mathbb{R}^n)$ in (5) on the other hand are of the same type. Furthermore we have $H_p = L_p = F^0_{p2}$ if $1 < p < \infty$, see (1.3.4/4), (1.3.3/13) and (1.3.2/2). The problem is whether this Littlewood–Paley assertion can be extended to $p \leqq 1$. We discuss this question in the next subsection.

Remark 3. Recall that we look at function spaces from the point of view of possible applications to PDE's and ψDE's. In particular a useful function space should have both enough Fourier multipliers and enough pointwise multipliers. Any $\psi \in S$ should be a pointwise multiplier, i.e., $f \to \psi f$ should be a linear and bounded operator from the space in question into itself, and S should be a linear subspace of this space. Unfortunately, $H_p(\mathbb{R}^n)$ with $0 < p \leqq 1$ has not this property, in contrast to $H_p(\mathbb{R}^n) = L_p(\mathbb{R}^n)$ with $1 < p < \infty$. More precisely: If $f \in S$ belongs to $H_p(\mathbb{R}^n)$ with $0 < p \leqq 1$ then we have $\hat{f}(0) = 0$, see e.g., [Gol2], [Tri5: p. 164]. But pointwise multiplication does not preserve this property. We ask for a modification and look more closely at (1.3.4/2–4). We may assume $\varphi_j(\xi) = \varphi(2^{-j}\xi)$ if $j \in \mathbb{N}$, see the discussion after (1.3.3/12). That means that we have in (1.3.4/2–4) only expressions of the type

$$\varphi_j(D)f = \varphi(tD)f \quad \text{with} \quad t = 2^{-j} \leqq 1. \tag{6}$$

This observation may suggest to replace $\sup_{t>0}$ in (5) by something like $\sup_{0<t<1}$. This will be done in the following subsection. The other difference in the formulation (1.3.4/2–4) on the one hand and (5) on the other hand is unimportant: One can replace, for example, $\sup_{t>0} |\varphi(tD)f|$ in (5) by $\sup_{j \,\text{integer}} |\varphi(2^{-j}D)f|$ (equivalent quasi-norms). Similarly one can replace the discrete variable $j \in \mathbb{N}$ in (1.3.4/2–4) in the sense of (6) by the continuous variable $0 < t < 1$. This is a matter of taste and will be described later on in detail.

1.4.4 Local Hardy spaces

The discussion in Remark 1.4.3/3 suggests the following modification of the Hardy spaces $H_p(\mathbb{R}^n)$ from (1.4.3/5).

Definition. Let $0 < p < \infty$ and $\varphi \in S(\mathbb{R}^n)$ with $\hat{\varphi}(0) = 1$. Then

$$h_p = \left\{ f \in S'(\mathbb{R}^n) \colon \left\| \sup_{0<t<1} |\varphi(tD)f(\cdot)| \mid L_p \right\| < \infty \right\}. \tag{1}$$

Remark 1. One can replace $\sup_{0<t<1} |\varphi(tD)f|$ in (1) by $\sup_{j\in\mathbb{N}_0} |\varphi(2^{-j}D)f|$ in the sense of equivalent quasi-norms. Now it is quite clear that the characterizations (1.3.4/2–4) of the spaces B^s_{pq}, $\mathcal{C}^s$ and F^s_{pq} on the one hand and h_p on the other hand are of the same type, see also the discussion in Remark 1.4.3/3. The definition (1) is due to D. Goldberg, see [Gol1,2] (1979). These spaces can also be described via harmonic extensions and maximal functions in the sense of 1.4.2 and 1.4.3: The function u in (1.4.2/8) need not be harmonic in $\mathbb{R}^{n+1}_+$, but only in the strip $\{(x,t) \in \mathbb{R}^{n+1}\colon\ x \in \mathbb{R}^n,\ 0 < t < 1\}$ and the suprema in the maximal functions (1.4.2/7) and (1.4.3/4) are restricted to the truncated cones $|x-y| < t < 1$. After these replacements we have (1.4.2/8) and (1.4.3/3) with h_p instead of H_p.

Theorem (of Littlewood–Paley type). Let $\{\varphi\}_{j=0}^{\infty}$ be a smooth dyadic resolution of unity in the sense of (1.3.3/10–12). Let $0 < p < \infty$ then

$$h_p = \left\{ f \in S'\colon \left\| \left(\sum_{j=0}^{\infty} |\varphi_j(D)f(\cdot)|^2 \right)^{1/2} \mid L_p \right\| < \infty \right\} \tag{2}$$

(equivalent quasi-norms).

Remark 2. This theorem generalizes Theorem 1.3.3 with $s = 0$. In particular, we have $h_p = L_p$ if $1 < p < \infty$. By Remark 1.4.3/2 we can complement this assertion as follows,

$$h_p = H_p = L_p = F^0_{p2} \quad \text{if} \quad 1 < p < \infty. \tag{3}$$

Remark 3. The Littlewood–Paley assertion (2) is due to H.-Q. Bui, see [Bui1] (1980), we refer also to [Mey2]. However the homogeneous counterpart was known several years before. For sake of completeness we give a formulation and references. Let φ be a C^∞ function on $\mathbb{R}^n$ with $\operatorname{supp} \varphi \subset \{\xi \in \mathbb{R}^n\colon\ |\xi| < 2\}$ and $\varphi(\xi) = 1$ if $|\xi| \leqq 1$. Let $\varphi_j(\xi) = \varphi(2^{-j}\xi) - \varphi(2^{-j+1}\xi)$ where j is an integer. Then $\{\varphi_j\}_{j=-\infty}^{\infty}$ is a resolution of unity in $\mathbb{R}^n \setminus \{0\}$. Let $H_p = H_p(\mathbb{R}^n)$ be the Hardy spaces introduced in 1.4.3. Then

$$\|f \mid H_p\| \sim \left\| \left(\sum_{j=-\infty}^{\infty} |\varphi_j(D)f(\cdot)|^2 \right)^{1/2} \mid L_p \right\|, \qquad 0 < p < \infty, \tag{4}$$

(equivalent quasi-norms at least for those $f \in S$ with supp $\hat{f}$ off the origin). This Littlewood–Paley assertion is due to J. Peetre [Pee4,5] (1975). Extensions of this result to anisotropic Hardy spaces in the sense of A.P. Calderón and A. Torchinsky, [CaT: II], may be found in [Tri8], see also [Tri5: 167–169].

1.4.5 A digression: Inhomogeneous and local spaces

We have spoken several times about inhomogeneous and local spaces. We use the opportunity to say a few words about these notations. Let $f \in H_p$ in the sense of (1.4.3/5) and let $\lambda > 0$. Then $f(\lambda\cdot) \in H_p$ and

$$\|f(\lambda\cdot) \mid H_p\| = \lambda^{-n/p}\|f \mid H_p\| \tag{1}$$

where we identified $\|\cdot \mid H_p\|$ with the quasi-norm in (1.4.3/5). The homogeneity relation (1) is an easy consequence of

$$[\varphi(tD)f(\lambda\cdot)^\wedge](x) = [\varphi(t\lambda D)\hat{f}](\lambda x). \tag{2}$$

In our context quasi-normed spaces A with $A \subset S'$ are called homogeneous if

$$\|f(\lambda\cdot) \mid A\| = \lambda^\kappa\|f \mid A\|, \qquad f \in A,\ \lambda > 0, \tag{3}$$

holds for some $\kappa \in \mathbb{R}$. By (1) the spaces H_p have this property, but by (2) and (1.4.4/1) the spaces h_p have not this property. This is the reason why the latter spaces are called inhomogeneous. In this book we prefer inhomogeneous spaces, but it comes out that all these spaces have homogeneous counterparts. The pairing (h_p, H_p) may serve as an example. To shed more light on this somewhat vague remark we return to the spaces $\mathcal{C}^s$ and B^s_{pq}, see (1.2.2/5) and (1.2.5/1,2). The dominating parts of the respective norms can be taken to define corresponding homogeneous spaces normed by

$$\|f \mid \mathcal{C}^s_{\text{hom}}\| = \sum_{|\alpha|=[s]^-} \sup_{0\neq h\in\mathbb{R}^n} |h|^{-\{s\}^+}\|\Delta_h^2 D^\alpha f \mid C\|, \tag{4}$$

where all symbols have the same meaning as in (1.2.2/5). Similarly one defines

$$\|f \mid B^s_{pq,\text{hom}}\| = \sum_{|\alpha|=[s]^-} \left(\int_{\mathbb{R}^n} |h|^{-\{s\}^+q}\|\Delta_h^2 D^\alpha f \mid L_p\|^q \frac{dh}{|h|^n}\right)^{1/q} \tag{5}$$

(modification if $q = \infty$). We have

$$\|f(\lambda\cdot) \mid \mathcal{C}^s_{\text{hom}}\| = \lambda^s\|f \mid \mathcal{C}^s_{\text{hom}}\|, \qquad \lambda > 0, \tag{6}$$

and

$$\|f(\lambda\cdot) \mid B^s_{pq,\mathrm{hom}}\| = \lambda^{s-\frac{n}{p}} \|f \mid B^s_{pq,\mathrm{hom}}\|, \qquad \lambda > 0. \tag{7}$$

If one uses (4) or (5) to introduce corresponding spaces then one faces a technical difficulty: these spaces must be considered modulo polynomials, see also (1.4.4/4) where the right-hand side vanishes for any polynomial. Otherwise homogeneous spaces are more elegant and they are preferred in particular by the American school. As far as the homogeneous counterparts of B^s_{pq} and F^s_{pq} are concerned we refer to [Pee6] and [Triß].

Next we explain what is meant by local spaces. First we remark that (1.4.3/2) can be rewritten as

$$\varphi(tD)f(x) = \int_{\mathbb{R}^n} \hat{\varphi}(y) f(x+ty)dy, \qquad t > 0,\ x \in \mathbb{R}^n,\ f \in S', \tag{8}$$

(appropriate interpretation). Next we choose in (1.4.4/1) a function $\varphi \in S$ with $\hat{\varphi}(0) = 1$, and, say, $\operatorname{supp}\hat{\varphi} \subset \{y \in \mathbb{R}^n\colon\ |y| < 1\}$. To calculate (8) in a given point $x \in \mathbb{R}^n$ one needs only a knowledge of f in a ball centered at x and of radius t. Because t in (1.4.4/1) is restricted by $0 < t < 1$, it is now clear what is meant by local spaces. One needs only a knowledge of f in a neighbourhood of a given point in order to calculate the needed means (8). As we shall see later on in 1.8.4, *all spaces* B^s_{pq} *and* F^s_{pq} treated in this book *are local spaces.* However at this moment this is not clear, see, for instance (1.3.4/4) where the φ_j's and not the $\hat{\varphi}_j$'s have compact supports. On the other hand, it is obvious that one can replace $h \in \mathbb{R}^n$ in (1.2.2/5) and (1.2.5/1,2) by $|h| \leqq 1$. Then it follows that at least all spaces W^k_p, $\mathcal{C}^s$ and B^s_{pq} are local spaces in the above sense.

1.5 The Fourier-analytical approach

1.5.1 The spaces B^s_{pq} and F^s_{pq}

First we summarize what has been done so far. We know the Besov spaces B^s_{pq} with $s > 0$, $1 < p < \infty$; $1 \leqq q \leqq \infty$, and the Hölder–Zygmund spaces $\mathcal{C}^s$ with $s > 0$, introduced in Definitions 1.2.5 and 1.2.2/2, respectively. By Theorem 1.3.4 these spaces can be redefined via spectral decompositions. Furthermore, we considered Sobolev spaces W^k_p with $1 < p < \infty$, $k \in \mathbb{N}_0$, and, more generally, fractional Sobolev spaces H^s_p with $1 < p < \infty$, $s \in \mathbb{R}$, see (1.2.3/1), (1.3.2/2,4). Theorem 1.3.3 describes the corresponding characterizations via spectral decompositions. Finally we used this technique to introduce the spaces F^s_{pq} with $1 < p < \infty$, $1 < q < \infty$,

$s \in \mathbb{R}$ in Definition 1.3.4. All these spaces are Banach spaces and all these spaces can be constructed by the same principle: $f \in S'$ is decomposed by

$$f = \sum_{j=0}^{\infty} \varphi_j(D) f \tag{1}$$

in entire analytic functions, where $\{\varphi_j\}_{j=0}^{\infty}$ is a smooth dyadic resolution of unity in $\mathbb{R}^n$ in the sense of (1.3.3/10–12) and $\varphi_j(D)$ is given by (1.3.3/9) (with φ_j instead of φ). Afterwards the analytic pieces $2^{js}\varphi_j(D)f$ are measured via $l_q(L_p)$ or $L_p(l_q)$. The description given so far represents the situation at the beginning of the seventies. In 1.4 we did a decisive step beyond the realm of Banach spaces and introduced the Hardy spaces. Of special interest is the Paley–Littlewood characterization (1.4.4/2) of h_p with $0 < p < \infty$. It is really surprising but this formula suggests the possibility to extend the definition of the spaces F^s_{pq} in (1.3.3/4) to values $p \leqq 1$ (and maybe $q \leqq 1$). If such an extension makes sense than even more, a corresponding extension of (1.3.4/2). The latter was done in 1973 and the former in 1975, both by J. Peetre. (The reader may have the impression that we again argue backwards in time because (1.4.4/2) was proved in 1980. But (1.4.4/2) is only a technical modification of its homogeneous counterpart (1.4.4/4) and this was known in the middle of the seventies).

Definition. Let $\Phi = \{\varphi_j\}_{j=0}^{\infty}$ be a smooth dyadic resolution of unity in $\mathbb{R}^n$ in the sense of (1.3.3/10–12) and let $\varphi_j(D)$ be given by (1.3.3/9) (with φ_j instead of φ).

(i) Let $0 < p \leqq \infty$, $0 < q \leqq \infty$ and $s \in \mathbb{R}$ then

$$B^s_{pq} = \left\{ f \in S' \colon \|f \mid B^s_{pq}\|_\Phi = \left(\sum_{j=0}^{\infty} 2^{jsq} \|\varphi_j(D) f \mid L_p\|^q \right)^{1/q} < \infty \right\} \tag{2}$$

(usual modification if $q = \infty$).

(ii) Let $0 < p < \infty$, $0 < q \leqq \infty$ and $s \in \mathbb{R}$ then

$$F^s_{pq} = \left\{ f \in S' \colon \|f \mid F^s_{pq}\|_\Phi = \left\| \left(\sum_{j=0}^{\infty} 2^{jsq} |\varphi_j(D) f(\cdot)|^q \right)^{1/q} \mid L_p \right\| < \infty \right\} \tag{3}$$

(usual modification if $q = \infty$).

Remark 1. First we remark that these spaces make sense: they are independent of the chosen system Φ and they are quasi-Banach spaces (Banach spaces if $p \geqq 1$ and

$q \geqq 1$). Furthermore, our notation is consistent with Theorem 1.3.4, in particular, $\mathcal{C}^s = B^s_{\infty\infty}$, and Definition 1.3.4.

Remark 2. The surprising extension of B^s_{pq} in the sense of Theorem 1.3.4(i) to $p \leqq 1$ is due to J. Peetre in 1973, see [Pee3]. In his book [Pee6: 225–227] one can find a vivid description of how he discovered that possibility. He emphasizes that he was inspired by T.M. Flett, see [Fle2] (based in turn, as so often in this field of research, on classical ideas of G.H. Hardy and J.E. Littlewood). The main point seems to be a psychological barrier which comes from the fact that the spaces L_p with $p < 1$ are really nasty from the point of view of harmonic analysis. The extension of F^s_{pq} in the sense of Definition 1.3.4 to $0 < p < \infty$, $0 < q \leqq \infty$ is also due to J. Peetre in 1975, see [Pee5]. The crucial point here is the technique of maximal functions developed in [FeS2] in connection with Hardy spaces and which can be carried over to the general F^s_{pq} spaces. (It is almost superfluous to remark that also the idea of maximal functions goes back to G.H. Hardy and J.E. Littlewood some 60 years ago, see [Tri21] for detailed references.) If one compares the two parts of the above definition then $p = \infty$ is missing in connection with the spaces F^s_{pq}. It comes out that a direct extension of the above definition of F^s_{pq} to $p = \infty$ is not possible. But we shall return to this problem in the next subsection.

Theorem. (i) Let $1 < p < \infty$, $1 \leqq q \leqq \infty$ and $s > 0$, then B^s_{pq} coincides with the spaces in Definition 1.2.5.

(ii) Let $s > 0$, then $B^s_{\infty\infty} = \mathcal{C}^s$.

(iii) Let $1 < p < \infty$ and $s \in \mathbb{R}$ then $F^s_{p2} = H^s_p$. In particular let $1 < p < \infty$ and $k \in \mathbb{N}_0$ then $F^k_{p2} = W^k_p$.

(iv) Let $0 < p < \infty$, then $F^0_{p2} = h_p$.

Remark 3. In this theorem we simply fixed what had been said before.

1.5.2 The spaces $F^s_{\infty q}$

An extension of (1.5.1/3) to $p = \infty$ and $0 < q < \infty$ is not reasonable. A corresponding space would depend on the chosen smooth dyadic resolution of unity in $\mathbb{R}^n$. However a proof of this claim is rather complicated and we refer for details to [Tri5: 2.1.4] where we proved a corresponding "non-multiplier" theorem. On the other hand, in the framework of a duality theory we have

$$(F^s_{pq})' = F^{-s}_{p'q'}, \quad 1 < p < \infty,\ 1 < q < \infty,\ \frac{1}{p} + \frac{1}{p'} = \frac{1}{q} + \frac{1}{q'} = 1. \tag{1}$$

In this book we shall not discuss duality theory and refer for details to [Trіß: 2.11]. It would be desirable to extend (1) to $p = 1$, in particular, the case $q = 2$ is of interest because $(F^0_{1,2})' = h'_1$ is the inhomogeneous version of BMO, see 1.7.1. In any case it is highly desirable to extend (1) to $p = 1$, i.e., $p' = \infty$ and to describe the corresponding spaces $F^\sigma_{\infty r}$.

Definition. Let $\Phi = \{\varphi_j\}_{j=0}^\infty$ be a smooth dyadic resolution of unity in $\mathbb{R}^n$ in the sense of (1.3.3/10–12) and let $\varphi_j(D)$ be given by (1.3.3/9) (with φ_j instead of φ). Let $1 < q \leqq \infty$ and $s \in \mathbb{R}$ then

$$F^s_{\infty q} = \Big\{ f \in S' \colon \text{there exists } \{f_j\}_{j=0}^\infty \subset L_\infty \text{ with } f = \sum_{j=0}^\infty \varphi_j(D) f_j \tag{2}$$
$$\text{in } S' \text{ and } \sup_{x\in\mathbb{R}^n} \Big(\sum_{j=0}^\infty 2^{jsq} |f_j(x)|^q \Big)^{1/q} < \infty \Big\}$$

(usual modification if $q = \infty$).

Remark 1. The norm in $F^s_{\infty q}$ is given by

$$\|f \mid F^s_{\infty q}\|_\Phi = \inf \sup_{x\in\mathbb{R}^n} \Big(\sum_{j=0}^\infty 2^{jsq} |f_j(x)|^q \Big)^{1/q} \tag{3}$$

where the infimum is taken over all possible representations of f of the above type. These spaces make sense, they are independent of Φ (equivalent norms) and they are Banach spaces. Furthermore, (1) can be extended to $1 \leqq p < \infty$ and $1 \leqq q < \infty$, see [Triß: 2.4.3].

Remark 2. The above spaces $F^s_{\infty q}$ have been introduced in [Tri8] (anisotropic case) and in [Tri5: p. 118]. In particular we proved a Littlewood–Paley theorem for BMO-spaces which will be explicitly formulated later on, see 1.7.1. Recently M. Frazier and B. Jawerth defined in [FrJ2] spaces $F^s_{\infty q}$ for all $0 < q \leqq \infty$, which cover the above spaces. They used atomic representations. See also [See2] in this connection.

1.5.3 Spaces of analytic functions as building blocks

We mentioned several times that the spaces B^s_{pq} are simpler than the spaces F^s_{pq} (at least at this stage in the game, later on, in connection with spaces on Riemannian manifolds and Lie groups, the situation will be different). Although we promised to avoid technicalities in this introductory chapter we wish to shed some light on the above claim. This may also be of self-contained interest and it provides a better understanding of what follows in the next subsection.

Let $\Omega = B(r)$ be the closed ball of radius $r > 0$ in $\mathbb{R}^n$ centered at the origin and let $0 < p \leqq \infty$, then

$$L^\Omega_p = \{f \in S' \colon \operatorname{supp} \hat{f} \subset \Omega, f \in L_p\}. \tag{1}$$

By the Paley–Wiener–Schwartz theorem L^Ω_p consists of entire analytic functions, furthermore, L^Ω_p is a quasi-Banach space. We formulate three typical assertions for these spaces:

(i) Let $\mathbb{Z}^n = \{k = (k_1, \ldots, k_n) \in \mathbb{R}^n \colon k_1, \ldots, k_n \text{ integers}\}$ be the usual lattice in $\mathbb{R}^n$. If $\kappa > 0$ is sufficiently small (in dependence on r) then

$$\Big(\sum_{k \in \mathbb{Z}^n} |f(\kappa k)|^p\Big)^{1/p} \sim \|f \mid L_p\|, \qquad f \in L^\Omega_p,\ 0 < p \leqq \infty, \tag{2}$$

(equivalent quasi-norms).

(ii) If $0 < p \leqq q \leqq \infty$ and if α is a given multi-index then

$$\|D^\alpha f \mid L_q\| \leqq c\|f \mid L_p\|, \qquad f \in L^\Omega_p, \tag{3}$$

where c depends on r, p, q, and α, but not on f.

(iii) Recall that the Hardy–Littlewood maximal function is given by

$$(Mg)(x) = \sup |B|^{-1} \int_B |g(y)| dy, \qquad x \in \mathbb{R}^n, \tag{4}$$

where the supremum is taken over all balls B in $\mathbb{R}^n$ centered at x. Let $\kappa > 0$ and let

$$f \in S', \qquad \operatorname{supp} \hat{f} \subset \Omega = B(r), \tag{5}$$

then

$$\sup_{z \in \mathbb{R}^n} \frac{|\nabla f(x-z)|}{1 + |z|^{n/\kappa}} \leqq c_1 \sup_{z \in \mathbb{R}^n} \frac{|f(x-z)|}{1 + |z|^{n/\kappa}} \leqq c_2 (M|f|^\kappa)^{1/\kappa}(x), \tag{6}$$

where c_1 and c_2 depend only on r and κ.

We return to the spaces B^s_{pq} given by (1.5.1/1,2). Then it is clear that f is decomposed into

$$\varphi_j(D)f \in L_p^\Omega \quad \text{with} \quad \Omega = B(2^{j+1}); \quad j \in \mathbb{N}_0. \tag{7}$$

Now one can believe why the theory of the spaces B^s_{pq} is simpler than the theory of the spaces F^s_{pq}. If $f \in B^s_{pq}$ then the pieces $\varphi_j(D)f$ can be separately treated within the theory of the beautiful spaces L_p^Ω, where (2,3,6) are only few highlights, and the dependence of the involved constants on j can be calculated easily by homogeneity arguments. Afterwards one has to sum up the resulting estimates in the sense of (1.5.1/2). For example, (2) and (1.5.1/2) provide a better understanding of the structure of B^s_{pq} mentioned at the end of 1.2.5, whereas (3) is the basis for the embedding theory of the spaces B^s_{pq}. Furthermore, (6) can be used to prove maximal inequalities for B^s_{pq} spaces. In this context we recall the maximal inequality

$$\|Mg \mid L_\lambda\| \leqq c_\lambda \|g \mid L_\lambda\|, \qquad 1 < \lambda \leqq \infty, \; g \in L_\lambda. \tag{8}$$

For the spaces F^s_{pq} the situation is more complicated. The pieces $\varphi_j(D)f$, see (7), cannot be treated one after the other, but they have to be considered simultaneously, see (1.5.1/3). For instance, maximal inequalities for F^s_{pq} spaces can be obtained on the basis of (6) and the vector-valued counterpart of (8),

$$\left\| \left(\sum_{k=1}^{\infty} |Mg_k(\cdot)|^\kappa \right)^{1/\kappa} \mid L_\lambda \right\| \leqq c \left\| \left(\sum_{k=1}^{\infty} |g_k(\cdot)|^\kappa \right)^{1/\kappa} \mid L_\lambda \right\|, \tag{9}$$

$1 < \lambda < \infty$, $1 < \kappa \leqq \infty$, $g_k \in L_\lambda$.

The above rough discussion should convince the reader that

(i) the spaces L_p^Ω deserve to be studied for their own sake,

(ii) the spaces L_p^Ω are building blocks for the spaces B^s_{pq} and

(iii) the spaces F^s_{pq} are more complicated than the spaces B^s_{pq}.

Remark. Although this subsection may be considered as a digression it seems to be appropriate to incorporate some (historical) references about the above mentioned facts. Inequalities of type (2) and (3) have a long history and play a crucial role in the theory of function spaces. The equivalence (2) is essentially due to M. Plancherel and G. Pólya, see [PlP] (1937) and [Boa: 197–199]. Inequality (3), restricted to $1 \leqq p \leqq q \leqq \infty$, is the famous Nikol'skij inequality, see [Nik1] (1951) and [Nik2]. A proof of (6) was given in [Tri4: 1.3.2] (1977) based on the technique of maximal functions developed in [FeS2] and [Pee5], see also [Triß: 1.3.2] and [ScT: 1.4.2]. The maximal inequality (8) is one of the outstanding theorems of functional analysis. The case $n = 1$ is due to G.H. Hardy and J.E. Littlewood [HaL1] (1930), it was extended by N. Wiener [Wie] (1939) to $n > 1$. The vector-valued version (9) goes back to C. Fefferman and E.M. Stein [FeS1] (1971). Systematic treatments on the theory of the spaces L_p^Ω including weighted and mixed-normed generalizations may be found in [Tri4, Triß, ScT].

1.5.4 Approximations by entire analytic functions

Since the fundamental thesis of D. Jackson (1911), it is well-known that smoothness of functions can also be measured by approximation schemes. Examples of distinguished systems of approximating functions are algebraic and trigonometric polynomials, and entire analytic functions in $\mathbb{R}^n$, where the latter may be considered as the natural counterpart of trigonometric polynomials in the case of the n-torus. It is beyond the scope of this book to discuss the diverse aspects of approximation theory in connection with function spaces. The interested reader may consult [Lor, BuB, BuN, Zyg2, Pee6, Nik2]. We restrict ourselves to few assertions which are directly connected with the theory of the spaces B^s_{pq} and F^s_{pq} and the consideration of the preceding subsection.

Let again $B(r)$ be the ball in $\mathbb{R}^n$ of radius $r > 0$ centered at the origin. Let

$$B^j = B(2^j); \qquad j \in \mathbb{N}_0, \tag{1}$$

and

$$A^0 = B^1, \;\; A^k = B^{k+1}/B^{k-1}; \;\; k \in \mathbb{N}. \tag{2}$$

Definition. Let $0 < p \leqq \infty$ then

$$\alpha_p = \{a = (a_j)_{j=0}^\infty : a_j \in S' \cap L_p, \; \operatorname{supp} \hat{a}_j \subset A^j\} \tag{3}$$

and

$$\beta_p = \{b = (b_j)_{j=0}^\infty : b_j \in S' \cap L_p, \; \operatorname{supp} \hat{b}_j \subset B^j\}. \tag{4}$$

Remark 1. In other words, both α_p and β_p are systems of entire analytic functions where the supports of the Fourier transforms of the involved distributions are restricted to dyadic annuli and dyadic balls, respectively. In the periodic case b_j corresponds to a trigonometrical polynomial of degree 2^j, see Remark 4 below.

Theorem. (i) Let $0 < p \leqq \infty$, $0 < q \leqq \infty$ and $s \in \mathbb{R}$, then

$$B^s_{pq} = \Big\{ f \in S'\colon \text{there exists } a \in \alpha_p \text{ with } f = \sum_{j=0}^{\infty} a_j \text{ in } S' \text{ and} \tag{5}$$

$$\|f \mid B^s_{pq}\|^a = \Big(\sum_{j=0}^{\infty} 2^{jsq} \|a_j \mid L_p\|^q \Big)^{1/q} < \infty \Big\}.$$

(ii) Let $0 < p \leqq \infty$, $0 < q \leqq \infty$ and $s > n(\frac{1}{p} - 1)_+$, then

$$B^s_{pq} = \Big\{ f \in S'\colon \text{there exists } b \in \beta_p \text{ with } f = \lim_{j\to\infty} b_j \text{ in } S' \text{ and} \tag{6}$$

$$\|f \mid B^s_{pq}\|^b = \|b_0 \mid L_p\| + \Big(\sum_{j=1}^{\infty} 2^{jsq} \|f - b_j \mid L_p\|^q \Big)^{1/q} < \infty \Big\}.$$

(iii) Let $0 < p < \infty$, $0 < q \leqq \infty$ and $s \in \mathbb{R}$, then

$$F^s_{pq} = \Big\{ f \in S'\colon \text{there exists } a \in \alpha_p \text{ with } f = \sum_{j=0}^{\infty} a_j \text{ in } S' \text{ and} \tag{7}$$

$$\|f \mid F^s_{pq}\|^a = \Big\| \Big(\sum_{j=0}^{\infty} 2^{jsq} \mid a_j(\cdot)|^q \Big)^{1/q} \mid L_p \Big\| < \infty \Big\}.$$

(iv) Let $0 < p < \infty$, $0 < q \leqq \infty$ and $s > n(\frac{1}{\min(p,q)} - 1)_+$, then

$$F^s_{pq} = \Big\{ f \in S'\colon \text{there exists } b \in \beta_p \text{ with } f = \lim_{j\to\infty} b_j \text{ in } S' \text{ and} \tag{8}$$

$$\|f \mid F^s_{pq}\|^b = \|b_0 \mid L_p\| + \Big\| \Big(\sum_{j=1}^{\infty} 2^{jsq} |(f - b_j)(\cdot)|^q \Big)^{1/q} \mid L_p \Big\| < \infty \Big\}.$$

Remark 2. If $q = \infty$ then one has to modify (5–8) in the usual way. Furthermore

$$\inf \|f \mid B^s_{pq}\|^a, \ \inf \|f \mid B^s_{pq}\|^b, \ \inf \|f \mid F^s_{pq}\|^a, \ \inf \|f \mid F^s_{pq}\|^b,$$

where the infima are taken with respect to the corresponding admissible systems of functions, are equivalent quasi-norms in the respective spaces B^s_{pq} and F^s_{pq}. In connection with (6) and (8), there arises the question whether $f \in L_p$ if $f \in B^s_{pq}$. This is clear if $p \geqq 1$, but it is also valid if $0 < p < 1$, but then one needs $s > n(\frac{1}{p} - 1)$, see the discussion in [Trіß: 2.5.3]. Furthermore, characterizations of type (ii) and (iv) cannot be expected if $s < n(\frac{1}{p} - 1)_+$, see again [Trіß: 2.5.3].

We wish to modify (i) somewhat. Let $f \in S' \cap L_p$, then

$$E_p^j(f) = \inf \|f - g \mid L_p\|, \tag{9}$$

where the infimum is taken over all $g \in S' \cap L_p$ with $\operatorname{supp} \hat{g} \subset B^j$. These are best approximations of f by entire analytic functions of a given order.

Proposition. Let $0 < p \leqq \infty$, $0 < q \leqq \infty$ and $s > n(\frac{1}{p} - 1)_+$, then

$$B_{pq}^s = \left\{ f \in S' \cap L_p \colon \|f \mid L_p\| + \left(\sum_{j=0}^{\infty} 2^{jsq} E_p^j(f)^q \right)^{1/q} < \infty \right\}. \tag{10}$$

(equivalent quasi-norms).

Remark 3. It is one of the striking discoveries of the first decades of our century that smoothness of functions can be expressed in terms of approximation schemes. In this context smoothness is described via derivatives and differences in the language of Hölder–Zygmund and Besov spaces, see Definitions 1.2.2/2 and 1.2.5. On the other hand, if the underlying domain is a cube or the n-torus then one studies approximations via algebraic or trigonometric polynomials. The $\mathbb{R}^n$-counterpart is the approximation of functions via entire analytic functions. After the pioneer work of D. Jackson (1911) many mathematicians contributed to this flourishing field of research. In particular the work of S.N. Bernštejn in the non-periodic case, i.e., approximation by entire analytic functions, influenced decisively the development of this theory in the Soviet Union since the late forties, see [Nik2: 5.5, 5.6 and the related note sections]. Characterizations of type (10) for $\mathcal{C}^s = B^s_{\infty\infty}$, $s > 0$, are due to A. Zygmund [Zyg1], whereas the corresponding characterizations for $B^s_{p\infty}$, $s > 0$, $1 < p < \infty$ and for B^s_{pq}, $s > 0$, $1 < p < \infty$, $1 \leqq q < \infty$, are due to S.M. Nikol'skij [Nik1] (1951) and O.V. Besov [Bes2] (1961), respectively. However many other mathematicians contributed to this field, both the periodic and the non-periodic case. Characterizations of type (5) (with $1 < p < \infty$, $1 \leqq q \leqq \infty$, $s \in \mathbb{R}$) and (7) (with $1 < p < \infty$, $1 < q < \infty$, $s \in \mathbb{R}$) had been used by the author to define the respective spaces, see [Tri2] (1973). The above theorem in its full generality may be found in [Triß: 2.5.2, 2.5.3], see also [Triα: 2.3] and [Tri5: 2.2.3] for further references and comments.

Remark 4. In connection with the above discussion it seems to be in order to say a word about periodic spaces. Let

$$\mathbb{T}^n = \{x = (x_1, \ldots, x_n) \in \mathbb{R}^n \colon |x_j| \leqq \pi\}$$

be the n-torus with the usual convention that opposite points are identified. Then $f \in D'(\mathbb{T}^n)$ can be represented as

$$f = \sum_{k \in \mathbb{Z}^n} \hat{f}(k) e^{ikx}, \tag{11}$$

where $\mathbb{Z}^n$ is again the lattice of all points $k \in \mathbb{R}^n$ with integer-valued components. It is quite clear how to define the periodic counterparts of $\mathcal{C}^s$ (Definition 1.2.2/2), W_p^k (Definition 1.2.3) and B^s_{pq} (Definition 1.2.5) in terms of differences and derivatives. There is also no difficulty to find the periodic counterparts of B^s_{pq} and F^s_{pq} in the sense of Definition 1.5.1. For examle, if p, q, s and $\{\varphi_j\}_{j=0}^{\infty}$ have the same meaning as in Definition 1.5.1(ii) then

$$F^s_{pq}(\mathbb{T}^n) = \left\{ f \in D'(\mathbb{T}^n) \colon \left\| \left(\sum_{j=0}^{\infty} 2^{jsq} \left| \sum_{k \in \mathbb{Z}^n} \varphi_j(k) \hat{f}(k) e^{ikx} \right|^q \right)^{1/q} |L_p(\mathbb{T}^n) \right\| < \infty \right\} \tag{12}$$

where f is given by (11). In this sense one can develop a theory of trigonometric series in the language of function spaces. In particular the periodic counterpart of what has been discussed in this subsection is the problem of approximation of periodic functions and distributions by trigonometric polynomials. We do not go into any detail, an extensive treatment of all these problems, including strong summability, has been given in [ScT: Chapter 3].

1.6 Interpolation methods

1.6.1 An introductory remark

We return to the situation around 1960. This time may be considered as a first culmination point in the theory of function spaces. Almost simultaneously, the Besov spaces B^s_{pq} from Definition 1.2.5, the fractional Sobolev spaces H^s_p from Definition 1.3.2, the real variable Hardy spaces $H_p(\mathbb{R}^{n+1}_+)$ from 1.4.2 and the space of functions of bounded mean oscillation BMO from 1.7.1 came into being. All these spaces had been defined by quite different concrete procedures. It arose the urgent need to find unifying methods. The first impressing major step in this direction was done around 1960 with the creation of abstract interpolation theory. Concrete interpolation theorems in spaces of L_p type have a longer history. In 1926 M. Riesz proved a first version of the nowadays celebrated Riesz–Thorin theorem. Afterwards, in the late thirties and, in particular in the fifties, many concrete interpolation theorems had been proven. Detailed references may be found in [Triα: 1.18]. In 1958–1961, the abstract interpolation theory had been established by J.L. Lions, E. Gagliardo,

A.P. Calderón, S.G. Krejn and N. Aronszajn. Comprehensive treatments of interpolation theory, including many historical references, may be found in [BeL, Triα, KPS, BKS]. The dominating figure in interpolation theory in the sixties and early seventies was J. Peetre. In particular, he created the so-called K-functional which is the crucial instrument in real interpolation up to our time. At the beginning, in the early sixties, abstract interpolation theory was closely connected with PDE's and function spaces. Even more, applications to PDE's and function spaces were the motivating force behind interpolation theory at this stage of the theory. Nowadays interpolation theory is a technically highly developed sub-branch of functional analysis; the book by V.I. Ovchinnikov [Ovc] gives an impression of some recent trends in interpolation theory; see also [Mal] where one finds on some 120 pages almost all references in interpolation theory from 1926 to 1988. Compared with this overwhelming flood of papers, our aim is rather modest. We restrict ourselves to a description of the two outstanding classical interpolation methods, the real one and the complex one, including some more recent extensions to quasi-Banach spaces, and their interplay with function spaces of the above type. There is only one major exception. We could not withstand the temptation to include the semi-group approach in real interpolation which shows in a striking way the intense interplay between concrete assertions in the theory of Sobolev–Besov spaces and their abstract counterparts.

1.6.2 Real interpolation

Let A_0 and A_1 be two complex quasi-Banach spaces, then $\{A_0, A_1\}$ is called an interpolation couple if there exists a linear complex Hausdorff space A such that both A_0 and A_1 are linearly and continuously embedded in A. Then $A_0 \cap A_1$ and

$$A_0 + A_1 = \{a \in A \colon a = a_0 + a_1 \text{ for some } a_0 \in A_0 \text{ and } a_1 \in A_1\} \tag{1}$$

make sense. We equip $A_0 \cap A_1$ with the quasi-norm $\|a \mid A_0 \cap A_1\| + \|a \mid A_0\| + \|a \mid A_1\|$ and get a quasi-Banach space. Let $t > 0$, then Peetre's K-functional is given by

$$K(t, a) = \inf(\|a_0 \mid A_0\| + t\|a_1 \mid A_1\|), \qquad a \in A, \tag{2}$$

where the infimum is taken over all representations $a = a_0 + a_1$ with $a_0 \in A_0$ and $a_1 \in A_1$. For fixed $t > 0$, the linear space $A_0 + A_1$, equipped with the quasi-norm $K(t, a)$, becomes a quasi-Banach space. It is quite obvious that, for different values of t, the quasi-norms $K(t, a)$ are mutually equivalent.

Definition (Real interpolation). Let $\{A_0, A_1\}$ be the above interpolation couple. Let $0 < \theta < 1$ and $0 < q < \infty$, then

$$(A_0, A_1)_{\theta,q} = \Big\{ a \in A_0 + A_1 \colon \|a \mid (A_0, A_1)_{\theta,q}\| = \Big(\int_0^\infty t^{-\theta q} K(t,a)^q \frac{dt}{t} \Big)^{1/q} < \infty \Big\} \tag{3}$$

and

$$(A_0, A_1)_{\theta,\infty} = \Big\{ a \in A_0 + A_1 \colon \|a \mid (A_0, A_1)_{\theta,\infty}\| = \sup_{t>0} t^{-\theta} K(t,a) < \infty \Big\}. \tag{4}$$

Remark 1. The above form of the real interpolation method for Banach spaces is due to J. Peetre. Its extension to quasi-Banach spaces is obvious. Historical references and detailed studies may be found in [Triα] and [BeL]. All spaces $(A_0, A_1)_{\theta,q}$ with $0 < \theta < 1$, $0 < q \leqq \infty$, are quasi-Banach spaces. If A_0 and A_1 are Banach spaces, then $(A_0, A_1)_{\theta,q}$ with $0 < \theta < 1$ and $1 \leqq q \leqq \infty$, are also Banach spaces.

Remark 2 (Interpolation property). The basic assertion in interpolation theory is the following: Let $\{A_0, A_1\}$ and $\{B_0, B_1\}$ be two interpolation couples and let T be a linear operator mapping from $A_0 + A_1$ into $B_0 + B_1$ such that its restriction to A_j is a linear and bounded operator from A_j into B_j, where $j = 0, 1$. Then the restriction of T to $(A_0, A_1)_{\theta,q}$ is a linear and bounded operator from $(A_0, A_1)_{\theta,q}$ into $(B_0, B_1)_{\theta,q}$. The proof of this crucial assertion follows immediately from the above definition.

1.6.3 Complex interpolation of Banach spaces

Let $\{A_0, A_1\}$ be an interpolation couple, where we now assume that both A_0 and A_1 are complex Banach spaces. Let $A = A_0 + A_1$ and

$$\sigma = \{ z \in \mathbb{C} \colon 0 < \operatorname{Re} z < 1 \} \tag{1}$$

be a strip in the complex plane. Then $F[A]$ is the collection of all functions $f(z)$, defined on $\overline{\sigma}$, with the following two properties:

(i) $f(z)$ is A-continuous in $\overline{\sigma}$ and A-analytic in σ with

$$\sup_{z \in \overline{\sigma}} \|f(z) \mid A\| < \infty, \tag{2}$$

(ii) $f(it) \in A_0$ and $f(1+it) \in A_1$, with $t \in \mathbb{R}$, are continuous in the respective Banach spaces and

$$\|f \mid F[A]\| = \sup_{t\in\mathbb{R}}(\|f(it) \mid A_0\| + \|f(1+it) \mid A_1\|) < \infty. \tag{3}$$

By the maximum principle $F[A]$ is a Banach space (here one needs that A_0 and A_1 are Banach spaces and not merely quasi-Banach spaces).

Definition. Let $\{A_0, A_1\}$ be an interpolation couple of Banach spaces and let $0 < \theta < 1$. Let $A = A_0 + A_1$ then

$$[A_0, A_1]_\theta = \{a \in A\colon \text{there exists } f(z) \in F[A] \text{ with } f(\theta) = a\}. \tag{4}$$

Remark. Let

$$\|a \mid [A_0, A_1]_\theta\| = \inf \|f \mid F[A]\| \tag{5}$$

where the infimum is taken over all $f \in F[A]$ with $f(\theta) = a$. Then $[A_0, A_1]_\theta$ becomes a Banach space which has the interpolation property from Remark 1.6.2/2. References and further properties may be found in [Triα: 1.9].

1.6.4 Fractional Sobolev spaces and Besov spaces as interpolation spaces

At the beginning, in the early sixties, abstract interpolation theory was closely connected with the theory of function spaces. Even more, it was one of the highlights at this time that Besov spaces and fractional Sobolev spaces could be interpreted as interpolation spaces of Sobolev spaces, see the theorem below. On the one hand, this shed new light on the theory of function spaces and on the other hand, it stimulated the further development of interpolation theory. Let W_p^k be the Sobolev spaces from (1.2.3/1), B_{pq}^s be the Besov spaces from Definition 1.2.5 and H_p^s be the fractional Sobolev spaces from (1.3.2/2).

Theorem. Let $1 < p < \infty$, $k_0 \in \mathbb{N}_0$ and $k_1 \in \mathbb{N}_0$ with $k_0 \neq k_1$. Let $1 \leqq q \leqq \infty$, $0 < \theta < 1$ and

$$s = k_0(1-\theta) + k_1\theta, \tag{1}$$

then

$$(W_p^{k_0}, W_p^{k_1})_{\theta,q} = B_{pq}^s \tag{2}$$

and

$$[W_p^{k_0}, W_p^{k_1}]_\theta = H_p^s. \tag{3}$$

Remark. Both formulas, including several generalizations, are known since the early sixties. One can find (2) in the celebrated survey [LiP] (1964) and (3) in [Cal3] (1963), see also [Triα: 2.4] for detailed references, further discussions and extensions. Some generalizations will be described below, see 1.6.7. The formulas (2) and (3) show in a striking way that the classical Besov spaces B_{pq}^s and the fractional Sobolev spaces H_p^s with $s > 0$ can be generated by interpolation of Sobolev spaces. If one looks at (1.3.2/2,4) then one gets the feeling that (3) comes from the "interpolation" of the weights $(1+|\xi|^2)^{k_0/2}$ and $(1+|\xi|^2)^{k_1/2}$. This impression is completely correct and it explains the following generalization of (3) which also may be found in [Cal3]. Let $1 < p < \infty$, $s_0 \in \mathbb{R}$ and $s_1 \in \mathbb{R}$ then

$$[H_p^{s_0}, H_p^{s_1}]_\theta = H_p^s \text{ with } 0 < \theta < 1 \text{ and } s = (1-\theta)s_0 + \theta s_1. \tag{4}$$

To provide a better understanding of (2) one has to explain how derivatives, real interpolation and differences are interrelated. This will be done in the next subsection in an abstract setting.

1.6.5 Semi-groups of operators and interpolation

In the classical theory of function spaces of the above type one studies, among other things, embeddings between different function spaces, traces on hyperplanes, duality assertions, equivalent norms connected with harmonic and thermic functions, etc. In the sixties and early seventies, the astonishing fact had been discovered that many of these concrete assertions have counterparts within abstract interpolation theory. In [Triα] we gave a thorough description of this close interrelation, including detailed references. In this subsection we restrict ourselves to a distinguished example of this abstract theory which provides a better understanding of (1.6.4/2) and its abstract counterparts.

Let A be a complex Banach space, then the family $\{G(t),\ 0 \leqq t < \infty\}$ of linear and bounded operators mapping A into itself is called a strongly continuous semi-group if

$$G(t_1)G(t_2) = G(t_1 + t_2) \text{ where } 0 \leqq t_1 < \infty \text{ and } 0 \leqq t_2 < \infty, \tag{1}$$

$$G(0) = id \quad (\text{identity in } A) \text{ and} \tag{2}$$

$$\lim_{\tau \to t} G(\tau)a = G(t)a, \text{ where } 0 \leqq t < \infty \text{ and } a \in A \tag{3}$$

(with $\tau \downarrow 0$ if $t = 0$). For sake of simplicity we assume

$$\|G(t)\| \leqq 1, \qquad 0 \leqq t < \infty. \tag{4}$$

The infinitesimal generator Λ of $\{G(t)\}$ is given by

$$\Lambda a = \lim_{t \downarrow 0} t^{-1}[G(t)a - a] \tag{5}$$

with the domain of definition

$$D(\Lambda) = \{a \in A\colon \text{the limit in (5) exists}\}. \tag{6}$$

Recall that Λ is a densely defined closed operator. The theory of semigroups of operators can be found, for example, in [Yos] or [BuB]. Let Λ^m with $m \in \mathbb{N}$ be the usual powers of Λ, then the domain of definition $D(\Lambda^m)$ equipped with the norm

$$\|a \mid D(\Lambda^m)\| = \|a \mid A\| + \|\Lambda^m a \mid A\| \tag{7}$$

becomes a Banach space. Of course $\{A, D(\Lambda^m)\}$ is an interpolation couple.

Theorem. Let $0 < \theta < 1$, $1 \leqq q \leqq \infty$ and $m \in \mathbb{N}$ then

$$(A, D(\Lambda^m))_{\theta,q} = \left\{ a \in A\colon \left(\int_0^\infty t^{-\theta m q} \|(G(t) - id)^m a \mid A\|^q \frac{dt}{t} \right)^{1/q} < \infty \right\} \tag{8}$$

(modification if $q = \infty$).

Remark 1. The connection between interpolation theory and semi-groups of operators had been discovered by J.L. Lions around 1960, see [Triα: 1.13] for references and further results. Furthermore,

$$\|a \mid A\| + \left(\int_0^\infty t^{-\theta m q} \|(G(t) - id)^m a \mid A\|^q \frac{dt}{t} \right)^{1/q} \tag{9}$$

is an (equivalent) norm in $(A, D(\Lambda^m))_{\theta,q}$.

Example. Let $A = L_p(\mathbb{R})$ with $1 < p < \infty$, then the translations

$$G(t)\colon f(x) \longrightarrow f(x+t), \qquad t \geqq 0,\ x \in \mathbb{R}, \tag{10}$$

generate a strongly continuous semi-group of operators in the above sense. Let Λ be the corresponding infinitesimal generator, then

$$\Lambda^m f(x) = \frac{d^m f}{dx^m}(x), \qquad D(\Lambda^m) = W_p^m(\mathbb{R}), \tag{11}$$

where $W_p^m(\mathbb{R})$ is the Sobolev space from (1.2.3/1) (with $n = 1$). We refer for details and extensions to $n > 1$ to [Triα: 2.5.1]. In other words, the interpolation spaces $(L_p, W_p^m)_{\theta,q}$ (one-dimensional case) coincide with the left-hand side of (8) and can be calculated by the right-hand side of (8) with (9) as an equivalent norm. We have

$$(G(t) - id)^m f(x) = \Delta_t^m f(x), \qquad x \in \mathbb{R},\ t \geqq 0, \tag{12}$$

where Δ_t^m are the usual differences (one-dimensional case), see (1.2.2/3). Then (9) coincides with

$$\|f \mid L_p\| + \left(\int_0^\infty t^{-sq} \|\Delta_t^m f \mid L_p\|^q \frac{dt}{t} \right)^{1/q}, \qquad s = \theta m. \tag{13}$$

However (13) is an equivalent norm in B_{pq}^s, see (1.2.5/3). In other words, we obtain a special case of (1.6.4/2). Extensions to several dimensions, further equivalent norms and references may be found in [Triα: 2.5.1].

Remark 2. One can ask for other possibilities to use (8) in connection with Sobolev–Besov spaces. The procedure seems to be quite clear. Let, for example, $A = L_p = L_p(\mathbb{R}^n)$ with $1 < p < \infty$. What about semi-groups such that the infinitesimal generator Λ coincides with $-\Delta$ or $(-\Delta)^k$, or closely connected differential operators, where $k > 0$, and Δ stands for the Laplacian. In such a case one can expect that $D(\Lambda^m)$ coincides with some Sobolev spaces W_p^l. Then (8) and (1.6.4/2) yield new equivalent norms for Besov spaces. There are two distinguished examples in this direction: The Cauchy–Poisson semi-group given by

$$G(t)f(x) = (e^{-t|\xi|} \hat{f}(\xi))^\vee(x), \qquad t \geqq 0, \tag{14}$$

and the Gauss–Weierstrass semi-group given by

$$G(t)f(x) = (e^{-t|\xi|^2} \hat{f}(\xi))^\vee(x), \qquad t \geqq 0. \tag{15}$$

We used the notations introduced in 1.2.4, see also (1.3.3/8,9). These two semi-groups describe harmonic and thermic extensions to $\mathbb{R}^{n+1}_+$ of functions defined on $\mathbb{R}^n$. We return to this problem in detail in 1.8. At this moment we remark that the corresponding infinitesimal generators are closely connected with $(-\Delta)^{1/2}$ and Δ, respectively, and that (8) gives in principle the possibility to characterize Besov spaces in terms of Cauchy–Poisson or Gauss–Weierstrass semi-groups. However, we have to add a comment. Both semi-groups are so-called analytic semi-groups and for analytic semi-groups one has further and even more handsome characterizations of $(A, D(\Lambda^m))_{\theta,q}$ on an abstract level. The specialization of these new characterizations to the two above semi-groups yields the well-known descriptions of Besov spaces in terms of harmonic and thermic extensions. Details about the underlying abstract theory and applications to the above semi-groups may be found in [Triα: 1.14.5, 2.5.2, 2.5.3]. Further information and precise formulations are also given in 1.8.

1.6.6 Complex interpolation of quasi-Banach spaces

An extension of the complex interpolation method for Banach spaces described in 1.6.3 to quasi-Banach spaces is not possible, in sharp contrast to the real method from 1.6.2. In recent times several proposals had been made to extend the complex interpolation method from Banach spaces to quasi-Banach spaces. But on an abstract level the problem is complicated. We shall give some references in the next subsection. The situation improves essentially if one specializes the quasi-Banach spaces in question. The complex interpolation of the Hardy spaces from 1.4.4 and their homogeneous counterparts are outstanding examples which attracted much attention and which stimulated the discussion about complex interpolation of quasi-Banach spaces, see the references in the next subsection. We describe here a method which is due to A.P. Calderón and A. Torchinsky, see [CaT], in connection with Hardy spaces and which was extended by L. Päivärinta, [Päi1], and the author to spaces of type B^s_{pq} and F^s_{pq}, see [Tri10,11] and [Trißβ: 2.4].

Let again S' be the collection of all complex tempered distributions in $\mathbb{R}^n$ and let D be the collection of all complex compactly supported C^∞ functions in $\mathbb{R}^n$. The strip σ in the complex plane has the same meaning as in (1.6.3/1). We deal with S'-valued functions $f(z)$ in the strip $\overline{\sigma}$. Such a function is called S'-analytic if for any $\varphi \in D$, $(x, z) \to (\varphi \hat{f})^\vee(x, z)$ is uniformly continuous and bounded in $\mathbb{R}^n \times \overline{\sigma}$ and $z \to (\varphi \hat{f})^\vee(x, z)$ is analytic in σ for any fixed $x \in \mathbb{R}^n$. Recall that by the Paley–Wiener–Schwartz theorem $x \to (\varphi \hat{f})^\vee(x, z)$ is also analytic for any $z \in \overline{\sigma}$.

Next we define the counterpart of $F[A]$ from 1.6.3. Let $A = \{A_0, A_1\}$ where A_k with $k = 0$ or $k = 1$ is given by

$$\text{either } B^{s_k}_{p_k q_k} \text{ with } s_k \in \mathbb{R},\ 0 < p_k \leqq \infty,\ 0 < q_k \leqq \infty,$$

or $F^{s_k}_{p_k q_k}$ with $s_k \in \mathbb{R}$, $0 < p_k < \infty$, $0 < q_k \leqq \infty$.

Then $F(A)$ is the collection of all S'-analytic functions $f(z)$ with $f(it) \in A_0$ and $f(1+it) \in A_1$ for all $t \in \mathbb{R}$ and

$$\|f \mid F(A)\| = \sup_{t\in\mathbb{R}} (\|f(it) \mid A_0\| + \|f(1+it) \mid A_1\|) < \infty. \tag{1}$$

The rest is now the same as in 1.6.3.

Definition. Let $A = \{A_0, A_1\}$ be the above couple and let $0 < \theta < 1$, then

$$(A_0, A_1)_\theta = \{g \in S' \colon \text{there exists an element } f(z) \in F(A) \text{ with } f(\theta) = g\}. \tag{2}$$

Remark. Let

$$\|g \mid (A_0, A_1)_\theta\| = \inf \|f \mid F(A)\| \tag{3}$$

where the infimum is taken over all $f(z) \in F(A)$ with $f(\theta) = g$. One can prove that both $F(A)$ and $(A_0, A_1)_\theta$ are quasi-Banach spaces, see [Trіß: 2.4.4] and the references given there. If A_0 and A_1 are Banach spaces, then we have two complex methods, $[A_0, A_1]_\theta$, see 1.6.3, and $(A_0, A_1)_\theta$. As we shall see it may happen in limiting cases that these two methods yield different results.

1.6.7 Interpolation of B^s_{pq} and F^s_{pq}

It was the aim of Theorem 1.6.4 to demonstrate the unifying power of interpolation methods. Now we add further interpolation results both for the real method from 1.6.2 and the two complex methods from 1.6.3 (restricted to Banach spaces) and from 1.6.6. The spaces B^s_{pq} and F^s_{pq} have the same meaning as in 1.5.1.

Theorem. (i) Let $s_0 \in \mathbb{R}$, $s_1 \in \mathbb{R}$, $0 < q_0 \leqq \infty$, $0 < q_1 \leqq \infty$, $0 < p_0 < \infty$ and $0 < p_1 < \infty$. Let $0 < \theta < 1$,

$$s = (1-\theta)s_0 + \theta s_1 \tag{1}$$

and

$$\frac{1}{p} = \frac{1-\theta}{p_0} + \frac{\theta}{p_1}, \qquad \frac{1}{q} = \frac{1-\theta}{q_0} + \frac{\theta}{q_1}, \tag{2}$$

then

$$(B^{s_0}_{p_0q_0}, B^{s_1}_{p_1q_1})_\theta = B^s_{pq} \tag{3}$$

and

$$(F^{s_0}_{p_0q_0}, F^{s_1}_{p_1q_1})_\theta = F^s_{pq}. \tag{4}$$

(ii) Let $s_0 \in \mathbb{R}$, $s_1 \in \mathbb{R}$, $1 < q_0 < \infty$, $1 < q_1 < \infty$, $1 < p_0 < \infty$ and $1 < p_1 < \infty$. Let $0 < \theta < 1$ and let s, p, q be given by (1,2), then

$$[B^{s_0}_{p_0q_0}, B^{s_1}_{p_1q_1}]_\theta = B^s_{pq} \tag{5}$$

and

$$[F^{s_0}_{p_0q_0}, F^{s_1}_{p_1q_1}]_\theta = F^s_{pq}. \tag{6}$$

(iii) Let $s_0 \in \mathbb{R}$ and $s_1 \in \mathbb{R}$ with $s_0 \neq s_1$; let $0 < q_0 \leqq \infty$ and $0 < q_1 \leqq \infty$. Let $0 < q \leqq \infty$, $0 < p < \infty$, $0 < \theta < 1$ and let s be given by (1) then

$$(B^{s_0}_{pq_0}, B^{s_1}_{pq_1})_{\theta,q} = (F^{s_0}_{pq_0}, F^{s_1}_{pq_1})_{\theta,q} = B^s_{pq}. \tag{7}$$

Remark 1. Formula (5), the special case (1.6.4/4) of (6) and the restriction of (7) to the case of Banach spaces and with the fractional Sobolev spaces instead of the F-spaces are known since the early sixties. Beside the above-mentioned papers [Cal3, LiP] we refer also to [Gri, Tai], but many other mathematicians contributed at this time to this flourishing field of research. In [Triα: Remark 2.4.1/5 and 2.4.2/3] we listed carefully (we hope) the relevant names and papers. Furthermore, the extension of (7) from the fractional Sobolev spaces to the F-spaces and in particular, from the case of Banach spaces to the full range of the involved parameters is known since 1973–78, we refer to [Pee3,6, Jaw1,2, Tri5: 2.2.10]. Formula (6) may be found in [Tri2]. Recall $h_p = F^0_{p2}$, see Theorem 1.5.1(iv). Then

$$(h_{p_0}, h_{p_1})_\theta = h_p, \quad 0 < p_0 < \infty, \quad 0 < p_1 < \infty, \quad \frac{1}{p} = \frac{1-\theta}{p_0} + \frac{\theta}{p_1}, \tag{8}$$

is a special case of (4). The homogeneous counterpart of (8) is due to [CaT: II] (1977). The extension of (8) to (4), and also (3), have been proved in 1980/81, see [Päi1, Tri10,11]. Further details may be found in [Triß: 2.4].

Remark 2. The extension of the real interpolation method from a Banach space setting to quasi-Banach spaces causes no serious problem. However, for the complex method the situation is quite different. There exist apparently two different proposals to introduce a complex interpolation method for quasi-Banach spaces. We described above the method which started in 1977 with the second part of [CaT]. On the other hand, N. Riviere suggested in 1966 (unpublished) another way which, at least in principle, is nearer to A.P. Calderón's original complex method for Banach spaces. But only recently substantial progress has been made. In our context the paper by M. Cwikel, M. Milman and Y. Sagher [CMS2] (1986) is of special interest. In particular, an interpolation theorem of type (8) is proved and seemingly it can be extended to an assertion of type (4). In this connection we refer also to [RVVW, CwS, Vig].

Remark 3. In the above theorem we collected some typical and interesting results. But in [Triα: 2.4] and [Triß: 2.4] one finds many other interpolation theorems. For example, some of the above results can be extended to limiting cases which, in turn, cover the Hölder–Zygmund spaces $\mathcal{C}^s = B^s_{\infty\infty}$. Furthermore, if $1 < p < \infty$, $s_0 \in \mathbb{R}$, $s_1 \in \mathbb{R}$, $0 < \theta < 1$ and if s is given by (1) then

$$(B^{s_0}_{p\infty}, B^{s_1}_{p\infty})_\theta = B^s_{p\infty}. \tag{9}$$

But, on the other hand, $[B^{s_0}_{p\infty}, B^{s_1}_{p\infty}]_\theta$ is the completion of D, the collection of all compactly supported C^∞ functions in $\mathbb{R}^n$, in $B^s_{p\infty}$, and this space is different from $B^s_{p\infty}$. In other words, in this limiting case the two complex methods yield different results, see [Triß: 2.4.7] for details.

1.7 Local approximation spaces

1.7.1 Functions of bounded mean oscillation

As so often in this survey we return to the situation around 1960. Let f be a locally Lebesgue-integrable function on $\mathbb{R}^n$, then

$$f_Q = \frac{1}{|Q|} \int_Q f(x)\, dx \tag{1}$$

stands for the mean value of f with respect to the cube Q. Then *bmo* is the collection of all complex-valued locally Lebesgue-integrable functions f such that

$$\|f \mid bmo\| = \sup_{|Q| \leqq 1} \frac{1}{|Q|} \int_Q |f(x) - f_Q| dx + \sup_{|Q|=1} \int_Q |f(x)| dx < \infty. \tag{2}$$

This is the local counterpart of the celebrated space BMO which consists of all functions such that

$$\|f \mid BMO\| = \sup \frac{1}{|Q|} \int_Q |f(x) - f_Q| dx < \infty, \tag{3}$$

where the supremum is now taken over all cubes in $\mathbb{R}^n$. This space had been introduced in 1961 by F. John and L. Nirenberg, see [JoN]. The space BMO influenced the further development of the theory of function spaces in at least two quite different directions. First, it may be considered as one of the forerunners of the theory of Campanato spaces and other local approximation spaces. Secondly, BMO was suddenly in the limelight when C. Fefferman in 1971 discovered the surprising fact that BMO is the dual space of the Hardy space H_1 defined in (1.4.3/5), see [Fef, FeS2]. This was the starting point of the remarkable (Fourier-analytical) career of BMO. We prefer local spaces. Then Fefferman's result reads as follows,

$$h_1' = bmo, \tag{4}$$

where h_1 is given by (1.4.4/1) and duality must be understood in the sense of the dual pairing (S, S').

Theorem (of Littlewood–Paley type). Let $0 < p < \infty$, then

$$h_p = F^0_{p2} \tag{5}$$

and

$$bmo = F^0_{\infty 2}. \tag{6}$$

Remark. Recall that h_p, F^0_{p2} and $F^0_{\infty 2}$ have been defined in (1.4.4/1), (1.5.1/3) and (1.5.2/2). Furthermore, (5) coincides with (1.4.4/3) and Theorem 1.5.1(iv), see Remark 1.4.4/3 for historical remarks. In other words, (6) complements (5) on the basis of (4). A proof of the homogeneous counterpart of (6) may be found in [Tri5: 3.2.2] with a reference to [Tri8], see also [Triß: 2.5.8].

1.7.2 Oscillations, sharp maximal functions and related function spaces

Let Ω be an arbitrary domain in $\mathbb{R}^n$, then

$$B(x,t) = \{y \in \Omega \colon |x-y| < t\}; \qquad x \in \Omega;\ t > 0. \tag{1}$$

Beside the collection $\mathbb{N}$ of all natural numbers and $\mathbb{N}_0 = \mathbb{N} \cup \{0\}$ it is convenient for us to introduce $\mathbb{N}_{-1} = \mathbb{N}_0 \cup \{-1\}$. Let $N \in \mathbb{N}_{-1}$ then $\mathbb{P}_N$ is the collection of all polynomials (with complex coefficients) of degree less than or equal to N, where we put $\mathbb{P}_{-1} = \{0\}$. Let $0 < u \leqq \infty$, then

$$osc_u^N f(x,t) = \inf\Big(|B(x,t)|^{-1} \int_{B(x,t)} |f(y) - P(y)|^u dy \Big)^{1/u}, \quad x \in \Omega,\ t > 0, \tag{2}$$

denotes the local oscillation where the infimum is taken over all $P \in \mathbb{P}_N$ with the usual modification if $u = \infty$, i.e.,

$$osc_\infty^N f(x,t) = \inf \sup_{y \in B(x,t)} |f(y) - P(y)|.$$

Of course the notation $osc_u^N f$ depends on Ω, but we shall not indicate this dependence. Furthermore, we assumed tacitly that $|f|^u$ is integrable in $B(x,t)$. Let again $N \in \mathbb{N}_{-1}$, $0 < u \leqq \infty$ and $s \in \mathbb{R}$, then

$$f_u^{N,s}(x) = \sup_{0<t<1} t^{-s} osc_u^N f(x,t) \tag{3}$$

denotes the sharp maximal function. In particular, we have

$$f_u^{-1,0}(x) = \Big(\sup_{0<t<1} |B(x,t)|^{-1} \int_{B(x,t)} |f(y)|^u dy \Big)^{1/u} = (M|f|^u)^{1/u}(x), \tag{4}$$

$0 < u < \infty$, where M is essentially the Hardy–Littlewood maximal function from (1.5.3/4). This connection explains the notation "sharp maximal function" for $f_u^{N,s}$.

Definition. Let Ω be a bounded C^∞ domain in $\mathbb{R}^n$.

(i) (Campanato spaces) Let $1 \leqq p < \infty$, $s \geqq -\frac{n}{p}$ and $N = \max(-1, [s])$, then

$$L_p^s(\Omega) = \Big\{ f \in L_p(\Omega) \colon \sup_{x \in \Omega} f_p^{N,s}(x) < \infty \Big\}. \tag{5}$$

(ii) Let $s > 0$, $0 < p \leqq \infty$ and $\overline{p} = \max(1,p)$, then

$$C_p^s(\Omega) = \Big\{ f \in L_{\overline{p}}(\Omega) \colon f_p^{[s],s} \in L_p(\Omega) \Big\}. \tag{6}$$

Remark 1. The spaces $L_p^s(\Omega)$ and $C_p^s(\Omega)$ are (quasi-)normed by

$$\|f \mid L_p^s(\Omega)\| = \|f \mid L_p(\Omega)\| + \sup_{x\in\Omega} f_p^{N,s}(x) \tag{7}$$

and

$$\|f \mid C_p^s(\Omega)\| = \|f \mid L_p(\Omega)\| + \|f_p^{[s],s} \mid L_p(\Omega)\|, \tag{8}$$

respectively. The history of the Campanato spaces begins in 1938 when C.B. Morrey [Mor] suggested to complement the scale of $L_p(\Omega)$-spaces by spaces normed via

$$\|f \mid L_p(\Omega)\| + \sup_{x\in\Omega}\sup_{0<t<1}\left(t^{-sp-n}\int_{B(x,t)}|f(y)|^p dy\right)^{1/p} \tag{9}$$

where $1 \leqq p < \infty$ and $s < 0$. Of course, if $s \leqq -\frac{n}{p}$, then these spaces coincide with $L_p(\Omega)$. If $-\frac{n}{p} \leqq s < 0$, then these Morrey spaces coincide with the above spaces $L_p^s(\Omega)$. If one extends (9) to $s = 0$ then one obtains $L_\infty(\Omega)$. The next decisive step was done by F. John and L. Nirenberg in 1961, see [JoN]. Let $f \in L_1(\Omega)$ and let f_B be the mean value of f in $B = B(x,t)$, see (1.7.1/1) and (1), then $bmo(\Omega)$ is the collection of all $f \in L_1(\Omega)$ such that

$$\|f \mid bmo(\Omega)\| = \|f \mid L_1(\Omega)\| + \sup_B |B|^{-1}\int_B |f(y) - f_B(y)|dy < \infty. \tag{10}$$

This is the appropriate substitute of (1.7.1/2) for bounded domains. Now it is easy to see (and had been observed in [JoN]) that we have $bmo(\Omega) = L_1^0(\Omega)$. The combination of the ideas of Morrey and John–Nirenberg led in the middle of the sixties to the study of the above Campanato spaces $L_p^s(\Omega)$. We refer to [Cam1–4, Mey1, Spa, Sta1,2]. The original notations are different. Further details and references may be found in [Pee2] and in [KJF: Chapter 4, in particular 4.10]. The spaces $C_p^s(\Omega)$ have been introduced by R.A. DeVore, R.C. Sharpley and M. Christ, see [DeS, Chr] in 1984. The difference between (7) and (8) is quite clear, one replaces the sup-norm by a L_p-quasi-norm. However, the idea to replace the sup-norm by some L_q-norms goes back to the mid-sixties, see [Sta2] and [BIN: §28] where the latter summarizes the work by V.P. I'lin at that time. Further discussion may be found in [Boj].

Remark 2. The restriction to bounded smooth domains Ω is convenient for us but not necessary. However, if Ω is unbounded, for example $\mathbb{R}^n$, then one should replace $L_p(\Omega)$ in (5) by $L_p^{\mathrm{loc}}(\Omega)$ and $\|f \mid L_p(\Omega)\|$ in (7) by $\sup_{x\in\mathbb{R}^n}\|f \mid L_p(B(x,1))\|$.

Next we summarize what is known about the interrelations of the spaces from the above definition with other function spaces. The Hölder–Zygmund spaces $\mathcal{C}^s$, the Sobolev spaces W_p^k and the spaces $F^s_{p\infty}$ have been defined in 1.2.2, 1.2.3 and 1.5.1, respectively. If Ω is a domain then $\mathcal{C}^s(\Omega)$, $W_p^k(\Omega)$ and $F^s_{p\infty}(\Omega)$ are restrictions of the corresponding spaces on $\mathbb{R}^n$ to Ω, (quasi-)normed in the usual way. Let, for brevity, $F^s_{\infty\infty}(\Omega) = \mathcal{C}^s(\Omega)$.

Theorem. Let Ω be a bounded C^∞ domain in $\mathbb{R}^n$.

(i) Let $1 \leqq p < \infty$ then

$$L_p^{-n/p}(\Omega) = L_p(\Omega) \quad \text{and} \quad L_p^0(\Omega) = bmo(\Omega). \tag{11}$$

(ii) Let $1 \leqq p < \infty$ and $s > 0$ then

$$L_p^s(\Omega) = \mathcal{C}^s(\Omega). \tag{12}$$

(iii) Let $0 < p \leqq \infty$ and $s > n(\frac{1}{p} - 1)_+$, then

$$C_p^s(\Omega) = F^s_{p\infty}(\Omega). \tag{13}$$

(iv) Let $k \in \mathbb{N}_0$ and $1 \leqq u < p < \infty$, then

$$W_p^k(\Omega) = \{f \in L_p(\Omega) \colon \|f \mid L_p(\Omega)\| + \|f_u^{k-1,k} \mid L_p(\Omega)\| < \infty\} \tag{14}$$

(equivalent norms).

Remark 3. The first part of (11) was mentioned above, whereas the second part generalizes $bmo(\Omega) = L_1^0(\Omega)$ and is known since 1961, see [JoN]. Formula (12) is a classical assertion in the theory of Campanato spaces and known since the middle of the sixties, see the references given above. Formula (13) may be found in [Tri23], but we wish to emphasize that it is also more or less a consequence of [See1: Corollary 1]. The interesting characterization (14) of Sobolev spaces is usually attributed to A.P. Calderón [Cal5], but see also [CaS, Chr, DeS]. If $k = 0$ then (14) coincides essentially with the celebrated Hardy–Littlewood maximal inequality

$$\|Mg \mid L_p\| \leqq c\|g \mid L_p\|, \qquad 1 < p < \infty, \tag{15}$$

see (4). This was the first maximal inequality in the history of function spaces. Its one-dimensional case is due to G.H. Hardy and J.E. Littlewood, see [HaL1], (1930) and [HLP: Theorem 398], and it was extended by N. Wiener to n dimensions, [Wie] (1939). In other words, (14) generalizes the Hardy–Littlewood maximal inequality to smooth functions. Some further comments will be given in the next subsection.

Remark 4. In 5.3 we return to this subject. Then we give a detailed proof of the above theorem.

1.7.3 Characterizations of B^s_{pq} and F^s_{pq} as local approximation spaces

We return to $\mathbb{R}^n$ as the underlying domain. Then B^s_{pq} and F^s_{pq} are the spaces from 1.5.1 and $osc^N_u f(x,t)$ has the same meaning as in (1.7.2/2), now with $\Omega = \mathbb{R}^n$.

Theorem. (i) Let $0 < p < \infty$, $0 < q \leqq \infty$, $1 \leqq r \leqq \infty$ and

$$s > n\left(\frac{1}{p} - \frac{1}{r}\right)_+ \quad \text{and} \quad s > n\left(\frac{1}{q} - \frac{1}{r}\right)_+ . \tag{1}$$

Let $0 < u \leqq r$ and $N \in \mathbb{N}_0$ with $N \geqq [s]$, then

$$F^s_{pq} = \left\{ f \in L_{\max(p,r)} : \|f \mid L_p\| + \left\| \left(\int_0^1 t^{-sq} osc^N_u f(\cdot,t)^q \frac{dt}{t} \right)^{1/q} \mid L_p \right\| < \infty \right\} \tag{2}$$

(modification if $q = \infty$) in the sense of equivalent quasi-norms.

(ii) Let $0 < p \leqq \infty$, $0 < q \leqq \infty$, $1 \leqq r \leqq \infty$ and

$$s > n\left(\frac{1}{p} - \frac{1}{r}\right)_+ . \tag{3}$$

Let $0 < u \leqq r$ and $N \in \mathbb{N}_0$ with $N \geqq [s]$, then

$$B^s_{pq} = \left\{ f \in L_{\max(p,r)} : \|f \mid L_p\| + \left(\int_0^1 t^{-sq} \|osc^N_u f(\cdot,t) \mid L_p\|^q \frac{dt}{t} \right)^{1/q} < \infty \right\} \tag{4}$$

(modification if $q = \infty$) in the sense of equivalent quasi-norms.

Remark 1. Characterizations of type (2) for the fractional Sobolev spaces $H^s_p = F^s_{p2}$ with $s > 0$ and $1 < p < \infty$ may be found in [Dor1]. Extensions to more general spaces F^s_{pq} have been given in [Dor2] and [See1], where the latter paper covers more or less (2) with $u = r$. The general case of (2) may be found in [Tri23] where the extension from $u = r$ to $0 < u \leqq r$ is crucial for some applications (e.g., in connection with intrinsic characterizations of some F^s_{pq} spaces on domains via differences). Characterizations of type (4) may be found in a somewhat hidden form in the work by V.P. Il'in in the sixties, see [BIN: §28]. We refer also to Ju.A. Brudnyj, [Bru], H. Wallin, [Wal: Theorem 7], and the references given in the latter paper to the work by Ju.A. Brudnyj, J.R. Dorronsoro, F. Ricci, M. Taibleson, A. Jonsson, H. Wallin, R.A. DeVore and V. Popov in this connection. We give a detailed proof of the above theorem in 3.5.1.

Remark 2. The above theorem and the assertions of the last subsections show that some spaces F^s_{pq} and B^s_{pq} and all spaces L^s_p and C^s_p can be based on the same principle: measurement of smoothness via approximation by algebraic polynomials. The restriction of this procedure to $\mathbb{R}^n$ or to bounded C^∞ domains is not necessary. Just on the contrary, one can use these methods in order to introduce and study, say, spaces of Besov type on arbitrary non-smooth sets. We refer to this connection to [JoW, Wal] and the papers mentioned there.

1.8 Local means, traces of harmonic and thermic functions

1.8.1 Gauss–Weierstrass and Cauchy–Poisson semi-groups

We return to 1.6.5, where we described the connection between semi-groups of operators and real interpolation. Let A be a complex Banach space and let $\{G(t), 0 \leqq t < \infty\}$ be a strongly continuous semi-group in the sense of (1.6.5/1–6). Such a semi-group is called analytic if $G(t)$ with $t > 0$ maps A into $D(\Lambda)$ and if

$$\sup_{t>0} t\|\Lambda G(t)\| < \infty. \tag{1}$$

Now Theorem 1.6.5 can be complemented in the following way.

Proposition 1. Let $\{G(t)\}$ be an analytic semi-group. Let $0 < \theta < 1$, $1 \leqq q \leqq \infty$ and $m \in \mathbb{N}$, then

$$(A, D(\Lambda^m))_{\theta,q} = \left\{ a \in A\colon \left(\int_0^\infty t^{m(1-\theta)q} \|\Lambda^m G(t)a \mid A\|^q \frac{dt}{t} \right)^{1/q} < \infty \right\} \tag{2}$$

(modification if $q = \infty$).

Remark 1. We refer for details, proofs and further informations to [Triα: 1.14.5]. In particular, the corresponding term in (2) complemented by $\|a \mid A\|$ is an equivalent norm in $(A, D(\Lambda^m))_{\theta,q}$. This result is known since 1967, see [Kom] and [BuB: §3.5].

We return to the Gauss–Weierstrass and Cauchy–Poisson semi-groups briefly mentioned in 1.6.5. Let $x \in \mathbb{R}^n$ and $t > 0$, then

$$W(t)f(x) = (4\pi t)^{-n/2} \int_{\mathbb{R}^n} e^{-\frac{|x-y|^2}{4t}} f(y)\, dt \tag{3}$$

(Gauss–Weierstrass semi-group) and

$$P(t)f(x) = c_n \int_{\mathbb{R}^n} \frac{t}{(|x-y|^2 + t^2)^{(n+1)/2}} f(y)\, dy \tag{4}$$

(Cauchy–Poisson semi-group) with $c_n \|(1+|x|^2)^{-(n+1)/2} \mid L_1\| = 1$ and complemented by $W(0) = P(0) = id$.

Proposition 2. Let $1 < p < \infty$, then both $\{W(t)\}$ and $\{P(t)\}$ are analytic semi-groups in L_p and we have

$$\Lambda_W^m f = \Delta^m f, D(\Lambda_W^m) = W_p^{2m} \tag{5}$$

and

$$\Lambda_P^{2m} f = (-\Delta)^m f, D(\Lambda_P^{2m}) = W_p^{2m}, \tag{6}$$

where Λ_W and Λ_P are the respective infinitesimal generators and $m \in \mathbb{N}$.

Remark 2. Of course, W_p^k stands for the Sobolev spaces on $\mathbb{R}^n$ from 1.2.3 and Δ denotes the Laplacian in $\mathbb{R}^n$. One can rewrite (3) and (4) by

$$u(x,t) = W(t)f(x) = (e^{-t|\xi|^2} \hat{f})^\vee, \qquad x \in \mathbb{R}^n,\ t \geqq 0 \tag{7}$$

and

$$v(x,t) = P(t)f(x) = (e^{-t|\xi|} \hat{f})^\vee, \qquad x \in \mathbb{R}^n,\ t \geqq 0, \tag{8}$$

see also 1.6.5. These two formulas provide a better understanding of the semi-group property of $W(t)$ and $P(t)$ and also of (5) and (6). Furthermore, we have (appropriate interpretation)

$$\frac{\partial u(x,t)}{\partial t} = \Delta u(x,t) \text{ in } \mathbb{R}^{n+1}_+ \text{ and } u(x,0) = f(x) \tag{9}$$

and

$$\frac{\partial^2 v(x,t)}{\partial t^2} + \Delta v(x,t) = 0 \text{ in } \mathbb{R}^{n+1}_+ \text{ and } u(x,0) = f(x), \tag{10}$$

where Δ is the Laplacian in $\mathbb{R}^n$, i.e., $\Delta u(x,t) = \sum_{j=1}^n \frac{\partial^2 u(x,t)}{\partial x_j^2}$. Hence $u(x,t)$ is a solution of the heat equation in $\mathbb{R}^{n+1}_+$ and $v(x,t)$ is a harmonic function in $\mathbb{R}^{n+1}_+$. In other words, $u(x,t)$ is a thermic extension of f from $\mathbb{R}^n$ to $\mathbb{R}^{n+1}_+$ and $v(x,t)$ is a harmonic extension of f from $\mathbb{R}^n$ to $\mathbb{R}^{n+1}_+$. We combine the two properties and obtain by (1.6.4/2) new characterizations of some Besov spaces.

Theorem. Let $0 < s < \infty$, $1 < p < \infty$ and $1 \leqq q \leqq \infty$. Let $m \in \mathbb{N}$ and $k \in \mathbb{N}$ with $m > s/2$ and $k > s$, then

$$B^s_{pq} = \left\{ f \in L_p \colon \|f \mid L_p\| + \left(\int_0^\infty (t^{(m-(s/2))q} \left\| \frac{\partial^m W(t) f}{\partial t^m} \mid L_p \right\|^q \frac{dt}{t} \right)^{1/q} < \infty \right\} \quad (11)$$

$$= \left\{ f \in L_p \colon \|f \mid L_p\| + \left(\int_0^\infty t^{(k-s)q} \left\| \frac{\partial^k P(t) f}{\partial t^k} \mid L_p \right\|^q \frac{dt}{t} \right) < \infty \right\}$$

(modification if $q = \infty$) in the sense of equivalent norms.

Remark 3. The above theorem is a consequence of (1.6.4/2), the two propositions and (9,10). We refer for details to [Triα: 2.5.2, 2.5.3]. First, comprehensive treatments of Besov spaces and fractional Sobolev spaces in the sense of the above theorem are due to M.H. Taibleson (1964) and T.M. Flett (1971), see [Tai, Fle1]. However, many partial results had been known since the late forties and the fifties, some of them can even be traced back to G.H. Hardy and J.E. Littlewood. We gave in [Triα: 2.5.2, 2.5.3] references to these older papers. In addition, we mention that corresponding results for Besov spaces and fractional Sobolev spaces have been proved by P.I. Lizorkin in the early sixties, see [NLL] and the references given there.

Remark 4. In 2.6.4 we return to this subject in detail and prove far-reaching generalizations of the above theorem. Some formulations and discussions of these generalizations beyond the above classical results will be given in the following two subsections.

1.8.2 A Fourier-analytical discussion

Recall our notation

$$\varphi(tD)f = (\varphi(t\cdot)\hat{f})^\vee, \qquad t > 0, \tag{1}$$

which we introduced in 1.4.3 and used afterwards several times, see also 1.3.3. By (1.8.1/7) and (1.8.1/8) we have

$$\varphi(\sqrt{t}\,D)f(x) = (-t)^m \frac{\partial^m W(t)f}{\partial t^m}(x) \ \text{ with } \ \varphi(\xi) = |\xi|^{2m} e^{-|\xi|^2} \tag{2}$$

and

$$\psi(tD)f(x) = (-t)^k \frac{\partial^k P(t)f}{\partial t^k}(x) \ \text{ with } \ \psi(\xi) = |\xi|^k \, e^{-|\xi|}. \tag{3}$$

We use Theorem 1.8.1, where we replace $\int_0^\infty$ by $\int_0^1$, and obtain that

$$\|f \mid L_p\| + \left(\int_0^1 t^{-sq} \|\varphi(tD)f \mid L_p\|^q \frac{dt}{t} \right)^{1/q} \tag{4}$$

and

$$\|f \mid L_p\| + \left(\int_0^1 t^{-sq} \|\psi(tD)f \mid L_p\|^q \frac{dt}{t} \right)^{1/q} \tag{5}$$

are equivalent norms in B^s_{pq}, where φ and ψ have the above meaning with $2m > s$ and $k > s$. Coming from Theorem 1.8.1 we have the same restrictions for s, p, q as above, i.e.,

$$s > 0, \quad 1 < p < \infty, \quad 1 \leqq q \leqq \infty. \tag{6}$$

On the other hand, by the discussion in front of Theorem 1.3.3 and the continuous version of (1.5.1/2) we have the following assertion: Let φ_0 be a C^∞ function on $\mathbb{R}^n$ with supp $\varphi_0 \subset \{\xi \in \mathbb{R}^n\colon\ |\xi| < 2\}$ and $\varphi_0(\xi) = 1$ if $|\xi| \leqq 1$, and let $\varphi(\xi) = \varphi_0(\xi/2) - \varphi_0(\xi)$, then

$$\|\varphi_0(D) \mid L_p\| + \left(\int_0^1 t^{-sq} \|\varphi(tD)f \mid L_p\|^q \frac{dt}{t} \right)^{1/q} \tag{7}$$

is an equivalent quasi-norm in B^s_{pq} where now

$$s \in \mathbb{R}, \quad 0 < p \leqq \infty, \quad 0 < q \leqq \infty. \tag{8}$$

Furthermore one has a corresponding reformulation of (1.5.1/3) for the spaces F^s_{pq}. If one compares (7,8) on the one hand with (4–6) on the other hand then one can ask several questions:

(i) What about an extension of the equivalent (quasi-)norms (4) and (5) for B^s_{pq} from (6) to (8)? The price to pay seems to be clear: One has to replace, in general, $\|f \mid L_p\|$ by $\|\varphi_0(D)f \mid L_p\|$, and one has to choose m and k in (2) and (3), respectively, in dependence on the parameters s, p, q.

(ii) Instead of the quite natural assumption that φ in (7) has a compact support off the origin, the functions φ and ψ in (2) and (3) do not have a compact support but they have a sufficiently strong decay near the origin and when $|\xi|$ tends to infinity. The question is whether this relaxed behaviour is connected with the special nature of (2) and (3) or whether these special functions are merely examples for a more general theorem waiting behind the scene to be discovered. Such a theorem will be one of the major results of Chapter 2. Of peculiar interest is the question whether one can replace the compactness of φ by the compactness of $\hat{\varphi}$. This problem will be discussed in 1.8.4, it leads us to characterizations of B^s_{pq} and F^s_{pq} spaces via local means.

(iii) Finally, one can ask whether possible characterizations of B^s_{pq} in the sense of (i) and (ii) can be extended to the spaces F^s_{pq}. A detailed study of these problems will be given in Chapter 2. Some distinguished aspects of this theory may be found in the following two subsections.

1.8.3 Harmonic and thermic extensions

We interpret Theorem 1.8.1 as a thermic or harmonic extension of a given function $f \in B^s_{pq}$ from $\mathbb{R}^n$ into $\mathbb{R}^{n+1}_+$. After the discussion in 1.8.2 there is some hope that this assertion can be generalized to other spaces B^s_{pq} and also to F^s_{pq} spaces. The operators $W(t)$ and $P(t)$ have the same meaning as in (1.8.1/3,4). The first formula makes sense for any $f \in S'$ whereas the second formula must be understood in the following theorem via a limiting procedure.

Theorem. Let $s \in \mathbb{R}$, $0 < q \leqq \infty$ and let φ_0 with $\varphi_0(0) \neq 0$ be a compactly supported C^∞ function.

(i) Let $0 < p \leqq \infty$. Let $m \in \mathbb{N}_0$ and $k \in \mathbb{N}_0$ with $m > s/2$ and $k > n(\frac{1}{p} - 1)_+ + \max(s, n(\frac{1}{p} - 1)_+)$, then

$$\|\varphi_0(D)f \mid L_p\| + \left(\int_0^1 t^{(m-(s/2))q} \left\| \frac{\partial^m W(t)f}{\partial t^m} \mid L_p \right\|^q \frac{dt}{t}\right)^{1/q} \tag{1}$$

and

$$\|\varphi_0(D)f \mid L_p\| + \left(\int_0^1 t^{(k-s)q} \left\| \frac{\partial^k P(t)f}{\partial t^k} \mid L_p \right\|^q \frac{dt}{t}\right)^{1/q} \tag{2}$$

(modification if $q = \infty$) are equivalent quasi-norms in B^s_{pq}. If $s > n(\frac{1}{p} - 1)_+$ then $\|\varphi_0(D)f \mid L_p\|$ in (1) and (2) can be replaced by $\|f \mid L_p\|$.

(ii) Let $0 < p < \infty$. Let $m \in \mathbb{N}_0$ and $k \in \mathbb{N}_0$ with $m > s/2$ and $k > \frac{n}{\min(p,q)} + \max(s, n(\frac{1}{p} - 1)_+)$, then

$$\|\varphi_0(D)f \mid L_p\| + \left\| \left(\int_0^1 t^{(m-(s/2))q} \left| \frac{\partial^m W(t)f}{\partial t^m}(\cdot) \right|^q \frac{dt}{t} \right)^{1/q} \mid L_p \right\| \tag{3}$$

and

$$\|\varphi_0(D)f \mid L_p\| + \left\| \left(\int_0^1 t^{(k-s)q} \left| \frac{\partial^k P(t)f}{\partial t^k}(\cdot) \right|^q \frac{dt}{t} \right)^{1/q} \mid L_p \right\| \tag{4}$$

(modification if $q = \infty$) are equivalent quasi-norms in F^s_{pq}. If $s > n(\frac{1}{p} - 1)_+$ then $\|\varphi_0(D)f \mid L_p\|$ in (3) and (4) can be replaced by $\|f \mid L_p\|$.

Remark. The theorem shows that all spaces B^s_{pq} and F^s_{pq} can be characterized via harmonic and thermic extensions from $\mathbb{R}^n$ into $\mathbb{R}^{n+1}_+$. However, the semigroup approach described in 1.8.1 breaks down for the spaces F^s_{pq} and also for the spaces B^s_{pq} with $p < 1$. The above theorem has been proved in [Tri16: Theorem 9, p. 199] (1988). However many special cases are known before. First of all we mention the classical results for Besov spaces (and fractional Sobolev spaces) which we described in 1.8.1, including some references. Further results of this type for some Besov spaces B^s_{pq} with $p < 1$ and also for some F^s_{pq} spaces have been proved in the late seventies and the early eighties, see G.A. Kaljabin [Kal1,2], H.-Q. Bui [Bui2,3] and [Tri12]. In 2.6.4 we give a proof of the above theorem.

1.8.4 Local means

We return to the discussion in 1.8.2 and rewrite (1.8.2/1) as a convolution

$$\varphi(tD)f(x) = c \int_{\mathbb{R}^n} \hat{\varphi}(y) f(x + ty) dy, \qquad x \in \mathbb{R}^n,\ t > 0, \tag{1}$$

where c is an uninteresting constant. Let us assume that $\hat{\varphi}$ is a compactly supported C^∞ function, then (1) makes sense for any $f \in S'$, and even for any $f \in D'$, appropriately interpreted, and $\varphi(tD)f(x)$ is a C^∞ function in $\mathbb{R}^n$. Furthermore, in order to calculate $\varphi(tD)f(x)$ at a given point $x \in \mathbb{R}^n$ one needs only a knowledge of f in some neighbourhood of x. This justifies to call (1) local means if supp $\hat{\varphi}$ is compact. See also the discussion in 1.4.5. The question is, under what conditions (1.8.2/7) is an equivalent quasi-norm in B^s_{pq} and what about the F^s_{pq}-counterpart. It is convenient for us to reformulate (1) somewhat. Let $B = \{y \in \mathbb{R}^n,\ |y| < 1\}$ be the unit ball in $\mathbb{R}^n$, and let k be a C^∞ function in $\mathbb{R}^n$ with supp $k \subset B$, then we introduce the local means

$$k(t,f)(x) = \int_{\mathbb{R}^n} k(y) f(x+ty)dy, \qquad x \in \mathbb{R}^n,\ t > 0, \tag{2}$$

which makes sense for any $f \in D'$ (appropriately interpreted). Recall that D' stands for the collection of all complex-valued distributions in $\mathbb{R}^n$. Let k_0 and k^0 be two C^∞ functions in $\mathbb{R}^n$ with

$$\operatorname{supp} k_0 \subset B, \qquad \operatorname{supp} k^0 \subset B \tag{3}$$

and $\hat{k}_0(0) \neq 0$, $\hat{k^0}(0) \neq 0$. Let

$$k(y) = \Delta^N k^0(y) = \left(\sum_{j=1}^n \frac{\partial^2}{\partial x_j^2} \right)^N k^0(y) \tag{4}$$

where $N \in \mathbb{N}$ is at our disposal.

Theorem. (i) Let $s \in \mathbb{R}$, $0 < p \leqq \infty$ and $0 < q \leqq \infty$. Let $N \in \mathbb{N}$ with $2N > \max(s, n(\frac{1}{p} - 1)_+)$, then

$$\|k_0(1,f) \mid L_p\| + \left(\int_0^1 t^{-sq} \|k(t,f) \mid L_p\|^q \frac{dt}{t} \right)^{1/q} \tag{5}$$

(modification if $q = \infty$) is an equivalent quasi-norm in B^s_{pq}.

(ii) Let $s \in \mathbb{R}$, $0 < p < \infty$ and $0 < q \leqq \infty$. Let $N \in \mathbb{N}$ with $2N > \max(s, n(\frac{1}{p} - 1)_+)$, then

$$\|k_0(1,f) \mid L_p\| + \left\| \left(\int_0^1 t^{-sq} |k(t,f)(\cdot)|^q \frac{dt}{t} \right)^{1/q} \mid L_p \right\| \tag{6}$$

(modification if $q = \infty$) is an equivalent quasi-norm in F^s_{pq}.

Remark 1. This theorem follows from [Tri16] (1988), although the above version has been explicitly stated by the author in later papers, see [Tri18, 20]. We return to this subject in detail in 2.4.6 and 2.5.3. Characterizations of this local type play a crucial role in this book. Most of our later considerations will be based on representations of this type, including function spaces on Riemannian manifolds and Lie groups and applications to pseudodifferential operators and Fourier integral operators. In some special cases, characterizations of the above type are known. For example, if one chooses the function φ in (1.4.4/1) in an appropriate way then one obtains a definition of the Hardy spaces h_p in the above sense. Furthermore, we refer to the survey [KaL] where one finds a description of what has been done in this direction in the Soviet Union, see also the recent paper [Liz6].

Remark 2. We return to the discussion in 1.8.2. We have, on the one hand, (5) and, on the other hand, the equivalent quasi-norm (1.8.2/7) where φ has the previous meaning. The characterization (1.8.2/2–5) are somewhat in between. As we shall see in Chapter 2 all these descriptions are special cases of a general theorem. Furthermore, we assumed in (2) and (1.8.2/7) that the underlying kernels and functions belong to C^∞. This is also valid for the function φ from (1.8.2/2) but not for the function ψ from (1.8.2/3). In other words, in a unifying theory one should carefully check what differentiability properties for the underlying kernels and functions are really needed. We discuss an extreme example which shows in a striking way what highly non-smooth kernels k in (2) should be covered. Let $n = 1$, $s > 0$, $1 < p < \infty$, $1 \leqq q \leqq \infty$, $m \in \mathbb{N}$ with $m > s$, then

$$\|f \mid L_p\| + \left(\int_0^1 t^{-sq} \|\Delta_t^m f \mid L_p\|^q \frac{dt}{t} \right)^{1/q} \tag{7}$$

is an equivalent norm in B^s_{pq}, see Theorem 1.2.5. We have

$$\Delta_t^m f(x) = \int_{\mathbb{R}} k(y) f(x+ty) dy \tag{8}$$

if $k = \sum_{j=0}^m (-1)^{m-j} \binom{m}{j} \delta_j$, where δ_j is the δ-distribution with j as off-point. In other words, a unifying theorem should also cover kernels of this type, in dependence on s, p, q. All this will be done in Chapter 2.

1.9 A new device: Atoms as building blocks

1.9.1 Atoms in Hardy spaces

In (1.5.1/1) we decomposed $f \in S'$ in a sum of entire analytic functions $\varphi_j(D)f$. Then we used this representation in order to introduce the spaces B^s_{pq} and F^s_{pq}, see Definition 1.5.1. Hence entire analytic functions may be considered as building blocks for the spaces B^s_{pq} and F^s_{pq} in the following sense: Let a problem be given, for example, mapping properties of PDE's or ψDE's between spaces of the above type, or traces on hyperplanes, etc. First one asks what happens when the problem is applied to entire analytic functions. Then the rest reduces to the question of convergence in the sense of (1.5.1/1) or (1.5.1/2,3). In other words, representations via building blocks break problems into two parts: application of the problem to the building blocks and a discussion of convergence. The aim of this section is to present another type of building blocks, so-called atoms. As far as the spaces B^s_{pq} and F^s_{pq} are concerned, more detailed considerations will be given in later chapters, including some striking applications.

Let $H_p = H_p(\mathbb{R}^n)$ be the Hardy spaces in the sense of 1.4.3 where $0 < p < \infty$. Atomic representations for H_1 are due to C. Fefferman and C. Herz, the latter for the martingale case. The extension from H_1 to H_p with $0 < p \leqq 1$ is due to R.R. Coifman (1974) in the one-dimensional case and R.H. Latter (1978) in the n-dimensional case, see [Coi, Lat] and also [Wil] for a more recent approach. We give a description. Let $0 < p \leqq 1$, then a function $a(x)$ on $\mathbb{R}^n$ is called a p-atom if there exists a cube Q in $\mathbb{R}^n$ with

$$\operatorname{supp} a \subset Q, \qquad |a(x)| \leqq |Q|^{-1/p} \tag{1}$$

and

$$\int_{\mathbb{R}^n} a(x)x^\alpha dx = 0, \quad \alpha \text{ multi-index}, \ |\alpha| \leqq n\left(\frac{1}{p}-1\right). \tag{2}$$

Theorem. Let $0 < p \leqq 1$, then $f \in H_p$ if and only if there exist p-atoms $a_j(x)$ and complex numbers λ_j such that

$$f = \sum_{j=1}^{\infty} \lambda_j a_j(x) \qquad (\text{in } D') \tag{3}$$

and

$$\sum_{j=1}^{\infty} |\lambda_j|^p < \infty. \tag{4}$$

Remark 1. Let $\|f \mid H_p\|$ be given by (1.4.3/3) or (1.4.3/5). Then

$$\inf\left(\sum_{j=1}^{\infty} |\lambda_j|^p\right)^{1/p}, \tag{5}$$

where the infimum is taken over all representations in the sense (3,4), is equivalent to $\|f \mid H_p\|$. This formulation coincides essentially with [Lat].

Remark 2. Compared with the historical way the Hardy spaces came to us (see 1.4) the above characterization is really astonishing. So it is quite clear that this new device attracted much attention, in particular, in the U.S.A., with the St. Louis School as a centre. Since the late seventies, many function spaces have been examined whether they admit atomic representations. Some new function spaces have been defined via atomic representations and more sophisticated atoms and compounds of atoms, called molecules, have been introduced. The papers [CoW, CoR, TaW, RiT] give a good impression of the stage of the art in the early eighties. We concentrate ourselves on those developments, mostly since the middle of the eighties, which are directly connected with atomic representations of B^s_{pq} and F^s_{pq} spaces. References will be given later on.

Remark 3. Recall $H_p = L_p$ if $1 < p < \infty$. What about representations of type (3–5) for these spaces? Let us assume that we have (3,4) for $f \in L_p$ with $1 < p < \infty$. Then

$$f = \sum_{j=1}^{\infty} \sum_{k=1}^{N} \frac{\lambda_j}{N} a_{j,k}(x) \quad \text{with} \quad a_{j,k}(x) = a_j(x) \tag{6}$$

is also a representation of type (3), where $N \in \mathbb{N}$. Instead of (4) we have

$$N^{1-p} \sum_{j=1}^{\infty} |\lambda_j|^p \tag{7}$$

and this expression tends to zero if N tends to infinity. In other words, atomic representations of the above type where (5) is equivalent to the (quasi-) norm $\|f \mid H_p\|$ are restricted to $p \leqq 1$. However the above consideration indicates also a possible cure of this shortcoming. If we require not only a rigid control on the size of the admitted atoms in the sense of (1,2) but also on the location of the support of the atom, then one has a good chance to extend atomic representations of the above type to spaces with $p > 1$ in (4). Some of the representations in the papers mentioned in Remark 2 are just of this type, but sometimes the counterpart of (4) looks more difficult. In what follows both types of atoms, with and without a rigid control on the location of the support, will make their appearance.

1.9.2 Atoms in Besov spaces

Let B^s_{pq} be the Besov spaces in $\mathbb{R}^n$ from 1.5.1. Let $\mathbb{Z}^n$ be the lattice of all points in $\mathbb{R}^n$ with integer-valued components. Then $\mathbb{Q}^n$ stands for the collection of all cubes $Q_{\nu k}$ in $\mathbb{R}^n$ with sides parallel to the axes, centered at $2^{-\nu}k$, and with side length $2^{-\nu}$, where $k \in \mathbb{Z}^n$ and $\nu \in \mathbb{N}_0$. Let Q be a cube in $\mathbb{R}^n$ and $r > 0$, then rQ is the cube in $\mathbb{R}^n$ concentric with Q and with side length r times the side length of Q. We write $(\nu, k) < (\nu', k')$ if $\nu \geqq \nu'$ and

$$Q_{\nu k} \subset 2Q_{\nu' k'} \quad \text{with} \quad Q_{\nu k} \in \mathbb{Q}^n \quad \text{and} \quad Q_{\nu' k'} \in \mathbb{Q}^n. \tag{1}$$

Let $c \in \mathbb{R}$, then $c_+ = \max(c, 0)$ and $[c]$ stands for the largest integer less than or equal to c.

Definition. Let $s \in \mathbb{R}$, $0 < p \leqq \infty$ and $0 < q \leqq \infty$. Let K and L be integers with

$$K \geqq ([s] + 1)_+ \quad \text{and} \quad L \geqq \max\left(\left[n \left(\frac{1}{p} - 1 \right)_+ - s \right], -1 \right). \tag{2}$$

(i) A function $a(x)$ is called a s-atom if

$$\operatorname{supp} a \subset 5Q \tag{3}$$

for some $Q = Q_{0k} \in \mathbb{Q}^n$ (side length 1) and

$$|D^\alpha a(x)| \leqq 1 \quad \text{for} \quad |\alpha| \leqq K. \tag{4}$$

(ii) Let $Q = Q_{\nu k} \in \mathbb{Q}^n$. The function $a(x)$ is called a (Q, s, p, q)-atom if (3) is satisfied,

$$|D^\alpha a(x)| \leqq |Q|^{-\frac{1}{q} + \frac{s}{n} - \frac{|\alpha|}{n}} \quad \text{for} \quad |\alpha| \leqq K \tag{5}$$

and

$$\int_{\mathbb{R}^n} x^\beta a(x) dx = 0 \quad \text{for} \quad |\beta| \leqq L. \tag{6}$$

(iii) A (Q, s, p, p)-atom is called a (Q, s, p)-atom.

Remark 1. Recall $x^\beta = x_1^{\beta_1} \cdots x_n^{\beta_n}$ if $\beta = (\beta_1, \ldots, \beta_n)$ is a multi-index and $x = (x_1, \ldots, x_n) \in \mathbb{R}^n$. Furthermore, in the above definition, it is tacitly assumed that $a(x)$ has classical derivatives up to order K. If $L = -1$ then (6) means that no moment conditions are required. Obviously, the above atoms depend on the chosen numbers K and L. The atoms in (i) and (iii) coincide with corresponding atoms in [FrJ1]. The slight generalization in (ii) is not necessary in this subsection, but it prepares atomic representations of F^s_{pq} described in 1.9.4.

Theorem. Let $s \in \mathbb{R}$, $0 < p \leqq \infty$ and $0 < q \leqq \infty$. Let K and L be fixed integers satisfying (2). Then $f \in S'$ is an element of B^s_{pq} if and only if it an be represented as

$$f = \sum_{k \in \mathbb{Z}^n} \left[\mu_k a_k(x) + \sum_{\nu=0}^{\infty} \lambda_{\nu k} a_{\nu k}(x) \right] \quad \text{(convergence in } S'\text{)} \tag{7}$$

where $a_k(x)$ is a s-atom related to the cube Q_{0k}, $a_{\nu k}(x)$ is a $(Q_{\nu k}, s, p)$-atom, μ_k and $\lambda_{\nu k}$ complex numbers with

$$\left(\sum_{k \in \mathbb{Z}^n} |\mu_k|^p \right)^{1/p} + \left(\sum_{\nu=0}^{\infty} \left(\sum_{k \in \mathbb{Z}^n} |\lambda_{\mu k}|^p \right)^{q/p} \right)^{1/p} < \infty \tag{8}$$

(modification if either $p = \infty$ and/or $q = \infty$). Furthermore, the infimum of (8) with respect to all admissible representations (for fixed integers K and L) is an equivalent quasi-norm in B^s_{pq}.

Remark 2. This theorem is due to M. Frazier and B. Jawerth, see [FrJ1: Theorem 7.1] (1985). The papers mentioned in Remark 1.9.1/2 cover among other things atomic representations of spaces of harmonic and thermic functions in $\mathbb{R}^{n+1}_+$ equipped with mixed quasi-norms of the same type as in Theorem 1.8.3. This suspicious situation was examined in detail by S. Sands in her unpublished Ph.D. thesis [San] (1981) and led to some atomic representations of Besov spaces. Furthermore, some partial results of the above type have been obtained independently by H.-Q. Bui in [Bui4], and by Ju.V. Netrusov in [Net1,2,4].

Remark 3. At the end of Remark 1.9.1/3 we discussed two types of atoms, with and without a rigid control on the location of supports of the atoms. For the above atomic representations, a control of the location of the supports of the atoms is crucial, in sharp contrast to atomic representations in Hardy spaces described in Theorem 1.9.1. The difference between these two theorems is quite clear: Theorem 1.9.1 is restricted to $0 < p \leqq 1$ and cannot be extended to $p > 1$, whereas the above theorem is valid for all values p and q with $0 < p \leqq \infty$ and $0 < q \leqq \infty$.

1.9.3 Atoms in tent spaces and characterizations of F^s_{pq} spaces in terms of Lusin functions

Our way to look for atomic representations in F^s_{pq} spaces will be based on corresponding assertions in so-called tent spaces. For this reason, we give here a description of tent spaces, including atomic representations, and their close connection with F^s_{pq} spaces. Let again $\mathbb{R}^{n+1}_+ = \{(x,t) \in \mathbb{R}^{n+1} \colon x \in \mathbb{R}^n,\, t > 0\}$ and

$$\Gamma(x) = \{(y,t) \in \mathbb{R}^{n+1}_+ \colon |x-y| < t\}, \tag{1}$$

be a cone of aperture 1 whose vertex is $x \in \mathbb{R}^n$. We need the functional

$$A_q f(x) = \left(\int_{\Gamma(x)} |f(y,t)|^q \frac{dy\, dt}{t^{n+1}} \right)^{1/q}, \qquad 0 < q < \infty, \quad x \in \mathbb{R}^n, \tag{2}$$

with the usual modification

$$A_\infty f(x) = \sup_{(y,t) \in \Gamma(x)} |f(y,t)|, \qquad x \in \mathbb{R}^n. \tag{3}$$

Definition. Let $0 < p < \infty$ and $0 < q < \infty$, then T^p_q is the collection of all Lebesgue-measurable complex-valued functions in $\mathbb{R}^{n+1}_+$ such that

$$\|f\|_{q,p} = \|A_q f \mid L_p\| < \infty. \tag{4}$$

Furthermore, T^p_∞ is the collection of all continuous functions in $\mathbb{R}^{n+1}_+$ such that

$$\|f\|_{\infty,p} = \|A_\infty f \mid L_p\| < \infty. \tag{5}$$

Remark 1. These spaces have been introduced in [CMS1] (1985), they are called tent spaces for reasons which will be clear later on. The additional assumption in [CMS1] in connection with T^p_∞ is not necessary, it is always satisfied.

Next we describe the connection with the space F^s_{pq}. Let

$$\gamma(x) = \{(y,t) \in \mathbb{R}^{n+1}_+ \colon |x-y| < t < 1\} \tag{6}$$

be a truncated cone of aperture 1 whose vertex is $x \in \mathbb{R}^n$, let $k(t,f)(x)$ be the local means from (1.8.4/2) and

$$k_q^s f(x) = \left(\int_{\gamma(x)} t^{-sq} |k(t,f)(y)|^q \frac{dy\,dt}{t^{n+1}} \right)^{1/q} \tag{7}$$

(usual modification if $q = \infty$).

Proposition. Let $s \in \mathbb{R}$, $0 < p < \infty$ and $0 < q \leqq \infty$. Let k_0 and k be the same functions as in Theorem 1.8.4, then

$$\|k_0(1,f) \mid L_p\| + \|k_q^s f \mid L_p\| \tag{8}$$

is an equivalent quasi-norm in F_{pq}^s.

Remark 2. The connection of (8) with tent spaces is obvious. A proof of this proposition has been given in [Tri16: p. 179–180] (1988). We return in 2.4.5 to this subject in detail. The functional $k_q^s f$ can be rewritten as

$$k_q^s f(x) = c \left(\int_0^1 t^{-sq} |B(x,t)|^{-1} \int_{B(x,t)} |k(t,f)(y)|^q dy \frac{dt}{t} \right)^{1/q} \tag{9}$$

(modification if $q = \infty$), where $B(x,t)$ stands for a ball of radius t in $\mathbb{R}^n$ centered at $x \in \mathbb{R}^n$ with $|B(x,t)|$ as its volume. Constructions of type (9) are called Lusin functions. Characterizations of F_{pq}^s spaces in terms of Lusin functions are given by L. Päivärinta, see [Päi2] (1982). In this connection we refer also to [CaT].

After this digression we return to the above tent spaces. Let Q be a cube in $\mathbb{R}^n$, then

$$\hat{Q} = \{(y,t) \in \mathbb{R}_+^{n+1} : y \in Q, 0 < t < \operatorname{dist}(y, \partial Q)\} \tag{10}$$

is the tent over Q. A complex-valued function $a(x,t)$ in $\mathbb{R}_+^{n+1}$ is called a T_q^p-atom if

$$\operatorname{supp} a(x,t) \subset \hat{Q} \quad \text{for some cube } Q \text{ in } \mathbb{R}^n \tag{11}$$

and

$$\left(\int_{\hat{Q}} |a(x,t)|^q \frac{dx\,dt}{t} \right)^{1/q} \leqq |Q|^{\frac{1}{q} - \frac{1}{p}}. \tag{12}$$

Theorem. Let $0 < p \leqq 1 < q \leqq \infty$, then $f \in T_q^p$ if and only if f can be represented as

$$f = \sum_{j=1}^{\infty} \lambda_j a_j \tag{13}$$

where a_j are T^p_q-atoms, λ_j complex numbers, and

$$\left(\sum_{j=1}^{\infty} |\lambda_j|^p \right)^{1/p} < \infty. \tag{14}$$

Furthermore, the infimum of (14) taken over all admissible representations is an equivalent quasi-norm in T^p_q.

Remark 3. A proof of this theorem may be found in [CMS1: Section 8]. A combination of this atomic decomposition and the above proposition is the basis to establish atomic representations for some F^s_{pq} spaces. A description of the result will be given in the next subsection. Furthermore we return to this subject in detail in 3.2.

1.9.4 Atoms in F^s_{pq} spaces

Let F^s_{pq} be the spaces in $\mathbb{R}^n$ from 1.5.1. First we complement Definition 1.9.2, where all notations have the same meaning as in 1.9.2.

Definition. Let $s \in \mathbb{R}$ and $0 < p \leqq 1 < q \leqq \infty$, then $g \in S'$ is called an (s,p,q)-atom if

$$g = \sum_{(\mu,l)<(\nu,k)} d_{\mu l} a_{\mu l}(x) \quad \text{for some } \nu \in \mathbb{N}_0 \text{ and } k \in \mathbb{Z}^n, \tag{1}$$

where $a_{\mu l}(x)$ is a $(Q_{\mu l}, s, p, q)$-atom and $d_{\mu l}$ are complex numbers with

$$\left(\sum_{(\mu,l)<(\nu,k)} |d_{\mu l}|^q \right)^{1/q} \leqq |Q_{\nu k}|^{\frac{1}{q}-\frac{1}{p}} \tag{2}$$

(modification if $q = \infty$).

Remark 1. The atom g belongs to F^s_{pq} and the convergence in (1) must be understood in this sense. These rather complicated atoms have been introduced in [Tri22] and announced in [Tri19]. No control on the location of (s,p,q)-atoms is required in contrast to its sub-atoms $a_{\mu l}(x)$. Hence, in the light of our discussion at the end of Remark 1.9.1/3, the above atoms have a hybrid structure.

Theorem. Let $s \in \mathbb{R}$, $0 < p \leqq 1 < q \leqq \infty$. Let K and L be fixed integers satisfying (1.9.2/2). Then $f \in S'$ is an element of F^s_{pq} if and only if it can be represented as

$$f = \sum_{j=1}^{\infty} (\mu_j a_j + \lambda_j g_j) \qquad \text{(convergence in } S'\text{)} \tag{3}$$

where a_j is a s-atom, g_j is a (s,p,q)-atom, μ_j and λ_j are complex numbers with

$$\left(\sum_{j=1}^{\infty} |\mu_j|^p + |\lambda_j|^p \right)^{1/p} < \infty. \tag{4}$$

Furthermore, the infimum of (4) with respect to all admissible representations (for fixed integers K and L) is an equivalent quasi-norm in F^s_{pq}.

Remark 2. The theorem has been proved in [Tri22], see also the announcement [Tri19] (1988). A definition of s-atoms may be found in 1.9.2. There is no control on the location of the supports of a_j and g_j. This brings the above atomic decomposition near to the original atomic decomposition in Hardy spaces, see Theorem 1.9.1. By the discussion in Remark 1.9.1/3 it is clear that the above theorem cannot be extended to values $p > 1$. Another type of atomic representations in F^s_{pq} spaces is due to M. Frazier and B. Jawerth, see [FrJ2,3] (1988) and [FJW]. Their decomposition applies to all parameters $s \in \mathbb{R}$, $0 < p < \infty$, $0 < q \leqq \infty$, but one needs a control on the location of the corresponding atoms and the coefficient spaces look more complicated. Nearest to us comes [FrJ3: Theorem 7.4]. These decompositions proved very useful and one can find many applications in the cited papers and [FHJW]. Apparently independently discovered Ju.V. Netrusov similar atomic representations in F^s_{pq} spaces as in the above-cited papers by M. Frazier and B. Jawerth, [Net2,4] (1987–89). He complemented the impressive list of striking applications of atomic representations by the study of capacities in F^s_{pq} spaces. A proof of the above theorem will be given in 3.2.3.

Remark 3. Recently, an atomic representation in function spaces was given a group-theoretical interpretation, mostly by the work of H.-G. Feichtinger and K.H.

Gröchenig, see [FeG1] (1988) for a first survey, and [FeG2, Grö1,2]. The basic idea reads as follows. Let B be a Banach space and G be a locally compact group which acts on B via a representation π. Then atoms should be of the form $\{\pi(x_i)g\colon i \in I\}$, where $g \in B$ is an appropriate element of B and $x_i \in G$. An outstanding example in this theory is the so-called n-dimensional $(ax + b)$-group $\mathbb{R}^n \times \mathbb{R}_+$ with the multiplication

$$(x,t)(y,\tau) = (x + ty, t\tau), \qquad x \in \mathbb{R}^n,\ y \in \mathbb{R}^n,\ t \in \mathbb{R}_+,\ \tau \in \mathbb{R}_+,$$

and the unitary representations $\pi(x,t)$,

$$\pi(x,t)f(z) = t^{-n/2} f\left(\frac{z-x}{t}\right), \tag{5}$$

which act on $L_2(\mathbb{R}^n)$. On this basis, H.-G. Feichtinger and K.H. Gröchenig shed new light on the atomic representations due to M. Frazier and B. Jawerth, see [FrJ1–3], in homogeneous spaces of B^s_{pq} and F^s_{pq} type (restricted to the Banach space case, i.e., $p \geqq 1$, $q \geqq 1$). They used a characterization of homogeneous F^s_{pq} spaces in terms of Lusin functions of type (1.9.3/8).

Remark 4. Closely connected with atomic representations, and in particular, with the group-theoretical interpretations described in the previous remark, is the concept of wavelets developed in the eighties. We do not go into details and refer to the papers by M. Frazier, B. Jawerth, H.-G. Feichtinger and K.H. Gröchenig mentioned in the previous remark and in particular, to the extensive recent surveys by I. Daubechies [Dau1,2]. It comes out that wavelets are more constructive than atoms, they enjoy properties like (5) and they form unconditional bases in some L_p spaces, $1 < p < \infty$, and some Besov–Sobolev spaces. Furthermore, wavelets seem to be a good instrument for numerical, physical, and technical problems, see [Dau1,2]. An extension of the concept of wavelets to F^s_{pq} spaces has been given recently by W. Sickel, see [Sic2].

1.10 Spaces on domains

1.10.1 Definitions and preliminaries

In 1.7.2 we introduced some function spaces on arbitary domains Ω in $\mathbb{R}^n$, among them the Hölder–Zygmund spaces $\mathcal{C}^s(\Omega)$ and the Sobolev spaces $W^k_p(\Omega)$. Now we extend this definition to all spaces of type B^s_{pq} and F^s_{pq}.

Definition. Let Ω be a domain in $\mathbb{R}^n$.

(i) Let $s \in \mathbb{R}$, $0 < p \leqq \infty$ and $0 < q \leqq \infty$, then

$$B^s_{pq}(\Omega) = \{f \in D'(\Omega)\colon \text{there exists a } g \in B^s_{pq} \text{ with } g \mid \Omega = f\}. \tag{1}$$

(ii) Let $s \in \mathbb{R}$, $0 < p < \infty$ and $0 < q \leqq \infty$, then

$$F^s_{pq}(\Omega) = \{f \in D'(\Omega)\colon \text{there exists a } g \in F^s_{pq} \text{ with } g \mid \Omega = f\}. \tag{2}$$

Remark. Recall that B^s_{pq} and F^s_{pq} stand for the spaces from 1.5.1, defined on $\mathbb{R}^n$. Furthermore, $g \mid \Omega$ denotes the restriction of $g \in S'(\mathbb{R}^n)$ to Ω in the sense of $D'(\Omega)$. Let $f \in B^s_{pq}(\Omega)$ then

$$\|f \mid B^s_{pq}(\Omega)\| = \inf \|g \mid B^s_{pq}\| \tag{3}$$

where the infimum is taken over all g in the sense of (1), similarly we define $\|f \mid F^s_{pq}(\Omega)\|$.

In connection with spaces on domains, there are two outstanding problems:

(i) the extension problem, and

(ii) the problem of intrinsic characterizations.

Let re be the restriction operator from $\mathbb{R}^n$ to Ω, i.e.,

$$\mathrm{re}\, g = g \mid \Omega. \tag{4}$$

Obviously, re is a linear and bounded operator from B^s_{pq} onto $B^s_{pq}(\Omega)$ and from F^s_{pq} onto $F^s_{pq}(\Omega)$. The question is whether there exists a linear and bounded operator ext from $B^s_{pq}(\Omega)$ into B^s_{pq} such that

$$\mathrm{re} \circ \mathrm{ext} = id \qquad (\text{identity in } B^s_{pq}(\Omega)), \tag{5}$$

and similarly for F^s_{pq} spaces. Intrinsic characterization means that we are looking for a description where one needs only a knowledge of $f \in B^s_{pq}(\Omega)$ or $f \in F^s_{pq}(\Omega)$ inside of Ω. We are interested in three types of domains: $\mathbb{R}^n$,

$$\mathbb{R}^n_+ = \{x = (x_1, \ldots, x_n) \in \mathbb{R}^n\colon x_n > 0\} \tag{6}$$

and bounded C^∞ domains. Non-smooth domains will not be considered, although many authors paid much attention to function spaces in non-smooth domains, mostly in connection with Sobolev spaces, see [Ada, EdE, Maz].

1.10.2 Extensions

Let $N \in \mathbb{N}$ and

$$1 < \lambda_1 < \lambda_2 < \cdots < \lambda_{2N+1}, \tag{1}$$

then there exist uniquely determined real numbers $a_1, \ldots, a_{2N+1}$ with

$$\sum_{l=1}^{2N+1} a_l (-\lambda_l)^j = 1 \quad \text{where} \quad j = -N, -N+1, \ldots, 0, \ldots, N-1, N \tag{2}$$

(Vandermonde's determinant). Let $f(x)$ be a function defined in $\mathbb{R}^n_+$, then we extend $f(x)$ from $\mathbb{R}^n_+$ to $\mathbb{R}^n$ by

$$\text{ext}_N f(x) = \begin{cases} f(x', x_n) & \text{if } x_n \geqq 0 \\ \sum_{k=1}^{2N+1} a_k f(x', -\lambda_k x_n) & \text{if } x_n < 0, \end{cases} \tag{3}$$

where $x = (x', x_n) \in \mathbb{R}^n$ with $x' \in \mathbb{R}^{n-1}$ and $x_n \in \mathbb{R}$. Obviously, the linear operator ext_N depends on the chosen numbers $\lambda_1, \ldots, \lambda_{2N+1}$, but this is immaterial. If $f(x)$ has classical derivatives in $\overline{\mathbb{R}^n_+}$ up to order N, then one checks easily that $\text{ext}_N f(x)$ has classical derivatives up to order N in $\mathbb{R}^n$. In other words, the differentiability properties of $f(x)$ in $\mathbb{R}^n_+$ are preserved after extension.

Theorem. Let $a > 0$ and $b > 0$, then there exists a natural number $N = N(a, b)$ such that ext_N is a (linear and bounded) extension operator from $F^s_{pq}(\mathbb{R}^n_+)$ into $F^s_{pq}(\mathbb{R}^n)$, where s, p, q are real numbers with

$$|s| \leqq a, \quad b \leqq p < \infty, \quad 0 < q \leqq \infty, \tag{4}$$

and from $B^s_{pq}(\mathbb{R}^n_+)$ into $B^s_{pq}(\mathbb{R}^n)$, where s, p, q are real numbers with

$$|s| \leqq a, \quad b \leqq p \leqq \infty, \quad 0 < q \leqq \infty. \tag{5}$$

Remark 1. Extension procedures of type (3) have a long history. The classical version, where one has only $N+1$ numbers $\lambda_1, \ldots, \lambda_{N+1}$ in (1) and the counterpart of (2) is restricted to $j = 0, \ldots, N$, is often attributed to G.M. Fichtengol'c. This version works pretty well for the above-mentioned classical situation and also for extensions of Sobolev spaces, see [Triα: 2.9.1]. However, for spaces B^s_{pq} and F^s_{pq}, in particular, if $s < 0$ and/or $p < 1$, the classical version is not sufficient. The above modification (1–3) is especially well adapted to cover all spaces B^s_{pq} and F^s_{pq}

for all admissible values of the parameters. It was first applied in [Triα: 2.9.3] (1978) to B^s_{pq} and $H^s_p = F^s_{p2}$ with $s \in \mathbb{R}$, $1 < p < \infty$, and $1 \leqq q \leqq \infty$. In [Triß: 2.9] (1983) we extended this method to all spaces B^s_{pq} and some spaces F^s_{pq}. There remained some gaps. The first full proof of the above theorem is due to J. Franke [Fra] (1986); it is a modification of [Triß] based on (1–3). We return to this subject in 4.5, especially 4.5.5, where we give a comparatively simple proof of this crucial extension property based on the newly developed techniques. Then we shall also explain how to understand (3) if $f(x)$ is not a (smooth) function, see also the quoted sections in [Triα, Triß] in this context.

Corollary. Let Ω be a bounded C^∞ domain in $\mathbb{R}^n$. Let $a > 0$ and $b > 0$, then there exists a common extension operator for all spaces $F^s_{pq}(\Omega)$ into F^s_{pq} with (4) and for all spaces $B^s_{pq}(\Omega)$ into B^s_{pq} with (5).

Remark 2. The extension operator in question can be reduced to the above operator ext_N by some standard procedures, see [Triß: 3.3.4] for details. This makes clear what is meant by the a little sloppy notation of a "common" extension operator. On the other hand, the situation is different if Ω is a bounded domain with minimally smooth boundary. Stein's extension method for Sobolev spaces in domains with non-smooth boundary has been used by G.A. Kaljabin in connection with some spaces of F^s_{pq} type, see [Ste1: VI,3] and [Kal4] (1985), we refer also to [DeS: §10] and [See1]. But a generalization of this method to all spaces F^s_{pq}, in particular, to those ones with $s < 0$, seems to be rather doubtful.

Remark 3. A further extension method goes back to A.P. Calderón, see [Cal1,2] (1960). It is characterized by integral representations of functions and an application of mapping properties for singular integrals in L_p-spaces with $1 < p < \infty$. In [Triα: 4.2.3] we gave a brief description. Furthermore, we refer to [BIN]. This book contains a lot of sophisticated integral representations of functions via differences and derivatives, and constructions of corresponding extension operators, including detailed references to the underlying, mostly Russian, papers.

1.10.3 Intrinsic characterizations of classical Sobolev, Hölder–Zygmund and Besov spaces

Let Ω be a bounded C^∞ domain in $\mathbb{R}^n$, then $W^k_p(\Omega)$ with $k \in \mathbb{N}_0$, $1 < p < \infty$, and $\mathcal{C}^s(\Omega)$ with $s > 0$, are the restrictions of W^k_p and $\mathcal{C}^s$, respectively, to Ω, see 1.7.2, 1.2.2 and 1.2.3. Recall $W^k_p = F^k_{p2}$ and $\mathcal{C}^s = B^s_{\infty\infty}$, see 1.5.1. Let $f(x)$ be a function defined in Ω, then

$$\Delta^m_h f(x,\Omega) = \begin{cases} \Delta^m_h f(x) \text{ if } x + jh \in \Omega \text{ for } j = 0, \ldots m \\ 0 \qquad\qquad \text{otherwise,} \end{cases} \tag{1}$$

where $\Delta^m_h f(x)$ with $m \in \mathbb{N}$, $h \in \mathbb{R}^n$ and $x \in \Omega$, are the differences introduced in (1.2.2/3).

Theorem. Let Ω be a bounded C^∞ domain.

(i) Let $1 < p < \infty$ and $k \in \mathbb{N}_0$, then

$$W^k_p(\Omega) = \left\{ f \in L_p(\Omega) \colon \left(\sum_{|\alpha| \leqq k} \| D^\alpha f \mid L_p(\Omega)\|^p \right)^{1/p} < \infty \right\}. \tag{2}$$

(ii) Let $s > 0$ and $m \in \mathbb{N}$ with $m > s$, then

$$\mathcal{C}^s(\Omega) = \left\{ f \in C(\Omega) \colon \sup_{\substack{x \in \Omega \\ 0 \neq h \in \mathbb{R}^n}} |h|^{-s} \mid \Delta^m_h f(x,\Omega)| < \infty \right\}. \tag{3}$$

(iii) Let $s > 0$, $1 < p < \infty$ and $1 \leqq q \leqq \infty$. Let $m \in \mathbb{N}$ with $m > s$, then

$$B^s_{pq}(\Omega) = \left\{ f \in L_p(\Omega) \colon \left(\int_{|h| \leqq 1} |h|^{-sq} \|\Delta^m_h f(\cdot,\Omega) \mid L_p(\Omega)\|^q \frac{dh}{|h|^n} \right)^{1/q} < \infty \right\} \tag{4}$$

(modification if $q = \infty$).

Remark. All these remarks are very classical ones. They make clear what is meant by intrinsic characterizations. Obviously, $L_p(\Omega)$ denotes the usual Lebesgue spaces, whereas $C(\Omega)$ stands for the collection of all continuous functions in $\overline{\Omega}$. This coincides with the definition of $C(\Omega)$ given at the beginning of 1.2.1. These spaces are normed in an obvious manner, denoted by $\|f \mid C(\Omega)\|$ and $\|f \mid L_p(\Omega)\|$, respectively. The $\mathbb{R}^n$-counterparts of the above theorem have been discussed in 1.2.2,

1.2.3 and 1.2.5. Sobolev, Hölder–Zygmund, and Besov spaces on domains are as old as the corresponding spaces on $\mathbb{R}^n$. At the first glance this observation sounds a little astonishing. But one has to have in mind that Hölder and Sobolev spaces, and at least partly, also Zygmund classes and Besov spaces, came into being in close connection with boundary value problems and initial boundary value problems for PDE's. We refer to the classical books [Mir, Sob4]. As far as the Besov spaces are concerned, things may be a little bit different, because also other fields, for example, approximation theory and Fourier analysis, gave a strong impetus to study Besov spaces. Nevertheless, characterizations of type (4) are well-known since the sixties, at least in the Russian literature, see [BIN] for detailed references and discussions. The concrete version (4) may be found in [Mur] (1971), extended to very general domains. Many further results and references have been given in [Triα: 4.2.3, 4.4.2, 4.6.2].

1.10.4 Intrinsic characterizations of $B^s_{pq}(\Omega)$ and $F^s_{pq}(\Omega)$

We incorporated the preceding subsection for historical reasons and to make clear that intrinsic characterizations for the treated classical spaces are known since ages. But only very recently, satisfactory intrinsic descriptions for the general spaces $B^s_{pq}(\Omega)$ and $F^s_{pq}(\Omega)$ have been obtained, in particular, if $p < 1$. We are interested in characterizations based on oscillations and differences in three types of domains: $\mathbb{R}^n$, $\mathbb{R}^n_+$, and bounded C^∞ domains. Let $N \in \mathbb{N}_{-1}$ and $0 < u \leqq \infty$, then $osc^N_u f(x,t)$ has the same meaning as in (1.7.2/2). As in 1.7.2 we shall not indicate the dependence of $osc^N_u f(x,t)$ on Ω.

Proposition. Let Ω be either $\mathbb{R}^n$, $\mathbb{R}^n_+$ or a bounded C^∞ domain in $\mathbb{R}^n$. After replacing F^s_{pq}, B^s_{pq} and L_ρ with $\rho = p$ or $\rho = \max(p,r)$ by $F^s_{pq}(\Omega)$, $B^s_{pq}(\Omega)$ and $L_\rho(\Omega)$, respectively, Theorem 1.7.3 remains valid, where now $osc^N_u f(x,t)$ has the above meaning.

Remark 1. In other words, this proposition extends Theorem 1.7.3 from $\mathbb{R}^n$ to $\mathbb{R}^n_+$ and to bounded C^∞ domains in the sense of intrinsic characterizations. This is by no means obvious and only few of the papers mentioned in Remark 1.7.3/1 deal with this case. As far as the spaces F^s_{pq} are concerned, the case $u = r$ is more or less covered by [See1], the general case is due to [Tri23]. We return in detail to this subject in 4.5.3 and 5.2.1.

Now we describe characterizations via differences. Let again Ω be either $\mathbb{R}^n$, or $\mathbb{R}^n_+$ or a bounded C^∞ domain in $\mathbb{R}^n$. Let $M \in \mathbb{N}$, $t > 0$ and $x \in \Omega$, then

$$V^M(x,t) = \{h \in \mathbb{R}^n : x + \tau h \in \Omega,\ |h| < t,\ 0 \leqq \tau \leqq M\} \tag{1}$$

is the largest truncated cone of radius t with vertex in the origin such that $x + MV^M(x,t) \subset \Omega$. We omit to indicate the dependence of $V^M(x,t)$ on Ω. Furthermore, $\Delta_h^M f(x)$ are the differences from (1.2.2/3). Let $f \in L_\rho(\Omega)$ with $\rho \geqq 1$ then the mean values

$$d_t^M f(x) = t^{-n} \int_{V^M(x,t)} |\Delta_h^M f(x)| dh, \qquad x \in \Omega,\ t > 0, \tag{2}$$

make sense. As we shall see later on, (2) and $osc_1^M f(x,t)$ from (1.7.2/2) are closely connected.

Theorem. Let Ω be either $\mathbb{R}^n$, or $\mathbb{R}^n_+$ or a bounded C^∞ domain in $\mathbb{R}^n$.

(i) Let $0 < p < \infty$, $0 < q \leqq \infty$ and

$$s > n\left(\frac{1}{p} - 1\right)_+, \qquad s > n\left(\frac{1}{q} - 1\right)_+. \tag{3}$$

Let $M \in \mathbb{N}$ with $M > s$, then

$$F^s_{pq}(\Omega) = \Bigg\{ f \in L_{\max(1,p)}(\Omega) : \|f \mid L_p(\Omega)\| \tag{4}$$

$$+ \left\| \left(\int_0^1 t^{-sq} d_t^M f(\cdot)^q \frac{dt}{t} \right)^{1/q} \mid L_p(\Omega) \right\| < \infty \Bigg\}$$

(modification if $q = \infty$) in the sense of equivalent quasi-norms.

(ii) Let $0 < p \leqq \infty$, $0 < q \leqq \infty$ and $s > n(\frac{1}{p} - 1)_+$. Let $M \in \mathbb{N}$ with $M > s$, then

$$B^s_{pq}(\Omega) = \Bigg\{ f \in L_{\max(1,p)}(\Omega) : \|f \mid L_p(\Omega)\| \tag{5}$$

$$+ \left(\int_0^1 t^{-sq} \|d_t^M f \mid L_p(\Omega)\|^q \frac{dt}{t} \right)^{1/q} < \infty \Bigg\}$$

(modification if $q = \infty$) in the sense of equivalent quasi-norms.

Remark 2. Characterizations of function spaces via differences have been extensively studied. In the preceding subsection we collected some really classical assertions. In the case $\Omega = \mathbb{R}^n$, characterizations of the fractional Sobolev spaces

$H^s_p = F^s_{p2}$ with $s > 0$, $1 < p < \infty$, via differences are known since the sixties and early seventies, see [Str1] (1967) and [Liz3,4] (1972), references to some forerunners may be found in [Ste1: V.6]. Characterizations of F^s_{pq} with $1 < p < \infty$ and $1 < q < \infty$ on $\mathbb{R}^n$ in terms of differences are due to G.A. Kaljabin, see [Kal1,2] (1977–80), extensions to values $p \leqq 1$ and/or $q \leqq 1$ may be found in [Tri9] (1980). In [Triß: 2.5.9–2.5.12] (1983) we discussed in detail what was known at that time about characterizations of F^s_{pq} and B^s_{pq} on $\mathbb{R}^n$ via differences, including many references. In particular (4) and (5) with $\Omega = \mathbb{R}^n$ coincide essentially with the Corollary in [Triß: 2.5.11]. The extension of these results from $\mathbb{R}^n$ to $\mathbb{R}^n_+$ and to more general domains caused a lot of trouble and around 1980 only very few results had been known. For example, let $0 < p \leqq 1$ and $n(\frac{1}{p} - 1) < s < 1$, then

$$\|f \mid L_p(\Omega)\| + \left(\int_{\Omega\times\Omega} \frac{|f(x) - f(y)|^p}{|x-y|^{n+sp}} dxdy \right)^{1/p} \tag{6}$$

is an equivalent quasi-norm in $B^s_{pp}(\Omega)$, where Ω is a bounded C^∞ domain, see [Tri7] (1979). The first major step was done by G.A. Kaljabin, see [Kal3,4] (1983–85), who proved (4) for bounded domains Ω (even more general than C^∞ domains) and spaces $F^s_{pq}(\Omega)$ with $s > 0$, $1 < p < \infty$, $1 < q < \infty$, see also [KaL] and [Bes3]. A proof of the full theorem has been given in [Tri23] (1989). We return to this subject in detail in 3.5.2, 3.5.3, 4.5.4 and 5.2.2, including some generalizations.

1.11 Spaces on Riemannian manifolds and Lie groups

1.11.1 Introduction

Function spaces on Riemannian manifolds and Lie groups of type F^s_{pq} and B^s_{pq} will be discussed in great detail in Chapter 7. This covers, in particular, Sobolev spaces, Hardy spaces and Hölder–Zygmund spaces on manifolds. On the other hand, it is the aim of this introductory chapter to describe the relevant parts of the theory of function spaces from a historical point of view. So we have the feeling that some material on function spaces of the above type should be included in this chapter.

Compared with the overwhelming flood of papers and books on function spaces on $\mathbb{R}^n$ or on domains Ω, there are only few papers on function spaces on Riemannian manifolds and only some papers on function spaces on Lie groups. Apparently the first comprehensive study of Sobolev spaces on Riemannian manifolds is due to T. Aubin, [Aub1] (1976), see also [Aub2] (1982) where he used his results in connection with non-linear PDE's on manifolds. The basic idea is simply to replace partial derivatives in $\mathbb{R}^n$ by covariant derivatives. However, a detailed discussion of the properties of these spaces in dependence on geometric data of the manifold is difficult. R.S. Strichartz introduced in [Str2] (1983) fractional Sobolev spaces on Riemannian manifolds. Let Δ be the Laplace–Beltrami operator on the underlying manifold, then, roughly speaking, Strichartz introduced the spaces $H^s_p = (id - \Delta)^{-s/2} L_p$ with $1 < p < \infty$ and $s \in \mathbb{R}$. But even the identification of H^s_p, where $s \in \mathbb{N}$, $1 < p < \infty$, with the corresponding Sobolev spaces defined via covariant derivatives caused some trouble. This spectral approach was used in [Sak2] to introduce Besov spaces. A closer look reveals that seemingly any deeper analysis on Riemannian manifolds is closely connected with a deep and thorough study of the Laplace–Beltrami operator. For example, it comes out that a satisfactory theory of function spaces on Riemannian manifolds needs far-reaching assertions on decay properties of the kernels of the heat equation depending on geometric data such as sectional curvature, etc. Sufficiently strong results had not been available up to the early eighties. At least from our point of view, the paper by J. Cheeger, M. Gromov and M. Taylor, [CGT] (1982), was the first one containing estimates for the heat kernels good enough for our purpose. In Chapter 7 we give further references to more recent papers. On this basis (and the local means from 1.8.4) we developed since 1986 a theory of function spaces on Riemannian manifolds which will be discussed in detail in Chapter 7 and outlined in the following subsections. Further information on Sobolev spaces on manifolds and many references may be found in [Eic1,2], see also [Skr1,2].

Function spaces on a Lie group G have been studied by several authors with rather different methods. Under strong restrictions for G (simply connected, nilpotent, stratifying), G.E. Folland, E.M. Stein, S.G. Krantz and K. Saka studied Hölder–Zygmund spaces (called Lipschitz spaces), Hardy spaces and Besov spaces, see [Fol1,2, FoS, Kra, Sak]. They used differences, derivatives, Gauss–Weierstrass semi-groups and, in particular, Cauchy–Poisson semi-groups, see also the recent papers [CoS1,2]. Another method was examined by I.Z. Pesenson, M. Geisler and D.W. Robinson, [Pes1–4, Gei1, Rob1,2]. Using invariant Haar measures and invariant vector fields they introduced the spaces $L_p(G)$, $1 \leqq p < \infty$, and the Sobolev spaces $W^m_p(G)$. They defined the Besov spaces $B^s_{pq}(G)$ via real interpolation. Based on abstract non-commutative interpolation theory for semi-groups of operators and representation theory for Lie groups they are able to characterize the Besov spaces $B^s_{pq}(G)$ via differences similar as in the Euclidean case. Our own approach uses the rather fashionable idea to combine the advantages of Lie group theory and Riemannian geometry. We equip G with a (left-invariant) Riemannian metric. A striking application of this idea in the completely different field of Eulerian motions of rigid bodies may be found in [Arn: Appendix 2]. We reduce function spaces on Lie groups to function spaces on Riemannian manifolds and characterize them afterwards in terms of Lie groups. This will be done in detail in Chapter 7, but some highlights will be outlined below.

1.11.2 Sobolev spaces, fractional Sobolev spaces, and Hölder–Zygmund spaces on M

The letter M is reserved for the underlying Riemannian manifolds. It is always supposed that M is a n-dimensional connected C^∞ manifold equipped with a smooth Riemannian metric. The two additional assumptions we need read as follows:

(i) M is complete and has a bounded geometry,

(ii) M has a positive injectivity radius r_0.

Detailed explanations of these technical assumptions will be given in 7.2.1, including some references, see also the following subsection where we add few comments. Let $P \in M$ and $X \in T_P M$, where the latter stands for the tangent space, then $c(P, X, t)$ with $t \in \mathbb{R}$ characterizes the geodesic with

$$c(P, X, 0) = P \quad \text{and} \quad \frac{dc}{dt}(P, X, 0) = X. \tag{1}$$

Furthermore, $t\|X\|$ is the arc length measured along $c(P, X, t)$ with P as the off-point. Let $m \in \mathbb{N}$, $P \in M$ and $X \in T_P M$, then

$$\Delta_X^m f(P) = \sum_{j=0}^{m} (-1)^{m-j} \binom{m}{j} f\left(c\left(P, X, \frac{j}{m}\right)\right) \tag{2}$$

is the natural Riemannian counterpart of the differences (1.2.2/3). Let ∇_α with $\alpha = 1, \ldots, n$ be the covariant derivative, ∇^α its contravariant version and

$$|\nabla^k f|^2 = \nabla^{\alpha_1} \cdots \nabla^{\alpha_k} f \cdot \nabla_{\alpha_1} \cdots \nabla_{\alpha_k} \overline{f}, \qquad k \in \mathbb{N}_0, \tag{3}$$

(summation convention), where f is a complex-valued C^∞-function. Let Δ be the Laplace–Beltrami operator. Furthermore, $L_p(M)$ with $0 < p \leqq \infty$ are the usual L_p-spaces on M with respect to the Riemannian volume element. Let $D(M)$ be the collection of all complex-valued C^∞ functions on M with compact support and $D'(M)$ be the dual space of all distributions. Under the above restrictions for M it comes out that $\rho\, id - \Delta$ with $\rho > 0$ and $D(M)$ as domain of definition is a positive-definite essentially self-adjoint operator in the Hilbert space $L_2(M)$. By spectral theory, $(\rho\, id - \Delta)^{-s/2}$ with $s > 0$, makes sense in $L_2(M)$. This can be extended to $L_p(M)$ with $1 < p < \infty$, see [Str2].

Definition. (i) (Sobolev spaces). Let $1 < p < \infty$ and $k \in \mathbb{N}_0$ then

$$\|f \mid W_p^k(M)\| = \sum_{l=0}^{k} \|\nabla^l f \mid L_p(M)\| \tag{4}$$

and $W_p^k(M)$ is the completion of $\{h \in C^\infty(M)\colon \|h \mid W_p^k(M)\| < \infty\}$ in $L_p(M)$ with respect to (4).

(ii) (Fractional Sobolev spaces). Let $1 < p < \infty$. Let $s > 0$, then $H_p^s(M)$ is the collection of all $f \in L_p(M)$ which can be represented as $f = (id - \Delta)^{-s/2} h$ with $h \in L_p(M)$, and

$$\|f \mid H_p^s(M)\| = \|h \mid L_p(M)\|. \tag{5}$$

Let $H_p^0(M) = L_p(M)$. Let $s < 0$, then $H_p^s(M)$ is the collection of all $f \in D'(M)$ which can be represented as $f = (id - \Delta)^k h$ with $h \in H_p^{s+2k}(M)$, $k \in \mathbb{N}$ and $2k + s > 0$. Furthermore,

$$\|f \mid H_p^s(M)\| = \|h \mid H_p^{2k+s}(M)\|. \tag{6}$$

(iii) (Hölder–Zygmund spaces). Let $s > 0$, then

$$\mathcal{C}^s(M) = \left\{ f \in L_\infty(M)\colon \sup_{P \in M} \sup_{X \in T_P M} \|X\|^{-s} \mid \Delta_X^{1+[s]} f(P)| < \infty \right\}. \tag{7}$$

Remark 1. Part (i) goes back to T. Aubin [Aub1,2], part (ii) is due to R.S. Strichartz [Str2] and part (iii) we took over from [Tri14]. Let $M = \mathbb{R}^n$, then $W_p^k(M)$, $H_p^s(M)$ and $\mathcal{C}^s(M)$ coincides with the spaces from 1.2.3, 1.3.2 and 1.2.2, respectively.

Proposition. Let M be the above Riemannian manifold. Then the spaces $W_p^k(M)$, $H_p^s(M)$ and $\mathcal{C}^s(M)$ are Banach spaces. Let $k \in \mathbb{N}$ and $1 < p < \infty$, then

$$H_p^k(M) = W_p^k(M). \tag{8}$$

Remark 2. The equivalence (8) is not obvious. As we shall see in 7.4.5, we employ all the geometric properties of the Riemannian manifold M in order to prove it.

1.11.3 The spaces $F_{pq}^s(M)$ and $B_{pq}^s(M)$; decompositions of F_{pq}^s on $\mathbb{R}^n$

Let M be the Riemannian manifold of the preceding subsection. Let $P \in M$ then the exponential map $\exp_P$ maps $T_P M$ into M:

$$\exp_P X = c(P, X, 1) \quad \text{with} \quad X \in T_P M, \tag{1}$$

where $c(P, X, t)$ is the geodesic introduced in (1.11.2/1). The injectivity radius r_0 of the preceding subsection may be defined as the largest positive number such that for any r with $0 < r < r_0$ the restriction of $\exp_P$ to $B_P(r) = \{X \in T_P M: \|X\| < r\}$ is a diffeomorphic map of $B_P(r)$ onto $\Omega_P(r) = \exp_P B_P(r)$. Let $\delta > 0$ be sufficiently small, then there exists an at most countable covering of M by geodesic balls $\Omega_{P_j}(\delta)$ in the above sense such that any given ball $\Omega_{P_j}(\delta)$ has a non-empty intersection with at most L of other balls of that family. Here $L \in \mathbb{N}$. This covering $M = \bigcup_{j=1}^{J} \Omega_{P_j}(\delta)$, where either $J \in \mathbb{N}$ or $J = \infty$, can be chosen such that there exists a family $\{\psi_j\}_{j=1}^{J} \subset C^\infty(M)$ with

(i) $0 \leqq \psi_j \leqq 1$, $\sum_{j=1}^{J} \psi_j = 1$ on M,

(ii) $\operatorname{supp} \psi_j \subset \Omega_{P_j}(\delta)$,

(iii) for any multi-index α there exists a number $b_\alpha > 0$ such that

$$|D^\alpha(\psi_j \circ \exp_{P_j})(x)| \leqq b_\alpha, \qquad x \in B_{P_j}(r),$$

where $0 < \delta < r < r_0$. We refer to 7.2.1 or [Tri13: Remark 6]. Decompositions of this type go back to E. Calabi, see also [Aub2: Lemmata 2.25, 2.26]. Several large classes of Riemannian manifolds are known which obey the above requirements, see 7.2.1 for more details.

Recall that F^s_{pq} and B^s_{pq} are the spaces from 1.5.1, defined on $\mathbb{R}^n$. Now it is convenient for us to put $F^s_{\infty\infty} = B^s_{\infty\infty}$. Let again $D'(M)$ be the collection of all complex-valued distributions in M.

Definition. (i) Let either $0 < p < \infty$, $0 < q \leqq \infty$ or $p = q = \infty$, let $s \in \mathbb{R}$, then

$$F^s_{pq}(M) = \left\{ f \in D'(M), \left(\sum_{j=1}^{J} \|\psi_j f \circ \exp_{P_j} \mid F^s_{pq}\|^p \right)^{1/p} < \infty \right\} \tag{2}$$

(modification if $p = \infty$).

(ii) Let $0 < p \leqq \infty$, $0 < q \leqq \infty$ and $-\infty < s_0 < s < s_1 < \infty$ with $s = (1-\theta)s_0 + \theta s_1$, then

$$B^s_{pq}(M) = (F^{s_0}_{pp}(M), F^{s_1}_{pp}(M))_{\theta,q}. \tag{3}$$

Remark 1. These spaces have been introduced in [Tri13, 14] (1986–87). We must add some explanations. First, let $0 < \delta < r_1 < r_0$, then we assumed tacitly in (2) that $\psi_j f \circ \exp_{P_j}$ is extended by zero outside of $B_{P_j}(r_1)$. Then the extended function belongs to S' and (1.5.1/3) makes sense, including $F^s_{\infty\infty} = B^s_{\infty\infty}$. Secondly, $B^s_{pq}(M)$ is defined via the real interpolation method from 1.6.2. This is the first time that the spaces $F^s_{pq}(M)$ have priority over $B^s_{pq}(M)$. Usually, B^s_{pq} spaces are simpler, and it is more ambitious to prove assertions in F^s_{pq} spaces, for example, mapping properties for PDE's and ψDE's. But now the situation is different. To provide a better understanding, we describe the following observation for the spaces F^s_{pq} on $\mathbb{R}^n$.

Proposition. Let either $0 < p < \infty$, $0 < q \leqq \infty$ or $p = q = \infty$ and let $s \in \mathbb{R}$. Let ψ be a compactly supported C^∞ function on $\mathbb{R}^n$ such that $\sum_{k\in\mathbb{Z}^n} \psi(x-k) = 1$ for $x \in \mathbb{R}^n$, then

$$\left(\sum_{k\in\mathbb{Z}^n} \| \psi(\cdot - k) f \mid F^s_{pq} \|^p \right)^{1/p} \tag{4}$$

is an equivalent quasi-norm in F^s_{pq}.

Remark 2. In other words, we took a property of the spaces F^s_{pq} on $\mathbb{R}^n$ in order to extend the definition of these spaces to manifolds M with positive injectivity radius. In particular, $F^s_{pq} = F^s_{pq}(M)$ if $M = \mathbb{R}^n$. If $p \neq q$ then the counterpart of (4) for B^s_{pq} is wrong. This can be seen by constructing counterexamples, see 2.4.7 where we return to this subject. Hence, the simple localization method (2), which works so nicely in the case of the F^s_{pq} spaces, cannot be applied to the B^s_{pq} spaces. Rescue comes from the second part of the real interpolation formula (1.6.7/7) which is now converted into the definition (3).

After this digression we return to the function spaces from the above definition on the Riemannian manifold M. First we remark that $F^s_{pq}(M)$ are quasi-Banach spaces which are independent of the distinguished local charts $(\Omega_{P_j}(r), \exp^{-1}_{P_j})$ and the resolution of unity $\{\psi_j\}$. Furthermore, $B^s_{pq}(M)$ are quasi-Banach spaces which are independent of s_0 and s_1. Finally we compare the Sobolev spaces, fractional Sobolev spaces and Hölder–Zygmund spaces introduced in the preceding subsection with the above spaces.

Theorem. (i) Let $1 < p < \infty$ and $k \in \mathbb{N}_0$ then

$$F^k_{p2}(M) = W^k_p(M). \tag{5}$$

(ii) Let $1 < p < \infty$ and $s \in \mathbb{R}$, then

$$F^s_{p2}(M) = H^s_p(M). \tag{6}$$

(iii) Let $s > 4$ then

$$F^s_{\infty\infty}(M) = \mathcal{C}^s(M). \tag{7}$$

Remark 3. Proofs may be found in [Tri13,14]. The restriction $s > 4$ in (iii) is awkward, it depends on our method, and (7) should be valid for all $s > 0$. See also (1.11.2/8) as a special case of (5,6). This theorem is the extension of previous assertions on $\mathbb{R}^n$, see (1.3.2/4) and Theorem 1.5.1. A detailed proof of the above theorem will be given in 7.4.5 and 7.5.3.

1.11.4 Intrinsic characterizations of $F^s_{pq}(M)$ and $B^s_{pq}(M)$

The definition of $F^s_{pq}(M)$ and, as a consequence, also of $B^s_{pq}(M)$ is based on the distinguished atlas $(\Omega_{P_j}(r), \exp^{-1}_{P_j})$. The problem is to find intrinsic descriptions. We discussed a little bit different but similar questions in connection with spaces on domains, see 1.10. First we look for a counterpart of Theorem 1.8.4. Let k_0, k^0 and k be the same functions as in 1.8.4, where we now assume, in addition, that k_0 and k^0, and hence also k, are rotation-invariant, i.e.,

$$k_0(x) = K_0(|x|) \quad \text{and} \quad k(x) = K(|x|), \quad x \in \mathbb{R}^n. \tag{1}$$

Recall that $T_P M$ stands for the tangent space in $P \in M$, and $c(P, X, t)$ for the geodesic with (1.11.2/1). Then the substitute of (1.8.4/2) reads as follows,

$$k(t, f)(P) = \int_{T_P M} K(\|X\|) f(c(P, X, t)) dX, \qquad P \in M, \quad t \in \mathbb{R}. \tag{2}$$

We may restrict t by $0 \leqq t < r_0$, where r_0 is the injectivity radius. Then (2) makes sense for any $f \in D'(M)$, see the beginning of 1.11.3.

Theorem. (i) Let either $0 < p < \infty$, $0 < q \leqq \infty$ or $p = q = \infty$. Let $0 < r < r_0$ and $s \in \mathbb{R}$. Let $\varepsilon > 0$ be sufficiently small and $N \in \mathbb{N}$ be sufficiently large (in dependence on s, p, q), then

$$\|k_0(\varepsilon, f) \mid L_p(M)\| + \left\| \left(\int_0^r t^{-sq} |k(t,f)(\cdot)|^q \frac{dt}{t} \right)^{1/q} \mid L_p(M) \right\| \tag{3}$$

(modification if $q = \infty$) is an equivalent quasi-norm in $F^s_{pq}(M)$.

(ii) Let $0 < p \leqq \infty$, $0 < q \leqq \infty$, $0 < r < r_0$ and $s \in \mathbb{R}$. Let $\varepsilon > 0$ be sufficiently small and $N \in \mathbb{N}$ be sufficiently large (in dependence on s, p, q) then

$$\|k_0(\varepsilon, f) \mid L_p(M)\| + \left(\int_0^r t^{-sq} \|k(t,f) \mid L_p(M)\|^q \frac{dt}{t} \right)^{1/q} \tag{4}$$

(modification if $q = \infty$) is an equivalent quasi-norm in $B^s_{pq}(M)$.

Remark 1. This theorem extends essentially Theorem 1.8.4 from $\mathbb{R}^n$ to M. It is one of the main assertions proved in [Tri13, 14]. We return to this subject in detail in 7.2.4 and 7.3.2.

In Remark 1.10.4/2 we discussed characterizations of F^s_{pq} spaces and B^s_{pq} spaces on $\mathbb{R}^n$ and on domains in terms of differences. See, in particular, Theorem 1.10.4. We ask for similar descriptions of the spaces $F^s_{pq}(M)$ and $B^s_{pq}(M)$ where M is the above Riemannian manifold. For this purpose we modify (1.10.4/2) in the following way. Let $d(x)$ be a rotation-invariant nonnegative C^∞ function in $\mathbb{R}^n$ with $d(0) > 0$ and $\operatorname{supp} d \subset B$, where B is the unit ball in $\mathbb{R}^n$. Let $d(x) = D(|x|)$, then we introduce the means

$$D^m(t,f)(P) = \int_{T_P M} D(\|X\|) \Delta^m_{tX} f(P) dX, \qquad P \in M,\ t \in \mathbb{R}, \tag{5}$$

where Δ^m_{tX} has been defined in (1.11.2/2) with $m \in \mathbb{N}$ and $X \in T_P M$. The means (5) modify both (2) and (1.10.4/2). Recall that $r_0 > 0$ stands for the injectivity radius.

Proposition. (i) Let either $0 < p < \infty$, $0 < q \leqq \infty$ or $p = q = \infty$. Let $0 < r < r_0$,

$$s > 2\left(\frac{n}{p} + 2\right), \quad s > 2n\left(\frac{1}{q} - 1\right)_+, \quad m \in \mathbb{N} \text{ with } m > \frac{3}{2}s, \tag{6}$$

then

$$\|f \mid L_p(M)\| + \left\| \left(\int_0^r t^{-sq} |D^m(t,f)(\cdot)|^q \frac{dt}{t} \right)^{1/q} \mid L_p(M) \right\| \tag{7}$$

(modification if $q=\infty$) is an equivalent quasi-norm in $F^s_{pq}(M)$.

(ii) Let $0<p\leqq\infty$, $0<q\leqq\infty$, $0<r<r_0$,

$$s>2\left(\frac{n}{p}+2\right),\quad m\in\mathbb{N} \text{ with } m>\frac{3}{2}s, \tag{8}$$

then

$$\|f \mid L_p(M)\|+\left(\int_0^r t^{-sq}\|D^m(t,f)\mid L_p(M)\|^q\frac{dt}{t}\right)^{1/q} \tag{9}$$

(modification if $q=\infty$) is an equivalent quasi-norm in $B^s_{pq}(M)$.

Remark 2. This proposition has been proved in [Tri14]. We return to this subject in 7.5.4, see also 7.5.2. The restrictions (6) and (8) are somewhat artificial. Let $M=\mathbb{R}^n$, then (6) can be replaced by

$$n\left(\frac{1}{\min(p,q)}-1\right)_+<s<m \tag{10}$$

and (8) can be replaced by

$$n\left(\frac{1}{p}-1\right)_+<s<m. \tag{11}$$

These restrictions are natural, details may be found in 2.6.3. Surprisingly also this $\mathbb{R}^n$-version seems to be quite new. The above formulation with $\mathbb{R}^n$ and (10,11) may be found in [Tri14] (1987), but it is a slight modification of a corresponding result in [Tri16] (1988).

1.11.5 Spaces on Lie groups

We discuss briefly spaces of type F^s_{pq} and B^s_{pq} on an arbitrary n-dimensional connected Lie group G. The basic idea has been mentioned in the introduction 1.11.1. We equip G with a left-invariant Riemannian metric and apply Definition 1.11.3. Let e be the neutral element in G and let $\hat{G}=T_eG$ be the corresponding Lie algebra. Let g be a positive-definite symmetric bilinear form on $\hat{G}$ and let $L_a\colon x\to ax$ with $a\in G$ and $x\in G$ be the left translation on G. The pull-back operation $g_a=(L_{a^{-1}})^*_a g$ with $a\in G$, generates a left-invariant Riemannian metric on G. The resulting Riemannian manifold has the properties required in the previous subsections. In particular, we can apply Definition 1.11.3 and denote now the corresponding spaces by $F^s_{pq}(G)$

and $B^s_{pq}(G)$. These spaces are independent of the chosen positive-definite symmetric bilinear form g on $\hat{G}$ (equivalent quasi-norms). Furthermore, the corresponding Riemannian volume element generates a left-invariant Haar measure on G. Then it is clear what is meant by $L_p(G)$ with $0 < p \leqq \infty$. Intrinsic descriptions of $F^s_{pq}(G)$ and $B^s_{pq}(G)$ can be given in the sense of Theorem 1.11.4 with the means (1.11.4/2) where $c(P, X, t)$ are the geodesics in the sense of the generated Riemannian metric. It would be highly desirable to replace the Riemannian geodesics $c(P, X, t)$ by their Lie group counterparts, i.e., by

$$x \cdot \exp tX \quad \text{with} \quad x \in G, X \in \hat{G} \text{ and } \quad t \in \mathbb{R}, \tag{1}$$

where exp has the usual meaning. (1) with $x = e$, $X \in \hat{G}$ fixed, and $t \in \mathbb{R}$ variable, are the one-dimensional subgroups of G. Then (1.11.4/2) must be replaced by

$$k(t, f)(x) = \int_{\hat{G}} K(|X|) f(x \cdot \exp tX) dX, \qquad x \in G, \ t \in \mathbb{R}. \tag{2}$$

In general, the Riemannian geodesics and the Lie group geodesics $x \cdot \exp tX$ are different. Nevertheless, Theorem 1.11.4 can be modified in the following way.

Theorem. Let G be the above connected Lie group equipped with a left-invariant Riemannian metric.

(i) Under the same restrictions for p, q, r, s, ε and N as in Theorem 1.11.4(i),

$$\|k_0(\varepsilon, f) \mid L_p(G)\| + \left\| \left(\int_0^r t^{-sq} |k(t, f)(\cdot)|^q \frac{dt}{t} \right)^{1/q} \mid L_p(G) \right\| \tag{3}$$

(modification if $q = \infty$) is an equivalent quasi-norm in $F^s_{pq}(G)$, where $k(t, f)$ stands either for the Riemannian version (1.11.4/2) or for the Lie group version (2).

(ii) Under the same restrictions for p, q, r, s, ε and N, as in Theorem 1.11.4(ii),

$$\|k_0(\varepsilon, f) \mid L_p(G)\| + \left(\int_0^r t^{-sq} \|k(t, f) \mid L_p(G)\|^q \frac{dt}{t} \right)^{1/q} \tag{4}$$

(modification if $q = \infty$) is an equivalent quasi-norm in $B^s_{pq}(G)$, where $k(t, f)$ stands either for the Riemannian version (1.11.4/2) or for the Lie group version (2).

Remark 1. The Riemannian version is completely covered by Theorem 1.11.4. As for the Lie group version, we gave several independent proofs of the above theorem, see [Tri15, 17, 20] (1987–88). We return to this subject in detail in 7.6. One can ask for further descriptions, but at the moment, we mention only a characterization of the Sobolev spaces $W_p^k(G) = F_{p2}^k(G)$ with $k \in \mathbb{N}_0$ and $1 < p < \infty$, in terms of Lie group theory.

Proposition. Let $k \in \mathbb{N}_0$ and $1 < p < \infty$. Let $X_1(x), \ldots, X_n(x)$ with $x \in G$ be linearly independent left-invariant vector fields on G, then

$$W_p^k(G) = \left\{ f \in L_p(G) \colon \sum_{l=0}^{k} \sum_{1 \leqq n_i \leqq n} \| X_{n_1} \cdots X_{n_l} f \mid L_p(G) \| < \infty \right\}. \tag{5}$$

Remark 2. Some care is necessary because the vector fields $X_1, \ldots, X_n$ are non-commutative, in general. We refer to the above-mentioned papers or to 7.6.

1.12 Epilogue

It is the main aim of this chapter to highlight some seminal ideas in the theory of function spaces which came into being during the last 50 years. We concentrated ourselves strictly on those developments which are directly connected with the theme of this book, that is, the theory of the spaces B^s_{pq} and F^s_{pq}. We did virtually nothing to illuminate other areas in the theory of function spaces. We even hardly touched on very neighbouring branches which grew out of the treated topics in the course of time. Even under these severe restrictions, it is hard to decide what subjects should be selected, what ideas should be emphasized, and for whom credit should be given for inventing these or those idea, device or space. Even worse, the latter is to often a matter of taste and knowledge, and sometimes it also depends on in which part of the world one lives. Our point of view is rather conventional: We quote those papers and authors which are usually quoted in connection with the just treated idea, well aware of the fact that almost each of these milestone papers has forerunners and that often contemporaries had similar ideas published in lesser known journals or were simply overlooked (because they are not members of self-contained recognized circles). We gave a conventional picture, but occasionally we highlighted some aspects which may be known to directly involved specialists, but which are more or less unknown to a wider audience interested in function spaces. A typical example is our attempt to emphasize the importance of the work of G.H. Hardy in the late twenties and thirties. We add a list of those spaces treated in this chapter accompanied by the years when they came into being. We say nothing on their mutual interrelations and almost nothing on the underlying techniques (there are two exceptions, as the reader will notice). Detailed considerations have been given in the corresponding sections of this chapter, which may be found easily by consulting the table of contents. At the same time, this is the end of our historically-orientated survey. The rest of this book deals with recent developments in the theory of the spaces B^s_{pq} and F^s_{pq} which took place in the eighties.

Problem: What is the year of birth of the Theory of Function Spaces as a self-contained branch of Functional Analysis?

Hint: Look at the table.

Answer: 1960 is apparently a good candidate.

year	
	L_p $(0 < p \leqq \infty)$; C^m $(m \in \mathbb{N})$; Hölder spaces C^s $(0 < s \neq \text{integer})$
36–38	Sobolev spaces W^k_p $(k \in \mathbb{N},\ 1 < p < \infty)$; Morrey spaces
45	Hölder–Zygmund spaces $\mathcal{C}^s$ $(s > 0)$
51	Nikol'skij spaces $B^s_{p\infty}$ $(s > 0;\ 1 < p < \infty)$
55–58	W^s_p $(0 < s \neq \text{integer},\ 1 < p < \infty)$
59–61	Besov spaces B^s_{pq} $(s > 0,\ 1 < p < \infty,\ 1 \leqq q \leqq \infty)$; fractional Sobolev spaces H^s_p $(s \in \mathbb{R},\ 1 < p < \infty)$; abstract interpolation; BMO; real n dimensional version of Hardy spaces H_p $(0 < p < \infty)$
60–64	B^s_{pq} and H^s_p as interpolation spaces
63–65	Campanato spaces
67	Fourier analytical version of B^s_{pq} $(s \in \mathbb{R},\ 1 \leqq p \leqq \infty,\ 1 \leqq q \leqq \infty)$
72	Fourier analytical version of H_p $(0 < p < \infty)$
72–74	F^s_{pq} $(s \in \mathbb{R},\ 1 < p < \infty,\ 1 < q < \infty)$
73	B^s_{pq} $(s \in \mathbb{R},\ 0 < p \leqq \infty,\ 0 < q \leqq \infty)$
75	F^s_{pq} $(s \in \mathbb{R},\ 0 < p < \infty,\ 0 < q \leqq \infty)$

Chapter 2

The Spaces B^s_{pq} and F^s_{pq}: Definitions and Characterizations

2.1 Introduction

It was the aim of the first chapter to look at function spaces from a historical point of view and to convince the reader that the two scales B^s_{pq} and F^s_{pq} occupy the very heart of the theory of function spaces. In the present chapter and the following one we develop systematically the technical part of the theory of the spaces B^s_{pq} and F^s_{pq} on $\mathbb{R}^n$ as it stands at the end of the eighties. The remaining chapters rest on these fundamentals. Our recent approach, compared with earlier ones, for example in [Triß], not to speak about [Triα], can be described as follows. We give rather general and, unfortunately, highly technical characterizations of the spaces F^s_{pq} and B^s_{pq} which more or less cover desirable concrete characterizations, for example, via differences or derivatives of functions, or via harmonic or thermic extensions, and which also provide the basis for later applications. This will be done in 2.4 for F^s_{pq} and in 2.5 for B^s_{pq}, always with a preference of the more complicated spaces F^s_{pq}. In this sense, 2.4 and 2.5 may be considered as the heart of the present chapter. Characterizations of some spaces F^s_{pq} and B^s_{pq} in terms of differences of functions, or as harmonic or thermic extensions are known, for example, we derived them in [Triß] by rather specific and sometimes quite tricky arguments. Now we return in 2.6 to this subject, but this time as consequences of the characterizations in 2.4 and 2.5.

2.2 Prerequisites

2.2.1 Basic notations

For the convenience of the reader we collect in 2.2 some known assertions on maximal inequalities and entire analytic functions.

H. Triebel, *Theory of Function Spaces II*, Modern Birkhäuser Classics,
DOI 10.1007/978-3-0346-0419-2_2, © Birkhäuser Verlag 1992

First we recall some standard notations. Let:

$\mathbb{R}^n$ be the n-dimensional real euclidean space,
$\mathbb{N}$ be the collection of all natural numbers, and
$\mathbb{N}_0$ be the collection of all non-negative integers.

The general point in $\mathbb{R}^n$ is denoted by $x = (x_1, \ldots, x_n)$. Furthermore $D^\alpha = \frac{\partial^{|\alpha|}}{\partial x_1^{\alpha_1} \cdots \partial x_n^{\alpha_n}}$ has the usual meaning: $\alpha = (\alpha_1, \ldots, \alpha_n)$ is a multi-index, $\alpha_j \in \mathbb{N}_0$ and $|\alpha| = \sum_{j=1}^n \alpha_j$.

Let $S = S(\mathbb{R}^n)$ be the Schwartz space of all complex-valued rapidly decreasing C^∞ functions on $\mathbb{R}^n$ equipped with the usual topology. We adopt here and in the sequel the following convention: If there is no danger of confusion we omit $\mathbb{R}^n$ in $S(\mathbb{R}^n)$ and in the other spaces below. Let S' be the collection of all tempered distributions on $\mathbb{R}^n$, i.e., the topological dual of S, equipped with the strong topology. If $\varphi \in S$, then

$$\hat{\varphi}(x) = F\varphi(x) = (2\pi)^{-n/2} \int_{\mathbb{R}^n} e^{-ix\xi} \varphi(\xi) d\xi, \qquad x \in \mathbb{R}^n, \tag{1}$$

denotes the Fourier transform $\hat{\varphi}$ or $F\varphi$ of φ. Here $x\xi = \sum_{j=1}^n x_j \xi_j$ is the scalar product in $\mathbb{R}^n$ of $x = (x_1, \ldots, x_n) \in \mathbb{R}^n$ and $\xi = (\xi_1, \ldots, \xi_n) \in \mathbb{R}^n$. The inverse Fourier transform of φ is given by

$$\check{\varphi}(x) = F^{-1}\varphi(x) = (2\pi)^{-n/2} \int_{\mathbb{R}^n} e^{ix\xi} \varphi(\xi) d\xi, \qquad x \in \mathbb{R}^n. \tag{2}$$

One extends F and F^{-1} in the usual way from S to S'. Recall that both F and F^{-1} yield isomorphic mappings from S onto itself and from S' onto itself.

The collection of all complex-valued C^∞ functions on $\mathbb{R}^n$ with compact support is denoted by D; and D' stands for the set of all complex distributions on $\mathbb{R}^n$.

Let A be a complex linear vector space, then $\|a \mid A\|$ is said to be a quasi-norm if $\|a \mid A\|$ satisfies the usual conditions of a norm with exception of the triangle inequality which is replaced by the assumption that there exists a constant $c \geqq 1$ such that

$$\|a_1 + a_2 \mid A\| \leqq c\|a_1 \mid A\| + c\|a_1 \mid A\| \tag{3}$$

holds for all $a_1 \in A$ and $a_2 \in A$. (If $c = 1$ is admissible then A is a normed space). A quasi-normed space A is said to be a quasi-Banach space if it is complete (i.e., any fundamental sequence in A with respect to $\|\cdot \mid A\|$ converges).

Let $0 < q \leqq \infty$, then l_q is the set of all sequences $b = (b_k)_{k \in \mathbb{N}_0}$ of complex numbers such that

$$\|b \mid l_q\| = \left(\sum_{k=0}^{\infty} |b_k|^q \right)^{1/q} < \infty \tag{4}$$

(modification if $q = \infty$ by $\sup_{k\in\mathbb{N}_0} |b_k|$). Of course, l_q is a quasi-Banach space (if $q \geqq 1$, a Banach space). Let $0 < p \leqq \infty$ and $0 < q \leqq \infty$, let $f = (f_k(x))_{k\in\mathbb{N}_0}$ be a sequence of complex-valued Lebesgue measurable functions on $\mathbb{R}^n$, then

$$\|f \mid L_p(l_q)\| = \left(\int_{\mathbb{R}^n} \left(\sum_{k=0}^{\infty} |f_k(x)|^q\right)^{p/q} dx\right)^{1/p} \tag{5}$$

and

$$\|f \mid l_q(L_p)\| = \left(\sum_{k=0}^{\infty} \left(\int_{\mathbb{R}^n} |f_k(x)|^p dx\right)^{q/p}\right)^{1/q} \tag{6}$$

(modification if $p = \infty$ by $\operatorname{ess\,sup}_{x\in\mathbb{R}^n}$ and/or $q = \infty$ by $\sup_k$). Let $L_p(l_q) = L_p(\mathbb{R}^n, l_q)$ be the set of all sequences f such that $\|f \mid L_p(l_q)\| < \infty$, and let $l_q(L_p) = l_q(L_p(\mathbb{R}^n))$ be the set of all sequences f such that $\|f \mid l_q(L_p)\| < \infty$. In the scalar case the corresponding space is denoted by L_p, quasi-normed by $\|\cdot \mid L_p\| = (\int_{\mathbb{R}^n} |f(x)|^p dx)^{1/p}$ (modification if $p = \infty$ by $\operatorname{ess\,sup}_{x\in\mathbb{R}^n} |f(x)|$). Again we omit $\mathbb{R}^n$ if there is no danger of confusion. $L_p(l_q)$, $l_q(L_p)$ and the scalar case L_p are quasi-Banach spaces (Banach spaces if $p \geqq 1$ and $q \geqq 1$).

2.2.2 Some maximal inequalities

Let $f(x)$ be a complex-valued locally Lebesgue-integrable function on $\mathbb{R}^n$, then

$$Mf(x) = \sup |B|^{-1} \int_B |f(y)| dy \tag{1}$$

is the Hardy–Littlewood maximal function, where the supremum is taken over all balls B centered at x. We have $Mf(x) \geqq |f(x)|$ almost everywhere.

Theorem. Let $1 < p < \infty$ and $1 < q \leqq \infty$, let $f = (f_k(x))_{k\in\mathbb{N}_0}$ and $Mf = (Mf_k(x))_{k\in\mathbb{N}_0}$, then there exists a constant c such that

$$\|Mf \mid L_p(l_q)\| \leqq c\|f \mid L_p(l_q)\| \tag{2}$$

holds for all $f \in L_p(l_q)$.

Remark. The scalar case is due to G.H. Hardy, J.E. Littlewood ($n = 1$) and N. Wiener ($n > 1$), see [HaL1, Wie] (1930/39). The above vector-valued version goes back to C. Fefferman, E.M. Stein, see [FeS1] (1971). Comments and further informations may be found in [Ste1].

2.2.3 Entire analytic functions

Let $\varphi \in S'$ and $\operatorname{supp} \hat{\varphi}$ be compact in $\mathbb{R}^n$, then $\varphi = \varphi(x)$ is an entire analytic function in $\mathbb{R}^n$. This is a well-known fact and by the Paley–Wiener–Schwartz theorem there is a close connection between the size of $\operatorname{supp} \hat{\varphi}$ and the growth of the analytic extension $\varphi(z)$ with $z \in \mathbb{C}^n$. We shall not need these assertions and refer to [Triß: 1.2.1] for details. More interesting for us is the following Fourier multiplier assertion.

Let $a_+ = \max(a, 0)$ where $a \in \mathbb{R}$. Let $0 < p \leqq \infty$ then

$$\sigma_p = n \left(\frac{1}{p} - 1 \right)_+ . \tag{1}$$

Let $H^s_2 = H^s_2(\mathbb{R}^n)$ be the fractional Sobolev spaces from 1.3.2 including the classical Sobolev spaces $W^k_2 = H^k_2$ if $k \in \mathbb{N}_0$.

Theorem. Let Ω be a compact subset of $\mathbb{R}^n$, $0 < p \leqq \infty$ and $s > \sigma_p + \frac{n}{2}$, then there exists a constant c such that

$$\|(M\hat{f})^\vee \mid L_p\| \leqq c\|M \mid H^s_2\| \, \|f \mid L_p\| \tag{2}$$

for all $f \in L_p \cap S'$ with $\operatorname{supp} \hat{f} \subset \Omega$ and all $M \in H^s_2$.

Remark. A proof may be found in [Triß: 1.5.2]. In comparison with [Triß] the number σ_p has now a different meaning.

2.2.4 The spaces $L^\Omega_p(l_q)$ of entire analytic functions

We need an extension of Theorem 2.2.3 from the scalar case to the case of vector functions, including corresponding maximal inequalities. Let $0 < p \leqq \infty$, $0 < q \leqq \infty$ and let $\Omega = (\Omega_k)_{k \in \mathbb{N}_0}$ be a collection of compact subsets of $\mathbb{R}^n$, then

$$L^\Omega_p(l_q) = \{ f = (f_k)_{k\in\mathbb{N}_0} \subset S' \text{ with } \operatorname{supp} \hat{f}_k \subset \Omega_k \tag{1}$$
$$\text{if } k \in \mathbb{N}_0 \text{ and } \|f \mid L_p(l_q)\| < \infty \},$$

see (2.2.1/5). We always assume that the diameters

$$d_k = \sup |x - y| > 0, \qquad k \in \mathbb{N}_0, \tag{2}$$

where the supremum in (2) is taken over all $x \in \Omega_k$ and $y \in \Omega_k$.

Theorem. Let $0 < p < \infty$, $0 < q \leqq \infty$, and let $\Omega = (\Omega_k)_{k \in \mathbb{N}_0}$ be a collection of compact subsets of $\mathbb{R}^n$ with $d_k > 0$ if $k \in \mathbb{N}_0$.

(i) Let $0 < r < \min(p,q)$, then there exists a constant c such that

$$\left\| \left(\sum_{k=0}^{\infty} \left(\sup_{t \in \mathbb{R}^n} \frac{|f_k(\cdot - z)|}{1 + |d_k z|^{n/r}} \right)^q \right)^{1/q} \mid L_p \right\| \leqq c \| f \mid L_p(l_q) \| \tag{3}$$

(modification if $q = \infty$) holds for all $f = (f_k)_{k \in \mathbb{N}_0} \in L_p^{\Omega}(l_q)$.

(ii) Let $\kappa > \frac{n}{2} + \frac{n}{\min(p,q)}$, then there exists a constant c such that

$$\left\| \left(\sum_{k=0}^{\infty} |(M_k \hat{f}_k)^\vee|^q \right)^{1/q} \mid L_p \right\| \leqq c \sup_{l \in \mathbb{N}_0} \| M_l(d_l \cdot) \mid H_2^\kappa \| \, \| f \mid L_p(l_q) \| \tag{4}$$

(modification if $q = \infty$) holds for all $f = (f_k)_{k \in \mathbb{N}_0} \in L_p^{\Omega}(l_q)$ and all $(M_k(x))_{k \in \mathbb{N}_0} \subset H_2^\kappa$.

Remark 1. A proof of the theorem may be found in [Trіß: 1.6.1–1.6.3]. The spaces H_2^κ have the same meaning as in the preceding subsection. For our later purposes it is useful to remark that (4) follows from (3) and

$$\sup_{z \in \mathbb{R}^n} \frac{|(M_k \hat{f}_k)^\vee (x - z)|}{1 + |d_k z|^{n/r}} \leqq c \sup_{z \in \mathbb{R}^n} \frac{|f_k(x - z)|}{1 + |d_k z|^{n/r}} \| M_k(d_k \cdot) \mid H_2^\kappa \|, \tag{5}$$

with $0 < r < \min(p,q)$, $\kappa > \frac{n}{2} + \frac{n}{r}$, and where c is independent of $x \in \mathbb{R}^n$, $k \in \mathbb{N}_0$, M_k, and f_k. The maximal inequality (5) coincides with [Trіß: (1.6.3/2)].

Remark 2. The restrictions for s in Theorem 2.2.3 are better than the scalar case of (4). However one can improve the restrictions for κ in (4). Let

$$\sigma_{pq} = n \left(\frac{1}{\min(p,q)} - 1 \right)_+ \tag{6}$$

be the counterpart of (2.2.3). Let $0 < p < \infty$, $0 < q < \infty$, and

$$\kappa > \frac{n}{2} + \sigma_{pq}, \tag{7}$$

then (4) holds for all $f \in L^{\Omega}_p(l_q)$ and all $(M_k(x))_{k\in\mathbb{N}_0} \subset H^{\kappa}_2$. A proof may be found in [Trïß: 2.4.9]. It is based on part (ii) of the above theorem and a rather sophisticated complex interpolation of quasi-Banach spaces, whereas the proof of (ii) is straightforward. Furthermore, for technical reasons, we have $q < \infty$, although there is no doubt that the just-mentioned improvement is also correct if $q = \infty$.

2.3 Definitions and basic properties

2.3.1 Definitions

Let φ be a C^∞ function on $\mathbb{R}^n$ with

$$\operatorname{supp}\varphi \subset \{\xi \in \mathbb{R}^n \colon |\xi| \leqq 2\}, \quad \varphi(\xi) = 1 \text{ if } |\xi| \leqq 1. \tag{1}$$

Let $j \in \mathbb{N}$ and

$$\varphi_j(\xi) = \varphi(2^{-j}\xi) - \varphi(2^{-j+1}\xi), \qquad \xi \in \mathbb{R}^n. \tag{2}$$

Then we have

$$\operatorname{supp}\varphi_j \subset \{\xi \in \mathbb{R}^n \colon 2^{j-1} \leqq |\xi| \leqq 2^{j+1}\}, \qquad j \in \mathbb{N}, \tag{3}$$

and, with $\varphi_0 = \varphi$,

$$\sum_{k=0}^{\infty} \varphi_k(\xi) = 1 \text{ if } \xi \in \mathbb{R}^n. \tag{4}$$

In other words, $(\varphi_k)_{k\in\mathbb{N}_0}$ is a resolution of unity with (1.3). Let $f \in S'$ then

$$\varphi_k(D)f(x) = (\varphi_k \hat{f})^\vee(x), \qquad k \in \mathbb{N}_0, \ x \in \mathbb{R}^n, \tag{5}$$

is an entire analytic function, see 2.2.3, and we have

$$f = \sum_{k=0}^{\infty} \varphi_k(D) f \qquad \text{(convergence in } S'). \tag{6}$$

In other words, we decompose f in entire analytic functions, and we introduce spaces by checking the behaviour of these analytic functions with respect to $x \in \mathbb{R}^n$ and $k \in \mathbb{N}_0$.

Definition. (i) Let $0 < p \leqq \infty$, $0 < q \leqq \infty$ and $s \in \mathbb{R}$ then

$$B^s_{pq} = \left\{ f \in S' : \|f \mid B^s_{pq}\|_\varphi = \left(\sum_{k=0}^{\infty} 2^{ksq} \|\varphi_k(D) f \mid L_p\|^q \right)^{1/q} < \infty \right\} \tag{7}$$

(modification if $q = \infty$).

(ii) Let $0 < p < \infty$, $0 < q \leqq \infty$ and $s \in \mathbb{R}$ then

$$F^s_{pq} = \left\{ f \in S' : \|f \mid F^s_{pq}\|_\varphi = \left\| \left(\sum_{k=0}^{\infty} 2^{ksq} |\varphi_k(D) f(\cdot)|^q \right)^{1/q} \mid L_p \right\| < \infty \right\} \tag{8}$$

(modification if $q = \infty$).

Remark. The history of these spaces has been described in 1.3.4 and 1.5.1, including their connections with classical function spaces. We return to the latter point later on. An extension of (8) to $p = \infty$ is not reasonable, see 1.5.2 where we discussed this point in some detail.

2.3.2 Basic properties

Let $(\varphi_k)_{k \in \mathbb{N}_0}$ be the same system of functions as in 2.3.1 with the generating function $\varphi = \varphi_0$. We introduce the maximal functions

$$(\varphi^*_k f)_a(x) = \sup_{y \in \mathbb{R}^n} \frac{|\varphi_k(D) f(x-y)|}{1 + |2^k y|^a}, \qquad f \in S', \quad a > 0, \tag{1}$$

where $x \in \mathbb{R}^n$ and $k \in \mathbb{N}_0$.

Theorem. (i) Let $0 < p \leqq \infty$, $0 < q \leqq \infty$, $a > \frac{n}{p}$, and $s \in \mathbb{R}$, then

$$B^s_{pq} = \left\{ f \in S' : \left(\sum_{k=0}^{\infty} 2^{ksq} \|(\varphi^*_k f)_a \mid L_p\|^q \right)^{1/q} < \infty \right\} \tag{2}$$

(modification if $q = \infty$) in the sense of equivalent quasi-norms. Furthermore, B^s_{pq} is a quasi-Banach space (Banach space if $p \geqq 1$, $q \geqq 1$) and it is independent of the generating function φ (equivalent quasi-norms).

(ii) Let $0 < p < \infty$, $0 < q \leqq \infty$, $a > \frac{n}{\min(p,q)}$, and $s \in \mathbb{R}$, then

$$F^s_{pq} = \left\{ f \in S' : \left\| \left(\sum_{k=0}^{\infty} 2^{ksq} |(\varphi^*_k f)_a(\cdot)|^q \right)^{1/q} \mid L_p \right\| < \infty \right\} \tag{3}$$

(modification if $q = \infty$) in the sense of equivalent quasi-norms. Furthermore, F^s_{pq} is a quasi-Banach space (Banach space if $p \geqq 1$, $q \geqq 1$) and it is independent of the generating function φ (equivalent quasi-norms).

Proof. Step 1. We apply Theorem 2.2.4(i) with $d_k = c2^k$ and $\frac{n}{r} = a > \frac{n}{\min(p,q)}$, see also (2.3.1/1,2,4). Then it follows that the quasi-norms in (3) can be estimated from above by the quasi-norm $\|f \mid F^s_{pq}\|_\varphi$ in (2.3.1/8), besides an unimportant factor. This proves (3). Let ψ be another generating function. Then we have $\psi_k(x) = \chi_k(x)\psi_k(x)$ with

$$\chi_k(x) = \sum_{r=-2}^{2} \varphi_{k+r}(x), \qquad k \in \mathbb{N}_0, \quad \varphi_l = 0 \ \text{ if } \ l < 0, \tag{4}$$

where ψ_k is constructed in the same way as φ_k in 2.3.1. Now (2.2.4/4) and

$$\psi_k(D)f = \sum_{r=-2}^{2} (\psi_k[\varphi_{k+r}(D)f]^\wedge)^\vee \tag{5}$$

prove

$$\|f \mid F^s_{pq}\|_\psi \leqq c\|f \mid F^s_{pq}\|_\varphi. \tag{6}$$

Hence, F^s_{pq} is independent of φ. In the same way one proves the corresponding assertions for B^s_{pq}.

Step 2. By (2.3.1/6) it follows that $\|f \mid F^s_{pq}\|_\varphi$ from (2.3.1/8) is a quasi-norm (norm if $p \geqq 1$, $q \geqq 1$). We prove that F^s_{pq} is complete, where we may assume $s = 0$, without restriction of generality. For that purpose we introduce the space $L^*_p(l_q)$ which consists, by definition, of all sequences $(f_k)_{k\in\mathbb{N}_0}$ of measurable functions with

$$\left\| \left(\sum_{k=0}^{\infty} \left(\sup_{y\in\mathbb{R}^n} \frac{|f_k(\cdot - y)|}{1 + |2^k y^a|} \right)^q \right)^{1/q} \mid L_p \right\| < \infty \tag{7}$$

for some fixed $a > \frac{n}{\min(p,q)}$ (modification if $q = \infty$). Then $L^*_p(l_q)$ is a quasi-Banach space, in particular it is complete. Let $\varphi_k(x)$ and $\chi_k(x)$ be the same functions as above, see (4). We construct two linear operators. First, by (3) and (7),

$$A\colon f \in F^0_{pq} \longrightarrow (\varphi_k(D)f)_{k\in\mathbb{N}_0} \tag{8}$$

is a continuous map from F^0_{pq} into $L^*_p(l_q)$. Secondly, we claim that

$$B\colon (f_k)_{k\in\mathbb{N}_0} \in L^*_p(l_q) \longrightarrow \sum_{k=0}^{\infty} \chi_k(D) f_k \tag{9}$$

is a continuous map from $L_p^*(l_q)$ into F^0_{pq}. Let $f_k^*(x)$ be the maximal function in (7), then we have

$$f_k^*(x) \leqq c f_k^*(z)(1+2^k|x-z|)^a, \qquad x \in \mathbb{R}^n, \quad z \in \mathbb{R}^n, \tag{10}$$

for some $c>0$, which follows from

$$1+|2^k u| \leqq (1+2^k|v|)(1+2^k|u-v|), \qquad u \in \mathbb{R}^n, \quad v \in \mathbb{R}^n.$$

In particular $f_k^*(0)<\infty$ and

$$|f_k(y)| \leqq c f_k^*(0)(1+|2^k y|^a) \tag{11}$$

which shows that $f_k(y)$ is of at most polynomial growth. Then $f_k \in S'$ and $\chi_k(D) f_k$ makes sense. Let us assume that the sum in (9) converges in S' to some $f \in S'$: we return to this point in Remark 1 below. At most 5 summands in (9) contribute to $\varphi_j(D)f$, that ones with $|j-k| \leqq 2$,

$$\varphi_j(D) f = \sum_{|j-k|\leqq 2} \varphi_j(D) \circ \chi_k(D) f_k = \varphi_j(D) f_j + \cdots \tag{12}$$

where we used $\varphi_j \chi_j = \varphi_j$, and $+\cdots$ stands for 4 similar terms. Let $j \in \mathbb{N}$ then by (2.3.1/2) we have

$$\begin{aligned}\varphi_j(D) f_j(x) &= \int_{\mathbb{R}^n} \check{\varphi}_j(y) f_j(x-y)dy = 2^{jn}\int_{\mathbb{R}^n} \hat{\varphi}(2^j y) f_j(x+y)dy \\ &= \int_{\mathbb{R}^n} \hat{\varphi}(y) f_j(x+2^{-j})dy + \end{aligned} \tag{13}$$

where "+" indicates a second term of the same type. It follows

$$|\varphi_j(D) f_j(x)| \leqq c \int_{\mathbb{R}^n} |\hat{\varphi}(y)|(1+|y|^a)dy \cdot f_j^*(x). \tag{14}$$

Now (14) and similar estimates for the 4 other terms in (12) prove $f \in F^0_{pq}$, including an estimate for the involved quasi-norms. Hence B is a continuous map from $L_p^*(l_q)$ into F^0_{pq}. By $\varphi_k \chi_k = \varphi_k$ we have

$$B \circ A = \mathrm{id} \qquad (\text{identity in } F^0_{pq}) \tag{15}$$

and $\operatorname{Im} A$, the image of A, consists of those $(f_k)_{k\in\mathbb{N}_0} \in L_p^*(l_q)$ with

$$A \circ B(f_k) = (f_k). \tag{16}$$

In particular, $\operatorname{Im} A$ is a closed subspace of the quasi-Banach space $L_p^*(l_q)$ and A is an isomorphic map from F^0_{pq} onto $\operatorname{Im} A$. Hence F^0_{pq} is complete. The proof of part (ii) of the theorem is complete. Similarly one proves part (i).

Remark 1. We prove the convergence in S' of the sum in (9). Let $\eta \in S$, then we have

$$\int_{\mathbb{R}^n} \varphi_k(D) f_k(x)\eta(x)dx = \int_{\mathbb{R}^n} (\varphi_k \hat{f}_k)(\xi)\check{\eta}(\xi)d\xi = \int_{\mathbb{R}^n} (\varphi_k \hat{f}_k)(\xi)\chi_k(\xi)\check{\eta}(\xi)d\xi \quad (17)$$
$$= \int_{\mathbb{R}^n} \varphi_k(D) f_k(x)(\chi_k\check{\eta})^\wedge(x)dx,$$

where χ_k has the same meaning as in (4). Recall the Plancherel–Polya–Nikol'skij inequality for entire analytic functions

$$|\varphi_k(D) f_k(x)| \leqq c\, 2^{k\frac{n}{p}} \|\varphi_k(D) f_k \mid L_p\|, \tag{18}$$

see [Trib: (1.3.2/5) and Theorem 1.4.1(ii)], where c is independent of $k \in \mathbb{N}_0$. Let C be the number on the left-hand side of (7), then (17), (18), and (14) yield

$$\left| \int_{\mathbb{R}^n} \varphi_k(D) f_k(x)\eta(x)dx \right| \leqq c_1 C\, 2^{c_2 k} \sup_{y\in\mathbb{R}^n} (1+|y|)^{c_3} |(\chi_k\check{\eta})^\wedge(y)|, \tag{19}$$

where c_1, c_2, c_3 are positive numbers which are independent of k. Let Δ be the Laplacian, then we have by standard arguments

$$\begin{aligned} 2^{c_2 k}(1+|y|)^{c_3}|(\chi_k\check{\eta})^\wedge(y)| &\leqq 2^{c_2 k}|((1-\Delta)^{N_1}\chi_k\check{\eta})^\wedge(y)| \\ &\leqq c\, 2^{c_2 k} \sup_{z\in\mathbb{R}^n} (1+|z|)^{N_2} |(1-\Delta)^{N_1}(\chi_k\check{\eta})(z)| \\ &\leqq c'\, 2^{-k} \sup_{z\in\mathbb{R}^n} (1+|z|)^{N_3} \sum_{|\alpha|\leqq N_4} |D^\alpha \eta(z)|, \end{aligned} \tag{20}$$

for some natural numbers N_1, N_2, N_3, N_4. We insert (20) in (19) and note that we have a corresponding estimate with χ_k instead of φ_k. But now it is clear that the sum in (9) converges in S'.

Remark 2. By the same arguments as in the last remark we get

$$S \subset B^s_{pq} \subset S' \quad \text{and} \quad S \subset F^s_{pq} \subset S', \tag{21}$$

for all admissible values of s, p, q, where "$\subset$" always stands for topological embedding. We complement (21) by some further embedding assertions:

(i) Let $s \in \mathbb{R}$, $0 < p < \infty$ and $0 < q \leqq \infty$, then

$$B^s_{pu} \subset F^s_{pq} \subset B^s_{pv}, \qquad u = \min(p,q), \quad v = \max(p,q). \tag{22}$$

(ii) Let $0 < p \leqq \infty$ and $0 < r \leqq \infty$, $0 < t \leqq \infty$, then

$$B^s_{pr} \subset B^\sigma_{pt}, \qquad -\infty < \sigma < s < \infty. \tag{23}$$

These elementary embeddings follow easily from Definition 2.3.1. However, we are not going into details and refer to [Trіß: 2.3.2, 2.3.3] which covers also (21), a different proof of the completeness of B^s_{pq} and F^s_{pq} and further elementary embeddings.

Remark 3. Let $0 < q \leqq \infty$, then

$$B^s_{pq} \subset L_p \quad \text{if} \quad 1 \leqq p \leqq \infty \quad \text{and} \quad s > 0 \tag{24}$$

follows immediately from (2.3.1/6,7), including a corresponding assertion for F^s_{pq} (with $p < \infty$). At first glance (24) seems also to be valid if $0 < p < 1$. But this is not correct and it sheds some light on a dangerous pitfall in the theory of function spaces with $p < 1$. If $1 \leqq p \leqq \infty$ then the convergence of the right-hand side of (2.3.1/6) in L_p is also a convergence in S' to the same limit element. This is the basis in order to prove (24). If $p < 1$ then this needs not to be the case. For example, let φ be the above function, then $2^{jn}\hat{\varphi}(2^j x) \to \delta$ in S' if $j \to \infty$ where δ stands for the δ-distribution, but

$$2^{jn}\hat{\varphi}(2^j x) \to 0 \quad \text{in} \quad L_p \quad \text{if} \quad j \to \infty, \quad \text{where } 0 < p < 1.$$

Rescue comes from the following embedding. Let $0 < q \leqq \infty$, then

$$B^s_{pq} \subset L_1 \quad \text{if} \quad 0 < p \leqq 1 \quad \text{and} \quad s > \sigma_p = n\left(\frac{1}{p} - 1\right) \tag{25}$$

and a corresponding embedding for F^s_{pq}. This assertion follows from (2.3.1/7) and the Plancherel–Polya–Nikol'skij inequality

$$\|\varphi_k(D)f \mid L_1\| \leqq c\, 2^{k\sigma_p} \|\varphi_k(D)f \mid L_p\|, \qquad k \in \mathbb{N}, \tag{26}$$

see [Trіß: (1.3.2/5) and Theorem 1.4.1(ii)].

2.3.3 Spaces with $s > \sigma_p$

Recall $a_+ = \max(0, a)$ where $a \in \mathbb{R}$ and

$$\sigma_p = n\left(\frac{1}{p} - 1\right)_+ \quad \text{where} \quad 0 < p \leqq \infty, \tag{1}$$

see (2.2.3/1). By (2.3.2/24, 25) all elements f of B^s_{pq} or F^s_{pq} with $s > \sigma_p$ are regular distributions, more precisely, $f \in L_{\max(p,1)}$. We wish to complement this observation. Let φ be the same C^∞ function as in 2.3.1, in particular we have (2.3.1/1). We extend the definition of φ_j from (2.3.1/2) to all integers j. It should be noted that φ_0 has now a different meaning as in 2.3.1.

Theorem. (i) Let $0 < p \leqq \infty$, $0 < q \leqq \infty$, and $s > \sigma_p$, then

$$\|\varphi(D)f \mid L_p\| + \left(\sum_{j=-\infty}^{\infty} 2^{jsq}\|\varphi_j(D)f \mid L_p\|^q\right)^{1/q} \tag{2}$$

and

$$\|f \mid L_p\| + \left(\sum_{j=-\infty}^{\infty} 2^{jsq}\|\varphi_j(D)f \mid L_p\|^q\right)^{1/q} \tag{3}$$

(modification if $q = \infty$) are equivalent quasi-norms in B^s_{pq}.

(ii) Let $0 < p < \infty$, $0 < q \leqq \infty$, and $s > \sigma_p$, then

$$\|\varphi(D)f \mid L_p\| + \left\|\left(\sum_{j=-\infty}^{\infty} 2^{jsq}|\varphi_j(D)f(\cdot)|^q\right)^{1/q} \mid L_p\right\| \tag{4}$$

and

$$\|f \mid L_p\| + \left\|\left(\sum_{j=-\infty}^{\infty} 2^{jsq}|\varphi_j(D)f(\cdot)|^q\right)^{1/q} \mid L_p\right\| \tag{5}$$

(modification if $q = \infty$) are equivalent quasi-norms in F^s_{pq}.

Proof. Step 1. In order to prove that (2) and (4) are equivalent quasi-norms in B^s_{pq} and F^s_{pq}, respectively, it is sufficient to show that there exists a constant $c > 0$ such that

$$\|\varphi_j(D)f \mid L_p\| \leqq c\,2^{-j\sigma_p}\|\varphi(D) \mid L_p\|, \qquad -j \in \mathbb{N}, \tag{6}$$

holds. For those j's we have $\varphi_j(\xi) = \varphi_j(\xi)\varphi(\xi)$ and hence

$$\begin{aligned} \|\varphi_j(D)f \mid L_p\| &= \|(\varphi_j(\varphi(D)f)^\wedge)^\vee \mid L_p\| \\ &\leqq c\|\check{\varphi}_j \mid L_r\|\,\|\varphi(D)f \mid L_p\|, \qquad r = \min(1,p), \end{aligned} \tag{7}$$

where the equality comes from (2.3.1/5) and the inequality is a Fourier multiplier assertion for entire analytic functions proved in [Trieβ: Proposition 1.5.1]. By $\check{\varphi}_j(x) = 2^{jn}\check{\varphi}_0(2^{jn}x)$ we obtain (6).

Step 2. We prove that (5) is an equivalent quasi-norm in F^s_{pq}. Because $s > \sigma_p$ we may assume that (2.3.1/6) converges not only in S', but also, say, almost everywhere in $\mathbb{R}^n$. Then we have

$$\|f \mid L_p\| \leqq c\|\varphi(D)f \mid L_p\| + c\left(\sum_{j=1}^{\infty}\|\varphi_j(D)f \mid L_p\|^p\right)^{1/p} \tag{8}$$

if $0 < p \leqq 1$ and a corresponding estimate if $1 < p < \infty$. Now, (4) and (8) prove that (5) can be estimated from above by $c\|f \mid F^s_{pq}\|$. We prove the reverse inequality. Because f is a regular distribution we have, say, a.e.

$$\varphi(D)f(x) = f(x) + ((1-\varphi(\cdot))\hat{f})^\vee(x) = f(x) + \sum_{j=0}^{\infty}((1-\varphi(\cdot))\varphi_j(\cdot)\hat{f})^\vee(x). \tag{9}$$

By the above-mentioned Fourier multiplier assertion we have

$$\|\varphi(D)f \mid L_p\| \leqq c\|f \mid L_p\| + c\left(\sum_{j=0}^{\infty}\|\varphi_j(D)f \mid L_p\|^p\right)^{1/p} \tag{10}$$

if $0 < p \leqq 1$ and a corresponding estimate if $1 < p < \infty$. Now (4) and (10) prove that $\|f \mid F^s_{pq}\|$ can be estimated from above by the quasi-norm in (5). In the same way one obtains that (3) is an equivalent quasi-norm in B^s_{pq}.

Remark. Later on, we shall see that any "discrete" quasi-norm of type (2–5) or also (2.3.1/7,8) has a "continuous" counterpart. To explain what this means we introduce

$$\rho(tD)f(x) = (\rho(t\cdot)\hat{f})^\vee(x), \qquad t > 0, \quad x \in \mathbb{R}^n$$

with $\rho(t\xi) = \varphi(t\xi) - \varphi(2t\xi)$, see (2.3.1/2). Then the continuous counterparts of (3) and (5) read as follows:

(i) Let $0 < p \leqq \infty$, $0 < q \leqq \infty$, and $s > \sigma_p$, then

$$\|f \mid L_p\| + \left(\int_0^\infty t^{-sq}\|\rho(tD)f \mid L_p\|^q \frac{dt}{t}\right)^{1/q} \tag{11}$$

(modification if $q = \infty$) is an equivalent quasi-norm in B^s_{pq}.

(ii) Let $0 < p < \infty$, $0 < q \leqq \infty$, and $s > \sigma_p$, then

$$\|f \mid L_p\| + \left\|\left(\int_0^\infty t^{-sq}|\rho(tD)f(\cdot)|^q \frac{dt}{t}\right)^{1/q} \mid L_p\right\| \tag{12}$$

(modification if $q = \infty$) is an equivalent quasi-norm in F^s_{pq}.

The quasi-norms (11) and (12) look more elegant than their discrete counterparts. Furthermore, both (11) and (12) is the sum of two homogeneous quasi-norms. Let $\lambda > 0$, then we have

$$\|f(\lambda\cdot) \mid L_p\| = \lambda^{-n/p}\|f \mid L_p\|, \tag{13}$$

and a corresponding assertion for the second terms in (11) and (12) with $s - \frac{n}{p}$ instead of $-\frac{n}{p}$. The latter claim follows from

$$[\rho(tD)f(\lambda\cdot)](x) = ((\rho(\lambda t\cdot)\hat{f})^\vee(\lambda x), \qquad \lambda > 0. \tag{14}$$

2.4 General characterizations for F^s_{pq}

2.4.1 The main theorem

Let $0 < p \leqq \infty$ and $0 < q \leqq \infty$, then we introduce the abbreviations

$$\sigma_p = n\left(\frac{1}{p} - 1\right)_+ \quad \text{and} \quad \sigma_{pq} = n\left(\frac{1}{\min(p,q)} - 1\right)_+ \tag{1}$$

which differ slightly from the corresponding notations in [Trіß: 2.5.3]. Recall $a_+ = \max(0, a)$ if $a \in \mathbb{R}$. Let $h(x) \in S$ and $H(x) \in S$ with

$$\operatorname{supp} h \subset \{y \in \mathbb{R}^n \colon |y| \leqq 2\}, \qquad \operatorname{supp} H \subset \left\{y \in \mathbb{R}^n \colon \frac{1}{4} \leqq |y| \leqq 4\right\}, \tag{2}$$

$$h(x) = 1 \;\text{ if }\; |x| \leqq 1 \;\text{ and }\; H(x) = 1 \;\text{ if }\; \frac{1}{2} \leqq |x| \leqq 2. \tag{3}$$

Recall our notation

$$\varphi(tD)f = (\varphi(t\cdot)\hat{f})^\vee, \qquad t > 0. \tag{4}$$

In the theorem below φ need not be an element of S, and then it is not immediately clear what is meant by (4). It will be defined via limiting procedures, see Step 3 of the proof and the following subsection.

Theorem. Let $0 < p < \infty$, $0 < q \leqq \infty$ and $s \in \mathbb{R}$. Let s_0 and s_1 be two real numbers with

$$s_0 + \sigma_{pq} < s < s_1 \quad \text{and} \quad s_1 > \sigma_p. \tag{5}$$

Let φ_0 and φ be two complex-valued C^∞ functions on $\mathbb{R}^n$ and $\mathbb{R}^n \setminus \{0\}$, respectively, which satisfy the Tauberian conditions

$$|\varphi_0(x)| > 0 \quad \text{if} \quad |x| \leqq 2 \tag{6}$$

and

$$|\varphi(x)| > 0 \quad \text{if} \quad \frac{1}{2} \leqq |x| \leqq 2. \tag{7}$$

Let $a > \frac{n}{\min(p,q)}$,

$$\int_{\mathbb{R}^n} \left| \left(\frac{\varphi(z)h(z)}{|z|^{s_1}} \right)^\vee (y) \right| (1+|y|)^a dy < \infty, \tag{8}$$

$$\sup_{m \in \mathbb{N}} 2^{-ms_0} \int_{\mathbb{R}^n} |(\varphi(2^m \cdot) H(\cdot))^\vee(y)| (1+|y|)^a dy < \infty, \tag{9}$$

and

$$\sup_{m \in \mathbb{N}} 2^{-ms_0} \int_{\mathbb{R}^n} |(\varphi_0(2^m \cdot) H(\cdot))^\vee(y)| (1+|y|)^a dy < \infty. \tag{10}$$

Let $\varphi_j(x) = \varphi(2^{-j}x)$ if $x \in \mathbb{R}^n \setminus \{0\}$ and $j \in \mathbb{N}$. Then

$$\left\| \left(\sum_{j=0}^\infty 2^{jsq} |\varphi_j(D) f(\cdot)|^q \right)^{1/q} \mid L_p \right\| \tag{11}$$

and

$$\|\varphi_0(D) f \mid L_p\| + \left\| \left(\int_0^1 t^{-sq} |\varphi(tD) f(\cdot)|^q \frac{dt}{t} \right)^{1/q} \mid L_p \right\| \tag{12}$$

(modification if $q = \infty$) are equivalent quasi-norms in F^s_{pq}.

Proof. Step 1. Let $f \in F^s_{pq}$. In the first two steps we prove that the quasi-norms in (11) can be estimated from above by $c\|f \mid F^s_{pq}\|$. Let $\{\rho_k(x)\}_{k \in \mathbb{N}_0}$ be a resolution of unity in the sense of (2.3.1/1–4) with ρ instead of φ. Then we have (2.3.1/8) with ρ instead of φ. Let $\rho_l(x) = 0$ if $-l \in \mathbb{N}$. Let $j \in \mathbb{N}_0$, then we obtain

$$2^{js}(\varphi_j \hat{f})^\vee(x) = \sum_{l=-\infty}^{\infty} 2^{js}(\varphi_j \rho_{l+1} \hat{f})^\vee(x) = \sum_{l=-\infty}^{K} \cdots + \sum_{l=K+1}^{\infty} \cdots \tag{13}$$

where $K \in \mathbb{N}$ will be chosen later on. Temporarily we take it for granted that $(\varphi_j \hat{f})^\vee \in S'$ is a regular distribution and that (13) coverges not only in S', but also pointwise a.e. We estimate the first sum where we have no problems of convergence. Let

$$\tilde{\rho}_j(x) = |2^{-j}x|^{s_1}\rho_j(x) \quad \text{if} \quad j \in \mathbb{N}_0. \tag{14}$$

Then we have

$$\left|\sum_{l=-\infty}^{K} 2^{js}(\varphi_j \rho_{l+j}\hat{f})^\vee(x)\right| \leqq \sum_{l=-\infty}^{K} 2^{l(s_1-s)} \left|\left(\frac{\varphi_j(z)}{|2^{-j}z|^{s_1}} 2^{s(j+l)}\tilde{\rho}_{j+l}(z)\hat{f}\right)^\vee(x)\right|. \tag{15}$$

One can replace $\varphi_j(z)$ on the right-hand side of (15) by $\varphi_j(z)h(c2^{-j}z)$, where c is an appropriate positive number which depends on K. Let $j \in \mathbb{N}$ then we have $\varphi_j(z) = \varphi(2^{-j}z)$ and $|(\ldots)^\vee(x)|$ on the right-hand side of (15) can be estimated from above by

$$\int_{\mathbb{R}^n} \left|\left(\frac{\varphi(2^{-j}z)}{|2^{-j}z|^{s_1}} h(c2^{-j}z)\right)^\vee(y)\right| |(2^{s(j+l)}\tilde{\rho}_{j+l}\hat{f})^\vee(x-y)|dy. \tag{16}$$

We apply $(\lambda(2^{-j}\cdot))^\vee(y) = 2^{jn}\lambda^\vee(2^j y)$ to the first factor in (16) and replace afterwards $2^j y$ by y. We use the maximal function from (2.3.2/1) with ρ instead of φ and obtain for $l \leqq K$

$$2^{s(j+l)}|(\tilde{\rho}_{j+l}\hat{f})^\vee(x-2^{-j}y)| \leqq c\, 2^{s(j+l)}(\tilde{\rho}^*_{j+l}f)_a(x)(1+|y|)^a, \tag{17}$$

where c depends on K, but not on x, y, j, and l. By (8) it follows that (16) can be estimated from above by $c2^{s(j+l)}(\tilde{\rho}^*_{j+l}f)_a(x)$. We return to (15) and have

$$\left|\sum_{l=-\infty}^{K} 2^{js}(\varphi_j \rho_{l+j}\hat{f})^\vee(x)\right| \leqq c \sum_{l=-\infty}^{K} 2^{l(s_1-s)}2^{s(j+l)}(\tilde{\rho}^*_{j+l}f)_a(x), \tag{18}$$

where $j \in \mathbb{N}$. We apply first the l_q-quasi-norm with respect to j and afterwards the L_p-quasi-norm with respect to x. Because $s_1 > s$ we obtain

$$\left\|\left(\sum_{j=1}^{\infty}\left|\sum_{l=-\infty}^{K} 2^{js}(\varphi_j\rho_{l+j}\hat{f})^\vee(\cdot)\right|^q\right)^{1/q} | L_p\right\| \leqq \tag{19}$$

$$c\left\|\left(\sum_{m=0}^{\infty} 2^{msq}(\tilde{\rho}^*_m f)^q_a(x)\right)^{1/q} | L_p\right\|.$$

Because $a > \frac{n}{\min(p,q)}$ we can use the vector-valued maximal inequality (2.2.4/3), only the term with $\tilde{\rho}_0(x) = |x|^{s_1}\rho_0(x)$ is critical. We return to this point later on in Remark 1, then we prove

$$\|(\tilde{\rho}_0 \hat{f})^\vee \mid L_p\| \leqq c\|(\rho_0 \hat{f})^\vee \mid L_p\|, \tag{20}$$

where we have to use $s_1 > \sigma_p$. Now the maximal inequality can be applied and (19) can be estimated from above by

$$c\left\|\left(\sum_{m=0}^{\infty} 2^{msq}|(\tilde{\rho}_m \hat{f})^\vee(\cdot)|^q\right)^{1/q} \mid L_p\right\|. \tag{21}$$

Finally, by (20) and (2.2.4/4), we can estimate (21), and consequently also (19), from above by

$$c\left\|\left(\sum_{m=0}^{\infty} 2^{msq}|(\rho_m \hat{f})^\vee(\cdot)|^q\right)^{1/q} \mid L_p\right\| = c\|f \mid F^s_{pq}\|_\rho. \tag{22}$$

The term with $j = 0$ can be incorporated afterwards. In other words, we have

$$\left\|\left(\sum_{j=0}^{\infty}\left|\sum_{\rho=-\infty}^{K} 2^{js}(\varphi_j \rho_{l+j} \hat{f})^\vee(\cdot)\right|^q\right)^{1/q} \mid L_p\right\| \leqq c\|f \mid F^s_{pq}\|, \tag{23}$$

where c depends on K.

Step 2. We estimate the second sum in (13) and we have to make sure now that (13) converges a.e. and also in some space L_r with $1 \leqq r \leqq \infty$. However the latter comes out as a by-product, see Step 3. Similar as in (14) we introduce

$$\rho'_j(x) = |2^{-j}x|^{s_0}\rho_j(x), \qquad j \in \mathbb{N}. \tag{24}$$

Instead of (15) we have now

$$\begin{aligned} &\left|\sum_{l=K+1}^{\infty} 2^{js}(\varphi_j \rho_{l+j} \hat{f})^\vee(x)\right| \\ &\quad \leqq \sum_{l=K+1}^{\infty} 2^{l(s_0-s)}\left|\left(\frac{\varphi_j(z)}{|2^{-j}z|^{s_0}} H(2^{-j-l}z) 2^{s(j+l)}\rho'_{j+l}(z)\hat{f}\right)^\vee(x)\right|. \end{aligned} \tag{25}$$

We have obvious counterparts of (16) and (17), where we replace $\tilde{\rho}$ and $x - 2^{-j}y$ by ρ' and $x - 2^{-j-l}y$, respectively. The crucial integral looks like

$$\int_{\mathbb{R}^n}\left|\left(\frac{\varphi(2^l z)}{|2^l z|^{s_0}} H(z)\right)^\vee(y)\right|(1+|y|)^a dy, \tag{26}$$

which in turn can be estimated from above by

$$c\,2^{-ls_0}\int_{\mathbb{R}^n}|(\varphi(2^l\cdot)H(\cdot))^\vee(y)|(1+|y|)^a dy, \tag{27}$$

see (2.2.3/2). Together with its φ_0-counterpart, (9) and (10), we arrive in the same way as in the first step at

$$\left|\sum_{l=K+1}^{\infty} 2^{js}(\varphi_j \rho_{l+j} \hat{f})^\vee(x)\right| \leqq c \sum_{l=K+1}^{\infty} 2^{l(s_0-s)} 2^{(j+l)s} (\rho'^*_{j+l} f)_a(x) \tag{28}$$

and

$$\left\| \left(\sum_{j=0}^{\infty} \left| \sum_{l=K+1}^{\infty} 2^{js}(\varphi_j \rho_{l+j} \hat{f})^\vee(x)\right|^q \right)^{1/q} \mid L_p \right\| \leqq c\, 2^{K(s-s_0)} \|f \mid F^s_{pq}\|, \tag{29}$$

where c is independent of K. We used $s_0 < s$. Now (13), (23), and (29) prove that the quasi-norm in (11) can be estimated from above by $c\|f \mid F^s_{pq}\|$.

Step 3. We postponed the technical discussion of $(\varphi_j \hat{f})^\vee$ with $f \in F^s_{pq}$. First we remark that there are no problems with $(\varphi_j \rho_m \hat{f})^\vee(x)$ if $j \in \mathbb{N}_0$ and $m \in \mathbb{N}$. Recall that φ need not be C^∞ near the origin, hence $(\varphi_j \rho_0 \hat{f})^\vee$ with $j \in \mathbb{N}$ must be treated separately. We have (18) term by term, in particular

$$|(\varphi_j \rho_0 \hat{f})^\vee(x)| \leqq c(\tilde{\rho}^*_0 f)_a(x). \tag{30}$$

Taking (20) for granted, see Remark 1 below, then the scalar case of Theorem 2.2.4 proves $(\varphi_j \rho_0 \hat{f})^\vee \in L_p \cap S'$ in any case. (A more detailed discussion of L_p spaces of entire analytic functions, related Fourier multipliers and maximal functions has been given in [Triß: Chapter 1]). We have to care about the convergence on the right-hand side of (13), both pointwise a.e. and in S'. We can rewrite (28) as follows:

$$\left|\sum_{l=K}^{L} 2^{js}(\varphi_j \rho_{l+j} \hat{f})^\vee(x)\right| \leqq c \sum_{l=K}^{L} 2^{l(s_0-s)} 2^{(j+l)s} (\rho'^*_{j+l} f)_a(x) \tag{31}$$

with $L > K$. If K is large then the right-hand side of (31) can be estimated from above by

$$\varepsilon \left(\sum_{l=K}^{\infty} 2^{lsq} (\rho'^*_l f)^q_a(x) \right)^{1/q}, \tag{32}$$

where $\varepsilon > 0$ is given. By Theorem 2.3.2 and the above considerations the last expression is finite a.e. This proves the desired pointwise convergence. We prove the S'-convergence and assume $0 < p < 1$. Let $\sigma = s - \sigma_p$, then we have the embedding $F^s_{pq} \subset F^\sigma_{1q}$, see [Triß: 2.7.1] and (1). We reformulate (31) by

$$\left|\sum_{l=K}^{L} 2^{j\sigma}(\varphi_j \rho_{l+j} \hat{f})^\vee(x)\right| \leqq c \sum_{l=K}^{L} 2^{l(s_0-\sigma)} 2^{(j+l)\sigma} (\rho'^*_{j+l} f)_a(x). \tag{33}$$

By (5) we have $\sigma = s-\sigma_p > s_0$. Via the counterpart of (32) and the L_1-convergence, resulting from the indicated embedding, it follows that (33), and hence (13), converges in L_1. If $1 \leqq p < \infty$, then we have by the same arguments a L_p-convergence. What has been done in the first two steps is now completely justified.

Step 4. Let again $f \in F^s_{pq}$. We prove that the quasi-norm in (12) can be estimated from above by $c\|f \mid F^s_{pq}\|$. We indicate the necessary modifications in the first two steps. Instead of $\varphi_j(x) = \varphi(2^{-j}x)$ we have now $\varphi(tx)$ with, say, $2^{-j} \leqq t \leqq 2^{-j+1}$. We use the same splitting as in (13). We follow (14–18), where we can replace now the left-hand side of (18) by

$$\sup_{2^{-j}\leqq t\leqq 2^{-j+1}} \left| \sum_{l=-\infty}^{K} 2^{js}(\varphi(t\cdot)\rho_{l+j}\hat{f})^\vee(x) \right| \tag{34}$$

with the corresponding counterpart in (19). Similarly in (28) and hence in (29). In other words, the estimates in the first two steps yield

$$\|\varphi_0(D)f \mid L_p\| + \left\| \left(\sum_{j=1}^{\infty} 2^{jsq} \sup_{2^{-j}\leqq t\leqq 2^{-j+1}} |\varphi(tD)f(\cdot)|^q \right)^{1/q} \mid L_p \right\| \leqq c\|f \mid F^s_{pq}\| \tag{35}$$

which is even stronger than the desired estimate.

Step 5. Let $f \in F^s_{pq}$. We prove that $\|f \mid F^s_{pq}\|_\rho$ from (22) can be estimated from above by the quasi-norm in (11). Let $\psi \in S$ be a function with $\operatorname{supp}\psi \subset \{y \in \mathbb{R}^n\colon |y| \leqq 2^{K+1}\}$ and $\psi(x) = 1$ if $|x| \leqq 2^K$, where we choose the natural number K later on. By (6,7) and the properties of φ_j and $\rho_j(x)$ we have

$$\begin{aligned} |(\rho_j\hat{f})^\vee(x)| &= |(\rho_j\psi(2^{-j}\cdot)\hat{f})^\vee(x)| \\ &\leqq c\int_{\mathbb{R}^n} \left| \left(\frac{\rho_j}{\varphi_j}\right)^\vee(y)(\varphi_j\psi(2^{-j}\cdot)\hat{f})^\vee(x-y) \right| dy, \quad j \in \mathbb{N}_0. \end{aligned} \tag{36}$$

For fixed $x \in \mathbb{R}^n$ the Fourier transform of the y-function in the integral in (36) has a support contained in a ball of radius $c2^{j+K}$, where c is independent of x, j, and K. Let $0 < r < \min(1, p, q)$. We use again an inequality of Plancherel–Polya–Nikol'skij type for entire analytic functions, see [Trіß: (1.3.2/5)], and obtain

$$|(\rho_j\hat{f})^\vee(x)|^r \leqq c\,2^{(j+K)n(1-r)} \int_{\mathbb{R}^n} \left| \left(\frac{\rho_j}{\varphi_j}\right)^\vee(y)(\varphi_j\psi(2^{-j}\cdot)\hat{f})^\vee(x-y) \right|^r dy. \tag{37}$$

Let $j \in \mathbb{N}$ then we have $\varphi_j(x) = \varphi(2^{-j}x)$ and $\rho_j(x) = \overline{\rho}(2^{-j}x)$ with $\overline{\rho}(x) = \rho(x) - \rho(2x)$, see (2.3.1/2), and

$$\left| \left(\frac{\rho_j}{\varphi_j}\right)^\vee(y) \right|^r = 2^{jnr} \left| \left(\frac{\overline{\rho}}{\varphi}\right)^\vee(2^jy) \right|^r \leqq c\,2^{jnr}(1+|2^jy|)^{-b}, \tag{38}$$

where $b > 0$ is at our disposal. A corresponding estimate holds for $j = 0$. We put (38) in (37) and obtain

$$|(\rho_j \hat{f})^\vee(x)|^r \tag{39}$$

$$\leqq c\, 2^{(j+K)n(1-r)+jnr} \sum_{l=0}^{\infty} 2^{-ld} \int_{\{y \in \mathbb{R}^n : |y| < 2^{-j+l}\}} |(\varphi_j \psi(2^{-j}\cdot)\hat{f})^\vee(x-y)|^r dy,$$

where $d > 0$ is at our disposal. The integrals in (39) can be estimated from above by

$$c\, 2^{-jn+ln}(M|(\varphi_j \psi(2^{-j}\cdot)\hat{f})^\vee|^r)(x), \tag{40}$$

where M stands for the Hardy–Littlewood maximal function. We put (40) in (39), choose $d > n$, and arrive at

$$|(\rho_j \hat{f})^\vee(x)|^r \leqq c\, 2^{Kn(1-r)}(M|(\varphi_j \psi(2^{-j}\cdot)\hat{f})^\vee|^r)(x). \tag{41}$$

Recall $1 < \frac{p}{r} < \infty$ and $1 < \frac{q}{r} \leqq \infty$. We multiply (41) with 2^{jsr}, apply the $l_{q/r}$-norm with respect to j and afterwards the $L_{p/r}$-norm with respect to x, then we obtain by Theorem 2.2.2

$$\left\| \left(\sum_{j=0}^{\infty} |2^{js}(\rho_j \hat{f})^\vee(\cdot)|^q \right)^{1/q} \mid L_p \right\|^r \leqq \tag{42}$$

$$c\, 2^{Kn(1-r)} \left\| \left[\sum_{j=0}^{\infty} (M|(2^{js}\varphi_j \psi(2^{-j}\cdot)\hat{f})^\vee|^r(\cdot))^{q/r} \right]^{r/q} \mid L_{p/r} \right\|$$

$$\leqq c'\, 2^{Kn(1-r)} \left\| \left(\sum_{j=0}^{\infty} |2^{js}(\varphi_j \psi(2^{-j}\cdot)\hat{f})^\vee|^q \right)^{1/q} \mid L_p \right\|^r,$$

where c and c' are independent of K. Because

$$\varphi_j \psi(2^{-j}\cdot) = \varphi_j - \varphi_j(1 - \psi(2^{-j}\cdot)) \tag{43}$$

the right-hand side of (42) can be estimated from above by the rth power of the quasi-norm in (11) (this is just what we want) and the additional term

$$c\, 2^{Kn(1-r)} \left\| \left(\sum_{j=0}^{\infty} 2^{jnq} |(\varphi_j(1 - \psi(2^{-j}\cdot))\hat{f})^\vee(\cdot)|^q \right)^{1/q} \mid L_p \right\|^r. \tag{44}$$

However this term can be treated in the same way as in the second step, in particular, we have a counterpart of the estimate in (29). Hence the term in (44) can be estimated from above by

$$c\, 2^{Kn(1-r)} 2^{-Kr(s-s_0)} \|f \mid F^s_{pq}\|^r. \tag{45}$$

By (5) we may assume $n(\frac{1}{r}-1)-(s-s_0)<0$. Recall that the natural number K is at our disposal. We choose K large. Then the term in (45) can be estimated from above, say, by $\frac{1}{2}\|f \mid F^s_{pq}\|^r_\rho$. Now (42) and the above splitting yield the desired result.

Step 6. Let $f \in F^s_{pq}$. We estimate $\|f \mid F^s_{pq}\|_\rho$ from (22) from above by the quasi-norm in (12). We follow Step 5 and indicate the necessary changes. First we replace $\varphi_j(\xi)=\varphi(2^{-j}\xi)$ in (36) with $j\in\mathbb{N}$ by $\varphi_j(\tau\xi)=\varphi(\tau 2^{-j}\xi)$ with $1 \leqq \tau \leqq 2$. This make sense, maybe after immaterial changes of ρ_j. Then we have (39) with $\varphi(\tau 2^{-j}\xi)$ instead of $\varphi(2^{-j}\xi)$. We integrate over τ. Besides the factor $c' 2^{Kn(1-r)}$ and the term with $j=0$ the modified right-hand side of (42) looks as follows,

$$\left\| \left[\sum_{j=1}^{\infty} \left(2^{jsr} \int_1^2 |(\varphi(\tau 2^{-j}\cdot)\hat{f})^\vee(\cdot)|^r d\tau \right)^{q/r} \right]^{r/q} \mid L_{p/r} \right\|. \tag{46}$$

We use (43) with $\varphi(\tau 2^{-j}\xi)$ instead of $\varphi(2^{-j}\xi)$. The remainder term can be treated as above, see (35). The main term, i.e., (46) without the factor $\psi(2^{-j}\cdot)$, can be estimated from above by

$$\left\| \left[\int_1^2 \sum_{j=1}^{\infty} 2^{jsq} |(\varphi(\tau 2^{-j}\cdot)\hat{f})^\vee(\cdot)|^q \right]^{r/q} \mid L_{p/r} \right\| \tag{47}$$

$$\leqq c \left\| \left(\int_0^1 t^{-sq} |(\varphi(t\cdot)\hat{f})^\vee(\cdot)|^q \right)^{1/q} \mid L_p \right\|,$$

which is the desired result. We used $r<q$.

Remark 1. We prove (20). This problem can be reduced to

$$\|(|x|^{s_1}\sigma(x)\hat{g})^\vee \mid L_p\| \leqq c\|g \mid L_p\| \tag{48}$$

where $\sigma \in S$ with $\sigma(x)=1$ if $|x|\leqq 1$ and $\sigma(x)=0$ if $|x|\geqq 2$, and $g\in L_p\cap S'$ is an arbitrary function with $\operatorname{supp}\hat{g}\subset\{y\in\mathbb{R}^n\colon |y|\leqq 1\}$. By (2.2.3/2) this estimate is valid if

$$|x|^{s_1}\sigma(x)\in H^s_2 \quad \text{with} \quad s>\sigma_p+\frac{n}{2}. \tag{49}$$

Because $s_1>\sigma_p$ we may assume $s_1+\frac{n}{2}>s>\sigma_p+\frac{n}{2}$. Let $\lambda(x)$ be a C^∞ function on $\mathbb{R}^n$ with $\lambda(x)=1$ if $|x|\geqq 2$ and $\lambda(x)=0$ if $|x|\leqq 1$. Then $\{|x|^{s_1}\sigma(x)\lambda(2^j x)\}_{j=1}^\infty$ is a fundamental sequence in H^s_2. This follows by straightforward calculations if $s\in\mathbb{N}$. For fractional numbers s it is a matter of interpolation or of the so-called multiplicative inequalities for fractional Sobolev spaces. This completes the proof of (48) and hence of (20).

Remark 2. The following observation will be of some use later on. In Steps 1, 2, and 4 of the above proof, we used only $s_0 < s < s_1$ and that (13) converges in S' and pointwise a.e. Under these assumptions the quasi-norms in (11) and (12) can be estimated from above by $c\|f \mid F^s_{pq}\|_\rho$.

It will be useful to give conditions (8–10) a more handsome reformulation. Recall that H^σ_2 are the fractional Sobolev spaces on $\mathbb{R}^n$ (Sobolev spaces if $\sigma \in \mathbb{N}_0$).

Corollary 1. Let p, q, s, s_0, s_1, and a be the same numbers as in the Theorem. Let $\sigma > a + \frac{n}{2}$. Let φ_0 and φ be two complex-valued C^∞ functions on $\mathbb{R}^n$ and $\mathbb{R}^n \setminus \{0\}$, respectively, which satisfy the Tauberian conditions (6), (7), and

$$\left\| \frac{\varphi(x)h(x)}{|x|^{s_1}} \mid H^\sigma_2 \right\| < \infty, \tag{50}$$

$$\sup_{m\in\mathbb{N}} 2^{-ms_0} \|\varphi(2^m \cdot)H(\cdot) \mid H^\sigma_2\| < \infty, \tag{51}$$

$$\sup_{m\in\mathbb{N}} 2^{-ms_0} \|\varphi_0(2^m \cdot)H(\cdot) \mid H^\sigma_2\| < \infty, \tag{52}$$

where $h(x)$ and $H(x)$ have the same meaning as in (2), (3). Let $\varphi_j(x) = \varphi(2^{-j}x)$ if $x \in \mathbb{R}^n \setminus \{0\}$ and $j \in \mathbb{N}$. Then (11) and (12) are equivalent quasi-norms in F^s_{pq}.

Proof. Let $0 < r \leqq 1$ and $b \in \mathbb{R}$, then

$$\|\hat{\lambda}(y)(1 + |y|)^b \mid L_r\| \leqq c\|\lambda \mid H^\delta_2\| \tag{53}$$

with $\delta > n(\frac{1}{r} - \frac{1}{2}) + b$. This is a well-known estimate, a proof may be found, e.g., in [ScT: 1.7.5], see also [TriB: 1.5.2, 1.5.4] for further informations. However, (53) with $r = 1$ and $a = b$ shows that the terms (8–10) can be estimated from above by the corresponding terms in (50–52).

Corollary 2. Let $0 < p < \infty$, $0 < q \leqq \infty$ and $s \in \mathbb{R}$. Let $a > \frac{n}{\min(p,q)}$. Let s_0 and s_1 be two real numbers with

$$s_0 + a < s < s_1 \quad \text{with} \quad s_1 > \sigma_p. \tag{54}$$

Let φ_0 and φ be two complex-valued C^∞ functions on $\mathbb{R}^n$ and $\mathbb{R}^n \setminus \{0\}$, respectively, which satisfy the Tauberian conditions (6), (7), and either (8–10) or (50–52). Let $\varphi_j(x) = \varphi(2^{-j}x)$ and

$$(\varphi^*_k f)_a(x) = \sup_{y \in \mathbb{R}^n} \frac{|\varphi_k(D) f(x-y)|}{1+|2^k y|^a}, \qquad k \in \mathbb{N}_0, \tag{55}$$

be the corresponding maximal functions, see (2.3.2/1). Then

$$\left\| \left(\sum_{k=0}^{\infty} 2^{ksq} (\varphi^*_k f)_a(\cdot)^q \right)^{1/q} \mid L_p \right\| \tag{56}$$

(modification if $q=\infty$) is an equivalent quasi-norm in F^s_{pq}.

Proof. One can follow the proof of the Theorem. We indicate the necessary modifications. We begin with (13) with $x - 2^{-j}z$ instead of x. Then we use (17) with $x - 2^{-j}y - 2^{-j}z$ instead of $x - 2^{-j}y$ and the additional factor $(1+|z|)^a$ on the right-hand side. We divide both sides of the modified estimate (18) by $(1+|z|)^a$ and take afterwards the supremum with respect to $z \in \mathbb{R}^n$. This yields (23) with

$$\sup_{z \in \mathbb{R}^n} \frac{|(\varphi_j \rho_{l+j} \hat{f})^\vee (x - 2^{-j}z)|}{(1+|z|)^a} \quad \text{instead of} \quad (\varphi_j \rho_{l+j} \hat{f})^\vee (x). \tag{57}$$

We modify the second step of the proof of the Theorem in the same way. We obtain (28) with $x - 2^{-j}z$ on the left-hand side and the additional factor $2^{la}(1+|z|)^a$ on the right-hand side. Because now $s_0 + a < s$ we obtain (29) with the same substitute as in (57). This proves that the quasi-norm in (56) can be estimated from above by $c\|f \mid F^s_{pq}\|_\rho$. The other direction follows from the Theorem because $\sigma_{pq} < a$.

Remark 3. In this subsection we followed closely [Tri16]. On the other hand, considerations of the above type are not new, both for the spaces F^s_{pq} and B^s_{pq}, and their homogeneous counterparts, see [Triß: 2.3.6, in particular, Remark 3] where we have given some references to preceding papers. Homogeneous spaces of type F^s_{pq} and B^s_{pq} will not be treated in this book, but in [Tri16] one finds detailed formulations. Conditions of type (5) or (54), and their even better counterparts for the spaces B^s_{pq}, see 2.5, improve essentially corresponding assertions in [Triß: 2.3.6]. Furthermore, the great service of the Tauberian conditions (6), (7) was simply overlooked in [Triß]. A first improvement was obtained in [Tri12], mostly for the spaces B^s_{pq}. Tauberian conditions have a long history. As far as the systematic use of ideas of the above type in the theory of function spaces and related problems in approximation theory is concerned, we refer to H.S. Shapiro, see [Sha1,2]. Furthermore, N.M. Rivière and W.R. Madych developed in [Riv, MaR] this method in great detail in order to study Hölder spaces. Some results in this connection for the homogeneous spaces of type $B^s_{p\infty}$ with $1 \leqq p \leqq \infty$ may also be found in [Pee6: Chapter 8], see also [Bom] for further informations and references. Recently G.A. Kaljabin obtained in [Ka15, 6] characterizations of spaces of F^s_{pq} type with $s > 0$, $1 < p < \infty$, $1 < q < \infty$ which are at least partly closely connected with the results in 2.4. But his method is rather different.

2.4.2 Characterizations

In Theorem 2.4.1 we described equivalent quasi-norms in F^s_{pq}. The question arises whether any $f \in S'$ such that (2.4.1/11) or (2.4.1/12) is finite belongs to F^s_{pq}. This is not covered by our method because we used in Step 5 of the proof of Theorem 2.4.1 that f is an element of F^s_{pq}. Now we discuss this problem in some detail where we restrict ourselves to the quasi-norm in (2.4.1/11). But corresponding assertions can be obtained for the quasi-norm in (2.4.1/12). First we recall how to understand

$$\varphi_j(D)f(x) = (\varphi_j \hat{f})^\vee(x), \qquad x \in \mathbb{R}^n, \ \ j \in \mathbb{N}_0, \tag{1}$$

where the φ_j's are the same functions as in Theorem 2.4.1 and $f \in F^s_{pq}$. Let again $\{\rho_k(x)\}_{k\in\mathbb{N}_0}$ be a resolution of unity in the sense of (2.3.1/1–4) with ρ instead of φ. Then $\|f \mid F^s_{pq}\|_\rho$ has the same meaning as in (2.3.1/8) with ρ instead of φ. As an abbreviation we introduce temporarily

$$\|f \mid F^s_{pq}\|^\varphi = \left\| \left(\sum_{j=0}^{\infty} 2^{jsq} |(\varphi_j \hat{f})^\vee(\cdot)|^q \right)^{1/q} \mid L_p \right\|. \tag{2}$$

In Step 3 of the proof of Theorem 2.4.1 we explained

$$(\varphi_j \hat{f})^\vee(x) = \sum_{l=0}^{\infty} (\varphi_j \rho_l \hat{f})^\vee(x), \qquad f \in F^s_{pq}, \ \ j \in \mathbb{N}_0, \tag{3}$$

where each summand on the right-hand side is well-defined, and (3) converges both pointwise a.e. and in $L_{\bar{p}}$ with $\bar{p} = \max(p, 1)$. In particular, (3) converges in S'. We extend these considerations somewhat and assume

$$s_0 < \sigma = s - \frac{n}{p} \tag{4}$$

in addition to the assumptions from Theorem 2.4.1. Recall the embedding assertion

$$F^s_{pq} \subset B^\sigma_{\infty\infty}, \tag{5}$$

see [Triß: 2.7.1]. By the reasoning given in Step 3 of the proof of Theorem 2.4.1 it follows that (3) converges uniformly, in particular $(\varphi_j \hat{f})^\vee \in L_\infty$. If $f \in S'$ then $(\varphi_j \hat{f})^\vee \in L_\kappa$ with $1 \leqq \kappa \leqq \infty$ must always be understood as convergence of (3) in L_κ. Finally let $\psi \in S$ with

$$\operatorname{supp}\psi \subset \{y \in \mathbb{R}^n, |y| < 2\} \quad \text{and} \quad \psi(x) = 1 \ \text{ if } |x| \leqq 1. \tag{6}$$

If $f \in S'$ then we put temporarily $f^k = (\psi(2^{-k}\cdot)\hat{f})^\vee$. We have

$$f^k \to f \ \text{ in } \ S' \ \text{ if } \ k \to \infty. \tag{7}$$

All other notations have the same meaning as in 2.4.1, in particular σ_p, σ_{pq}, and a. Furthermore we assume in the theorem below that φ_0 and φ are two complex-valued C^∞ functions on $\mathbb{R}^n$ and $\mathbb{R}^n \setminus \{0\}$, respectively, which satisfy (2.4.1/6–10) (or (2.4.1/50–52) instead of (2.4.1/8–10), respectively) with the indicated values of s_0 and s_1. The maximal function $(\varphi_k^* f)_a$ has been defined in (2.4.1/55).

Theorem. (i) Let $1 < p < \infty$, $1 < q \leqq \infty$ and $s \in \mathbb{R}$. Let s_0 and s_1 be two real numbers with

$$s_0 < s < s_1 \quad \text{and} \quad s_1 > \sigma_p, \tag{8}$$

then

$$F^s_{pq} = \{f \in S' \colon \|f \mid F^s_{pq}\|^\varphi < \infty\} \tag{9}$$

(equivalent norm).

(ii) Let $0 < p < \infty$, $0 < q \leqq \infty$ and $s \in \mathbb{R}$. Let s_0 and s_1 be two real numbers with

$$s_0 + \sigma_{pq} < s < s_1 \quad \text{and} \quad s_1 > \sigma_p, \tag{10}$$

let $\overline{p} = \max(p, 1)$, then

$$F^s_{pq} = \left\{f \in S' \colon (\varphi_j \hat{f})^\vee \in L_{\overline{p}} \ \text{ for } \ j \in \mathbb{N}_0 \ \text{ and } \ \varliminf_{k\to\infty} \|f^k \mid F^s_{pq}\|^\varphi < \infty\right\}. \tag{11}$$

Furthermore, both

$$\varliminf_{k\to\infty} \|f^k \mid F^s_{pq}\|^\varphi \quad \text{and} \quad \|f \mid F^s_{pq}\|^\varphi \tag{12}$$

are equivalent quasi-norms in F^s_{pq}.

(iii) Let $0 < p < \infty$, $0 < q \leqq \infty$ and $s \in \mathbb{R}$. Let s_0 and s_1 be two real numbers with

$$s_0 + a < s < s_1, \qquad s_1 > \sigma_p, \tag{13}$$

then

$$F^s_{pq} = \left\{f \in S' \colon (\varphi_j \hat{f})^\vee \in L_\infty \ \text{ for } \ j \in \mathbb{N}_0 \ \text{ and} \right. \tag{14}$$

$$\left\| \left(\sum_{k=0}^{\infty} 2^{ksq} (\varphi_k^* f)_a(\cdot)^q \right)^{1/q} \mid L_p \right\| < \infty \Big\}$$

(equivalent quasi-norm).

Proof. Step 1. Let $f \in F^s_{pq}$, then we have the desired convergence in (3), including $(\varphi_j \hat{f})^\vee \in L_\infty$ under the hypotheses of (iii) because $a > \frac{n}{p}$. Furthermore we have $f^j \in L_p$ with $j \in \mathbb{N}_0$, and

$$\|f^j \mid F^s_{pq}\|_\rho \leqq c\|f \mid F^s_{pq}\|_\rho, \tag{15}$$

where the latter follows from the definition (2.3.1/8), with ρ instead of φ, and the Fourier multiplier assertion (2.2.4/4), see also [Trіß: 2.3.7] for a more explicit version of a Fourier multiplier theorem in F^s_{pq}. The constant c in (15) is independent of j. Hence, we have

$$\underline{\lim}_{k\to\infty} \|f^k \mid F^s_{pq}\|_\rho \leqq c\|f \mid F^s_{pq}\|_\rho. \tag{16}$$

The reverse estimate follows from $(\rho_l \hat{f^k})^\vee(x) \to (\rho_l \hat{f})^\vee(x)$ if $k \to \infty$ pointwise and Fatou's lemma. Hence the quasi-norms in (12) are equivalent. Consequently, if $f \in F^s_{pq}$ then the right-hand sides of (9,11) and the equivalence in (12) are covered by Theorem 2.4.1 and the above observations, under the respective hypotheses for the involved parameters. Corollary 2.4.1/2 covers the right-hand side of (14).

Step 2. Let the hypotheses of part (i) be satisfied. Then $\|f \mid F^s_{pq}\|^\varphi < \infty$ includes $(\varphi_j \hat{f})^\vee \in L_p$ in the sense of the L_p-convergence of (3). Let $\{\rho_k\}_{k\in\mathbb{N}_0}$ be the above system of functions. Similarly as in (2.4.1/36) we have

$$|(\rho_j \hat{f})^\vee(x)| \leqq c \int_{\mathbb{R}^n} \left| \left(\frac{\rho_j}{\varphi_j}\right)^\vee (y) \right| |(\varphi_j \hat{f})^\vee(x-y)| dy, \qquad j \in \mathbb{N}_0. \tag{17}$$

By the same reasoning as after (2.4.1/36) we arrive at (2.4.1/41) with $r = 1$ and without the factor $\psi(2^{-j}\cdot)$. We apply Theorem 2.2.2 to this modified estimate and obtain

$$\left\| \left(\sum_{j=0}^{\infty} 2^{jsq} |(\rho_j \hat{f})^\vee(\cdot)|^q \right)^{1/q} \mid L_p \right\| \leqq c\|f \mid F^s_{pq}\|^\varphi \tag{18}$$

(modification if $q = \infty$), and, consequently, $f \in F^s_{pq}$, see (2.3.1/8). The proof of (i) is complete.

Step 3. We prove (ii). Let $f \in S'$ such that the right-hand side of (11) holds. Then we have

$$(\varphi_j \hat{f^k})^\vee(x) = (\psi(2^{-k}\cdot)\varphi_j(\cdot)\hat{f}(\cdot))^\vee(x) = 2^{kn}\int_{\mathbb{R}^n} \psi^\vee(2^k y)(\varphi_j \hat{f})^\vee(x-y)dy, \quad (19)$$

and, in particular, $(\varphi_j \hat{f^k})^\vee \in L_{\bar{p}}$. Let $j \geqq k$, then the same argument as in (2.4.1/37) yields

$$|(\rho_j \hat{f^k})^\vee(x)|^r \leqq c\, 2^{jn(1-r)} \int_{\mathbb{R}^n} \left| \left(\frac{\rho_j}{\varphi_j}\right)^\vee (y)(\varphi_j \hat{f^k})^\vee(x-y)\right|^r dy \quad (20)$$

with $0 < r < \min(1,p,q)$, and we arrive at the counterpart of (2.4.1/41),

$$|(\rho_j \hat{f^k})^\vee(x)|^r \leqq c(M|(\varphi_j \hat{f^k})^\vee|^r)(x). \quad (21)$$

This estimate can be extended to $k < j$, where c may depend on k. Now by our assumption it follows $f^k \in F^s_{pq}$, in the same way as after (2.4.1/41). Hence, Theorem 2.4.1 yields

$$\left\| \left(\sum_{j=0}^{N} 2^{jsq} |(\rho_j \hat{f^k})^\vee(\cdot)|^q \right)^{1/q} \mid L_p \right\| \leqq c \|f^k \mid F^s_{pq}\|^\varphi, \quad (22)$$

where c is independent of N and k. We used that both sides of (22) are equivalent quasi-norms in F^s_{pq} at least if N is large. By $(\rho_j \hat{f^k})^\vee(x) \to (\rho_j \hat{f})^\vee(x)$ and Fatou's lemma it follows $f \in F^s_{pq}$. Together with Step 1 this completes the proof of (ii).

Step 4. We prove (iii). Let $f \in S'$ be such that the right-hand side of (14) holds. By (17) and (2.4.1/38) we have

$$|(\rho_j \hat{f})^\vee(x)| \leqq c \int_{\mathbb{R}^n} \left| \left(\frac{\bar{\rho}}{\varphi}\right)^\vee (y)\right| |(\varphi_j \hat{f})^\vee(x - 2^{-j}y)| dy \quad (23)$$

where $j \in \mathbb{N}$. The second factor in the integral can be estimated from above by $c(1+|y|)^a(\varphi_j^* f)_a(x)$. Then we obtain

$$|(\rho_j \hat{f})^\vee(x)| \leqq c(\varphi_j^* f)_a(x), \qquad x \in \mathbb{R}^n, \quad j \in \mathbb{N}_0. \quad (24)$$

By our assumption we have $f \in F^s_{pq}$. The proof is complete.

Remark. Parts (i) and (iii) of the theorem are quite satisfactory. Based on the following interpretation one can even omit the assumption $(\varphi_j \hat{f})^\vee \in L_\infty$ for $j \in \mathbb{N}_0$ in (14). We assume that $(\varphi_j \hat{f})^\vee$ with $j \in \mathbb{N}_0$ is a regular distribution and that the quasi-norm in (14) is finite. In particular we have $(\varphi_j^* f)_a \in L_p$. By definition (2.4.1/55) and the counterpart of (2.3.2/10) we have

$$(\varphi_j^* f)_a(x) \leqq c(\varphi_j^* f)_a(y) \;\text{ if }\; |x-y| \leqq 2^{-j}. \quad (25)$$

Let $(\varphi_j \hat{f})^\vee(x)$ be unbounded, then there exists for any $A \in \mathbb{R}$ a point $x^A \in \mathbb{R}^n$ with

$$|(\varphi_j^* f)_a(y)| \geqq A \quad \text{if} \quad |y - x^A| \leqq 2^{-j}. \tag{26}$$

We used (2.4.1/55) and (25). However (26) contradicts $(\varphi_j^* f)_a \in L_p$. In other words, we have $(\varphi_j \hat{f})^\vee \in L_\infty$ if $j \in \mathbb{N}_0$. Let $p \geqq 1$, then we can omit the assumption $(\varphi_j \hat{f})^\vee \in L_p$ for $j \in \mathbb{N}_0$ in (11). In this case $\|f^k \mid F^s_{pq}\|^\varphi < \infty$ yields $(\varphi_j \hat{f^k})^\vee \in L_p$ and it was just this assertion which we needed in Step 3 of the above proof. However the situation seems to be different if $p < 1$. Furthermore it would be desirable to replace $\underline{\lim}_{k\to\infty} \|f^k \mid F^s_{pq}\|^\varphi < \infty$ in (11) by $\|f \mid F^s_{pq}\|^\varphi < \infty$. In general such a possibility seems to be somewhat doubtful. Later on we discuss this point under further restrictions for φ, see 2.6.4 in connection with thermic and harmonic extensions.

2.4.3 Modifications of the main theorem

Some modifications of Theorem 2.4.1 will be useful later on. Our formulations will be given in terms of equivalent quasi-norms. Although possible, we do not try to rephrase some of these assertions in the sense of Theorem 2.4.2. To give a feeling of our intensions we begin with a somewhat heuristic discussion. Let $\varphi(x) = (e^{i\gamma x} - 1)^M$, where γx stands for the scalar product of the variable $x \in \mathbb{R}^n$ and $\gamma \in \mathbb{R}^n$, and $M \in \mathbb{N}$. Then $(\varphi \hat{f})^\vee(x) = \Delta^M_\gamma f(x)$, where Δ^M_γ are the usual differences, see (1.2.2/3). Characterizations of some spaces of type F^s_{pq} and B^s_{pq} via differences Δ^M_γ are very desirable. For apropriate numbers s_0 and s_1 hypotheses of type (2.4.1/8,9) for the above function $\varphi(x)$ are fulfilled, but not the Tauberian condition (2.4.1/7). On the other hand if one does not deal with a single function $(e^{i\gamma x} - 1)^M$ but with an appropriate finite set of these functions $\{(e^{i\gamma^k x} - 1)^M\}_{k=1}^N$ or families of these functions of type $\{(e^{i\gamma x} - 1)^M\}_{|\gamma|=1}$ or $\{(e^{i\gamma x} - 1)^M\}_{1\leqq|\gamma|\leqq 2}$ then the Tauberian condition (2.4.1/7) is satisfied in some sense and one can expect equivalent quasi-norms of type (2.4.1/11,12). In this subsection we describe the necessary modifications compared with 2.4.1. We are not interested in most general formulations but we restrict our attention to those functions φ which cover the examples which we have in mind. We use the same notations as in 2.4.1, in particular, the numbers σ_p and σ_{pq} are given by (2.4.1/1), and $h(x) \in S$ and $H(x) \in S$ are the same functions as at the beginning of 2.4.1 with (2.4.1/2,3). Furthermore, $(\varphi(h\cdot)\hat{f})^\vee$ must be interpreted similarly as in 2.4.1 and 2.4.2.

Theorem 1. Let $0 < p < \infty$, $0 < q \leqq \infty$, and $s \in \mathbb{R}$. Let s_0 and s_1 be two real numbers with (2.4.1/5). Let $\varphi_0(x)$ be the same function as in Theorem 2.4.1, including (2.4.1/6,10) with $a > \frac{n}{\min(p,q)}$. Let $\varphi(t)$ be a complex-valued C^∞ function on $\mathbb{R} \setminus \{0\}$ which satisfies

$$|\varphi(t)| > 0 \quad \text{if} \quad \frac{1}{8} < t < 8 \tag{1}$$

(Tauberian condition) and

$$\sup_{1 \leqq |\gamma| \leqq 2} \int_{\mathbb{R}^n} \left| \left(\frac{\varphi(\gamma z) h(z)}{|z|^{s_1}} \right)^\vee (y) \right| (1 + |y|)^a dy < \infty, \tag{2}$$

$$\sup_{1 \leqq |\gamma| \leqq 2} \sup_{m \in \mathbb{N}} 2^{-ms_0} \int_{\mathbb{R}^n} |(\varphi(2^m \gamma z) H(z))^\vee (y)| (1 + |y|)^a dy < \infty. \tag{3}$$

Then

$$\|\varphi_0(D) f \mid L_p\| + \left\| \left(\int_{|h| \leqq 1} |h|^{-sq} |(\varphi(h \cdot) \hat{f})^\vee (\cdot)|^q \frac{dh}{|h|^n} \right)^{1/q} \mid L_p \right\| \tag{4}$$

and

$$\|\varphi_0(D) f \mid L_p\| + \left\| \left(\int_0^1 t^{-sq} \sup_{\frac{t}{2} \leqq |h| \leqq t} |(\varphi(h \cdot) \hat{f})^\vee (\cdot)|^q \frac{dt}{t} \right)^{1/q} \mid L_p \right\| \tag{5}$$

(modification if $q = \infty$) are equivalent quasi-norms in F^s_{pq}.

Proof. Step 1. We follow Step 1 of the proof of Theorem 2.4.1 with $\varphi(2^{-j}\gamma x)$ instead of $\varphi_j(x)$, where $j \in \mathbb{N}$ and $1 \leqq |\gamma| \leqq 2$. Then we obtain (2.4.1/18). We take the supremum with respect to γ and proceed afterwards in the same way as in (2.4.1/19–23). We obtain

$$\left\| \left[\sum_{j=1}^{\infty} \left(\sum_{l=-\infty}^{K} 2^{js} \sup_{1 \leqq |\gamma| \leqq 2} |(\varphi(2^{-j}\gamma z) \rho_{l+j}(z) \hat{f})^\vee (\cdot)| \right)^q \right]^{1/q} \mid L_p \right\| \leqq c \|f \mid F^s_{pq}\| \tag{6}$$

and a coresponding estimate with φ_0. Next we use the same arguments as in Step 2 of the proof of Theorem 2.4.1. We arrive at (2.4.1/28) with $\varphi(2^{-j}\gamma x)$ instead of $\varphi_j(x)$, where $j \in \mathbb{N}$ and $1 \leqq |\gamma| \leqq 2$. We take the supremum with respect to γ and obtain the following counterpart of (2.4.1/29)

$$\left\| \left[\sum_{j=1}^{\infty} \left(\sum_{l=K+1}^{\infty} 2^{js} \sup_{1 \leqq |\gamma| \leqq 2} |(\varphi(2^{-j}\gamma z) \rho_{l+j}(z) \hat{f})^\vee (\cdot)| \right)^q \right]^{1/q} \mid L_p \right\| \tag{7}$$

$$\leqq c\, 2^{-K(s-s_0)} \| f \mid F^s_{pq} \|$$

and a corresponding estimate with φ_0. However the quasi-norm in (4) can be estimated from above by the quasi-norm in (5), which, in turn, can be estimated from above by the sum of the left-hand sides of (6) and (7) and corresponding terms with φ_0, and hence by $c\|f \mid F^s_{pq}\|_\rho$.

Step 2. In order to prove the reverse inequality we must modify Step 5 of the proof of Theorem 2.4.1. Let γ with $1 \leqq |\gamma| \leqq 2$ be given. Then

$$\left\{x \in \mathbb{R}^n \colon \frac{1}{2} \leqq |x| \leqq 2,\ |\varphi(\gamma x)| > 0\right\}$$

covers a sectorial set

$$\Omega_\gamma = \left\{y \in \mathbb{R}^n \colon \frac{1}{2} \leqq |y| \leqq 2,\ \left|\frac{y}{|y|} - \frac{\gamma}{|\gamma|}\right| \leqq b\right\}$$

for some $b > 0$. Let again $\{\rho_k(x)\}_{k\in\mathbb{N}_0}$ be a resolution of unity in the sense of (2.3.1/1–4) with ρ instead of φ. Let $\overline{\rho}(x) = \rho(x) - \rho(2x)$, $x \in \mathbb{R}^n$, see (2.3.1/2), then $\rho_j(x) = \overline{\rho}(2^{-j}x)$ if $j \in \mathbb{N}$. We decompose the basic function $\overline{\rho}(x)$ by $\overline{\rho}(x) = \sum_{k=1}^N \rho^{(k)}(x)$ such that for any γ with $1 \leqq |\gamma| \leqq 2$ we find a number k such that $\operatorname{supp} \rho^{(k)} \subset \Omega_\gamma$. This is always possible if one chooses N large enough. Let k be given and let $1 \leqq |\gamma| \leqq 2$ with $\operatorname{supp} \rho^{(k)} \subset \Omega_\gamma$. We follow the arguments of Step 5 of the proof of Theorem 2.4.1 with $\rho^{(k)}(2^{-j}x)$ and $\varphi(2^{-j}\gamma x)$ instead of ρ_j and φ_j, respectively. We arrive at the counterparts of (2.4.1/37,38), where in the latter inequality we may assume that the corresponding right-hand side is independent of k and γ. We substitute this inequality in the just described modification of (2.4.1/37) and integrate over those γ's which are connected with the given number k in the above sense. Afterwards this integration can be extended to all admissible γ's. Then the corresponding right-hand sides are independent of k. Summation over k yields (2.4.1/39) with

$$\left(\int_{1\leqq|\gamma|\leqq 2} |(\varphi(2^{-j}\gamma z)\psi(2^{-j}z)\hat{f})^\vee(x-y)|^q d\gamma\right)^{1/q} \tag{8}$$

instead of $|(\varphi_j\psi(2^{-j}\cdot)\hat{f})^\vee(x-y)|$, where we used that

$$\int_\gamma |\cdots|^r d\gamma \leqq c\left(\int_\gamma |\cdots|^q d\gamma\right)^{r/q}. \tag{9}$$

We have the counterparts of (2.4.1/42,44). We use the same splitting (2.4.1/43). The substitute of $|(\varphi_j\hat{f})^\vee|$ is

$$\left(\int_{1\leqq|\gamma|\leqq 2} |(\varphi(2^{-j}\gamma\cdot)\hat{f})^\vee|^q d\gamma\right)^{1/q},$$

and the corresponding term yields (4). The remaining term, i.e., the counterpart of (2.4.1/44), can be estimated with the help of (7) in the same way as after (2.4.1/44). This shows that $\|f \mid F^s_{pq}\|_\rho$ can be estimated from above by the quasi-norm (4) and, consequently, also by the quasi-norm (5). The proof is complete.

Remark 1. We followed closely [Tri16: 175–178]. Under our specific assumptions for $\varphi(x)$ one can omit $\sup_{1\leqq|\gamma|\leqq 2}$ in (2) and (3), because it is suficient to know corresponding estimates for a fixed $\gamma \neq 0$. However, if one replaces $\varphi(\gamma x)$ by more general families of functions $\varphi_\gamma(x)$, then one needs formulations of type (2) and (3). Furthermore, in Step 1 we need only $s_0 < s < s_1$ and that $(\varphi(\gamma\cdot)\hat{f})^\vee$ is well-defined in the sense of the beginning of 2.4.2, see Remark 2.4.1/2. In other words, under these assumptions, the quasi-norms in (4) and (5) can be estimated from above by $c\|f \mid F^s_{pq}\|_\rho$.

A second modification of the main theorem in 2.4.1 will be useful later on, especially in connection with spaces on Riemannian manifolds. Let

$$l \in S'(\mathbb{R}) \quad \text{with} \quad \operatorname{supp} l \subset [-1,1]. \tag{10}$$

Then the (one-dimensional) Fourier transform $\psi = F_1 l$ is an analytic function. Let

$$0 < |\psi(\lambda)| \leqq c\lambda^m \ \text{ if } \ 0 < \lambda < \delta \tag{11}$$

for some $\delta > 0$ and some $m \in \mathbb{N}$. We introduce the (essentially one-dimensional) means

$$l(h,f)(x) = \int_{\mathbb{R}} l(\lambda) f(x + h\lambda) d\lambda, \qquad x \in \mathbb{R}^n, \ \ 0 \neq h \in \mathbb{R}^n, \tag{12}$$

with $f \in S' = S'(\mathbb{R}^n)$, as far as this expression makes sense: in the sequel this always will be ensured when these means are applied, maybe via limiting procedures, where one begins with smooth functions f. As in Theorem 1, we use the same notations as in 2.4.1, in particular the numbers σ_p and σ_{pq} are given by (2.4.1/1), and $H(x) \in S$ is the same function as at the beginning of 2.4.1 with (2.4.1/2,3). We shall now not need the function $h(x)$ from (2.4.1/2,3).

Theorem 2. Let $0 < p < \infty$, $0 < q \leqq \infty$, and $s \in \mathbb{R}$. Let $s_0 \in \mathbb{R}$ and $m \in \mathbb{N}$ with

$$s_0 + \sigma_{pq} < s < m, \qquad m > \sigma_p, \tag{13}$$

and (11) for some $\delta > 0$. Let

$$\sup |h|^{-s_0} \int_{\mathbb{R}^n} |(\psi(hz)H(z))^\vee(y)|(1+|y|)^a dy < \infty \tag{14}$$

with $a > \frac{n}{\min(p,q)}$, where the supremum is taken over all $h \in \mathbb{R}^n$ with $|h| \geqq 1$. Let $\varphi_0 \in S$ with $\varphi_0(0) \neq 0$, then

$$\|\varphi_0(D)f \mid L_p\| + \left\| \left(\int_{|h|\leqq 1} |h|^{-sq} |l(h,f)(\cdot)|^q \frac{dh}{|h|^n} \right)^{1/q} \mid L_p \right\| \tag{15}$$

and

$$\|\varphi_0(D)f \mid L_p\| + \left\| \left(\int_0^1 t^{-sq} \sup_{\frac{t}{2}\leqq |h| \leqq t} |l(h,f)(\cdot)|^q \frac{dt}{t} \right)^{1/q} \mid L_p \right\| \tag{16}$$

(modification if $q = \infty$) are equivalent quasi-norms in F^s_{pq}.

Proof. The function $\varphi_0 \in S$ satisfies the modified Tauberian condition (2.4.1/6), and also (2.4.1/10), see (2.4.1/53). Otherwise, the above theorem is a modification of Theorem 1. First we observe

$$(\psi(h\cdot)\hat{f})^\vee(x) = \int_{\mathbb{R}} (F_1^{-1}\psi)(\lambda) f(x - \lambda h) d\lambda = l(-h,f)(x). \tag{17}$$

Now we identify $\psi(\lambda)$ with $\varphi(\lambda)$ in Theorem 1. Then (11) modifies (1) which compensates the above modification of (2.4.1/6). Furthermore, (14) coincides with (3). We have to check (2). For this purpose we wish to replace $|z|^{s_1}$ in (2) by $(z\gamma)^m$, where m has the above meaning. We substitute (2.4.1/14) by

$$\tilde{\rho}_j(x) = (2^{-j}\gamma x)^m \rho_j(x) \quad \text{if} \quad j \in \mathbb{N}_0. \tag{18}$$

Because $(\gamma x)^m$ is a polynomial one has no difficulty to follow the reasoning after (2.4.1/14) with $\varphi(2^{-j}\gamma z)$ instead of $\varphi_j(z)$ up to the end of Step 1 of the proof of Theorem 2.4.1. The rest remains unchanged. In other words we may replace $|z|^{s_1}$ in (2) by $(z\gamma)^m$. However, under the above assumptions we have $\varphi(\gamma z) = \psi(\gamma z)$ where $\psi(\lambda)$ is an analytic function with (11). Hence (2) is satisfied. Now by (17) the quasi-norms in (4) and (5) coincide with the quasi-norms in (15) and (16), respectively. The proof is complete.

Remark 2. An interesting choice of the above function $\psi(\lambda)$ is given by $\psi(\lambda) = (e^{i\nu\lambda} - 1)^m$, where $\nu > 0$ is an appropriate number, e.g., $\nu = 1/m$. Then (15) and (16) are quasi-norms in terms of differences, see the beginning of this subsection. Details will be given later on.

2.4.4 Complements

We complement the two theorems of the last subsection. First we ask whether $\sup_{\frac{t}{2}\leqq|h|\leqq t}$ in (2.4.3/5,16) can be replaced by $\sup_{0<|h|\leqq t}$. This is of interest in connection with classical descriptions.

Proposition 1. (i) Let all the hypotheses of Theorem 2.4.3/1 be fulfilled and let in addition $s>0$. Then

$$\|\varphi_0(D)f \mid L_p\| + \left\| \left(\int_0^1 t^{-sq} \sup_{0<|h|\leqq t} |(\varphi(h\cdot)\hat{f})^\vee(\cdot)|^q \frac{dt}{t} \right)^{1/q} \mid L_p \right\| \tag{1}$$

(modification if $q=\infty$) is an equivalent quasi-norm in F^s_{pq}.

(ii) Let all the hypotheses of Theorem 2.4.3/2 be fulfilled and let in addition $s>0$. Then

$$\|\varphi_0(D)f \mid L_p\| + \left\| \left(\int_0^1 t^{-sq} \sup_{0<|h|\leqq t} |l(h,f)(\cdot)|^q \frac{dt}{t} \right)^{1/q} \mid L_p \right\| \tag{2}$$

(modification if $q=\infty$) is an equivalent quasi-norm in F^s_{pq}.

Proof. We have to prove that the quasi-norm in (1) can be estimated from above by the quasi-norms in (2.4.3/5). For this purpose we estimate $\sup_{0<|h|\leqq t}$ from above by

$$\sum_{j=0}^{\infty} \sup_{2^{-j-1}t\leqq|h|\leqq 2^{-j}t} .$$

Then we have

$$\int_0^1 t^{-sq} \sup_{0<|h|\leqq t} |(\varphi(h\cdot)\hat{f})^\vee(\cdot)|^q \frac{dt}{t} \tag{3}$$

$$\leqq \sum_{j=0}^{\infty} \int_0^1 t^{-sq} \sup_{2^{-j-1}t\leqq|h|\leqq 2^{-j}t} |(\varphi(h\cdot)\hat{f})^\vee(\cdot)|^q \frac{dt}{t}$$

$$\leqq \left(\sum_{j=0}^{\infty} 2^{-jsq} \right) \int_0^1 t^{-sq} \sup_{\frac{t}{2}\leqq|h|\leqq t} |(\varphi(h\cdot)\hat{f})^\vee(\cdot)|^q \frac{dt}{t}.$$

This is just the desired estimate. In the same way one estimates the quasi-norm in (2) from above by the quasi-norm in (2.4.3/16).

Proposition 2. Let $0 < p < \infty$, $0 < q \leqq \infty$ and $s \in \mathbb{R}$. Let s_0 and s_1 be two real numbers with (2.4.1/5). Let $\varphi_0(x)$ be the same function as in Theorem 2.4.1, including (2.4.1/6,10), where $a > \frac{n}{\min(p,q)}$. Let $\varphi^1(x), \ldots, \varphi^N(x)$ be N complex-valued C^∞ functions on $\mathbb{R}^n \setminus \{0\}$ which satisfy the Tauberian condition

$$\sum_{k=1}^{N} |\varphi^k(x)| > 0 \text{ if } \frac{1}{2} \leqq |x| \leqq 2 \tag{4}$$

and (2.4.1/8,9) with φ^k instead of φ. Then

$$\|\varphi_0(D)f \mid L_p\| + \sum_{k=1}^{N} \left\| \left(\int_0^1 t^{-sq} |\varphi^k(tD)f(\cdot)|^q \frac{dt}{t} \right)^{1/q} \mid L_p \right\| \tag{5}$$

and

$$\|\varphi_0(D)f \mid L_p\| + \sum_{k=1}^{N} \left\| \left(\int_0^1 t^{-sq} \sup_{\frac{t}{2} \leqq \tau \leqq t} |\varphi^k(\tau D)f(\cdot)|^q \frac{dt}{t} \right)^{1/q} \mid L_p \right\| \tag{6}$$

(modification if $q = \infty$) are equivalent quasi-norms in F^s_{pq}. If, in addition, $s > 0$ then

$$\|\varphi_0(D)f \mid L_p\| + \sum_{k=1}^{N} \left\| \left(\int_0^1 t^{-sq} \sup_{0 < \tau \leqq t} |\varphi^k(\tau D)f(\cdot)|^q \frac{dt}{t} \right)^{1/q} \mid L_p \right\| \tag{7}$$

(modification if $q = \infty$) is also an equivalent quasi-norm in F^s_{pq}.

Proof. This proposition is simply the discrete version of Theorem 2.4.3/1. The proof is the same. Furthermore, (7) is covered by the same estimate as in (3).

2.4.5 Characterizations via Lusin functions

In 1.9.3 we gave a brief description of tent spaces and their connections with F^s_{pq} spaces via Lusin functions. Now we return to this subject in the above context. Descriptions of F^s_{pq} spaces in terms of Lusin functions go back to L. Päivärinta [Päi2] (1982), see also [CaT] and [Trіß: 2.12.1]. Recently, H.G. Feichtinger and K.H. Gröchenig gave a group-theoretical interpretation of atomic representations in function spaces, where they used in a decisive way characterizations of F^s_{pq} spaces in terms of Lusin functions, see Remark 1.9.4/3 for references and some details. In other words, a detailed description of this possibility seems to be of some interest. We follow [Tri16: 179/180]. Let

$$\gamma(x) = \{(y,t) \in \mathbb{R}^{n+1}_+ : |x-y| < t < 1\} \tag{1}$$

be a truncated cone of aperture 1 whose vertex is $x \in \mathbb{R}^n$. Let φ be the same function as in Theorem 2.4.1. We use (2.4.1/4) and introduce

$$\varphi^s_q f(x) = \left(\int_{\gamma(x)} t^{-sq} |\varphi(tD)f(y)|^q \frac{dy\,dt}{t^{n+1}} \right)^{1/q}, \qquad x \in \mathbb{R}^n, \tag{2}$$

where $0 < q \leqq \infty$ (modification if $q = \infty$) and $s \in \mathbb{R}$. We have

$$\varphi^s_q f(x) = c \left(\int_0^1 t^{-sq} \frac{1}{|B(x,t)|} \int_{B(x,t)} |\varphi(tD)f(y)|^q dy \frac{dt}{t} \right)^{1/q} \tag{3}$$

for some $c > 0$, where $B(x,t)$ stands for a ball of radius t in $\mathbb{R}^n$ centered at $x \in \mathbb{R}^n$ with $|B(x,t)|$ as its volume. However (3) is a function of Lusin type, see [Päi1: p. 123] and [Triß: 2.12.1].

Theorem. Let $0 < p < \infty$, $0 < q \leqq \infty$, and $s \in \mathbb{R}$. Let $a > \frac{n}{\min(p,q)}$ and let s_0 and s_1 be two real numbers with (2.4.1/54). Let φ_0 and φ be the same two functions as in Theorem 2.4.1 with (2.4.1/6–10). Then

$$\|\varphi_0(D)f \mid L_p\| + \|\varphi^s_q f \mid L_p\| \tag{4}$$

(modification if $q = \infty$) is an equivalent quasi-norms in F^s_{pq}.

Proof. For fixed $x \in \mathbb{R}^n$ and, say, $t = 2^{-j}$ with $j \in \mathbb{N}$ we have

$$|\varphi(tD)f(x+y)| \leqq c(\varphi^*_j f)_a(x), \qquad |y| \leqq t, \tag{5}$$

where $(\varphi^*_j f)_a(x)$ is the maximal function introduced in (2.4.1/55). If, say, $2^{-j} \leqq t \leqq 2^{-j+1}$ with $j \in \mathbb{N}$, then one has an obvious counterpart of (5) and also of the equivalent quasi-norm (2.4.1/56) with $\int_0^1 \cdots \frac{dt}{t}$ instead of $\sum_{k=1}^\infty \cdots$. To justify the latter claim one has to modify the proof of Corollary 2.4.1/2 slightly. We insert (5) in (3), then the just indicated modification of (2.4.1/56) shows that the quasi-norm in (4) can be estimated from above by $\|f \mid F^s_{pq}\|$. In order to prove the reverse estimate we use (2.4.1/37) with $y+z$ instead of y (integration over y), where $|z| \leqq t \sim 2^{-j}$. We have (2.4.1/38) with $y+z$ on the left-hand side and y on the right-hand side. We put this estimate in the just-described modified version of (2.4.1/37) and take afterwards the $(\int_{|z|<t} |\cdots|^{q/r} dz)^{r/q}$-norm on both sides. Now the rest is the same as at the end of Step 2 of the proof of Theorem 2.4.3/1.

2.4.6 Local means

Let $B = \{y \in \mathbb{R}^n \colon |y| < 1\}$ be the unit ball in $\mathbb{R}^n$, and let k be a C^∞ function in $\mathbb{R}^n$ with $\operatorname{supp} k \subset B$, then we introduce the local means

$$k(t,f)(x) = \int_{\mathbb{R}^n} k(y) f(x+ty)dy, \qquad x \in \mathbb{R}^n, \quad t > 0, \tag{1}$$

which makes sense for any $f \in S'$ (appropriately interpreted). Let k_0 and k^0 be two C^∞ functions in $\mathbb{R}^n$ with

$$\operatorname{supp} k_0 \subset B, \qquad \operatorname{supp} k^0 \subset B, \tag{2}$$

and

$$\hat{k}_0(0) \neq 0, \qquad \hat{k^0}(0) \neq 0. \tag{3}$$

Let

$$k(y) = \Delta^N k^0(y) = \left(\sum_{j=1}^n \frac{\partial^2}{\partial y_j^2} \right)^N k^0(y), \qquad y \in \mathbb{R}^n, \tag{4}$$

where $N \in \mathbb{N}$ is at our disposal. Let again $\sigma_p = n\left(\frac{1}{p} - 1\right)_+$, see (2.4.1/1).

Theorem. Let $0 < p < \infty$, $0 < q \leqq \infty$, and $s \in \mathbb{R}$. Let $N \in \mathbb{N}$ with

$$2N > \max(s, \sigma_p), \tag{5}$$

then

$$\|k_0(1,f) \mid L_p\| + \left\| \left(\sum_{j=0}^\infty 2^{jsq} |k(2^{-j}, f)(\cdot)|^q \right)^{1/q} \mid L_p \right\|, \tag{6}$$

$$\|k_0(1,f) \mid L_p\| + \left\| \left(\int_0^1 t^{-sq} |k(t, f)(\cdot)|^q \frac{dt}{t} \right)^{1/q} \mid L_p \right\|, \tag{7}$$

and

$$\|k_0(1,f) \mid L_p\| + \left\| \left(\int_0^1 t^{-sq} \sup_{\frac{t}{2} \leqq \tau \leqq t} |k(\tau, f)(\cdot)|^q \frac{dt}{t} \right)^{1/q} \mid L_p \right\| \tag{8}$$

(modification if $q = \infty$) are equivalent quasi-norms in F^s_{pq}. If, in addition, $s > 0$ then

$$\|k_0(1,f) \mid L_p\| + \left\| \left(\int_0^1 t^{-sq} \sup_{0<\tau\leqq t} |k(\tau,f)(\cdot)|^q \frac{dt}{t} \right)^{1/q} \mid L_p \right\| \tag{9}$$

(modification if $q=\infty$) is also an equivalent quasi-norms in F^s_{pq}.

Proof. We identify $k_0^\vee$ and $k^\vee$ with φ_0 and φ in Theorem 2.4.1, respectively. Then (3) and $k^\vee(x) = |x|^{2N} k^{0\vee}$ fulfill immaterial modifications of the Tauberian conditions (2.4.1/6,7). We choose $s_1 = 2N$, then we have (2.4.1/8–10), see also (2.4.1/53). Hence, (2.4.1/11,12) are equivalent quasi-norms in F^s_{pq}. We have

$$\varphi(tD)f(x) = (\varphi(t\cdot)\hat{f})^\vee(x) = c \int_{\mathbb{R}^n} (\varphi(t\cdot))^\vee(y) f(x-y)dy. \tag{10}$$

We observe

$$(\varphi(t\cdot))^\vee(y) = t^{-n}\varphi^\vee\left(\frac{y}{t}\right) = t^{-n}\hat{\varphi}\left(-\frac{y}{t}\right) = t^{-n}k\left(-\frac{y}{t}\right),$$

which shows that (10) coincides with (1). Hence, (6) and (7) are equivalent quasi-norms in F^s_{pq}. The corresponding assertion for (8) follows from the arguments in the proof of Theorem 2.4.3/1. Finally, as far as (9) is concerned we refer to the proof of Proposition 2.4.4/1.

Remark. The above theorem covers in particular Theorem 1.8.4(ii), see Remark 1.8.4/1 for some references. The advantage of (1) compared, for instance, with $\varphi_k(D)$ from Definition 2.3.1 is its strictly local nature: In order to calculate $k(t,f)(x)$ in a given point $x \in \mathbb{R}^n$ one needs only a knowledge of $f(z)$ in a ball of radius t centered at x. This observation will be of great service for us in the sequel.

Proposition. Let k_0 and k^0 be the above functions. Let $0<p<\infty$, $0<q\leqq\infty$, and $s \in \mathbb{R}$. Let $M \in \mathbb{N}$ with

$$M > \max(s, \sigma_p), \tag{11}$$

then the counterpart of (7),

$$\|k_0(1,f) \mid L_p\| + \sum_{|\alpha|=M} \left\| \left(\int_0^1 t^{(M-s)q} \mid k^0(t, D^\alpha f)(\cdot)|^q \frac{dt}{t} \right)^{1/q} \mid L_p \right\| \tag{12}$$

(modification if $q = \infty$) and the corresponding counterparts of (6,8), and, if $s > 0$, of (9), are equivalent quasi-norms in F^s_{pq}.

Proof. We have

$$k^0(t, D^\alpha f)(x) = c\, t^{-M} \int_{\mathbb{R}^n} D^\alpha k^0(y) f(x + ty) dy. \tag{13}$$

We wish to apply Proposition 2.4.4/2 and put $k_0^\vee = \varphi_0$ and $(D^\alpha k^0)^\vee = \varphi^\alpha$ with $|\alpha| = M$. Recall $\varphi^\alpha(x) = x^\alpha (k^0)^\vee(x)$. By (3), obvious counterparts of (2.4.1/6) and (2.4.4/4) are fulfilled, similarly for (2.4.1/9,10). As for (2.4.1/8) we recall the slight modification described in the proof of Theorem 2.4.3/2 which shows that we can replace $\frac{\varphi(z)}{|z|^{s_1}} h(z)$ in (2.4.1/8) by $\frac{\varphi^\alpha(z)}{z^\alpha} h(z)$ and hence by $(k^0)^\vee(z) h(z)$.

2.4.7 Localization principle

Recall that F^s_{pq} and B^s_{pq} have been introduced in Definition 2.3.1. Let ψ be a compactly supported C^∞ function in $\mathbb{R}^n$ and let

$$\psi_k(y) = \psi(y - k), \qquad k \in \mathbb{Z}^n, \quad x \in \mathbb{R}^n, \tag{1}$$

where $\mathbb{Z}^n$ stands for the lattice of all points $x = (x_1, \ldots, x_n) \in \mathbb{R}^n$ with integer-valued components x_j. We assume that

$$\sum_{k \in \mathbb{Z}^n} \psi_k(x) = 1, \qquad x \in \mathbb{R}^n, \tag{2}$$

is a resolution of unity.

Theorem. (i) Let $0 < p < \infty$, $0 < q \leqq \infty$, and $s \in \mathbb{R}$, then

$$\left(\sum_{k \in \mathbb{Z}^n} \|\psi_k f \mid F^s_{pq}\|^p \right)^{1/p} \tag{3}$$

is an equivalent quasi-norm in F^s_{pq}.

(ii) Let $0 < p \leqq \infty$, $0 < q \leqq \infty$, and $s \in \mathbb{R}$, then

$$\left(\sum_{k \in \mathbb{Z}^n} \|\psi_k f \mid B^s_{pq}\|^r \right)^{1/r} \tag{4}$$

with $0 < r \leqq \infty$ (modification if $r = \infty$) is an equivalent quasi-norm in B^s_{pq} if and only if $p = q = r$.

Proof. Step 1. Let $0 < p < \infty$, $0 < q \leqq \infty$, and $s \in \mathbb{R}$. By the equivalent quasi-norm (2.4.6/7), based on (2.4.6/1), and

$$k(t, f)(x) = \sum_{m\in\mathbb{Z}^n} k(t, \psi_m f)(x), \qquad x \in \mathbb{R}^n, \tag{5}$$

it follows that $\|f \mid F^s_{pq}\|$ can be estimated from above by the quasi-norm in (3).

Step 2. Let $0 < p < \infty$, $0 < q \leq \infty$, and $s > n/p$. We prove that the quasi-norm in (3) can be estimated from above by $\|f \mid F^s_{pq}\|$. We use again (2.4.6/7). Let $M \in \mathbb{N}$ with $M > s$. Then we have

$$\begin{aligned} k(t, \psi_m f)(x) &= \int_{\mathbb{R}^n} k(y)\psi_m(x + ty)f(x + ty)dy \\ &= \sum_{|\alpha|\leqq M-1} \frac{1}{\alpha!} D^\alpha \psi_m(x) t^{|\alpha|} \int_{\mathbb{R}^n} k(y) y^\alpha f(x + ty)dy \\ &\quad + \sum_{|\alpha|=M} t^M \int_{\mathbb{R}^n} k(y) y^\alpha R_\alpha(x, y) f(x + ty)dy \end{aligned} \tag{6}$$

where $R_\alpha(x, y)$ are remainder terms of the type $D^\alpha \psi_m(x + \vartheta^\alpha ty)$ with $0 \leqq \vartheta^\alpha \leqq 1$. By (2.4.2/4,5) and $s > n/p$ it follows that $f \in F^s_{pq}$ is a bounded continuous function. In Remark 1 below we shall modify this assertion by

$$\| \sup_{|x-y|\leqq c} |f(y)| \mid L_p\| \leqq c\|f \mid F^s_{pq}\| \tag{7}$$

where $c > 0$ is an arbitrary number and the integration in (7) is taken with respect to x. Let $k_\alpha(y) = y^\alpha k(y)$, then we have by (6) and $M > s$

$$\begin{aligned} \left(\int_0^1 t^{-sq} |k(t, \psi_m f)(x)|^q \frac{dt}{t}\right)^{1/q} &\leqq c_1 \sum_{|\alpha|\leqq M-1} \left(\int_0^1 t^{-sq} |k_\alpha(t, f)(x)|^q \frac{dt}{t}\right)^{1/q} \\ &\quad + c_1 \sup_{|x-y|\leqq c_2} |f(y)| \end{aligned} \tag{8}$$

if $|x - m| \leqq c_2$ and

$$k(t, \psi_m f)(x) = 0 \quad \text{for} \quad |x - m| > c_2 \quad \text{and} \quad 0 < t \leqq 1, \tag{9}$$

for some positive numbers c_1 and c_2. We use (2.4.6/7) with $\psi_m f$ instead of f, where the corresponding first term in (2.4.6/7) can be estimated by the last term on the right-hand side of (8) with a counterpart of (9). We take the pth power of (8) (and its counterpart with k_0), sum over $m \in \mathbb{Z}^n$, integrate over $\mathbb{R}^n$, and obtain

$$\sum_{m\in\mathbb{Z}^n} \|\psi_m f \mid F^s_{pq}\|^p \leqq c_1 \sum_{|\alpha|\leqq M-1} \left\| \left(\int_0^1 t^{-sq} |k_\alpha(t,f)(\cdot)|^q \frac{dt}{t} \right)^{1/q} \mid L_p \right\|^p \tag{10}$$

$$+ c_1 \left\| \sup_{|x-y|\leqq c_2} |f(y)| \mid L_p \right\|^p$$

for some numbers c_1 and c_2. We choose N in (2.4.6/5) large and have

$$k^\vee_\alpha(x) = (y^\alpha k(y))^\vee(x) = |x|^{2K} g_\alpha(x), \qquad x\in\mathbb{R}^n, \tag{11}$$

with $0 \leqq |\alpha| \leqq M-1$, $K\in\mathbb{N}$, $2K > \max(s,\sigma_p)$, and $g_\alpha\in S$, see the beginning of the proof of Theorem 2.4.6. Then the corresponding counterparts of (2.4.1/8,9) are fulfilled. This is sufficient to estimate (2.4.1/12), and consequently the first terms on the right-hand side of (10), from above by $\|f \mid F^s_{pq}\|$ (Tauberian conditions are only needed for the reverse estimate). Together with (7) we arrive at the desired estimate

$$\sum_{m\in\mathbb{Z}^n} \|\psi_m f \mid F^s_{pq}\|^p \leqq c\|f \mid F^s_{pq}\|^p. \tag{12}$$

Step 3. Let $0<p<\infty$ and $0<q\leqq\infty$. Let (12) be valid for all s with $s\geqq\sigma$. We wish to prove that (12) holds also for $\sigma-1$ instead of s. Let $f\in F^{\sigma-1}_{pq}$, then we have

$$f = (\Delta - id)g, \quad g\in F^{\sigma+1}_{pq} \quad \text{and} \quad \|f \mid F^{\sigma-1}_{pq}\| \sim \|g \mid F^{\sigma+1}_{pq}\| \tag{13}$$

where Δ stands for the Laplacian and id is the identity. (13) is a lifting property for F^s_{pq} spaces, see [Trіß: 2.3.8], however, it is also a simple consequence of $\hat{f} = -(1+|x|^2)\hat{g}$, Definition 2.3.1(ii), and the same arguments as in Step 1 of the proof of Theorem 2.3.2. We have

$$\psi_m f = (\Delta - id)(\psi_m g) - (\Delta\psi_m)g - 2\sum_{j=1}^n \frac{\partial\psi_m}{\partial x_j}\frac{\partial g}{\partial x_j}, \tag{14}$$

and consequently

$$\|\psi_m f \mid F^{\sigma-1}_{pq}\| \leqq c\|\psi_m g \mid F^{\sigma+1}_{pq}\| + c\|(\Delta\psi_m)g \mid F^\sigma_{pq}\| + c\sum_{j=1}^n \left\| \frac{\partial\psi_m}{\partial x_j}\frac{\partial g}{\partial x_j} \mid F^\sigma_{pq} \right\| \tag{15}$$

where we used the equivalence in (13) and the monotonicity properties (2.3.2/22,23). By assumption we have (12) with $s=\sigma$ and $s=\sigma+1$, where we can replace ψ_m by $\Delta\psi_m$ or $\frac{\partial\psi_m}{\partial x_j}$. We apply this estimate to (15) and obtain

$$\left(\sum_{m\in\mathbb{Z}^n} \|\psi_m f \mid F^{\sigma-1}_{pq}\|^p \right)^{1/p} \leqq c\|g \mid F^{\sigma+1}_{pq}\| + c\|g \mid F^\sigma_{pq}\| + c\sum_{j=1}^n \left\| \frac{\partial g}{\partial x_j} \mid F^\sigma_{pq} \right\| \tag{16}$$

$$\leqq c' \| g \mid F^{\sigma+1}_{pq} \| \sim c' \| f \mid F^{\sigma+1}_{pq} \|$$

where we used that $\| \frac{\partial g}{\partial x_j} \mid F^{\sigma}_{pq} \|$ can be estimated from above by $c\| g \mid F^{\sigma+1}_{pq} \|$. This follows from $(\frac{\partial g}{\partial x_m})^{\wedge}(\xi) = i\xi_m \hat{g}(\xi)$ and the estimates in Step 1 of the proof of Theorem 2.3.2, see also [Triß: 2.3.8]. Now (16), Step 2, and mathematical induction complete the proof of part (i) of the theorem.

Step 4. We prove (ii). By (i) it follows that (4) with $r = q = p < \infty$ is an equivalent quasi-norm in $B^s_{pp} = F^s_{pp}$. The arguments from the Steps 1–3 can be extended to the case $r = q = p = \infty$, hence to the spaces $B^s_{\infty\infty}$, see Theorem 2.5.3.

Step 5. We assume that (4) is an equivalent quasi-norm in B^s_{pq}. Let $s = 0$, otherwise some immaterial modifications have to be made. Let χ be a non-vanishing C^∞ function on $\mathbb{R}^n$ with $\operatorname{supp} \hat{\chi} \subset B$, where the latter stands for the unit ball. Let

$$f(x) = \sum_{m=1}^{N} \sum_{l=1}^{N} c_{ml} e^{ix\sigma_m} \chi(x - x_l), \tag{17}$$

where c_{ml} are arbitrary complex numbers, $x_l \in \mathbb{R}^n$, $\sigma_m \in \mathbb{R}^n$ with $|\sigma_m| = 2^m$. We wish to use (2.3.1/7). First we observe that $(e^{i\sigma_m x}\chi(x - x_l))^{\wedge}$ is supported near σ_m. Hence we may assume

$$\varphi_m(D) f(x) = (\varphi_m \hat{f})^{\vee}(x) = \sum_{l=1}^{N} c_{ml} e^{ix\sigma_m} \chi(x - x_l). \tag{18}$$

We may choose x_l such that

$$\frac{1}{2} \| \varphi_m(D) f \mid L_p \|^p \leqq \| \chi \mid L_p \|^p \sum_{l=1}^{N} |c_{ml}|^p \leqq 2 \| \varphi_m(D) f \mid L_p \|^p. \tag{19}$$

By (2.3.1/7) we have

$$\| f \mid B^0_{pq} \| \sim \left[\sum_{m=1}^{N} \left(\sum_{l=1}^{N} |c_{ml}|^p \right)^{q/p} \right]^{1/q} \tag{20}$$

where "$\sim$" is independent of $N \in \mathbb{N}$ and the numbers c_{ml}. Let $x_j = k \in \mathbb{Z}^n$, then

$$\psi_k(x) f(x) = \sum_{m=1}^{N} \sum_{l=1}^{N} c_{ml} e^{ix\sigma_m} \psi(x - x_j) \chi(x - x_l). \tag{21}$$

We choose x_l in an appropriate way and obtain by the same arguments as above

$$\| \psi_k f \mid B^0_{pq} \| \sim \left(\sum_{m=1}^{N} |c_{mj}|^q \right)^{1/q} \tag{22}$$

where "$\sim$" is independent of $N \in \mathbb{N}$, $k \in \mathbb{Z}^n$, and the numbers c_{mj}. In other words, if (4) with $s = 0$ is an equivalent quasi-norm in B^0_{pq} then (20) and (22) yield

$$\left[\sum_{m=1}^{N}\left(\sum_{l=1}^{N}|c_{ml}|^p\right)^{q/p}\right]^{1/q} \sim \left[\sum_{l=1}^{N}\left(\sum_{m=1}^{N}|c_{ml}|^q\right)^{r/q}\right]^{1/r} \tag{23}$$

where "$\sim$" is independent of $N \in \mathbb{N}$ and the numbers c_{ml}. Let $c_{ml} = a_m b_l$, then it follows $r = p$. Afterwards the choice $c_{ml} = \delta_{ml} a_l$ yields $p = q$. Hence, if (23) holds then we have $p = q = r$. The proof is complete.

Remark 1. Let $0 < p < \infty$, $0 < q \leqq \infty$, and $s > n/p$. We prove (7). Let f be given by (2.3.1/6), then we have

$$\left\| \sup_{|x-y|\leqq c} |f(y)| \mid L_p \right\| \leqq c \sum_{k=0}^{\infty} 2^{k\varepsilon} \left\| \sup_{|x-y|\leqq c} |\varphi_k(D)f(y)| \mid L_p \right\| \tag{24}$$

where $\varepsilon > 0$ is an arbitrary number. Now we use the scalar case of (2.2.4/3) with

$$\sup_{|x-y|\leqq c} |\varphi_k(D)f(y)| \leqq c' 2^{ka} \sup_{z\in\mathbb{R}^n} \frac{|\varphi_k(D)f(x-z)|}{1+|2^k z|^a} \tag{25}$$

and $a > n/p$. Then the left-hand side of (24) can be estimated from above by

$$c\sum_{k=0}^{\infty} 2^{(a+\varepsilon)k} \|\varphi_k(D)f \mid L_p\| = c\|f \mid B^{a+\varepsilon}_{p1}\|, \tag{26}$$

where we used (2.3.1/7). Since a and ε are arbitrary numbers with $a > n/p$ and $\varepsilon > 0$, we can estimate the right-hand side of (26) from above by $c\|f \mid B^s_{pq}\|$, see (2.3.2/23). The proof is complete.

Remark 2. We denote part (i) of the theorem as the localization principle for the spaces F^s_{pq}. Part (ii) indicates that there is no counterpart for the spaces B^s_{pq}. The proof shows that we did not use the special structure of the functions $\psi_k(x) = \psi(x-k)$, it is sufficient to have a good control of the supports of ψ_k and of some of its derivatives (in dependence on s, p, q). The localization principle for F^s_{pq}, combined with the local means from the preceding subsections, are the basis for our considerations in the following chapters.

Remark 3. Instead of an "additive" localization based on $\psi_k(x) = \psi(x-k)$ one can ask for a corresponding "multiplicative" localization based on $\varphi_k(x) = \varphi(2^k x)$, where φ is a suitable function. This problem has been studied in [Bou2, You1,2], where the authors prefer the homogeneous counterparts of F^s_{pq}, see 1.4.5 as far as homogeneous spaces are concerned.

Remark 4. Actually, (3) is a characterization: If $f \in S'$ such that (3) is finite, then $f \in F^s_{pq}$, see Proposition 7.2.2.

2.4.8 Fourier multipliers and maximal inequalities

In connection with pseudodifferential operators we need some Fourier multiplier assertions and maximal inequalities for F^s_{pq} spaces. Although the needed properties are more or less covered by previous considerations we formulate them separately in this subsection. Let $a(\eta, \xi)$ be a complex-valued function defined for $\eta \in \mathbb{R}^n$ and $\xi \in \mathbb{R}^n$ such that $\xi \to a(\eta, \xi)$ for any fixed $\eta \in \mathbb{R}^n$ is a C^∞ function in $\mathbb{R}^n$ with

$$\sup_{\eta \in \mathbb{R}^n} (1 + |\xi|)^k \sum_{|\alpha| = k} |D^\alpha_\xi a(\eta, \xi)| < \infty, \qquad k \in \mathbb{N}_0. \tag{1}$$

We extend the notation (2.3.1/5) to

$$a(\eta, D) f(x) = (a(\eta, \cdot) \hat{f}(\cdot))^\vee (x), \qquad x \in \mathbb{R}^n, \tag{2}$$

where η must be considered as a parameter. It is convenient for us to write temporarily $F^s_{\infty\infty} = B^s_{\infty\infty}$ and to assume that Theorem 2.4.6 can be extended to $p = q = \infty$. This will be justified later on, see 2.5.3. Let $k(t, f)$ be the same means as in (2.4.6/1) and let k_0, k^0, and k be the same functions as in (2.4.6/2,4), where $N \in \mathbb{N}$ is at our disposal. The Tauberian conditions (2.4.6/3) need not be satisfied now.

Proposition. (i) Let either $0 < p < \infty$, $0 < q \leqq \infty$ or $p = q = \infty$, let $s \in \mathbb{R}$. Let $N \in \mathbb{N}$ with $2N > \frac{n}{2} + \frac{n}{\min(p,q)} + \max(s, \sigma_p)$, then there exists a positive number c such that

$$\left\| \sup_{\eta \in \mathbb{R}^n} |a(\eta, D) k_0(1, f)| \mid L_p \right\| \tag{3}$$

$$+ \left\| \left(\int_0^1 t^{-sq} \sup_{\eta \in \mathbb{R}^n} |a(\eta, D) k(t, f)(\cdot)|^q \frac{dt}{t} \right)^{1/q} \mid L_p \right\| \leqq c \|f \mid F^s_{pq}\|$$

(modification if $q = \infty$) for all $f \in F^s_{pq}$.

(ii) Let either $0 < p < \infty$, $0 < q \leqq \infty$ or $p = q = \infty$, let $s > n/p$ and $b > 0$, then there exists a positive number c such that

$$\left\| \sup_{|\cdot - y| \leqq b} \sup_{\eta \in \mathbb{R}^n} |a(\eta, D) f(y)| \mid L_p \right\| \leqq c \|f \mid F^s_{pq}\| \tag{4}$$

for all $f \in F^s_{pq}$.

Proof. Step 1. We prove (i) by combining the proof of Theorem 2.4.6 and the two first steps of the proof of Theorem 2.4.1, where we assume $0 < p < \infty$, $0 < q \leqq \infty$. The case $p = q = \infty$ can be treated in the same way based on Theorem 2.5.3. By (2) and the arguments from the proof of Theorem 2.4.6 we have

$$a(\eta, D) k(t, f)(x) = (a(\eta, \xi) t^{2N} |\xi|^{2N} k^{0\vee}(t\xi) \hat{f}(\xi))^\vee(x) \tag{5}$$

and similarly for $a(\eta, D) k_0(1, f)$. Now we follow the arguments from Step 1 of the proof of Theorem 2.4.1 and replace $\varphi_j(\xi) = \varphi(2^{-j}\xi)$ by $a(\eta, \xi) t^{2N} |\xi|^{2N} k^{0\vee}(t\xi)$ with $t = 2^{-j}$. We have the counterparts of (2.4.1/13,15,16,18). The critical first factor in the integral in (2.4.1/16) led to the hypotheses (2.4.1/8) or better (2.4.1/50) which now must be replaced by

$$\left\| a(\eta, 2^j x) |x|^{2N} \frac{k^{0\vee}(x) h(x)}{|x|^{s_1}} \mid H^\sigma_2 \right\| < \infty \tag{6}$$

with $\sigma > \frac{n}{2} + a$. Let $\sigma \in \mathbb{N}$, then we can estimate the factors connected with $a(\eta, 2^j x)$ by

$$|D^\alpha_x a(\eta, 2^j x)| \leqq c\, 2^{j|\alpha|} (1 + 2^j |x|)^{-|\alpha|} \leqq \frac{c}{|x|^{|\alpha|}}, \quad 0 \leqq |\alpha| \leqq \sigma, \tag{7}$$

where c is independent of $\eta \in \mathbb{R}^n$ and $j \in \mathbb{N}$ (and, of course, of $x \in \mathbb{R}^n$). In other words, if $2N$ is larger than $\sigma + s_1$ with $\sigma > \frac{n}{2} + \frac{n}{\min(p,q)}$ and $s_1 > \max(s, \sigma_p)$, then (6) is fulfilled uniformly with respect to η and j. Hence under our assumptions for N we obtain the counterpart of (2.4.1/18), where the corresponding constant c is independent of $\eta \in \mathbb{R}^n$ and $j \in \mathbb{N}$. We take the supremum with respect to η and proceed afterwards as in Step 1 of the proof of Theorem 2.4.1. Next we follow Step 2 of the proof of Theorem 2.4.1 with the same replacement of φ_j as above. The crucial integral is the counterpart of (2.4.1/26) which led to the hypotheses (2.4.1/9,10) or better (2.4.1/51,52), which look similar as (6). In (7) we have now $|x| \sim 1$, which shows that the left-hand side of (7) can be estimated by a constant uniformly with respect to $\eta \in \mathbb{R}^n$ and $j \in \mathbb{N}_0$. We arrive at the counterpart of (2.4.1/28), take the supremum with respect to $\eta \in \mathbb{R}^n$ and obtain finally the discrete version of (3). However by the same arguments as in Step 4 of the proof of Theorem 2.4.1 we have also a corresponding estimate for the continuous version. The proof of (i) is complete.

Step 2. We prove (ii) by modifying the arguments from Remark 2.4.7/1. The counterparts of (2.4.7/24,25) read as follows,

$$\left\| \sup_{|x-y|\le b} \sup_{\eta\in\mathbb{R}^n} |a(\eta,D)f(y)| \mid L_p \right\| \tag{8}$$

$$\le c \sum_{k=0}^{\infty} 2^{k\varepsilon} \left\| \sup_{|x-y|\le b} \sup_{\eta\in\mathbb{R}^n} |a(\eta,D)\varphi_k(D)f(y)| \mid L_p \right\|$$

and

$$\sup_{|x-y|\le b} \sup_{\eta\in\mathbb{R}^n} |a(\eta,D)\varphi_k(D)f(y)| \le c\, 2^{ka} \sup_{z\in\mathbb{R}^n,\eta\in\mathbb{R}^n} \frac{|a(\eta,D)\varphi_k(D)f(x-z)|}{1+|2^k z|^a} \tag{9}$$

with $a > n/p$. Next we remark that the proof of (2.2.4/4) is based on

$$\sup_{z\in\mathbb{R}^n} \frac{|(M_k \hat{f}_k)^\vee(x-z)|}{1+|d_k z|^a} \le c\|M_k(d_k\cdot) \mid H_2^\kappa\| \sup_{z\in\mathbb{R}^n} \frac{|f_k(x-z)|}{1+|d_k z|^a} \tag{10}$$

with $\kappa > a + n/2$, see [Triß: (1.6.3/2)]. We estimate the right-hand side of (9) by using (10) with $d_k \sim 2^k$, $M_k(\xi) = a(\eta,\xi)\varphi_k(\xi)$ and with $\sum_{r=-2}^{2} \varphi_{k+r}(D)f$ instead of f in (9), see (2.3.2/4) as far as the latter replacement is concerned. By (1) the first factor on the right-hand side of (10) can be estimated from above independently of $k \in \mathbb{N}_0$ and $\eta \in \mathbb{R}^n$. Taking the supremum over $\eta \in \mathbb{R}^n$ the indicated modification of (10) shows that the right-hand side of (9) can be estimated from above by

$$c\, 2^{ka} \sup_{z\in\mathbb{R}^n} \sum_{r=-2}^{2} \frac{|\varphi_{k+r}(D)f(x-z)|}{1+|2^k z|^a}. \tag{11}$$

The rest is the same as in Remark 2.4.7/1 and the proof of (ii) is complete.

2.5 General characterizations for B^s_{pq}

2.5.1 The main theorem

In 2.3 we introduced and studied the spaces B^s_{pq} and F^s_{pq} simultaneously. This will also be done later on. Sections 2.4 and 2.5 are exceptions: we treated the more complicated spaces F^s_{pq} in 2.4 and collect now in 2.5 the respective counterparts for the simpler spaces B^s_{pq}, where we refer mostly to 2.4 and indicate the necessary changes in the proofs. Recall

$$\sigma_p = n\left(\frac{1}{p} - 1\right)_+, \qquad 0 < p \le \infty, \tag{1}$$

see (2.4.1/1). Furthermore the functions h and H as well as $\varphi(tD)f$ have the same meaning as at the beginning of 2.4.1.

Theorem. Let $0 < p \leqq \infty$, $0 < q \leqq \infty$, and $s \in \mathbb{R}$. Let s_0 and s_1 be two real numbers with

$$s_0 + \sigma_p < s < s_1 \quad \text{and} \quad s_1 > \sigma_p. \tag{2}$$

Let φ_0 and φ be two complex-valued C^∞ functions on $\mathbb{R}^n$ and $\mathbb{R}^n \setminus \{0\}$, respectively, which satisfy the Tauberian conditions

$$|\varphi_0(x)| > 0 \quad \text{if} \quad |x| \leqq 2 \tag{3}$$

and

$$|\varphi(x)| > 0 \quad \text{if} \quad \frac{1}{2} \leqq |x| \leqq 2.$$

Let $\overline{p} = \min(1, p)$ and let

$$\int_{\mathbb{R}^n} \left| \left(\frac{\varphi(z)h(z)}{|z|^{s_1}} \right)^\vee (y) \right|^{\overline{p}} dy < \infty, \tag{4}$$

$$\sup_{m \in \mathbb{N}} 2^{-ms_0\overline{p}} \int_{\mathbb{R}^n} |(\varphi(2^m \cdot)H(\cdot))^\vee(y)|^{\overline{p}} dy < \infty \tag{5}$$

and

$$\sup_{m \in \mathbb{N}} 2^{-ms_0\overline{p}} \int_{\mathbb{R}^n} |(\varphi_0(2^m \cdot)H(\cdot))^\vee(y)|^{\overline{p}} dy < \infty. \tag{6}$$

Let $\varphi_j(x) = \varphi(2^{-j}x)$ if $x \in \mathbb{R}^n \setminus \{0\}$ and $j \in \mathbb{N}$. Then

$$\left(\sum_{j=0}^{\infty} 2^{jsq} \|\varphi_j(D)f \mid L_p\|^q \right)^{1/q} \tag{7}$$

and

$$\|\varphi_0(D)f \mid L_p\| + \left(\int_0^1 t^{sq} \|\varphi(tD)f \mid L_p\|^q \frac{dt}{t} \right)^{1/q} \tag{8}$$

(modification if $q = \infty$) are equivalent quasi-norms in B^s_{pq}.

Proof. This is the counterpart of Theorem 2.4.1 and we modify its proof. We have again the splitting (2.4.1/13), the estimate (2.4.1/15), and the expression (2.4.1/16). Let $1 \leqq p \leqq \infty$. Then we apply the L_p-norm to (2.4.1/15,16), use (4) with $\bar{p} = 1$, and apply afterwards the l_q-quasi-norm. Then we obtain the following counterpart of (2.4.1/19)

$$\left(\sum_{j=1}^{\infty} 2^{jsq} \left\| \sum_{l=-\infty}^{K} 2^{js} (\varphi_j \rho_{l+j} \hat{f})^\vee \mid L_p \right\|^q \right)^{1/q} \leqq c \left(\sum_{m=0}^{\infty} 2^{msq} \| (\tilde{\rho}_m \hat{f})^\vee \mid L_p \|^q \right)^{1/q} . \quad (9)$$

Let $0 < p < 1$. We use again an inequality of Plancherel–Polyá–Nikol'skij type, see [Triß: (1.3.2/5)] and estimate the integral in (2.4.1/16) from above by

$$c\, 2^{(j+K)\sigma_p} \left(\int_{\mathbb{R}^n} \left| \left(\frac{\varphi(2^{-j}z)}{|2^{-j}z|^{s_1}} h(c2^{-j}z) \right)^\vee (y) |^p | (2^{s(j+l)} \tilde{\rho}_{j+l} \hat{f} \right)^\vee (x-y) \right|^p dy \right)^{1/p} \quad (10)$$

where c is independent of j. We put this estimate in (2.4.1/15), apply the L_p-quasi-norm, use (4) with $\bar{p} = p$, apply the l_q-quasi-norm and obtain (9). Recall that $(\Sigma b_k)^p \leqq \Sigma b_k^p$ for non-negative b_k's. As in the first step of the proof of Theorem 2.4.1 we arrive at the following counterpart of (2.4.1/23)

$$\left(\sum_{j=0}^{\infty} 2^{jsq} \left\| \sum_{l=-\infty}^{K} (\varphi_j \rho_{l+j} \hat{f})^\vee \mid L_p \right\|^q \right)^{1/q} \leqq c \| f \mid B^s_{pq} \|. \quad (11)$$

In precisely the same way the second step of the proof of Theorem 2.4.1 can be modified. We have to use (5) and (6), and obtain

$$\left(\sum_{j=0}^{\infty} 2^{jsq} \left\| \sum_{l=K+1}^{\infty} (\varphi_j \rho_{l+j} \hat{f})^\vee \mid L_p \right\|^q \right)^{1/q} \leqq c\, 2^{-K(s-s_0)} \| f \mid B^s_{pq} \| \quad (12)$$

as the counterpart of (2.4.1/29). The constant c is independent of K. However (11,12) prove that the quasi-norm in (7) can be estimated from above by $c\|f \mid B^s_{pq}\|$. One can carry over the technical remarks from Step 3 of the proof of Theorem 2.4.1 in a slightly modified way. This can also be done with the arguments from Step 4 of the proof of Theorem 2.4.1 which proves that the quasi-norm in (8) can be estimated from above by $c\|f \mid B^s_{pq}\|$. In order to prove the reverse inequalities we modify Step 5 of the proof of Theorem 2.4.1. We have (2.4.1/41) where now $0 < r < \bar{p}$ is sufficient. We use the usual (scalar) Hardy–Littlewood maximal inequality with respect to the $L_{p/r}$-norm and apply afterwards the $l_{q/r}$-quasi-norm. By the same arguments as in Step 5 of the proof of Theorem 2.4.1 we obtain that $\|f \mid B^s_{pq}\|$ can be estimated from above by the quasi-norm in (7). Finally we modify Step 6 of the proof of Theorem 2.4.1 which shows that $\|f \mid B^s_{pq}\|$ can also be estimated from above by the quasi-norm in (8).

Remark 1. We followed closely [Tri16: pp. 180–182]. In contrast to Theorem 2.4.1 we avoided the technique of maximal functions completely. The advantages of

(4–6) compared with (2.4.1/8–10) will be clear later on when we discuss equivalent quasi-norms where differences Δ_h^M are involved.

Corollary 1. Let p, q, s, s_0, and s_1 be the same numbers as in the Theorem. Let $\sigma > \sigma_p + \frac{n}{2}$, see (1). Let φ_0 and φ be two complex-valued C^∞ functions on $\mathbb{R}^n$ and $\mathbb{R}^n \setminus \{0\}$, respectively, which satisfy the Tauberian conditions (3), (3′), and

$$\left\| \frac{\varphi(x)h(x)}{|x|^{s_1}} \mid H_2^\sigma \right\| < \infty, \tag{13}$$

$$\sup_{m\in\mathbb{N}} 2^{-ms_0} \|\varphi(2^m \cdot) H(\cdot) \mid H_2^\sigma\| < \infty, \tag{14}$$

$$\sup_{m\in\mathbb{N}} 2^{-ms_0} \|\varphi_0(2^m \cdot) H(\cdot) \mid H_2^\sigma\| < \infty, \tag{15}$$

where h and H have the same meaning as at the beginning of 2.4.1, and H_2^σ are the fractional Sobolev spaces. Then (7) and (8) are equivalent quasi-norms in B^s_{pq}, where $\varphi_j(x) = \varphi(2^{-j}x)$ and $\varphi(tD)f$ have the above meaning.

Proof. The assertion is an immediate consequence of the Theorem and (2.4.1/53).

Corollary 2. Let $0 < p \leqq \infty$, $0 < q \leqq \infty$, and $s \in \mathbb{R}^n$. Let $a > n/p$. Let s_0 and s_1 be two real numbers with

$$s_0 + a < s < s_1 \quad \text{and} \quad s_1 > \sigma_p. \tag{16}$$

Let φ_0 and φ be two complex-valued C^∞ functions on $\mathbb{R}^n$ and $\mathbb{R}^n \setminus \{0\}$, respectively, which satisfy the Tauberian conditions (3), (3′), and either (2.4.1/8–10) with the above number a of (13–15) with $\sigma > a + n/2$. Then

$$\left(\sum_{j=0}^{\infty} 2^{jsq} \|(\varphi_j^* f)_a \mid L_p\|^q \right)^{1/q} \tag{17}$$

(modification if $q = \infty$), with the maximal functions from (2.4.1/55), is an equivalent quasi-norm in B^s_{pq}.

Proof. Let (2.4.1/8–10) with $a > n/p$ be fulfilled, then one can follow the proof of Theorem 2.4.1 with the modifications described in Corollary 2.4.1/2. The replacement of (2.4.1/8–10) by (13–15) with $\sigma > a + \frac{n}{2}$ is covered by (2.4.1/53).

Remark 2. The above Theorem and the Corollaries 1 and 2 are the B^s_{pq}-counterparts of Theorem 2.4.1 and the Corollaries 2.4.1/1 and 2.4.1/2. We remark that the conditions (4–6) are less restrictive than (2.4.1/8–10) with $a > n/p$. This is clear if $p \geqq 1$ and it follows from Hölder's inequality if $p < 1$. Furthermore the characterizations of F^s_{pq} described in 2.4.2 can be carried over to B^s_{pq} spaces.

2.5.2 Modifications of the main theorem

We describe the counterparts of the two theorems in 2.4.3 for the spaces B^s_{pq}. We use the same notations as in 2.4.1 and 2.5.1, in particular σ_p is given by (2.5.1/1), and $h(x) \in S$ and $H(x) \in S$ are the same functions as at the beginning of 2.4.1 with (2.4.1/2,3). Furthermore, $(\varphi(h\cdot)\hat{f})^\vee$ must be interpreted similarly as in 2.4.1 and 2.4.2.

Theorem 1. Let $0 < p \leqq \infty$, $0 < q \leqq \infty$, and $s \in \mathbb{R}$. Let s_0 and s_1 be two real numbers with (2.5.1/2). Let $\varphi_0(x)$ be the same function as in Theorem 2.5.1, including (2.5.1/3,6), where again $\bar{p} = \min(1, p)$. Let $\varphi(t)$ be a complex-valued C^∞ function on $\mathbb{R} \setminus \{0\}$ which satisfies

$$|\varphi(t)| > 0 \quad \text{if} \quad \frac{1}{8} < t < 8 \tag{1}$$

(Tauberian condition) and

$$\sup_{1 \leqq |\gamma| \leqq 2} \int_{\mathbb{R}^n} \left| \left(\frac{\varphi(\gamma z) h(z)}{|z|^{s_1}} \right)^\vee (y) \right|^{\bar{p}} dy < \infty, \tag{2}$$

$$\sup_{1 \leqq |\gamma| \leqq 2} \sup_{m \in \mathbb{N}} 2^{-m s_0 \bar{p}} \int_{\mathbb{R}^n} |(\varphi(2^m \gamma z) H(z))^\vee (y)|^{\bar{p}} dy < \infty. \tag{3}$$

Then

$$\|\varphi_0(D) f \mid L_p\| + \left(\int_{|h| \leqq 1} |h|^{-sq} \|(\varphi(h\cdot)\hat{f})^\vee \mid L_p\|^q \frac{dh}{|h|^n} \right)^{1/q} \tag{4}$$

and

$$\|\varphi_0(D) f \mid L_p\| + \left(\int_0^1 t^{-sq} \sup_{\frac{t}{2} \leqq |h| \leqq t} \|(\varphi(h\cdot)\hat{f})^\vee \mid L_p\|^q \frac{dt}{t} \right)^{1/q} \tag{5}$$

(modification if $q = \infty$) are equivalent quasi-norms in B^s_{pq}. If, in addition, $s > 0$ then

$$\|\varphi_0(D)f \mid L_p\| + \left(\int_0^1 t^{-sq} \sup_{0<|h|\leq t} \|(\varphi(h\cdot)\hat{f})^\vee \mid L_p\|^q \frac{dt}{t} \right)^{1/q} \tag{6}$$

(modification if $q = \infty$) is also an equivalent quasi-norm in B^s_{pq}.

Proof. Recall that the above assertion, at least as far as (4) and (5) are concerned, is the counterpart of Theorem 2.4.3/1, which we proved by modifying the proof of Theorem 2.4.1. Now one can do the same with the above assertions starting with Theorem 2.5.1 and using the arguments from the proof of Theorem 2.4.3/1. We omit the details. Then we obtain that (4) and (5) are equivalent quasi-norms in B^s_{pq}. If, in addition, $s > 0$, then we can use the arguments from the proof of Proposition 2.4.4/1 in order to show that (6) is also an equivalent quasi-norm.

Remark. As we mentioned the above theorem is the counterpart both of Theorem 2.4.3/1 and Proposition 2.4.4/1(i). There is no problem to prove a corresponding counterpart of Proposition 2.4.4/2, where the latter one was of great service for us in connection with Proposition 2.4.6. We fix this observation.

Proposition. Let $0 < p \leqq \infty$, $0 < q \leqq \infty$ and $s \in \mathbb{R}$. Let s_0 and s_1 be two real numbers with (2.5.1/2). Let $\varphi_0(x)$ be the same function as in Theorem 2.5.1, including (2.5.1/3,6), where again $\overline{p} = \min(1, p)$. Let $\varphi^1(x), \ldots, \varphi^N(x)$ be N complex-valued C^∞ functions on $\mathbb{R}^n \setminus \{0\}$ which satisfy the Tauberian condition (2.4.4/4), and also (2.5.1/4,5) with φ^k instead of φ. Then

$$\|\varphi_0(D)f \mid L_p\| + \sum_{k=1}^N \left(\int_0^1 t^{-sq} \|\varphi^k(tD)f \mid L_p\|^q \frac{dt}{t} \right)^{1/q} \tag{7}$$

and

$$\|\varphi_0(D)f \mid L_p\| + \sum_{k=1}^N \left(\int_0^1 t^{-sq} \sup_{\frac{t}{2}\leqq\tau\leqq t} \|\varphi^k(\tau D)f \mid L_p\|^q \frac{dt}{t} \right)^{1/q} \tag{8}$$

(modification if $q = \infty$) are equivalent quasi-norms in B^s_{pq}. If, in addition, $s > 0$, then

$$\|\varphi_0(D)f \mid L_p\| + \sum_{k=1}^{N} \left(\int_0^1 t^{-sq} \sup_{0<\tau\leqq t} \|\varphi^k(\tau D)f \mid L_p\|^q \frac{dt}{t} \right)^{1/q} \tag{9}$$

(modification if $q=\infty$) is also an equivalent quasi-norm in B^s_{pq}.

Our next task is to describe the counterparts of Theorem 2.4.3/2 and Proposition 2.4.4/1(ii). Let again l be a distribution on the real line satisfying (2.4.3/10,11). Let $l(h,f)(x)$ be given by (2.4.3/12) if it makes sense. Let σ_p be given by (2.5.1/1), and let $H(x) \in S$ be the same function as at the beginning of 2.4.1 with (2.4.1/2,3). Again the function $h(x)$ from (2.4.1/2,3) will not be needed now.

Theorem 2. Let $0<p\leqq\infty$, $0<q\leqq\infty$, and $s\in\mathbb{R}$. Let $s_0\in\mathbb{R}$ and $m\in\mathbb{N}$ with

$$s_0+\sigma_p<s<m, \qquad m>\sigma_p, \tag{10}$$

and (2.4.3/11) for some $\delta>0$. Let

$$\sup |h|^{-s_0} \int_{\mathbb{R}^n} |(\psi(hz)H(z))^\vee(y)|^{\bar{p}} dy < \infty \tag{11}$$

with $\bar{p}=\min(1,p)$, where the supremum is taken over all $h\in\mathbb{R}^n$ with $|h|\geqq 1$. Let $\varphi_0\in S$ with $\varphi_0(0)\neq 0$, then

$$\|\varphi_0(D)f \mid L_p\| + \left(\int_{|h|\leqq 1} |h|^{-sq} \|l(h,f) \mid L_p\|^q \frac{dh}{|h|^n} \right)^{1/q} \tag{12}$$

and

$$\|\varphi_0(D)f \mid L_p\| + \left(\int_0^1 t^{-sq} \sup_{\frac{t}{2}\leqq|h|\leqq t} \|l(h,f) \mid L_p\|^q \frac{dt}{t} \right)^{1/q} \tag{13}$$

(modification if $q=\infty$) are equivalent quasi-norms in B^s_{pq}. If, in addition, $s>0$, then

$$\|\varphi_0(D)f \mid L_p\| + \left(\int_0^1 t^{-sq} \sup_{0<|h|\leqq t} \|l(h,f) \mid L_p\|^q \frac{dt}{t} \right)^{1/q} \tag{14}$$

(modification if $q=\infty$) is an equivalent quasi-norm in B^s_{pq}.

Proof. The proof is the same as the proof of Theorem 2.4.3/2, at least as far as the equivalent quasi-norms in (12) and (13) are concerned. It is reduced to Theorem 1 by the same scheme as Theorem 2.4.3/2 is reduced to Theorem 2.4.3/1. As far as (14) is concerned we refer to the proof of Proposition 2.4.4/1.

2.5.3 Local means

We formulate the counterparts of Theorem 2.4.6 and Proposition 2.4.6. Let $k(t, f)$ be the same means as in 2.4.6 and let k_0, k^0, and k be the same functions as in (2.4.6/2–4). Let again $\sigma_p = n(\frac{1}{p} - 1)_+$, see (2.4.1/1).

Theorem. Let $0 < p \leqq \infty$, $0 < q \leqq \infty$, and $s \in \mathbb{R}$. Let $N \in \mathbb{N}$ with

$$2N > \max(s, \sigma_p), \tag{1}$$

then

$$\|k_0(1, f) \mid L_p\| + \left(\sum_{j=0}^{\infty} 2^{jsq} \|k(2^{-j}, f) \mid L_p\|^q\right)^{1/q}, \tag{2}$$

$$\|k_0(1, f) \mid L_p\| + \left(\int_0^1 t^{-sq} \|k(t, f) \mid L_p\|^q \frac{dt}{t}\right)^{1/q}, \tag{3}$$

and

$$\|k_0(1, f) \mid L_p\| + \left(\int_0^1 t^{-sq} \sup_{\frac{t}{2} \leqq \tau \leqq t} \|k(\tau, f) \mid L_p\|^q \frac{dt}{t}\right)^{1/q} \tag{4}$$

(modification if $q = \infty$) are equivalent quasi-norms in B^s_{pq}. If, in addition, $s > 0$, then

$$\|k_0(1, f) \mid L_p\| + \left(\int_0^1 t^{-sq} \sup_{0 < \tau \leqq t} \|k(\tau, f) \mid L_p\|^q \frac{dt}{t}\right)^{1/q} \tag{5}$$

(modification if $q = \infty$) is also an equivalent quasi-norm in B^s_{pq}.

Proof. We use precisely the same arguments as in the proof of Theorem 2.4.6, now with Theorems 2.5.1 and 2.5.2/1 as basis.

Proposition. Let k_0 and k^0 be the above functions. Let $0 < p \leqq \infty$, $0 < q \leqq \infty$, and $s \in \mathbb{R}$. Let $M \in \mathbb{N}$ with

$$M > \max(s, \sigma_p), \tag{6}$$

then the counterpart of (3), i.e.,

$$\|k_0(1,f) \mid L_p\| + \sum_{|\alpha|=M} \left(\int_0^1 t^{(M-s)q} \|k^0(t, D^\alpha f) \mid L_p\|^q \frac{dt}{t} \right)^{1/q}, \tag{7}$$

(modification if $q = \infty$) and the corresponding counterparts of (2,4) and, if $s > 0$, of (5) are equivalent quasi-norms in B^s_{pq}.

Proof. We use precisely the same arguments as in the proof of Proposition 2.4.6, now with Proposition 2.5.2 as basis.

Remark. If one compares 2.4 with 2.5 then it is quite clear that we prefer F^s_{pq} to B^s_{pq}. By rule of thumb, assertions for B^s_{pq} can be proved easier than corresponding assertions for F^s_{pq}. The theorems, corollaries, and propositions in 2.5 can serve as examples, but this is not so clear by the way we came. The above claim is better reflected by the proofs given in 2.3. More important, occasionally, conditions in theorems for B^s_{pq} spaces have final character compared with corresponding conditions in related theorems for F^s_{pq} spaces. A typical example is given by the two main theorems if one compares (2.5.1/4–6) with (2.4.1/8–10). Technically speaking, this difference is caused by the fact that we had been forced in 2.4.1 to use the technique of maximal functions in contrast to the proof of Theorem 2.5.1. As a consequence, in some applications we obtain immediately final results for the spaces B^s_{pq}, whereas for the spaces F^s_{pq} few minor additional considerations are needed. On the other hand, there is a crucial exception of the above rule of thumb: the localization principle from Theorem 2.4.7 is correct only for the spaces F^s_{pq}. This observation will be of decisive importance in connection with the introduction of spaces of type F^s_{pq} and B^s_{pq} on general structures, such as Riemannian manifolds and Lie groups. We return to this subject in detail in Chapter 7.

2.6 Concrete characterizations

2.6.1 Differences and derivatives: the spaces B^s_{pq}

In 2.6 we shall describe concrete characterizations both for B^s_{pq} and F^s_{pq} which follow by specialization from 2.4 and 2.5. First we discuss equivalent quasi-norms in B^s_{pq} connected with derivatives D^α and differences Δ^M_h. This is a model case and for some assertions we give two independent proofs which show clearly both the power and the shortcomings of the general characterizations from 2.4 and 2.5. We recall some notations. If $\alpha = (\alpha_1, \ldots, \alpha_n)$ is a multi-index, i.e., $\alpha_j \in \mathbb{N}_0$, then

$$D^\alpha = \frac{\partial^{|\alpha|}}{\partial x_1^{\alpha_1} \cdots \partial x_n^{\alpha_n}} \quad \text{with} \quad |\alpha| = \alpha_1 + \cdots + \alpha_n$$

stands for the derivatives. Let

$$\Delta_h^M f(x) = \sum_{j=0}^{M} (-1)^{M-j} \binom{M}{j} f(x + hj), \qquad M \in \mathbb{N},\ h \in \mathbb{R}^n,\ x \in \mathbb{R}^n, \tag{1}$$

where $\binom{M}{j}$ are the binomial coefficients, see also (1.2.2/4). If $h = (t, 0, \ldots, 0)$ with $t \in \mathbb{R}$ then we write $\Delta_h^M = \Delta_{t,1}^M$ for the differences with respect to the first direction of the coordinates. Similarly $\Delta_{t,k}^M$ with $k = 2, \ldots, n$. Recall $\sigma_p = n(\frac{1}{p} - 1)_+$.

Theorem. Let $0 < p \leqq \infty$, $0 < q \leqq \infty$, and $\sigma_p < s < M$ where $M \in \mathbb{N}$. Then

$$\|f \mid L_p\| + \left(\int_{|h| \leqq 1} |h|^{-sq} \|\Delta_h^M f \mid L_p\|^q \frac{dh}{|h|^n} \right)^{1/q}, \tag{2}$$

$$\|f \mid L_p\| + \left(\int_0^1 t^{-sq} \sup_{0 < |h| \leqq t} \|\Delta_h^M f \mid L_p\|^q \frac{dt}{t} \right)^{1/q}, \tag{3}$$

$$\|f \mid L_p\| + \sum_{k=1}^{n} \left(\int_0^1 t^{-sq} \|\Delta_{t,k}^M f \mid L_p\|^q \frac{dt}{t} \right)^{1/q}, \tag{4}$$

and

$$\|f \mid L_p\| + \sum_{k=1}^{n} \left(\int_0^1 t^{-sq} \sup_{0 < \tau \leqq t} \|\Delta_{\tau,k}^M f \mid L_p\|^q \frac{dt}{t} \right)^{1/q} \tag{5}$$

(modification if $q = \infty$) are equivalent quasi-norms in B^s_{pq}.

Proof. Step 1. We use Theorem 2.5.2/1 with $\varphi_0(x) = 1$ and $\varphi(t) = (e^{it} - 1)^M$. Then we have (2.5.1/3,6) with $s_0 = 0$, and an immaterial modification of (2.5.2/1). Recall $[(e^{i\gamma z} - 1)^M g]^\vee = \Delta_\gamma^M g^\vee$, then it follows that (2.5.2/3) with $s_0 = 0$ is fulfilled. Finally we fix s_1 with $s_1 > s$. If M is large then (2.5.2/2) is fulfilled, see also

(2.4.1/53). Now (2.5.2/4,6) coincides with (2,3). This result can be extended to any $M \in \mathbb{N}$ with $M > s_1$, what will be done in Remark 2 below.

Step 2. We give a second proof that the expressions in (2,3) are equivalent quasi-norms in B^s_{pq}. This time we use Theorem 2.5.2/2. Under the above restrictions for s the function $\varphi_0 \in S$, $\varphi_0(0) \neq 0$, in this theorem can be replaced by the above choice $\varphi_0(x) = 1$. Let $\psi(\lambda) = (e^{i\lambda} - 1)^M$, then we have (2.4.3/11) and (2.5.2/10) with $m = M$. Furthermore (2.5.2/11) with $s_0 = 0$ is fulfilled, see Step 1. Finally we have an immaterial modification of (2.4.3/10) with $l = F_1^{-1}\psi$, and $l(h, f)(x) = c\Delta_h^M f(x)$, where c is an unimportant number. Then (2.5.2/12,14) coincides with (2,3).

Step 3. In order to prove that (4) and (5) are equivalent quasi-norms in B^s_{pq} we choose $\varphi_0(x) = 1$ and $\varphi^k(x) = (e^{ix_k} - 1)^M$ with $x = (x_1, \ldots, x_n)$ and $k = 1, \ldots, n$. We apply Proposition 2.5.2 where we assume that M is large. Then it follows by the same arguments as in Step 1 that the expressions in (2.5.2/7,9), and hence in (4) and (5), are equivalent quasi-norms in B^s_{pq}. The fact that even $M > s_1$ is sufficient follows from Remark 2 below.

Remark 1. We gave two proofs that the expressions in (2) and (3) are equivalent quasi-norms in B^s_{pq} which show advantages and disadvantages of the general characterizations from 2.5. In Step 1 we used Theorem 2.5.2/1 and obtained the desired result for large values of M. In Step 2 we used the rather sophisticated Theorem 2.5.2/2 which gave a final result. This discussion shows that the general characterizations are strong enough to cover a lot of concrete characterizations, but that one has to add some minor additional arguments from case to case.

Remark 2. By the Steps 1 and 3 we know that the expressions in (2–5) are equivalent quasi-norms in B^s_{pq} if M is large. We have to extend this assertion to all $M \in \mathbb{N}$ with $M > s$. In Step 2 we discussed one possibility. We describe a second possibility, where we restrict ourselves to (2). First we recall the identity

$$(\Delta_h^M f)(x) = 2^{-M}(\Delta_{2h}^M f)(x) + \Delta_h^{M+1}(\Sigma\, a_l f(x + lh)), \tag{6}$$

where Σ stands for a finite sum and a_l are real numbers, which is well known and which may be found, e.g., in [Trieß: (2.5.9/45)]. We denote the expression in (2) by $\|f \mid B^s_{pq}\|_M$. Then we obtain

$$\|f \mid B^s_{pq}\|_M \leqq c\|f \mid B^s_{pq}\|_{M+1} + 2^{-M}\left(\int_{|h|\leqq 1} |h|^{-sq}\|\Delta_{2h}^M f \mid L_p\|^q \frac{dh}{|h|^n}\right)^{1/q} \tag{7}$$

$$\leqq c\|f \mid B^s_{pq}\|_{M+1} + c'\|f \mid L_p\| + 2^{-M+s}\left(\int_{|h|\leqq 1} |h|^{-sq}\|\Delta_h^M f \mid L_p\|^q \frac{dh}{|h|^n}\right)^{1/q},$$

see also Remark 4 below. Because $M > s$ we obtain

$$\|f \mid B^s_{pq}\|_M \leqq c\|f \mid B^s_{pq}\|_{M+1}. \tag{8}$$

The reverse inequality is obvious. Hence, $\|f \mid B^s_{pq}\|_M$ and $\|f \mid B^s_{pq}\|_{M+1}$ are equivalent. Starting with large values of M we obtain that the expression in (2) is an equivalent quasi-norm for all $M \in \mathbb{N}$ with $M > s$. By the same argument one proves a corresponding assertion for the expressions in (3–5).

We prefer formulations in terms of equivalent quasi-norms. In 2.4.2 we discussed the problem under which conditions quasi-norms characterize the spaces F^s_{pq}. One can do the same for the spaces B^s_{pq} on the abstract level of 2.4.2. However we restrict ourselves at the moment to an example connected with the above concrete situation. Let again $\|f \mid B^s_{pq}\|_M$ be the quasi-norm in (2).

Corollary 1. Let $1 \leqq p \leqq \infty$, $0 < q \leqq \infty$, and $0 < s < M$ where $M \in \mathbb{N}$, then

$$B^s_{pq} = \{f \in S' \colon \|f \mid B^s_{pq}\|_M < \infty\}. \tag{9}$$

Proof. We return to the abstract situation where (2.4.2/23) can be taken as a starting point. In the case of B^s_{pq} one can apply first the L_p-norm and then the l_q-quasi-norm which yields the desired estimate under the respective hypothesis for φ. Immaterial modifications show that the arguments can also be applied to the situation described in Theorem 2.4.3/1 or, as far as the B^s_{pq} spaces are concerned, in Theorem 2.5.2/1. But (9) is simply a special case of these estimates.

Remark 3. If $p < 1$ then the above abstract argument does not work. However one can use rather special tools from approximation theory in order to prove the following assertion: Let $0 < p \leqq \infty$, $0 < q \leqq \infty$, and $\sigma_p < s < M$ where $M \in \mathbb{N}$, then

$$B^s_{pq} = \{f \in L_p \colon \|f \mid B^s_{pq}\|_M < \infty\}. \tag{10}$$

See [Triß: 2.5.12, in particular, Remark 2.5.12/3] where we gave details and references to related papers by E.A. Storoženko, P. Oswald and coworkers.

Remark 4. We add a technical observation which we used occasionally, for example in (7). But it seems to be useful to formulate it explicitly. Let $\nu > 0$ then one can replace the integration over $|h| \leqq 1$ in (2) by $|h| \leqq \nu$ or by $\mathbb{R}^n$. This is an immediate consequence of $\|\Delta_h^M f \mid L_p\| \leqq c\|f \mid L_p\|$. Similarly one can replace the integration over the interval $(0,1)$ in (3–5) by an integration over the interval $(0,\nu)$ or over $\mathbb{R}$.

Corollary 2. Let $0 < p \leqq \infty$, $0 < q \leqq \infty$, and $\sigma_p < s - L < M$, where $L \in \mathbb{N}$ and $M \in \mathbb{N}$, then

$$\|f \mid L_p\| + \sum_{|\alpha|=L} \left(\int_{|h|\leqq 1} |h|^{-(s-L)q} \|\Delta_h^M D^\alpha f \mid L_p\|^q \frac{dh}{|h|^n} \right)^{1/q} \tag{11}$$

and

$$\|f \mid L_p\| + \sum_{k=1}^{n} \left(\int_0^1 t^{-(s-L)q} \left\| \Delta_{t,k}^M \frac{\partial^L f}{\partial x_k^L} \mid L_p \right\|^q \frac{dt}{t} \right)^{1/q} \tag{12}$$

(modification if $q = \infty$) are equivalent quasi-norms in B_{pq}^s.

Proof. Step 1. We apply Proposition 2.5.2 with $\varphi_0(x) = 1$ and $\varphi^k(x) = x_k^L(e^{ix_k} - 1)^M$ where $k = 1, \ldots, n$. Let $s_0 = L$ and $s_1 = L + M$, then (2.5.1/2,3,6) are fulfilled. Furthermore we have an immaterial modification of (2.4.4/4); and (2.5.1/5) with φ^k instead of φ holds. As far as (2.5.1/4) is concerned we use the replacement described in connection with Theorems 2.4.3/2 and 2.5.2/2: the above functions φ^k take over the role of ψ in these theorems. Now (12) coincides with (2.5.2/7) and after these modifications it follows that (12) is an equivalent quasi-norm in B_{pq}^s.

Step 2. Now we complement the above functions $\varphi^k(x)$ by all functions $x_1^{L_1} \cdots x_n^{L_n}(e^{ix_k} - 1)^M$ with $L_1 + \cdots + L_n = L$. By the same arguments as above it follows that (12) with

$$\sum_{|\alpha|=L} \|\Delta_{t,k}^M D^\alpha f \mid L_p\| \quad \text{instead of} \quad \sum_{k=1}^{n} \left\| \Delta_{t,k}^M \frac{\partial^L f}{\partial x_k^L} \mid L_p \right\| \tag{13}$$

is also an equivalent quasi-norm in B_{pq}^s. We apply the above theorem to B_{pq}^{s-L}. Then it folows that the just indicated modifications of (12) and (11) are equivalent quasi-norms.

2.6.2 Differences and derivatives: the spaces F_{pq}^s

We formulate the F_{pq}^s-counterpart of Theorem 2.6.1, where the differences Δ_h^M have the same meaning as in (2.6.1/1).

Theorem. Let $0 < p < \infty$, $0 < q \leqq \infty$, and $\frac{n}{\min(p,q)} < s < M$ where $M \in \mathbb{N}$. Then

$$\|f \mid L_p\| + \left\| \left(\int_{|h|\leqq 1} |h|^{-sq} |\Delta_h^M f(\cdot)|^q \frac{dh}{|h|^n} \right)^{1/q} \mid L_p \right\|, \tag{1}$$

$$\|f \mid L_p\| + \left\| \left(\int_0^1 t^{-sq} \sup_{0<|h|\leqq t} |\Delta_h^M f(\cdot)|^q \frac{dt}{t} \right)^{1/q} \mid L_p \right\|, \tag{2}$$

$$\|f \mid L_p\| + \sum_{k=1}^n \left\| \left(\int_0^1 t^{-sq} |\Delta_{t,k}^M f(\cdot)|^q \frac{dt}{t} \right)^{1/q} \mid L_p \right\|, \tag{3}$$

and

$$\|f \mid L_p\| + \sum_{k=1}^n \left\| \left(\int_0^1 t^{-sq} \sup_{0<\tau\leqq t} |\Delta_{\tau,k}^M f(\cdot)|^q \frac{dt}{t} \right)^{1/q} \mid L_p \right\| \tag{4}$$

(modification if $q = \infty$) are equivalent quasi-norms in F^s_{pq}.

Proof. We apply Theorem 2.4.3/1. We use the same functions $\varphi_0(x) = 1$ and $\varphi(t) = (e^{it} - 1)^M$ as in the proof of Theorem 2.6.1. Let, in addition,

$$\sigma_{pq} + \frac{n}{\min(p,q)} < s. \tag{5}$$

We choose $s_0 = a > \frac{n}{\min(p,q)}$ with $s_0 + \sigma_{pq} < s$ and $s_1 = M$. Then (2.4.1/5,6,10) are fulfilled. Furthermore an immaterial modification of (2.4.3/1) holds. (2.4.3/3) is satisfied where we need now $s_0 = a$. If M is large then we have also (2.4.3/2), see (2.4.1/53). However one can modify this argument by using the modification described in the proof of Theorem 2.4.3/2. Then $M > s$ is sufficient, see also Step 2 of the proof of Theorem 2.6.1. Now the quasi-norms in (2.4.3/4) and (2.4.4/1) coincide with the quasi-norms in (1,2). As far as the quasi-norms in (3,4) are concerned, we argue in the same way as in Step 3 of the proof of Theorem 2.6.1 with Proposition 2.4.4/2 as basis. Then we obtain that the expressions in (3,4) are equivalent quasi-norms in F^s_{pq} at least if M is large. Again by the modification described in the proof of Theorem 2.4.3/2 it follows that $M > s$ is sufficient. This completes the proof under the additional assumption (5). If we have only $s > \frac{n}{\min(p,q)}$ and $a = s_0 < s$, then it follows by Remark 2.4.3/1 and the above modifications that the quasi-norms in (1–4) can be estimated from above by $c\|f \mid F^s_{pq}\|$. We postpone the converse inequality to Remark 2.6.3/3.

Remark 1. We follow closely in 2.6.1 and also in this subsection [Tri16]. However the results themselves are known and proved by different reasoning in [Triß: 2.5.10, 2.5.12] based on rather specific arguments directly connected with differences Δ_h^M. Our aim in 2.6 is different. We try to convince the reader that the known concrete characterizations of B^s_{pq} and F^s_{pq} grow out in a natural way from general abstract characterizations. However as the proofs in 2.6.1 and also the above proof show some minor specific complements will be necessary from case to case.

We formulate the counterpart of Corollary 2.6.1/1 and denote the quasi-norm in (1) by $\|f \mid F^s_{pq}\|_M$.

Corollary 1. Let $1 < p < \infty$, $1 < q \leqq \infty$, and $\frac{n}{\min(p,q)} < s < M$ where $M \in \mathbb{N}$ then

$$F^s_{pq} = \{f \in S' \colon \|f \mid F^s_{pq}\|_M < \infty\}. \tag{6}$$

Proof. The assertion follows from Theorem 2.4.2(i) or, even better, from Step 2 of the proof of that theorem.

Remark 2. The restrictions for s in Theorem 2.6.1 and Corollary 2.6.1/1 are natural. If $q \geqq p$ then we have also natural restrictions for s in the above theorem and the above corollary. For example, equivalent quasi-norms of type (2) cannot be expected if $s < n/p$, because in that case F^s_{pq} contains essentially unbounded functions for which the corresponding expression is infinite.

Remark 3. We prove the counterpart of Remark 2.6.1/4. Let $\nu > 0$ then one can replace the integration over $|h| \leqq 1$ in (1) by $|h| \leqq \nu$ or by $\mathbb{R}^n$. For this purpose we generalize (2.4.7/7) as follows. Let $\frac{n}{p} < a < s$, then there exists a positive number c such that

$$\sup_{t>1} t^{-a} \left\| \sup_{|x-y| \leqq t} |f(y)| \mid L_p \right\| \leqq c \|f \mid F^s_{pq}\|. \tag{7}$$

We fix $t > 1$ and follow the proof given in Remark 2.4.7/1, where (2.4.7/25) can be strengthened by

$$t^{-a} \sup_{|x-y| \leqq t} |\varphi_k(D) f(y)| \leqq c\, 2^{ka} \sup_{z \in \mathbb{R}^n} \frac{|\varphi_k(D) f(x-z)|}{1 + |2^k z|^a}, \tag{8}$$

where c is independent of t. Let $|h| = t > 1$, then

$$|h|^{-s}|\Delta_h^M f(x)| \leqq c\, t^{-s} \sup_{|x-y|\leqq Mt} |f(y)|. \tag{9}$$

This inequality and (7) show that (1), with $\mathbb{R}^n$ instead of $|h| \leqq 1$, can be estimated from above by $c\|f \mid F^s_{pq}\|$. Conversely, if $\nu > 0$ is a small number then (1) with $|h| \leqq \nu$ instead of $|h| \leqq 1$ is an equivalent quasi-norm in F^s_{pq}. This claim is covered by the proof of the theorem. Hence we obtain the desired assertion as far as (1) is concerned. Similarly one proves corresponding assertions for (2–4).

Corollary 2. Let $0 < p < \infty$, $0 < q \leqq \infty$, and $\frac{n}{\min(p,q)} < s - L < M$ where $L \in \mathbb{N}$ and $M \in \mathbb{N}$, then

$$\|f \mid L_p\| + \sum_{|\alpha|=L} \left\| \left(\int_{|h|\leqq 1} |h|^{-(s-L)q} |\Delta_h^M D^\alpha f(\cdot)|^q \frac{dh}{|h|^n} \right)^{1/q} \mid L_p \right\| \tag{10}$$

and

$$\|f \mid L_p\| + \sum_{k=1}^n \left\| \left(\int_0^1 t^{-(s-L)q} |\Delta_{t,k}^M \frac{\partial^L f}{\partial x_k^L}(\cdot)|^q \frac{dt}{t} \right)^{1/q} \mid L_p \right\| \tag{11}$$

(modification if $q = \infty$) are equivalent quasi-norms in F^s_{pq}.

Proof. We apply Proposition 2.4.4/2 and use the same arguments as in the proof of Corollary 2.6.1/2 and in the proof of the above theorem, see also Remark 2.6.3/3.

2.6.3 Weighted means of differences

The equivalent quasi-norms for B^s_{pq} and F^s_{pq} in the Theorems 2.6.1 and 2.6.2 are of interest for their own sake and from a historical point of view. This is the main reason why we included these results in this book. However for our later purposes we need some substantial modifications which we discuss now and which may also be considered as a modification of the local means introduced in 2.4.6, see also 2.5.3.

Let $K \in S$ be non-negative and rotation-invariant (i.e., $K(y)$ depends only on $|y|$) with a compact support. We assume that K does not vanish identically. Let $M \in \mathbb{N}$, then

$$K_M(t,f)(x) = \int_{\mathbb{R}^n} K(h) \Delta_{th}^M f(x)\, dh, \qquad x \in \mathbb{R}^n,\ t > 0, \tag{1}$$

may be interpreted as weighted means of differences. Let σ_p and σ_{pq} be the same numbers as above, see e.g., (2.4.1/1).

Theorem. (i) Let $0 < p < \infty$, $0 < q \leqq \infty$, and $\sigma_{pq} < s < M$ with $M \in \mathbb{N}$. Let $0 < \nu \leqq \infty$, then

$$\|f \mid L_p\| + \left\| \left(\int_0^\nu t^{-sq} |K_M(t,f)(\cdot)|^q \frac{dt}{t} \right)^{1/q} \mid L_p \right\| \tag{2}$$

(modification if $q = \infty$) is an equivalent quasi-norm in F^s_{pq}.

(ii) Let $0 < p \leqq \infty$, $0 < q \leqq \infty$, and $\sigma_p < s < M$ with $M \in \mathbb{N}$. Let $0 < \nu \leqq \infty$, then

$$\|f \mid L_p\| + \left(\int_0^\nu t^{-sq} \|K_M(t,f) \mid L_p\|^q \frac{dt}{t} \right)^{1/q} \tag{3}$$

(modification if $q = \infty$) is an equivalent quasi-norm in B^s_{pq}.

Proof. Step 1. We prove (i) with $\nu > 0$ small. We apply Theorem 2.4.1 with $s_0 = 0$, $s_1 = M$ if M is even and $s_1 = M + 1$ if M is odd, $\varphi_0(x) = 1$, and

$$\varphi(x) = \int_{\mathbb{R}^n} K(h)(e^{i\nu xh} - 1)^M dh, \qquad x \in \mathbb{R}^n, \tag{4}$$

where xh stands for the scalar product of $x \in \mathbb{R}^n$ and $h \in \mathbb{R}^n$. Then (2.4.1/5,6,10) are satisfied. If M is even then we have

$$\varphi(x) = \int_{\mathbb{R}^n} K(h)(i\nu xh)^M (1 + o(1)) dh \neq 0, \qquad 0 < |x| \leqq 2, \tag{5}$$

if M is odd then we have (5) with $M + 1$. In both cases we used that $K(h)$ is rotation-invariant and $\int_{\mathbb{R}^n} K(h)dh = \hat{K}(0) > 0$. Hence, (2.4.1/7) is fulfilled. We check (2.4.1/8). First we remark that $\varphi(x)$ is an analytic rotation-invariant function. If M is even then (5) shows that $\frac{\varphi(x)}{|x|^M}$ is also an analytic function. If M is odd then $\frac{\varphi(x)}{|x|^{M+1}}$ is an analytic function. In both cases we have (2.4.1/8) with the above choice of s_1. Finally we have to check (2.4.1/9) with $s_0 = 0$. First we remark that

$$\varphi(x) = (-1)^M \int_{\mathbb{R}^n} K(h)dh + \sum_{k=1}^{M} a_k \hat{K}(\nu kx) \tag{6}$$

holds, where a_k are appropriate numbers. We put (6) in (2.4.1/9) with $s_0 = 0$. The typical term we have to estimate reads as follows,

$$\int_{\mathbb{R}^n} (1+|y|)^a |(\hat{K}(2^m \cdot)H(\cdot))^\vee(y)| dy. \tag{7}$$

We use (2.4.1/53) which shows that this term can be estimated from above by a constant, uniformly with respect to $m \in \mathbb{N}$. Finally we remark $\varphi(tD)f = K_M(\nu t, f)$. Hence the quasi-norms in (2.4.1/12) and (2) coincide. The proof of (i) is complete provided that $\nu > 0$ is small.

Step 2. We extend (i) to arbitrary values of $\nu > 0$ and to ∞. By Remark 2.3.3 we have

$$\|f(\lambda \cdot) \mid F^s_{pq}\| \leqq c\lambda^{s-n/p} \|f \mid F^s_{pq}\|, \qquad \lambda \geqq 1, \tag{8}$$

where c is independent of λ. Futhermore we observe

$$K_M(t, f(\lambda \cdot))(x) = K_M(\lambda t, f)(\lambda x). \tag{9}$$

The second term of (2) with $\nu > 0$ small and with $f(\lambda x)$ instead of $f(x)$ can be estimated from above by $\|f(\lambda \cdot) \mid F^s_{pq}\|$ where $\lambda \geqq 1$. Then (9) and (8) prove

$$\left\| \left(\int_0^{\lambda \nu} t^{-sq} |K_M(t,f)(\cdot)|^q \frac{dt}{t} \right)^{1/q} \mid L_p \right\| \leqq c \|f \mid F^s_{pq}\| \tag{10}$$

where c is independent of $\lambda \geqq 1$. Now $\lambda \to \infty$ yields the desired result.

Step 3. The proof of (ii) is the same. Now we have to use Theorem 2.5.1 and again Remark 2.3.3. The proof is complete.

Remark 1. Part (i) of the theorem will be of some use for us in connection with spaces of type F^s_{pq} on Riemannian manifolds. For this purpose it will be useful to reformulate (1) somewhat. Let l and $l(h,f)$ be given by (2.4.3/10–12) and let $K(y)$ be the above function. Then we introduce the means

$$K^l(t,f)(x) = \int_{\mathbb{R}^n} K(h) l(th, f) dh \tag{11}$$

$$= \varphi^l(tD)f(x) = (\varphi^l(t\cdot)\hat{f})^\vee(x), \qquad t > 0, \quad x \in \mathbb{R}^n,$$

with

$$\varphi^l(x) = \int_{\mathbb{R}^n} K(h)\psi(hx) dh, \tag{12}$$

see (2.4.3/17). This is the generalization of $\varphi(tD)f = K_M(\nu t, f)$, where K_M and φ are given by (1) and (4), respectively. One can try to replace $l(h,f)$ in Theorem 2.4.3/2 by the means (11) and to generalize the above theorem. However for our purpose the above theorem and the just given reformulation will be sufficient.

Remark 2. The proof shows that it is not necessary to assume $K \in S$. However we shall not discuss the differentiability properties for K which are really needed. In [Triß: 2.5.11] we proved via rather specific means that even the characteristic function of the unit ball is an admissible choice of K, at least for F^s_{pq}, but this can be extended to B^s_{pq}. In other words, one can replace $K_M(t,f)$ in (2) and (3) by

$$\int_{|h|\leqq 1} \Delta^M_{th} f(x)dh.$$

See also 3.5.3 for further characterizations via differences.

Remark 3. We return to Theorem 2.6.2 and fill the gap at the end of the proof which simultaneously finishes the proof of Corollary 2.6.2/2. Recall that we have to prove that $\|f \mid F^s_{pq}\|$ can be estimated from above by each of the quasi-norms in (2.6.2/1–4). We restrict ourselves to (2.6.2/1,2) with $\int_{|h|\leqq\nu}$ and $\int_0^\nu$ instead of $\int_{|h|\leqq 1}$ and $\int_0^1$, respectively, where $0<\nu\leqq\infty$. The proof for the remaining quasi-norms is the same. Let $0<p<\infty$, $0<q\leqq\infty$, and $\frac{n}{\min(p,q)}<s<M$ with $M\in\mathbb{N}$. We use the equivalent quasi-norm (2). We have

$$|K_M(t,f)(x)| \leqq c\sum_{l=0}^{\infty} 2^{-rl} \sup_{0<|h|\leqq t2^l} |\Delta^M_h f(x)| \tag{13}$$

where $r>0$ is at our disposal. Then (2) with $r=\infty$ yields

$$\|f \mid F^s_{pq}\| \leqq c\|f \mid L_p\| + c\sum_{l=0}^{\infty} 2^{-r'l} \left\| \left(\int_0^\infty t^{-sq} \sup_{0<|h|\leqq t2^l} |\Delta^M_h f(x)|^q \frac{dt}{t}\right)^{1/q} \mid L_p \right\|, \tag{14}$$

where $r'>0$ is at our disposal. We substitute $\tau = t2^l$ in the respective terms on the right-hand side of (14). Then we have an additional factor 2^{ls}. We choose $r'>s$. It follows that $\|f \mid F^s_{pq}\|$ can be estimated from above by the quasi-norm in (2.6.2/2) with $\int_0^\infty$ instead of $\int_0^1$, and the proof is complete as far as this special case is concerned. Next we extend the proof to the quasi-norm (2.6.2/2) with $\int_0^\nu$ instead of $\int_0^1$ where $0<\nu<\infty$. By Remark 2.6.2/3 we have

$$\left\| \left(\int_\nu^\infty t^{-sq} \sup_{0<|h|\leqq t} |\Delta^M_h f(\cdot)|^q \frac{dt}{t}\right)^{1/q} \mid L_p \right\| \leqq c\|f \mid F^{\bar{s}}_{pq}\| \tag{15}$$

with $n/p<\bar{s}<s$. However the right-hand side can be estimated from above by $\varepsilon\|f \mid F^s_{pq}\| + c_\varepsilon\|f \mid L_p\|$ where $\varepsilon>0$ is at our disposal, see Remark 4 below. Now it follows that $\|f \mid F^s_{pq}\|$ can be estimated from above by the quasi-norm in (2.6.2/2) with $\int_0^\nu$ instead of $\int_0^1$. It remains to show that $\|f \mid F^s_{pq}\|$ can be estimated from above by the quasi-norm in (2.6.2/1) with $\int_{|h|\leqq \nu}$ instead of $\int_{|h|\leqq 1}$. We may assume $\nu = \infty$ because the cases with $\nu < \infty$ can be treated afterwards on the basis of (15). We use (2) with $\nu = \infty$. Let $1 \leqq q \leqq \infty$, then we have by Hölder's inequality (modification if $q = \infty$)

$$|K_M(t,f)(x)|^q \leqq c \int_{\mathbb{R}^n} K(h) |\Delta^M_{th} f(x)|^q dh = ct^{-n} \int_{\mathbb{R}^n} K\left(\frac{h}{t}\right) |\Delta^M_h f(x)|^q dh \quad (16)$$

and

$$\int_0^\infty t^{-sq} |K_M(t,f)(x)|^q \frac{dt}{t} \leqq c \int_{\mathbb{R}^n} |h|^{-sq} |\Delta^M_h f(x)|^q \frac{dh}{|h|^n}. \quad (17)$$

We put (17) in (2) and obtain the desired estimate in the case $1 \leqq q \leqq \infty$. Let $0 < q < 1$. Then we modify (13) by

$$|K_M(t,f)(x)| \leqq c \sum_{l=0}^{\infty} 2^{-rl} \sup_{0<|h|\leqq t2^l} |\Delta^M_h f(x)|^{1-q} \int_{\mathbb{R}^n} \overline{K}(h) |\Delta^M_{th} f(x)|^q dh, \quad (18)$$

where $\overline{K}(h)$ is a non-negative function on $\mathbb{R}^n$ with $\sup_{h\in\mathbb{R}^n} |h|^N \overline{K}(h) < \infty$ for any $N \in \mathbb{N}$. We take the qth power in (18) term by term, multiply afterwards with t^{-sq-1}, and integrate over $0 < t < \infty$. Then we apply Hölder's inequality, based on $q + (1-q) = 1$, and obtain

$$\int_0^\infty t^{-sq} |K_M(t,f)(x)|^q \frac{dt}{t} \leqq c \sum_{l=0}^{\infty} 2^{-rlq} \left(\int_0^\infty t^{-sq} \sup_{0<|h|\leqq t2^l} |\Delta^M_h f(x)|^q \frac{dt}{t} \right)^{1-q} \quad (19)$$
$$\times \left(\int_0^\infty t^{-sq} \int_{\mathbb{R}^n} \overline{K}(h) |\Delta^M_{th} f(x)|^q dh \frac{dt}{t} \right)^q.$$

The second factors are independent of l and they can be estimated in the same way as in (16) and (17). The first factors can be treated as above, i.e., we substitute $\tau = 2^l t$ and choose $r > s$. Now we apply the L_p-quasi-norm and we use again Hölder's inequality with respect to $q + (1-q) = 1$. By (2) with $\int_0^\infty$ instead of $\int_0^\nu$ and by (2.6.2/2) with $\int_0^\infty$ instead of $\int_0^1$ we arrive at

$$\|f \mid F^s_{pq}\| \leqq c \|f \mid F^s_{pq}\|^{1-q} \Big(\|f \mid L_p\| \quad (20)$$
$$+ \left\| \left(\int_{\mathbb{R}^n} |h|^{-sq} |\Delta^m_h f(\cdot)|^q \frac{dh}{|h|^n} \right)^{1/q} \mid L_p \right\| \Big)^q$$

which yields finally the desired estimate in the case $0 < q < 1$.

Remark 4. Let $\sigma_p < \overline{s} < s$ and let $\varepsilon > 0$, then there exists a number c_ε such that

$$\|f \mid B^{\bar{s}}_{p1}\| \leqq c_\varepsilon \|f \mid L_p\| + \varepsilon \|f \mid B^s_{p1}\|. \tag{21}$$

This estimate covers what we needed after (15), see the elementary embeddings mentioned in 2.3.2. We use (2.3.1/7) where φ and φ_k have the same meaning as in 2.3.1. Let $K \in \mathbb{N}$ and $k \in \mathbb{N}_0$ with $k \leqq N-1$, then a refined version of (2.2.3/2), see [Trifi: (1.5.2/13)], yields

$$\begin{aligned} \|(\varphi_k \hat{f})^\vee \mid L_p\| = \|(\varphi_k \varphi(2^{-K}\cdot)\hat{f})^\vee \mid L_p\| &\leqq c\, 2^{(K-k)\bar{s}} \|(\varphi(2^{-K}\cdot)\hat{f})^\vee \mid L_p\| \\ &\leqq c\, 2^{(K-k)\bar{s}} \|f \mid L_p\| + c'\, 2^{(K-k)\bar{s}} \|[(1-\varphi(2^{-K}\cdot))\hat{f}]^\vee \mid L_p\|, \end{aligned} \tag{22}$$

where c and c' are independent of K and k. It follows

$$2^{k\bar{s}} \|(\varphi_k \hat{f})^\vee \mid L_p\| \leqq c\, 2^{K\bar{s}} \|f \mid L_p\| + c\, 2^{-K\delta} \|f \mid B^s_{p1}\| \tag{23}$$

for some $\delta > 0$, where c is independent of K. We sum over $k = 0, \ldots, K-1$, choose K large and estimate the remaining terms with $k \geqq K$ in an obvious way. Then we obtain (21).

2.6.4 Harmonic and thermic extensions

In 1.8.1 we discussed harmonic and thermic extensions from a historical point of view, and described in 1.8.3 more recent results, including relevant references. Now we prove that Theorem 1.8.3 is more or less a special case of the above general characterizations. In order to be self-contained we recall some basic notations. Let $x \in \mathbb{R}^n$ and $t > 0$, then

$$W(t)f(x) = (4\pi t)^{-n/2} \int_{\mathbb{R}^n} e^{-\frac{|x-y|^2}{4t}} f(y)dy \tag{1}$$

is the Gauss–Weierstrass semi-group and

$$P(t)f(x) = c_n \int_{\mathbb{R}^n} \frac{t}{(|x-y|^2 + t^2)^{(n+1)/2}} f(y)dy \tag{2}$$

is the Cauchy–Poisson semi-group with $c_n \|(1+|x|^2)^{-(n+1)/2} \mid L_1\| = 1$, complemented by $W(0) = P(0) = id$. One can rewrite (1) and (2) by

$$u(x,t) = W(t)f(x) = (e^{-t|\xi|^2}\hat{f})^\vee, \qquad x \in \mathbb{R}^n, \quad t \geqq 0, \tag{3}$$

and

$$v(x,t) = P(t)f(x) = (e^{-t|\xi|}\hat{f})^\vee, \qquad x \in \mathbb{R}^n, \quad t \geqq 0. \tag{4}$$

We mention that $u(x,t)$ is a solution of the heat equation in $\mathbb{R}^{n+1}_+$ whereas $v(x,t)$ is a harmonic function in $\mathbb{R}^{n+1}_+$. We have $u(x,0) = v(x,0) = f(x)$ what explains to call $u(x,t)$ a thermic and $v(x,t)$ a harmonic extension of f. See 1.6.5 and, in particular, 1.8.1 for further details and references. The classical background, including proofs of (3), (4), and further references may also be found in [Triα: 2.5.2, 2.5.3]. Recall $\sigma_p = n(\frac{1}{p} - 1)_+$.

Theorem. Let $s \in \mathbb{R}$, $0 < q \leqq \infty$, and let φ_0 be a C^∞ function with a compact support and $\varphi_0(0) \neq 0$.

(i) Let $0 < p \leqq \infty$. Let $m \in \mathbb{N}_0$ and $k \in \mathbb{N}_0$ with

$$m > \frac{s}{2} \quad \text{and} \quad k > \sigma_p + \max(s, \sigma_p), \tag{5}$$

then

$$\|f \mid B^s_{pq}\|_W = \|\varphi_0(D)f \mid L_p\| + \left(\int_0^1 t^{(m-\frac{s}{2})q} \left\| \frac{\partial^m W(t)f}{\partial t^m} \mid L_p \right\|^q \frac{dt}{t}\right)^{1/q} \tag{6}$$

and

$$\|f \mid B^s_{pq}\|_P = \|\varphi_0(D)f \mid L_p\| + \left(\int_0^1 t^{(k-s)q} \left\| \frac{\partial^k P(t)f}{\partial t^k} \mid L_p \right\|^q \frac{dt}{t}\right)^{1/q} \tag{7}$$

(modification if $q = \infty$) are equivalent quasi-norms in B^s_{pq}. Furthermore

$$B^s_{pq} = \{f \in S' \colon \|f \mid B^s_{pq}\|_W < \infty\} = \{f \in S' \colon \|f \mid B^s_{pq}\|_P < \infty\}. \tag{8}$$

If $s > \sigma_p$ then $\|\varphi_0(D)f \mid L_p\|$ in (6–8) can be replaced by $\|f \mid L_p\|$.

(ii) Let $0 < p < \infty$. Let $m \in \mathbb{N}_0$ and $k \in \mathbb{N}_0$ with

$$m > \frac{s}{2} \quad \text{and} \quad k > \frac{n}{\min(p,q)} + \max(s, \sigma_p) \tag{9}$$

then

$$\|f \mid F^s_{pq}\|_W = \|\varphi_0(D)f \mid L_p\| + \left\| \left(\int_0^1 t^{(m-\frac{s}{2})q} \left| \frac{\partial^m W(t)f}{\partial t^m}(\cdot) \right|^q \frac{dt}{t}\right)^{1/q} \mid L_p \right\| \tag{10}$$

and

$$\|f \mid F^s_{pq}\|_P = \|\varphi_0(D)f \mid L_p\| + \left\| \left(\int_0^1 t^{(k-s)q} \left| \frac{\partial^k P(t)f}{\partial t^k}(\cdot) \right|^q \frac{dt}{t}\right)^{1/q} \mid L_p \right\| \tag{11}$$

(modification if $q = \infty$) are equivalent quasi-norms in F^s_{pq}. Furthermore

$$F^s_{pq} = \{f \in S' \colon \|f \mid F^s_{pq}\|_W < \infty\} = \{f \in S' \colon \|f \mid F^s_{pq}\|_P < \infty\}. \tag{12}$$

If $s > \sigma_p$ then $\|\varphi_0(D)f \mid L_p\|$ in (10–12) can be replaced by $\|f \mid L_p\|$.

Proof. Step 1. We prove that (10,11) are equivalent quasi-norms in F^s_{pq}. We use Theorem 2.4.1 with the above function φ_0 and $\varphi(x) = |x|^k e^{-|x|}$. Then immaterial modifications of (2.4.1/6,7) are satisfied. Furthermore (2.4.1/9,10), or better their substitutes (2.4.1/51,52), are fulfilled for any s_0. Finally (2.4.1/50) can be reduced to the question whether $e^{-|x|}|x|^{k-s_1}$ with $s_1 > \max(s, \sigma_p)$ belongs to H^σ_2 with $\sigma > \frac{n}{\min(p,q)} + \frac{n}{2}$. By the arguments in Remark 2.4.1/1 this property holds if $k - s_1 + \frac{n}{2} > \sigma$. By (9) we may assume that this is satisfied. Hence we can apply Theorem 2.4.1. We have

$$\begin{aligned}\varphi(tD)f = (\varphi(t\cdot)\hat{f})^\vee &= t^k(|y|^k e^{-t|y|}\hat{f}(y))^\vee \\ &= t^k \frac{\partial^k}{\partial t^k}(e^{-t|y|}\hat{f})^\vee = t^k \frac{\partial^k}{\partial t^k} P(t)f,\end{aligned} \tag{13}$$

see (4), which proves that $\|f \mid F^s_{pq}\|_P$ in (11) is an equivalent quasi-norm in F^s_{pq}. In order to prove the corresponding assertion for $\|f \mid F^s_{pq}\|_W$ we use again Theorem 2.4.1 with φ_0 and $\varphi(x) = |x|^{2m} e^{-|x|^2}$. Then all conditions, including (2.4.1/50–52) with $s_1 = 2m > s$, are satisfied, also (2.4.1/49) which shows that the assumption $s_1 > \sigma_p$ in Theorem 2.4.1 is not necessary now. The counterpart of (13) reads as follows,

$$\begin{aligned}\varphi(\sqrt{t}\,D)f = (\varphi(\sqrt{t}\cdot)\hat{f})^\vee &= t^m(|y|^{2m} e^{-t|y|^2}\hat{f})^\vee \\ &= t^m \frac{\partial^m}{\partial t^m}(e^{-t|y|^2}\hat{f})^\vee = t^m \frac{\partial^m}{\partial t^m} W(t)f,\end{aligned} \tag{14}$$

see (3). Then it follows that $\|f \mid F^s_{pq}\|_W$ is an equivalent quasi-norm in F^s_{pq}, where one has to take into consideration that we substituted t by $\sqrt{t}$. Let $s > \sigma_p$. Then we have (2.3.3/8,10) which prove that we can replace $\|\varphi_0(D)f \mid L_p\|$ in (10,11) by $\|f \mid L_p\|$.

Step 2. Let $1 < p < \infty$ and $1 < q \leqq \infty$, then (12) with the discrete versions of (10,11) follows from Theorem 2.4.2(i). However the argument in Step 2 of the proof of Theorem 2.4.2 are based on (2.4.2/17), which can easily be modified in order to replace the discrete version by the continuous one. Then we obtain (12) with $1 < p < \infty$ and $1 < q \leqq \infty$. To extend this assertion to arbitrary values of p and q we need some specific tools. Recall that $P(t)f(x)$ and hence also $v(x,t) = \frac{\partial^k P(t)f(x)}{\partial t^k}$ are harmonic functions in $\mathbb{R}^{n+1}_+$. We use the sub-mean value property

$$|v(x,t)|^r \leqq c_r(\operatorname{vol}\omega)^{-1} \int_\omega |v(y,\tau)|^r dy d\tau, \qquad 0 < r < \infty, \tag{15}$$

where ω is an arbitrary ball in $\mathbb{R}^{n+1}_+$ centered at (x,t). A proof of (15) may be found in [FeS2: Lemma 2 on p. 172] or in a more general version in [HiK]. We replace t in (15) by $2t$ and ω by $B \times [t,3t]$, where B is a ball in $\mathbb{R}^n$ of radius t, centered at x. Afterwards by (13) we can substitute $v(x,2t)$ by $\varphi(2tD)f(x)$. We arrive at

$$|\varphi(2tD)f(x)|^r \leqq c(\operatorname{vol} B)^{-1} \int_B \int_{-1}^1 |\varphi((2t+\tau t)D)f(z)|^r d\tau dz. \tag{16}$$

We use (2.4.2/17) with, say, $2t = 2^{-j}$ and $j \in \mathbb{N}$. We obtain the following counterpart of (2.4.1/39)

$$|(\rho_j \hat{f})^\vee(x)| \leqq c\, 2^{jn} \sum_{l=0}^\infty 2^{-ld} \int_{\{y\in\mathbb{R}^n: |y|<2^{-j+l}\}} |\varphi(2tD)f(x-y)|dy. \tag{17}$$

Let $0 < r < \min(1,p,q)$. We use (16) with $x-y$ instead of x where $|y| < 2^{-j+l}$. We replace the ball B of radius 2^{-j+1} and centered at $x-y$ by a ball of radius $c2^{-j+l}$ centered at x. Then we obtain

$$|\varphi(2tD)f(x-y)|^r \leqq c\, 2^{jn} 2^{(l-j)n} M\left(\int_{-1}^1 |\varphi((2t+\tau t)D)f|^r d\tau\right)(x), \tag{18}$$

where M stands again for the Hardy–Littlewood maximal function. We put (18) in (17), choose $d > n(1+\frac{1}{r})$ and arrive at

$$|(\rho_j \hat{f})^\vee(x)|^r \leqq c\, M\left(\int_{-1}^1 |\varphi((2t+\tau t)D)f|^r d\tau\right)(x). \tag{19}$$

Now we are in the same position as in (2.4.1/41). We apply Theorem 2.2.2 and obtain

$$\|f \mid F^s_{pq}\| \leqq \|\varphi_0(D)f \mid L_p\| \tag{20}$$

$$+c\left\|\left(\sum_{j=1}^\infty 2^{jsq}\left(\int_{-1}^1 |\varphi((2^{-j+1}+\tau 2^{-j})D)f|^r d\tau)^{q/r}\right)^{1/q} \mid L_p\right\|.$$

Because $q/r \geqq 1$ one can change the order of summation with respect to j and of integration with respect to τ. This shows that the right-hand side of (20) can be estimated from above by $c\|f \mid F^s_{pq}\|_P$. The proof of the second assertion in (12) is complete. As far as the first assertion is concerned we use the following counterpart of (15). Let $u(x,t)$ be a solution of the heat equation in $\mathbb{R}^{n+1}_+$ and let

$$\omega = \left\{(y,\tau) \in \mathbb{R}^{n+1} : |y_j - x_j| \leqq \frac{\lambda}{2},\ t-\lambda^2 \leqq \tau \leqq t\right\} \subset \mathbb{R}^{n+1}_+,$$

then

$$|u(x,t)|^r \leqq c_r(\operatorname{vol}\omega)^{-1} \int_\omega |u(y,\tau)|^r dy d\tau, \qquad 0 < r < \infty. \tag{21}$$

This inequality is a consequence of a more subtle sub-mean value property for temperatures due to H.-Q. Bui, see [Bui2: Lemma 2]. Now one obtains the first assertion in (12) in the same way as above. Finally let $s > \sigma_p$ and let $f \in S \cap L_p$ such that the second term on the right-hand side of (10) is finite. Let φ_0 be the above function with $\varphi_0(x) = 1$ near the origin. Then it follows by the above arguments $((1-\varphi_0)\hat{f})^\vee \in F^s_{pq}$ and by the discussion given in Step 2 of the proof of Theorem 2.3.3

$$(\varphi_0 \hat{f})^\vee = f + ((1-\varphi_0)\hat{f})^\vee \in L_p.$$

Hence $f \in F^s_{pq}$. By the end of Step 1 we can replace $\|\varphi_0(D)f \mid L_p\|$ in (12) by $\|f \mid L_p\|$. The proof of (ii) is complete.

Step 3. The proof of (i) is the same. Instead of Theorem 2.4.1 we have to use Theorem 2.5.1 and Corollary 2.5.1/1.

Remark. In connection with Theorem 1.8.1 the question is of interest whether $\int_0^1$ in (6,7,10,11) can be replaced by $\int_0^\nu$ with $0 < \nu \leqq \infty$. Corresponding assertions in connection with differences and means of differences may be found in Remarks 2.6.1/4 and 2.6.2/3 and Theorem 2.6.3. By the proof and the fact that $\varphi(x) \neq 0$ if $x \neq 0$ in both cases, see Step 1, it follows that $\int_0^1$ in (6,7,10,11) can be replaced by $\int_0^\nu$ with $0 < \nu < \infty$. Let $s > \sigma_p$ then we can argue in the same way as in Step 2 of the proof of Theorem 2.6.3 based on corresponding counterparts of (2.6.3/9). In other words, under the additional restriction $s > \sigma_p$ one can replace $\int_0^1$ in (6–8, 10–12) by $\int_0^\infty$.

2.6.5 The classical spaces

In Chapter 1 we introduced several classical function spaces on $\mathbb{R}^n$, discussed some properties, and described their connections with the spaces B^s_{pq} and F^s_{pq}, see Theorem 1.5.1 as far as the latter aspect is concerned. Now we return briefly to this subject and comment previous assertions on the basis of the results obtained in this chapter.

Hölder–Zygmund and Besov spaces. In 1.2.2 we defined the Hölder spaces C^s with $0 < s \neq$ integer and the Zygmund spaces $\mathcal{C}^s$ with $s > 0$. We have the classical assertions (1.2.2/6,7), where the norm $\|f \mid \mathcal{C}^s\|_{0,m}$ is of special interest now. The Besov spaces B^s_{pq} with $s > 0$, $1 < p < \infty$, $1 \leqq q \leqq \infty$ had been introduced in 1.2.5. Here we have the classical assertion (1.2.5/3) where again the norm $\|f \mid B^s_{pq}\|_{0,m}$ is of interest now. On the other hand, we have Corollary 2.6.1/1 with the quasi-norm

(2.6.1/2). By Remark 2.6.1/4 we can replace $\int_{|h|\leq 1}$ in (2.6.1/2) by $\int_{\mathbb{R}^n}$. Then this modified quasi-norm coincides with the above norms $\|f \mid \mathcal{C}^s\|_{0,m} = \|f \mid B^s_{\infty\infty}\|_{0,m}$ and $\|f \mid B^s_{pq}\|_{0,m}$. In other words, Corollary 2.6.1/1 proves that the classical Besov spaces B^s_{pq} from 1.2.5 coincide with the spaces B^s_{pq} introduced in (2.3.1/7), and that

$$\mathcal{C}^s = B^s_{\infty\infty}, \qquad s > 0, \tag{1}$$

holds. These assertions cover Theorem 1.5.1(i,ii).

Fractional Sobolev spaces. Although we have nothing new to say we recall for the sake of completeness, the relevant assertions for fractional Sobolev spaces. The Sobolev spaces W^k_p with $1 < p < \infty$ and $k \in \mathbb{N}_0$ have been introduced in 1.2.3. They are special cases of the fractional Sobolev spaces H^s_p from Definition 1.3.2, where $1 < p < \infty$ and $s \in \mathbb{R}$. The relevant assertion (1.3.2/4) is classical, it is based on L_p-Fourier multipliers which we shall not discuss in this book again. The connection of H^s_p with F^s_{p2} is based on Littlewood–Paley theorems for L_p spaces, see 1.3.3, in particular (1.3.3/13). Compared with (2.3.1/8) one arrives at

$$H^s_p = F^s_{p2}, \qquad s \in \mathbb{R}, \quad 1 < p < \infty, \tag{2}$$

what coincides with Theorem 1.5.1(iii).

Hardy spaces. We discussed in 1.4 Hardy spaces H_p and h_p, $0 < p < \infty$, rather extensively, see in particular (1.4.4/1). In Theorem 1.4.4 we formulated a Littlewood–Paley theorem which results in

$$h_p = F^0_{p2}, \qquad 0 < p < \infty, \tag{3}$$

which, in turn, generalizes (2) with $s = 0$. In particular $h_p = L_p$ if $1 < p < \infty$.

Remark. Only the above assertions on Hölder–Zygmund spaces and classical Besov spaces are covered by the results of this chapter. We did not prove any Littlewood–Paley assertions and mentioned (2,3) for the sake of completeness. More detailed discussions and proofs may be found in [Triß: 2.5.6–2.5.8, 2.5.12] and in the corresponding references given in the first chapter of this book.

Local methods. In contrast to [Triα: Triß], this book is characterized by the systematic use of local assertions and local methods. In Chapter 1 we described these new tools in 1.8.4 and in Proposition 1.11.3. These crucial assertions are now fully covered by 2.4.6 and Theorem 2.4.7.

Harmonic and thermic extensions. Characterizations of function spaces on $\mathbb{R}^n$ as trace spaces of certain spaces of harmonic and thermic functions on $\mathbb{R}^{n+1}_+$ have a long history. Classical results for Besov spaces may be found in 1.8.1. Corresponding results for Hardy spaces have been formulated in 1.4.2. In 1.8.3 we collected more recent results, including some references. Theorem 1.8.3 is completely covered by Theorem 2.6.4 where the latter one has a rather final character. In Theorem 2.6.4 we assumed that the function φ_0 has a compact support. This was helpful in connection with the characterizations (2.6.4/8,12). On the other hand, if one looks only for equivalent quasi-norms then such a restriction is not necessary.

Chapter 3

Atoms, Oscillations, and Distinguished Representations

3.1 Introduction

We recall the structure of this book. The first chapter served as a historically-oriented self-contained survey of relevant function spaces. In Chapters 2 and 3 we start again, now in full details. We develop the technical instruments which are the basis for the rest of the book. Whereas Chapter 2 dealt mainly with Fourier-analytical methods and their far-reaching consequences, we concentrate in the present chapter on some recent local techniques which are crucial for the rest of this book. First steps in this direction have already been done in the previous chapter in connection with local means, see 2.4.6 and 2.5.3. On this basis we study in 3.2 characterizations of F^s_{pq} via atoms. In 3.3 we ask for distinguished kernels and related representations in the sense of 2.4.6 and 2.5.3. In 3.4 we give a self-contained introduction into the L_p-theory of oscillations as far as it is of interest in our context. Based on the results obtained in 3.3 and 3.4, we describe in 3.5 characterizations of F^s_{pq} and B^s_{pq} in terms of oscillations and (again) of differences. It is a well-known fact that several types of function spaces are closely connected with oscillations, for example, Morrey–Campanato spaces. We wish to incorporate these spaces and to study their relations to F^s_{pq} and B^s_{pq}. But we postpone this task to Chapter 5 which deals with spaces on bounded smooth domains, see in particular, 5.3. The present chapter concentrates on spaces on $\mathbb{R}^n$ (as the previous chapter). However the technical studies on oscillations in 3.4 and on distinguished kernels in 3.3 prepare also our later intentions.

H. Triebel, *Theory of Function Spaces II*, Modern Birkhäuser Classics,
DOI 10.1007/978-3-0346-0419-2_3, © Birkhäuser Verlag 1992

3.2 Atomic decompositions of F^s_{pq}

3.2.1 Basic notations and an introductory remark

We complement the basic notations introduced in 2.2.1. Let

$\mathbb{Z}^n$ be the lattice of all points in $\mathbb{R}^n$ with integer-valued components,

$\mathbb{Z} = \mathbb{Z}^1$, and $\mathbb{R} = \mathbb{R}^1$ (real line).

Let $\mathbb{Q}^n$ be the collection of all cubes $Q_{\nu k}$ in $\mathbb{R}^n$ with sides parallel to the axes, centered at $2^{-\nu}k$, and with side length $2^{-\nu}$, where $k \in \mathbb{Z}^n$ and $\nu \in \mathbb{N}_0$.

Let Q be a cube in $\mathbb{R}^n$ and $r > 0$, then rQ is the cube in $\mathbb{R}^n$ concentric with Q and with side length r times the side length of Q. We write $(\nu, k) < (\nu', k')$ if $\nu \geqq \nu'$ and

$$Q_{\nu k} \subset 2Q_{\nu k'} \quad \text{with} \quad Q_{\nu k} \in \mathbb{Q}^n \quad \text{and} \quad Q_{\nu' k'} \in \mathbb{Q}^n.$$

Remark. Our point of view in 3.2 is the following. Based on [Tri22] we derive a new atomic decomposition of F^s_{pq}. We shall not deal with atomic decompositions of B^s_{pq}. As far as the latter spaces are concerned we refer to the results by M. Frazier and B. Jawerth which we described in 1.9.2 and which we shall use in the sequel. Furthermore, atomic decompositions in tent spaces will be of great service for us, see Theorem 1.9.3. Finally we wish to emphasize that there exist other types of atomic decompositions in F^s_{pq}, see [FrJ2,3] and [Net2,4], which will not be treated here. Nearest to us comes Theorem 7.4 in [FrJ3]. More information and detailed references about this subject may be found in 1.9.2 – 1.9.4. The interested reader should consult [FrJ3] where one can find many modifications and applications, as well as comprehensive references and historical remarks extending what had been said in 1.9.2 – 1.9.4.

3.2.2 Atoms

Let $a \in \mathbb{R}$, then $a_+ = \max(a, 0)$, and $[a]$ stands for the largest integer less than or equal to a.

Definition. Let $s \in \mathbb{R}$ and $0 < p \leqq 1 < q \leqq \infty$. Let K and L be integers with

$$K \geqq ([s]+1)_+ \quad \text{and} \quad L \geqq \max\left(\left[n\left(\frac{1}{p}-1\right)-s\right], -1\right). \tag{1}$$

(i) The (complex-valued) function $a(x)$ is called an s-atom if

$$\operatorname{supp} a \subset 5Q \tag{2}$$

for some $Q = Q_{0k} \in \mathbb{Q}^n$ (side length 1) and

$$|D^\alpha a(x)| \leqq 1 \quad \text{for} \quad |\alpha| \leqq K. \tag{3}$$

(ii) Let $Q = Q_{\nu k} \in \mathbb{Q}^n$. The (complex-valued) function $a(x)$ is called a (Q, s, p, q)-atom if (2) is satisfied,

$$|D^\alpha a(x)| \leqq |Q|^{-\frac{1}{q}+\frac{s}{n}-\frac{|\alpha|}{n}} \quad \text{for} \quad |\alpha| \leqq K, \tag{4}$$

and

$$\int_{\mathbb{R}^n} x^\beta a(x) dx = 0 \quad \text{for} \quad |\beta| \leqq L. \tag{5}$$

(iii) The distribution $g \in S'$ is called an (s, p, q)-atom if

$$g = \sum_{(\mu,l)<(\nu,k)} d_{\mu l} a_{\mu l}(x) \quad \text{for some} \quad \nu \in \mathbb{N}_0 \quad \text{and} \quad k \in \mathbb{Z}^n, \tag{6}$$

where $a_{\mu l}(x)$ is a $(Q_{\mu l}, s, p, q)$-atom and $d_{\mu l}$ are complex numbers with

$$\left(\sum_{(\mu,l)<(\nu,k)} |d_{\mu l}|^q\right)^{1/q} \leqq |Q_{\nu k}|^{\frac{1}{q}-\frac{1}{p}} \tag{7}$$

(usual modification if $q = \infty$).

Remark 1. The parts (i) and (ii) coincide essentially with Definition 1.9.2 (i,ii), whereas part (iii) repeats Definition 1.9.4. (Recall that by our intention both the first chapter and the rest of the book should be independently readable.) We add some technical explanations. As usual $x^\beta = x_1^{\beta_1} \cdots x_n^{\beta_n}$ if $\beta = (\beta_1, \ldots, \beta_n)$ is a multi-index and $x = (x_1, \ldots, x_n) \in \mathbb{R}^n$. Furthermore, in (i) and (ii) it is tacitly assumed that $a(x)$ has classical derivatives up to order K. If $L = -1$ then (5) means that no moment conditions are required. Obviously, the above atoms depend on K and L. As we shall see, the atom g in (6) belongs to F^s_{pq} and the convergence in (6) must be understood in this sense.

Remark 2. The atoms in (i) and (ii) coincide essentially with corresponding atoms in [FrJ1], whereas the atoms in (iii) go back to [Tri19,22]. Further remarks may be found in 1.9.1, 1.9.2, 1.9.4, and 3.2.1. In particular by Remark 1.9.1/3 the above restriction to $p \leqq 1$ in (iii) is natural.

3.2.3 The main theorem

The spaces F^s_{pq} have been introduced in 2.3.1. Furthermore the other notations we use have the same meaning as in the two preceding subsections. If s, p, q are given then the numbers K and L in (3.2.2/1) can be chosen as large as one wants. This observation will be of great use for us later on, for example, in connection with applications to pseudodifferential operators.

Theorem. Let $s \in \mathbb{R}$ and $0 < p \leqq 1 < q \leqq \infty$. Let K and L be fixed integers satisfying (3.2.2/1). Then $f \in S'$ is an element of F^s_{pq} if and only if it can be represented as

$$f = \sum_{j=1}^{\infty} (\mu_j a_j + \lambda_j g_j) \qquad \text{(convergence in } S'\text{)} \tag{1}$$

where a_j are s-atoms, g_j are (s,p,q)-atoms, μ_j and λ_j are complex numbers with

$$\left(\sum_{j=1}^{\infty} |\mu_j|^p + |\lambda_j|^p \right)^{1/p} < \infty. \tag{2}$$

Furthermore, the infimum of (2) with respect to all admissible representations (where K and L are fixed) is an equivalent quasi-norm in F^s_{pq}.

Proof. Step 1. First we construct a useful resolution of unity. Let $\varphi = k$ be the same function as in Theorem 2.4.6. We may assume, in addition, that φ is real and rotation-invariant, i.e., $\varphi(x) = \varphi(y)$ if $|x| = |y|$. Let $x^0 = (1, 0, \ldots, 0)$. Then $\hat{\varphi}(x) = |x|^{2N} k^{0\wedge}(x) \in S$ is a non-trivial real rotation-invariant function which can be normalized as

$$\int_0^\infty \hat{\varphi}^2(tx^0) \frac{dt}{t} = 1.$$

By dilation and rotation we have

$$\int_0^\infty \hat{\varphi}^2(tx) \frac{dt}{t} = 1 \quad \text{for} \quad x \neq 0, \quad x \in \mathbb{R}^n. \tag{3}$$

We define $\Phi(x)$ by

$$\hat{\Phi}(x) = 1 - \int_0^1 \hat{\varphi}^2(tx) \frac{dt}{t}, \qquad x \in \mathbb{R}^n. \tag{4}$$

We have $\hat{\Phi}(0) = 1$ and

$$\hat{\Phi}(x) = \int_1^\infty \hat{\varphi}^2(tx)\frac{dt}{t} \quad \text{if } \ x \neq 0. \tag{5}$$

By (4) and (5) it follows $\hat{\Phi} \in S$ and hence $\Phi \in S$. Furthermore, $\hat{\varphi}^2$ is an analytic function of exponential type. By (4) we have a corresponding assertion for $\hat{\Phi}$, and hence, by the Paley–Wiener–Schwartz theorem, Φ has a compact support. We arrive at the continuous resolution of unity

$$\hat{\Phi}(x) + \int_0^1 \hat{\varphi}^2(tx)\frac{dt}{t} = 1 \quad \text{for all } \ x \in \mathbb{R}^n. \tag{6}$$

Step 2. We need a preparation. Let $g(x,t)$ be a continuous function in

$$\mathbb{R}^{n+1}_+ = \{(x,t)\colon x \in \mathbb{R}^n, t > 0\}.$$

We assume that the projection from $\mathbb{R}^{n+1}_+$ into $\mathbb{R}^n$ given by

$$\pi_\varphi(g)(x) = \int_0^1 \int_{\mathbb{R}^n} \varphi(y) g(x+ty,t)\frac{dy\,dt}{t}, \qquad x \in \mathbb{R}^n, \tag{7}$$

makes sense, where φ has the above meaning. By $\varphi(y) = \varphi(-y)$ we obtain

$$\pi_\varphi(g)^\wedge(x) = \int_0^1 \hat{\varphi}(tx)\hat{g}(x,t)\frac{dt}{t}, \tag{8}$$

where $\hat{g}(x,t)$ is the n-dimensional Fourier transform with respect to the x-variable. (Again we assume that (8) makes sense.) Now we specialize $g(x,t) = \varphi(t,f)(x)$, where the latter is given by (2.4.6/1) with $\varphi = k$ and $f \in S'$. We have $\hat{g}(x,t) = \hat{\varphi}(tx)\hat{f}(x)$ and

$$\pi_\varphi(\varphi(t,f))^\wedge(x) = \hat{f}(x) \int_0^1 \hat{\varphi}^2(tx)\frac{dt}{t}. \tag{9}$$

We use (6) and obtain the representation formulas

$$\hat{f}(x) = \hat{f}(x)\hat{\Phi}(x) + \pi_\varphi(\varphi(t,f))^\wedge(x)$$

and

$$f(x) = f * \Phi(x) + \pi_\varphi(\varphi(t,f))(x), \qquad x \in \mathbb{R}^n. \tag{10}$$

Afterwards it is easy to see that the above calculation makes sense or any $f \in S'$ (appropriate interpretation).

Step 3. We need a second preparation. Let T^p_q be the tent spaces introduced in Definition 1.9.3 where we now assume $0 < p \leqq 1 < q \leqq \infty$. Then we have the atomic decomposition described in Theorem 1.9.3. Let Q be a cube in $\mathbb{R}^n$, then we denoted

$$\hat{Q} = \{(y,t) \in \mathbb{R}^{n+1}_+\colon y \in Q, 0 < t < \operatorname{dist}(y, \partial Q)\} \tag{11}$$

as the tent over Q, see (1.9.3/10). Atoms in tent spaces are characterized by (1.9.3/11,12). We need a technical modification of the atomic decomposition (1.9.3/13,14). Let $g(x,t) \in T^p_q$ with $g(x,t) = 0$ if $t > 1$. Let

$$g(x,t) = \sum_{j=1}^{\infty} \lambda_j a_j(x,t) \tag{12}$$

be the atomic decomposition (1.9.3/13) with

$$\sum_{j=1}^{\infty} |\lambda_j|^2 \leqq c \|g\|^p_{q,p} \tag{13}$$

where the constant c is independent of g. Let $a(x,t)$ be one of the atoms in (12) based on a cube Q with $|Q| > 1$. We may assume $a(x,t) = 0$ if $t > 1$. Then we have

$$\operatorname{supp} a(x,t) \subset \{(y,\tau) \in \mathbb{R}^{n+1}_+ : y \in Q, 0 \leqq \tau \leqq 1\}. \tag{14}$$

We cover this set by $(n+1)$-dimensional cubes Q_k with side length 1 and decompose

$$a(x,t) = \sum_{k=1}^{N} a^k(x,t), \qquad \operatorname{supp} a^k \subset Q_k, \tag{15}$$

with $N \sim |Q|$. Let

$$c_k = \left(\int_{Q_k} |a^k(x,t)|^q \frac{dx\,dt}{t} \right)^{1/q} \tag{16}$$

where we may assume $c_k > 0$. Besides immaterial constants $A_k(x,t) = \frac{a^k(x,t)}{c_k}$ can be interpreted as a T^p_q-atom based on a n-dimensional cube with side length, say, 3, see (1.9.3/11,12). We have

$$\left(\sum_{k=1}^{N} c_k^q \right)^{1/q} = \left(\int_Q |a(x,t)|^q \frac{dx\,dt}{t} \right)^{1/q} \leqq |Q|^{\frac{1}{q} - \frac{1}{p}}$$

and

$$\sum_{k=1}^{N} c_k^p \leqq \left(\sum_{k=1}^{N} c_k^q \right)^{p/q} N^{1-\frac{p}{q}} \leqq |Q|^{\frac{p}{q}-1} N^{1-\frac{p}{q}} \leqq c, \tag{17}$$

where c is independent of $a(x,t)$. We replace the considered atom $a(x,t)$ in (12) by $\sum_{k=1}^{N} c_k A_k(x,t)$. By (17) we have an immediate counterpart of (13). In other words: If $g(x,t) \in T^p_q$ with $0 < p \leqq 1 < q \leqq \infty$ and $g(x,t) = 0$ if $t \geqq 1$, then it is sufficient to restrict the decomposition (12) to atoms a_j based on cubes with side length less than 3.

Step 4. Let $s \in \mathbb{R}$ and $0 < p \leqq 1 < q \leqq \infty$, and let K and L be the same numbers as in the theorem. In this step we prove that any $f \in F^s_{pq}$ can be represented by (1) with

$$\left(\sum_{j=1}^{\infty} |\mu_j|^p + |\lambda_j|^p\right)^{1/p} \leqq c\|f \mid F^s_{pq}\|, \tag{18}$$

where c is independent of f. The functions φ and Φ and the means $\varphi(t,f)$ have the same meaning as in Steps 1 and 2. Let $f \in F^s_{pq}$ and

$$g(x,t) = t^{-s}\varphi(t,f)(x) \text{ if } 0 < t < 1 \text{ and } g(x,t) = 0 \text{ if } t \geqq 1. \tag{19}$$

By Theorem 2.4.5, (19), and Definition 1.9.3 we have $g(x,t) \in T^p_q$ and

$$\|\Phi * f \mid L_p\| + \|g\|_{q,p} \leqq c\|f \mid F^s_{pq}\|, \tag{20}$$

where c is independent of f. We decompose $g(x,t)$ in the sense of Step 3. By (12) and (13) we have

$$\varphi(t,f)(x) = \sum_{j=1}^{\infty} \lambda_j b_j(x,t), \qquad x \in \mathbb{R}^n, \quad 0 < t < 1 \tag{21}$$

with

$$\sum_{j=1}^{\infty} |\lambda_j|^p \leqq c\|f \mid F^s_{pq}\|^p \tag{22}$$

and $b_j(x,t) = t^s a_j(x,t)$, where $a_j(x,t)$ has the same meaning as in (12). In particular we have

$$\operatorname{supp} b_j \subset \hat{Q}_j \text{ and } \left(\int_{\hat{Q}_j} t^{-sq}|b_j(x,t)|^q \frac{dx\,dt}{t}\right)^{1/q} \leqq |Q_j|^{\frac{1}{q}-\frac{1}{p}}, \tag{23}$$

see (1.9.3/11,12). By Step 3 we may assume that the side length of Q_j is at most 3. By (10) we arrive at the decomposition

$$f = f * \Phi + \sum_{j=1}^{\infty} \lambda_j g_j \tag{24}$$

with

$$\begin{aligned} g_j(x) = \pi_\varphi b_j(x) &= \int_0^1 \int_{\mathbb{R}^n} \varphi(y) b_j(x+ty,t) \frac{dy\,dt}{t} \\ &= \int_0^1 t^{-n} \int_{\mathbb{R}^n} \varphi\left(\frac{x-y}{t}\right) b_j(y,t) \frac{dy\,dt}{t}. \end{aligned} \tag{25}$$

Next we decompose the slice $\hat{Q}_j \cap \{2^{-l} \leqq t \leqq 2^{-l+1}\}$ in cubes with side length 2^{-l}. By the above restriction of the side lengths of the cubes we may assume $l \in \mathbb{N}_0$, without any loss of generality. Let $\tilde{Q}$ be one of these cubes and let Q be its projection on $\mathbb{R}^n$. Let

$$a(x) = \int_{\tilde{Q}} t^{-n} \varphi\left(\frac{x-y}{t}\right) b_j(y,t) \frac{dy\,dt}{t} \tag{26}$$

where we temporarily omit to indicate Q and j on the left-hand side of (26). Recall that φ is supported by the unit ball, see (2.4.6/2). Then we have

$$\operatorname{supp} a(x) \subset 5Q, \qquad \operatorname{supp} a(x) \subset Q_j. \tag{27}$$

Let χ_Q be the characteristic function of Q. Then we have

$$\hat{a}(x) = \int_{2^{-l}}^{2^{-l+1}} \hat{\varphi}(tx)(b_j(\cdot,t)\chi_Q)^\wedge(x)\frac{dt}{t}, \tag{28}$$

where $(b_j(\cdot,t)\chi_Q)^\wedge(x)$ is an analytic function. Recall $\hat{\varphi}(x) = |x|^{2N} k^{0\wedge}(x)$, where we may assume that $N \in \mathbb{N}$ is a large number which is at our disposal. It follows

$$(-i)^{|\alpha|} \int_{\mathbb{R}^n} x^\alpha a(x)dx = (D^\alpha \hat{a})(0) = 0, \qquad |\alpha| \leqq L, \tag{29}$$

where $L \in \mathbb{N}$ is at our disposal. Furthermore, with $\frac{1}{q} + \frac{1}{q'} = 1$, we have

$$\begin{aligned} |a(x)| &\leqq \left(\int_{\tilde{Q}} t^{-nq'} \left|\varphi\left(\frac{x-y}{t}\right)\right|^{q'} \frac{dy\,dt}{t}\right)^{1/q'} \left(\int_{\tilde{Q}} |b_j(y,t)|^q \frac{dy\,dt}{t}\right)^{1/q} \\ &\leqq c\, 2^{ln-\frac{ln}{q'}} \left(\int_{\tilde{Q}} |b_j(y,t)|^q \frac{dy\,dt}{t}\right)^{1/q} \\ &\leqq c'\, |Q|^{-\frac{1}{q}+\frac{s}{n}} \left(\int_{\tilde{Q}} |b_j(y,t)|^q t^{-sq} \frac{dy\,dt}{t}\right)^{1/q}. \end{aligned} \tag{30}$$

Similarly we estimate $D^\alpha a(x)$ and obtain

$$|D^\gamma a(x)| \leqq c\,|Q|^{-\frac{1}{q}+\frac{s}{n}-\frac{|\gamma|}{n}} \left(\int_{\tilde{Q}} |b_j(y,t)|^q t^{-sq} \frac{dy\,dt}{t}\right)^{1/q}. \tag{31}$$

It is no restriction of generality to asumme that all involved n-dimensional cubes belong to $\mathbb{Q}^n$, see 3.2.1, and that the side length of the above cube Q_j is at most 1. Now we use the notations introduced in 3.2.1 and identify Q_j with $Q_{\nu_j k_j}$ for some $\nu_j \in \mathbb{N}_0$ and $k_j \in \mathbb{Z}^n$. Then (25) and the above consideration yield

$$g_j(x) = \sum_{(\mu,l)<(\nu_j,k_j)} d^j_{\mu l} a^j_{\mu l}(x) \tag{32}$$

where $a^j_{\mu l}(x)$ is a $(Q_{\mu l}, s, p, q)$-atom in the sense of Definition 3.2.2(ii), where the natural numbers K and L may be chosen in the above way. Of course, $a^j_{\mu l}(x)$ is the normalized function $a(x)$ from (26) with the properties (27), (29), and (31), and the normalizing factor is given by

$$d^j_{\mu l} = \left(\int_{\tilde{Q}^j_{\mu l}} |b_j(y,t)|^q t^{-sq} \frac{dy\,dt}{t} \right)^{1/q}$$

where $\tilde{Q}^j_{\mu l}$ is the related cube. By (23) we have

$$\left(\sum_{(\mu,l)<(\nu_j,k_j)} |d^j_{\mu l}|^q \right)^{1/q} \leqq c|Q_j|^{\frac{1}{q}-\frac{1}{p}} \tag{33}$$

where c is independent of f. We may assume $c = 1$, otherwise we put this c in (24) in the numbers λ_j which does not change (22). After this immaterial modification g_j becomes a (s,p,q)-atom in the sense of Definition 3.2.2(iii). This assertion and (22) cover the main part of (1) and (18). By (24) it remains to care about the term $f * \Phi$. Because $f \in F^s_{pq}$ it follows

$$((1+|\xi|^2)^{\kappa/2} \hat{\Phi} \hat{f})^\vee \in F^s_{pq} \quad \text{for any} \quad \kappa \in \mathbb{R}.$$

This simple Fourier multiplier assertion follows more or less from the definition of the spaces F^s_{pq}, see 2.4.8 for much more complicated assertions. In particular $\Phi * f \in F^\sigma_{pp} = B^\sigma_{pp}$ for any $\sigma \in \mathbb{R}$. We choose $\sigma \geqq s$ and apply Theorem 1.9.2. The resulting atomic decomposition of $\Phi * f$ can be incorporated in (1) and (18).

Step 5. We have to prove the reverse assertion. Let f be given by (1) with (2). Then we wish to prove $f \in F^s_{pq}$ and

$$\|f \mid F^s_{pq}\| \leqq c \left(\sum_{j=1}^{\infty} (|\mu_j|^p + |\lambda_j|^p) \right)^{1/p} \tag{34}$$

for some $c > 0$ which is independent of f. Let us assume temporarily

$$a_j \in F^s_{pq} \quad \text{with} \quad \|a_j \mid F^s_{pq}\| \leqq c \tag{35}$$

and

$$g_j \in F^s_{pq} \quad \text{with} \quad \|g_j \mid F^s_{pq}\| \leqq c, \tag{36}$$

where a_j and g_j have the same meaning as in (1) and c in (35,36) is independent of a_j and g_j. Then we have

$$\left\| \sum_{j=1}^{T} (\mu_j a_j + \lambda_j g_j) \mid F^s_{pq} \right\|^p \leqq c \sum_{j=1}^{T} (|\mu_j|^p + |\lambda_j|^p)$$

with the same c as in (35,36). A limiting argument yields $f \in F^s_{pq}$ with (34). We prove (35) and recall that $a_j(x)$ has classical derivatives up to order $K \geqq ([s]+1)_+$. We use Theorem 2.4.6 and denote temporarily the quasi-norm in (2.4.6/7) by $\|\cdot\|^*$. By (2.4.6/1) and partial integration it follows that $k(t, a_j)(x)$ is supported by a cube of side length 7 and that

$$|k(t, a_j)(x)| \leqq c\, t^K, \qquad 0 < t \leqq 1,$$

where c is independent of a_j. Then we obtain

$$\|a_j\|^* \leqq c \tag{37}$$

which does not mean automatically $a_j \in F^s_{pq}$. However we can mollify a_j, for example by Sobolev's method, then we have (37) uniformly for the mollified functions, what means, by Theorem 2.4.6, a corresponding uniform estimate for the mollified functions in the F^s_{pq}-quasi-norm. Then it follows by (2.3.1/8) and a Fatou argument $a_j \in F^s_{pq}$. Now (35) is a consequence of (37). In other words, it remains to prove (36).

Step 6. We prove (36). Let g be a (s,p,q)-atom in the sense of Definition 3.2.2(iii) based on a cube Q with side length $2^{-\nu}$, where $\nu \in \mathbb{N}_0$. By the arguments at the end of the last step any finite partial sum of (3.2.2/6) is an element of F^s_{pq}. Let us assume that we have a uniform estimate of type (37) for these partial sums. By the arguments below it follows that these partial sums converge at least in B^s_{qq}, and hence in S', to g. Again by Fatou's lemma and (2.3.1/8) we have $g \in F^s_{pq}$. In other words, we may concentrate ourselves on the proof of the second part of (36) and we may use (2.4.6/6) as an equivalent quasi-norm in F^s_{pq} for g, i.e.,

$$\|g \mid F^s_{pq}\| \sim \|k_0(1,g) \mid L_p\| + \left\| \left(\sum_{j=0}^{\infty} 2^{jsq} |k(2^{-j}, g)(\cdot)|^q \right)^{1/q} \mid L_p \right\|. \tag{38}$$

We split $\sum_{j=0}^{\infty}$ in $\sum_{j=0}^{\nu}$ and $\sum_{j=\nu+1}^{\infty}$. We have by Hölder's inequality

$$\begin{aligned} &\left\| \left(\sum_{j=\nu+1}^{\infty} 2^{jsq} |k(2^{-j}, g)|^q \right)^{1/q} \mid L_p \right\| \\ &\qquad \leqq c\, |Q|^{\frac{1}{p}-\frac{1}{q}} \left\| \left(\sum_{j=\nu+1}^{\infty} 2^{jsq} |k(2^{-j}, g)|^q \right)^{1/q} \mid L_q \right\| \\ &\qquad \leqq c'\, |Q|^{\frac{1}{p}-\frac{1}{q}} \|g \mid B^s_{qq}\| \leqq c''\, |Q|^{\frac{1}{p}-\frac{1}{q}} |Q|^{\frac{1}{q}-\frac{1}{p}} = c'', \end{aligned} \tag{39}$$

where we used Theorem 1.9.2 and (3.2.2/7). In order to estimate the term with $\sum_{j=0}^{\nu}$ we begin with a technical modification. Let

$$\tilde{g}(x) = 2^{\nu(s-\frac{n}{p})} g(2^{-\nu}x) = \sum_{(\mu,l)<(\nu,k)} 2^{\nu n(\frac{1}{q}-\frac{1}{p})} d_{\mu l} 2^{\nu(s-\frac{n}{q})} a_{\mu l}(2^{-\nu x}) \tag{40}$$

$$= \sum_{(\mu,l)<(0,\tilde{k})} \tilde{d}_{\mu l} \tilde{a}_{\mu l}(x) \quad \text{with} \quad \tilde{a}_{\mu l}(x) = 2^{\nu(s-\frac{n}{q})} a_{\mu l}(2^{-\nu}x).$$

Then $\tilde{a}_{\mu l}(x)$ is an atom in the sense of Definition 3.2.2(ii) and $\tilde{g}$ is a (s,p,q)-atom based on a cube with side length 1. Furthermore by (2.4.6/1) we have

$$k(2^{-j}, \tilde{g})(x) = 2^{\nu(s-\frac{n}{p})} k(2^{-j-\nu}, g)(2^{-\nu}x)$$

and consequently

$$k(2^{-j}, g)(x) = 2^{\nu(\frac{n}{p}-s)} k(2^{\nu-j}, \tilde{g})(2^{\nu}x). \tag{41}$$

Now we obtain

$$\left\| \left(\sum_{j=0}^{\nu} 2^{jsq} |k(2^{-j}, g)|^q \right)^{1/q} \mid L_p \right\| \leqq \left\| \left(\sum_{j=-\infty}^{0} 2^{jsq} |k(2^{-j}, \tilde{g})|^q \right)^{1/q} \mid L_p \right\|. \tag{42}$$

In other words, we may assume that the above (s,p,q)-atom g is based on a cube centered at the origin and with side length 1, and that we have to estimate the right-hand side of (42) with g instead of $\tilde{g}$. In order to do this we assume temporarily

$$|k(2^{-j}, g)(x)| \leqq c \frac{2^{jM+jn}}{1+|2^j x|^b}, \qquad -j \in \mathbb{N}_0, \tag{43}$$

where M is a real number with $M > n(\frac{1}{p}-1)-s$ and $b>0$ is at our disposal. The constant c in (43) is independent of j, x, and g. We choose $b > n/p$ and obtain

$$\int_{\mathbb{R}^n} |k(2^{-j}, g)|^p dx \leqq c\, 2^{jnp+jMp-jn}$$

and consequently

$$\sum_{j=-\infty}^{0} 2^{jsp} \int_{\mathbb{R}^n} |k(2^{-j}, g)|^p dx \leqq c \sum_{j=-\infty}^{0} 2^{jp(s+n+M-\frac{n}{p})} \leqq c', \tag{44}$$

where we used $M > n(\frac{1}{p}-1)-s$. By (44) and $l_p \subset l_q$ we have

$$\left\| \left(\sum_{j=-\infty}^{0} 2^{jsq} |k(2^{-j}, g)|^q \right)^{1/q} \mid L_p \right\| \leqq c. \tag{45}$$

This is the desired result. Together with (39) it follows that the second term on the right-hand side in (38) can be estimated from above by a constant which is independent of g. The first term on the right-hand side of (38) can be treated in the same way: We have (42) with $\nu = 0$ and with k_0 instead of k and, as we shall see, (43) with k_0 and $j = 0$ instead of k and $-j \in \mathbb{N}_0$.

Step 7. It remains to prove (43), where the calculations will show that we can also replace k by k_0. First we recall the one-dimensional representation formula

$$h(t) = \sum_{l=0}^{M-1} \frac{t^l}{l!} h^{(l)}(0) + \frac{1}{(M-1)!} \int_0^t (t-\tau)^{M-1} h^{(M)}(\tau) d\tau. \tag{46}$$

Let g be a (s,p,q)-atom based on a cube centered at the origin and with side length 1 given by (3.2.2/6,7) with $(\nu,k) = (0,0)$ and $|Q_{\nu k}| = 1$. Let $a = a_{\mu l}$ be one of the related (Q^a, s, p, q)-atoms, in particular the side length of Q^a is at most 2. Let

$$M = \begin{cases} [n(\frac{1}{p} - 1) - s] + 1 & \text{if } s \geqq n(\frac{1}{p} - 1) \\ 0 & \text{otherwise.} \end{cases} \tag{47}$$

Of course if $M = 0$ then (46) means that we work with the function $h(t)$ itself instead of any expansion. First we prove that there exists a positive number ε such that

$$|k(2^{-j}, a)(x)| \leqq c\, 2^{jM+jn} \frac{|Q^a|^{\frac{1}{p} - \frac{1}{q} + \varepsilon}}{1 + |2^j x|^b}, \qquad -j \in \mathbb{N}_0, \tag{48}$$

holds where $b > 0$ is at our disposal and c is independent of the chosen atom, $x \in \mathbb{R}^n$ and j. We may assume that Q^a is centered at the origin. By (2.4.6/1) and $k(y) = k(-y)$ we have

$$\begin{aligned} k(2^{-j}, a)(x) &= 2^{jn} \int_{\mathbb{R}^n} k(2^j y) a(x-y) dy \\ &= 2^{jn} \int_{\mathbb{R}^n} k(2^j x - 2^j y) a(y) dy. \end{aligned} \tag{49}$$

We apply (46) with $t = 1$ to $h(\tau) = k(2^j x - 2^j \tau y)$ and obtain

$$\begin{aligned} k(2^j x - 2^j y) = &\sum_{|\alpha| \leqq M-1} \frac{(-1)^{|\alpha|}}{\alpha!} 2^{j|\alpha|} y^\alpha (D^\alpha k)(2^j x) \\ &+ (-1)^M M \int_0^1 (1-\tau)^{M-1} \sum_{|\alpha| = M} \frac{1}{\alpha!} 2^{jM} y^\alpha (D^\alpha k)(2^j x - 2^j \tau y) d\tau. \end{aligned} \tag{50}$$

We insert (50) in (49). By (47) and (3.2.2/5) with (3.2.2/1) we have

$$k(2^{-j},a)(x) \tag{51}$$
$$= c\int_0^1 (1-\tau)^{M-1}2^{jM+jn}\sum_{|\alpha|=M}\int_{\mathbb{R}^n} y^\alpha D^\alpha k(2^jx-2^j\tau y)a(y)dyd\tau.$$

If $M = 0$ then (51) simply means (49). Because Q^a is centered at the origin (3.2.2/2,4) yield

$$|k(2^{-j},a)(x)| \leqq c\,2^{jM+jn}|Q^a|^{\frac{M}{n}-\frac{1}{q}+\frac{s}{n}+1}, \qquad x\in\mathbb{R}^n. \tag{52}$$

This is also valid if $M=0$. We have in any case

$$M = n\left(\frac{1}{p}-1\right) - s + \varepsilon n \quad \text{for some } \varepsilon > 0. \tag{53}$$

Now, (52) yields

$$|k(2^{-j},a)(x)| \leqq c\,2^{jM+jn}|Q^a|^{\frac{1}{p}-\frac{1}{q}+\varepsilon}, \qquad x\in\mathbb{R}^n. \tag{54}$$

On the other hand if $k(2^{-j},a)(x)\neq 0$, then by (49) we have $|2^jx-2^jy|\leqq 1$ for some $y\in Q^a$. Hence $k(2^{-j},a)(x)=0$ if $|2^jx|\geqq c$ for some positive number c which is independent of $-j\in\mathbb{N}_0$. Now (48) is a crude consequence of (54) and the last observation. Now we prove (43). By (3.2.2/6) with $(\nu,k)=(0,0)$ and (48) we have

$$|k(2^{-j},g)(x)| \leqq c\frac{2^{jM+jn}}{1+|2^jx|^b}\sum_{(\mu,l)<(0,0)} |d_{\mu l}|\,|Q_{\mu l}|^{1-\frac{1}{q}}2^{-\mu n(\frac{1}{p}-1+\varepsilon)} \tag{55}$$

$$\leqq c\frac{2^{jM+jn}}{1+|2^jx|^b}\sum_{\mu=0}^{\infty}2^{-\mu n(\frac{1}{p}-1+\varepsilon)}\left(\sum_{l\in\mathbb{Z}^n}|d_{\mu l}|^q\right)^{1/q}\left(\sum_{(\mu,l)<(0,0)}|Q_{\mu l}|\right)^{1-\frac{1}{q}}.$$

For fixed μ the union of the $Q_{\mu l}$ is essentially a partion of $2Q$, and the sum over $|d_{\mu l}|^q$ can be estimated from above by 1. Furthermore we have $\frac{1}{p}-1+\varepsilon>0$ in any case. Now (43) (also with k_0 and $j=0$ instead of k and $-j\in\mathbb{N}_0$) is a consequence of (55). The proof is complete.

3.2.4 Complements

It will be of some use for us to complement Theorem 3.2.3 by the following two assertions. All notations have the same meaning as in the previous subsections.

Proposition. Let $s\in\mathbb{R}$ and $0<p\leqq 1<q\leqq\infty$. Let K and L be fixed integers satisfying (3.2.2/1).

(i) Then $f \in S'$ is an element of F^s_{pq} if and only if it can be represented by (3.2.3/1,2), where the s-atom a_j are based on pairwise different cubes Q_{0k} and the (s,p,q)-atoms g_j are based on pairwise different cubes $Q_{\nu k}$. Furthermore, the infimum of (3.2.3/2) with respect to all these representations (where K and L are fixed) is an equivalent quasi-norm in F^s_{pq}.

(ii) Let $f \in S'$ with $\operatorname{supp} f \subset Q$, where Q is a cube in $\mathbb{R}^n$ with sides parallel to the axes and a side-length larger than 1. Then f is an element of F^s_{pq} if and only if it can be represented by (3.2.3/1,2) where both the s-atoms a_j and the (s,p,q)-atoms g_j are based on cubes $Q_{\nu k}$ with $Q_{\nu k} \subset 10Q$. Furthermore, the infimum of (3.2.3/2) with respect to all these representations (where K and L are fixed) is an equivalent quasi-norm in F^s_{pq}.

Proof. Step 1. Part (ii) is completely covered by the construction in Step 4 of the proof of Theorem 3.2.3.

Step 2. We prove (i). Let $g^1, g^2, \ldots$ be some of the (s,p,q)-atoms in the representation (3.2.3/1) which are based on the same fixed cube $Q_{\nu k}$, and let $\lambda^1, \lambda^2, \ldots$ be the related complex numbers in this representation. Let

$$\rho^r = \lambda^r \left(\sum_m |\lambda^m| \right)^{-1}$$

which makes sense because we may assume

$$0 < \sum |\lambda^j| \leqq \left(\sum |\lambda^j|^p \right)^{1/p} < \infty.$$

Then, in obvious notations, see in particular (3.2.2/6),

$$\sum_r \rho^r g^r = \sum_{(\mu,l)<(\nu,k)} \sum_r \rho^r d^r_{\mu l} a^r_{\mu l}(x) \tag{1}$$

$$= \sum_{(\mu,l)<(\nu,k)} \sum_m |\rho^m d^m_{\mu l}| \left[\sum_r \frac{\rho^r d^r_{\mu l}}{\sum_m |\rho^m d^m_{\mu l}|} a^r_{\mu l}(x) \right]$$

is a representation in the sense of (3.2.2/6,7). Furthermore we have

$$\sum \lambda^r g^r = \sum |\lambda^m| \sum \rho^r g^r \quad \text{and} \quad \left(\sum |\lambda^m| \right)^p \leqq \sum |\lambda^m|^p, \tag{2}$$

which completes the proof as far as the (s,p,q)-atoms are concerned. Of course, the same arguments work also for s-atoms.

3.3 Distinguished representations

3.3.1 A preparation

In this section we ask for distinguished kernels and related representations in the sense of 2.4.6 and 2.5.3. As often in this book we concentrate ourselves on the more complicated case, i.e., on the spaces F^s_{pq}. It will be clear how the B^s_{pq}-counterpart looks like. In this subsection we begin with a elementary preparation.

Lemma. Let $M \in \mathbb{N}$. There exist two C^∞ functions φ and ψ on $\mathbb{R}$ with $\operatorname{supp} \varphi \subset (0,1)$, $\operatorname{supp} \psi \subset (0,1)$,

$$\int_{\mathbb{R}} \varphi(\tau)d\tau = 1 \quad \text{and} \quad \varphi(t) - \frac{1}{2}\varphi\left(\frac{t}{2}\right) = \psi^{(M)}(t) \quad \text{for} \quad t \in \mathbb{R}. \tag{1}$$

Proof. Let $0 < \rho_0 < \rho_1 < \cdots < \rho_{M-1} < \frac{1}{2}$ and let $\varphi_0(t), \ldots, \varphi_{M-1}(t)$ be C^∞ functions on the real line with

$$\int_{\mathbb{R}} \varphi_j(t)dt = 1 \quad \text{if} \quad j = 0, \ldots, M-1, \tag{2}$$

and with supports concentrated near $\rho_0, \ldots, \rho_{M-1}$, respectively. Let $\varphi(t) = \sum_{j=0}^{M-1} \mu_j \varphi_j(t)$, where μ_j are coefficients such that

$$\int_{\mathbb{R}} t^k \varphi(t)dt = \sum_{j=0}^{M-1} \mu_j \int_{\mathbb{R}} t^k \varphi_j(t)dt = a_k \quad \text{with} \quad k = 0, \ldots, M-1, \tag{3}$$

$a_0 = 1$ and $a_1 = \cdots = a_{M-1} = 0$. The coefficient determinant of (3) is approximately Vandermonde's determinant $\det(\rho_j^k)_{j,k=0}^{M-1}$ and hence we may assume that it is different from zero. This shows that (3) has a unique solution. Let

$$\psi(t) = \frac{1}{(M-1)!} \int_0^t (t-\tau)^{M-1} \left(\varphi(\tau) - \left(\frac{1}{2}\right)\varphi\left(\frac{\tau}{2}\right)\right) d\tau \tag{4}$$

then φ and ψ have the required properties.

3.3.2 Distinguished kernels

First we extend Lemma 3.3.1 to n dimensions. The function φ and the number $M \in \mathbb{N}$ have the same meaning as in Lemma 3.3.1. Let

$$\Phi(x) = \prod_{l=1}^{n} \varphi(x_l), \qquad x = (x_1, \ldots, x_n) \in \mathbb{R}^n, \tag{1}$$

$$k_0(x) = \frac{(-1)^{M+1}}{M!} \sum_{r=1}^{M} \sum_{m=1}^{M} (-1)^{r+m} \binom{M}{r} \binom{M}{m} r^{-n} m^{M-n} \Phi\left(\frac{x}{rm}\right), \tag{2}$$

and

$$k(x) = k_0(x) - 2^{-n} k_0\left(\frac{x}{2}\right). \tag{3}$$

Let $f \in S'$ and $j \in \mathbb{N}$, then we put

$$f_j(x) = k(2^{-j}, f)(x) = \int_{\mathbb{R}^n} k(y) f(x + 2^{-j} y) dy, \qquad x \in \mathbb{R}^n, \tag{4}$$

where we used the notations introduced in (2.4.6/1). We complement (4) by

$$f_l(x) = k_0(2^{-l}, f)(x) = \int_{\mathbb{R}^n} k_0(y) f(x + 2^{-l} y) dy, \qquad x \in \mathbb{R}^n, \tag{5}$$

if $-l \in \mathbb{N}_0$. The functions k_0 and k are the distinguished kernels we are looking for. In any case, $f_j(x)$ and $f_l(x)$, defined in (4) and (5), respectively, are C^∞ functions on $\mathbb{R}^n$. Finally we recall what is meant by differences of functions and distributions. Let $M \in \mathbb{N}$ and $h \in \mathbb{R}^n$, then we put

$$\Delta_h^M f(x) = \sum_{r=0}^{M} (-1)^{M-r} \binom{M}{r} f(x + rh), \qquad x \in \mathbb{R}^n, \tag{6}$$

which makes sense for any $f \in S'$ (appropriate interpretation).

Proposition. (i) The functions k_0 and k from (2) and (3) are compactly supported C^∞ functions on $\mathbb{R}^n$ with

$$\operatorname{supp} k_0 \subset \mathbb{R}^n_+ \quad \text{and} \quad \operatorname{supp} k \subset \mathbb{R}^n_+ \tag{7}$$

where $\mathbb{R}^n_+ = \{x = (x_1, \ldots, x_n) \in \mathbb{R}^n : x_n > 0\}$. Furthermore

$$k_0^\wedge(0) = 1 \quad \text{and} \quad k^\wedge(\xi) = O(|\xi|^M) \tag{8}$$

near the origin.

(ii) Any $f \in S'$ can be represented as

$$f = \sum_{j=0}^{\infty} f_j \qquad (\text{convergence in } S') \tag{9}$$

where f_0 is given by (5) and f_j with $j \in \mathbb{N}$ by (4).

(iii) Let $f \in S'$, then f_j with $j \in \mathbb{N}$ can be represented as

$$f_j(x) = \sum_{|\alpha|=M} \int_{\mathbb{R}^n} D^\alpha k_\alpha(y) \Delta^M_{2^{-j}y} f(x)dy, \qquad x \in \mathbb{R}^n, \tag{10}$$

where k_α are compactly supported C^∞ functions in $\mathbb{R}^n$ with

$$\operatorname{supp} k_\alpha \subset \mathbb{R}^n_+. \tag{11}$$

(iv) Let $f \in S'$ and let β be a multi-index with $|\beta| = M$, then $D^\beta f_j$ with $j \in \mathbb{N}$ can be represented as

$$D^\beta f_j(x) = 2^{jM} \sum_{|\alpha|=M} \int_{\mathbb{R}^n} D^\alpha k_{\alpha\beta}(y) \Delta^M_{2^{-j}y} f(x)dy, \qquad x \in \mathbb{R}^n, \tag{12}$$

and $D^\beta f_l$ with $-l \in \mathbb{N}_0$ as

$$D^\beta f_l(x) = 2^{lM} \sum_{|\alpha|=M} \int_{\mathbb{R}^n} D^\alpha \tilde{k}_{\alpha\beta}(y) \Delta^M_{2^{-l}y} f(x)dy, \qquad x \in \mathbb{R}^n, \tag{13}$$

where $k_{\alpha\beta}$ and $\tilde{k}_{\alpha\beta}$ are compactly supported C^∞ functions in $\mathbb{R}^n$ with

$$\operatorname{supp} k_{\alpha\beta} \subset \mathbb{R}^n_+ \quad \text{and} \quad \operatorname{supp} \tilde{k}_{\alpha\beta} \subset \mathbb{R}^n_+. \tag{14}$$

Proof. Step 1. Let again φ and ψ be the two functions of Lemma 3.3.1 and let Φ be given by (1). Then we have

$$\int_{\mathbb{R}^n} \Phi(x)dx = 1, \qquad \operatorname{supp} \Phi \subset \mathbb{R}^n_+, \tag{15}$$

and

$$\begin{aligned} \Phi(x) - 2^{-n}\Phi\left(\frac{x}{2}\right) &= \prod_{j=1}^n \varphi(x_j) - 2^{-n} \prod_{j=1}^n \varphi\left(\frac{x_j}{2}\right) \\ &= \left[\varphi(x_1) - \frac{1}{2}\varphi\left(\frac{x_1}{2}\right)\right] \varphi(x_2) \cdots \varphi(x_n) \\ &+ \frac{1}{2}\varphi\left(\frac{x_1}{2}\right)\left[\varphi(x_2)\cdots\varphi(x_n) - 2^{-n+1}\varphi\left(\frac{x_2}{2}\right)\cdots\varphi\left(\frac{x_n}{2}\right)\right]. \end{aligned} \tag{16}$$

By (3.3.1/1) and an iteration argument it follows

$$\Phi(x) - 2^{-n}\Phi\left(\frac{x}{2}\right) = \sum_{|\alpha|=M} D^\alpha \varphi_\alpha(x), \qquad \operatorname{supp} \varphi_\alpha \subset \mathbb{R}^n_+, \tag{17}$$

where φ_α are appropriate compactly supported C^∞ functions on $\mathbb{R}^n$.

Step 2. We prove (i). By (2) and (3) we have (7) and, if we use in addition (17), the second assertion in (8). To prove the first assertion in (8) we begin with

$$k_0^\wedge(0) = \int_{\mathbb{R}^n} k_0(x)dx = \frac{(-1)^{M+1}}{M!} \sum_{r=1}^{M} \sum_{m=1}^{M} (-1)^{r+m} \binom{M}{r} \binom{M}{m} m^M, \tag{18}$$

where we used (15). Recall

$$0 = (1-1)^M = \sum_{r=0}^{M} (-1)^r \binom{M}{r}. \tag{19}$$

Furthermore we calculate the Mth derivative of

$$(e^t - 1)^M = \sum_{m=0}^{M} (-1)^{M-m} \binom{M}{m} e^{mt}$$

at the point $t = 0$ and obtain

$$M! = \sum_{m=1}^{M} (-1)^{M-m} \binom{M}{m} m^M. \tag{20}$$

We plug (19) and (20) in (18) and arrive at $k_0^\wedge(0) = 1$.

Step 3. We prove (ii). First we remark

$$2^{jn} k(2^j \cdot)^\wedge(\xi) = k^\wedge(2^{-j}\xi) = k_0^\wedge(2^{-j}\xi) - k_0^\wedge(2^{-j+1}\xi), \tag{21}$$

where $j \in \mathbb{N}$ and $\xi \in \mathbb{R}^n$, see (3). Let $g \in S'$, then (21) and (8) yield

$$g \left[\sum_{j=1}^{J} 2^{jn} k(2^j \cdot)^\wedge(\xi) + k_0^\wedge(\xi) \right] = g k_0^\wedge(2^{-J}\xi) \longrightarrow g \tag{22}$$

in S' if $J \to \infty$. Let $f \in S'$ and let f_j be given by (4) where $j \in \mathbb{N}$. Then we have

$$f_j(x) = \int_{\mathbb{R}^n} 2^{jn} k(-2^j y) f(x-y) dy, \qquad x \in \mathbb{R}^n, \tag{23}$$

and

$$f_j^\wedge(\xi) = f^\wedge 2^{jn} k(2^j \cdot)^\wedge(-\xi). \tag{24}$$

We apply (22) with $g = f^\wedge$ and arrive at

$$\sum_{j=0}^{J} f_j^\wedge \longrightarrow f^\wedge \quad \text{in} \quad S' \quad \text{if} \quad J \to \infty$$

which completes the proof of (9).

Step 4. We prove (iii). Let $j \in \mathbb{N}$, then we have

$$\int_{\mathbb{R}^n} k_0(y) f(x + 2^{-j}y) dy \tag{25}$$

$$= \frac{(-1)^{M+1}}{M!} \sum_{m=1}^{M} (-1)^{M-m} \binom{M}{m} m^{M-m} \int_{\mathbb{R}^n} \Phi\left(\frac{y}{m}\right) \sum_{r=1}^{M} (-1)^{M-r} \binom{M}{r} f(x+r2^{-j}y) dy$$

and a corresponding formula with $2^{-n}k_0\left(\frac{y}{2}\right)$ instead of $k_0(y)$ and with $2^{-n}\Phi\left(\frac{y}{2m}\right)$ instead of $\Phi\left(\frac{y}{m}\right)$. We take the difference of these two formulas and obtain

$$\int_{\mathbb{R}^n} k(y) f(x + 2^{-j}y) dy = \frac{(-1)^{M+1}}{M!} \sum_{m=1}^{M} (-1)^{M-m} \binom{M}{m} m^{M-n} \tag{26}$$

$$\times \int_{\mathbb{R}^n} \left[\Phi\left(\frac{y}{m}\right) - 2^{-n}\Phi\left(\frac{y}{2m}\right)\right] \sum_{r=0}^{M} (-1)^{M-r} \binom{M}{r} f(x + r2^{-j}y) dy,$$

where we added the term with $r = 0$ which is justified by (17). Now (10) and (11) is a consequence of (6), (17), and (15).

Step 5. We prove (iv). Let $|\beta| = M$ and let l be an integer, then we have

$$D^\beta \int_{\mathbb{R}^n} k_0(y) f(x + 2^{-l}y) dy = 2^{ln} D^\beta \int_{\mathbb{R}^n} k_0(2^l y - 2^l x) f(y) dy \tag{27}$$

$$= (-1)^M 2^{lM} \int_{\mathbb{R}^n} D^\beta k_0(y) f(x + 2^{-l}y) dy$$

$$= -\frac{2^{lM}}{M!} \sum_{r=1}^{M} (-1)^{M-r} \binom{M}{r} r^{-n-M} \int_{\mathbb{R}^n} D^\beta \Phi\left(\frac{y}{r}\right) \sum_{m=1}^{M} (-1)^{M-r} \binom{M}{m} f(x + m2^{-l}y) dy.$$

We can add the term with $m = 0$ on the right-hand side of (27). By (6) we obtain (13) with $-l \in \mathbb{N}_0$. Finally (3) and (27) prove (12) with $j \in \mathbb{N}$.

3.3.3 Distinguished equivalent quasi-norms

It is the aim of this subsection to modify and specify Theorem 2.4.6, where we now identify the means (2.4.6/1) with (3.3.2/4) if $t = 2^{-j}$ and $j \in \mathbb{N}$ and with (3.3.2/5) if $t = 2^{-l}$ and $-l \in \mathbb{N}_0$. In other words, the kernels $k_0(x)$ and $k(x)$, as well as the means $f_j(x)$, $j \in \mathbb{N}$, and $f_l(x)$, $-l \in \mathbb{N}_0$, have the same meaning as in (3.3.2/2–5).

Theorem. Let $0 < p < \infty$, $0 < q \leqq \infty$, and $s \in \mathbb{R}$. Let $M \in \mathbb{N}$ with $M > \max(s, n(\frac{1}{p} - 1)_+)$ and let $k_0(x)$ and $k(x)$ be given by (3.3.2/2,3). Then there exists a number $K \in \mathbb{N}_0$ such that

$$\left\| \left(\sum_{j=-K}^{\infty} 2^{jsq} |f_j(\cdot)|^q \right)^{1/q} \mid L_p \right\| \tag{1}$$

(modification if $q = \infty$) is an equivalent quasi-norm on F^s_{pq}.

Proof. Step 1. Let $f \in F^s_{pq}$. Recall that f_j with $j \in \mathbb{N}$ is given by (3.3.2/4) and f_j with $-j \in \mathbb{N}_0$ is given by (3.3.2/5). To clarify the role of the number $K \in \mathbb{N}$ we introduce

$$k'(x) = \sum_{r=0}^{K} k(2^{-r}x)2^{-nr} = k_0(x) - 2^{-(K+1)n} k_0(2^{-K-1}x), \tag{2}$$

see (3.3.2/3). Now we choose $K \in \mathbb{N}$ in such a way that we have for some $\varepsilon > 0$

$$k'^{\wedge}(\xi) = k_0^{\wedge}(\xi) - k_0^{\wedge}(2^{K+1}\xi) \neq 0 \quad \text{if} \quad \varepsilon \leqq |\xi| \leqq 2\varepsilon \tag{3}$$

and

$$k_0^{\wedge}(\xi) \neq 0 \quad \text{if} \quad |\xi| \leqq 4\varepsilon. \tag{4}$$

Step 2. To prove the theorem we replace the functions k_0 and k in the proof of Theorem 2.4.6 by the above function k_0 and k', respectively. Then (3) and (4) cover a modified version of the Tauberian conditions (2.4.1/6,7). In other words, by the same arguments as in the proof of Theorem 2.4.6 we reduce the above theorem to Theorem 2.4.1.

Remark. The quasi-norm (1) and the decomposition (3.3.2/9) will be of great service for us in the sequel. The functions $f_j(x)$ given by (3.3.2/10) have a useful structure. The kernels are local and one can incorporate as many derivatives and differences as one wants. This will be used later on in order to find intrinsic descriptions of some spaces $F^s_{pq}(\Omega)$ on bounded smooth domains Ω in terms of differences. The price to pay is the relatively complicated structure of the underlying kernels k_0 and k, see (3.3.2/2,3). Constructions of this type go back to G.A. Kaljabin [Kal4], see also Lizorkin's appendix D.2.3 of the Russian edition of [Triß].

For the sake of completeness we mention the B^s_{pq}-counterpart of the above theorem.

Proposition. Let $0 < p \leqq \infty$, $0 < q \leqq \infty$ and $s \in \mathbb{R}$. Let $M \in \mathbb{N}$ with $M > \max(s, n(\frac{1}{p} - 1)_+)$ and let $k_0(x)$ and $k(x)$ be given by (3.3.2/2,3). Let $K \in \mathbb{N}_0$ be the same number as in the above theorem, see (2) – (4), then

$$\left(\sum_{j=-K}^{\infty} 2^{jsq} \|f_j \mid L_p\|^q \right)^{1/q} \tag{5}$$

(modification if $q = \infty$) is an equivalent quasi-norm in B^s_{pq}.

Proof. The proof is the same as the proof of the above theorem. We have to use Theorem 2.5.3 instead of Theorem 2.4.6 with a reference to Theorem 2.5.1 instead of Theorem 2.4.1.

3.4 Oscillations

3.4.1 Definitions and preparations

In this section we give a self-contained introduction into the L_p-theory of oscillations as far as it is of interest in our context.

Let Ω be a domain in $\mathbb{R}^n$ and let

$$B(x,t) = \{y \in \Omega \colon |x-y| < t\}, \qquad x \in \Omega, \quad t > 0, \tag{1}$$

be that part of the ball centered at x and of radius t which is contained in Ω. Let $|B(x,t)|$ be the volume of $B(x,t)$. For sake of brevity we write

$$⨍_{B(x,t)} g(y)dy = |B(x,t)|^{-1} \int_{B(x,t)} g(y)dy \tag{2}$$

for the mean value. Recall that $\mathbb{N}$ stands for the collection of all natural numbers and $\mathbb{N}_0 = \mathbb{N} \cup \{0\}$. We complement these notations by $\mathbb{N}_{-1} = \mathbb{N}_0 \cup \{-1\}$, i.e., the collection of all integers larger than or equal to -1. Let $M \in \mathbb{N}_{-1}$ then $\mathbb{P}_M$ stands for the collection of all polynomials (with complex coefficients) of degree less than or equal to M, where we put $\mathbb{P}_{-1} = \{0\}$.

Definition. Let Ω be a domain in $\mathbb{R}^n$, let $M \in \mathbb{N}_{-1}$ and $0 < u \leqq \infty$. Then

$$\mathrm{osc}_u^M f(x,t) = \inf \left(\fint_{B(x,t)} |f(y) - P(y)|^u dy \right)^{1/u}, \qquad x \in \Omega, t > 0, \tag{3}$$

is called (local) oscillation, where the infimum is taken over all $P \in \mathbb{P}_M$ (usual modification if $u = \infty$).

Remark. Of course, $\mathrm{osc}_u^M f$ and also $B(x,t)$ depend on Ω. In general, we shall not indicate this dependence. Usually Ω is a given fixed domain. If several types of oscillations are involved then we indicate the dependence of the oscillations on the underlying domain. Furthermore in (3) it is tacitly assumed that $|f|^u$ is integrable in $B(x,t)$.

Domains. In this book we are interested in three types of domains: $\mathbb{R}^n$, $\mathbb{R}^n_+$, and bounded C^∞ domains in $\mathbb{R}^n$. In the latter case we always assume, without restriction of generality, $0 \in \partial\Omega$ and that $\partial\Omega$ can be described near the origin as

$$x_n = \psi(x'), \qquad x' = (x_1, \ldots, x_{n-1}) \in \omega, \tag{4}$$

where ψ is a C^∞ function in some neighbourhood ω of the origin in $\mathbb{R}^{n-1}$. Besides $\psi(0) = 0$ we may assume $\frac{\partial \psi}{\partial x_j}(0) = 0$ if $j = 1, \ldots, n-1$. Finally, let k be a compactly supported C^∞ kernel in $\mathbb{R}^n$ with

$$\operatorname{supp} k \subset \mathbb{R}^n_+. \tag{5}$$

Then the means

$$k(t,f)(x) = \int_{\mathbb{R}^n} k(y) f(x+ty) dy, \tag{6}$$

see (2.4.6/1), with, say, $f \in L_p(\Omega)$ where $1 \leqq p \leqq \infty$, makes sense at least for any $x \in \Omega$ near the origin and $t > 0$ small. To avoid awkward formulations we shall always assume that (6) makes sense for any $x \in \Omega$ near the origin and $0 < t \leqq 1$.

3.4.2 Optimal polynomials

The kernels k_0 and k have the same meaning as in (3.3.2/2,3). Furthermore Ω stands either for $\mathbb{R}^n$ or $\mathbb{R}^n_+$ or the domain described at the end of the last subsection. Let $f \in L_p(\Omega)$ with $1 \leqq p \leqq \infty$, then $f_j(x)$ with $j \in \mathbb{N}$ and $f_0(x)$ are given by (3.3.2/4) and (3.3.2/5), respectively. By our agreement at the end of the last subsection this makes sense at least for $x \in \Omega$ near the origin, see (3.3.2/7). Let

$$f^j(x) = \sum_{k=0}^{j} f_k(x) \quad \text{where } j \in \mathbb{N}_0 \text{ and } x \in \Omega \text{ near the origin.} \tag{1}$$

Proposition (Optimal polynomials). Let Ω be either $\mathbb{R}^n$ or $\mathbb{R}^n_+$ or a bounded C^∞ domain in the sense of 3.4.1. Let $1 \leqq p \leqq \infty$ and $M \in \mathbb{N}_0$, then there exist positive numbers c and c' such that

$$\begin{aligned} \operatorname{osc}_p^{M-1} f(x, 2^{-j}) &\leqq \left(\oint_{B(x,2^{-j})} \left| f(y) - \sum_{|\alpha| \leqq M-1} \frac{1}{\alpha!} (D^\alpha f^j)(x)(y-x)^\alpha \right|^p dy \right)^{1/p} \\ &\leqq c' \operatorname{osc}_p^{M-1} f(x, c2^{-j}) \end{aligned} \tag{2}$$

for all $f \in L_p(\Omega)$, $j \in \mathbb{N}_0$ and all $x \in \Omega$ (near the origin).

Proof. The left-hand side of (2) is obvious by definition. We prove the right-hand side. By (3.3.2/3–5) we have

$$\begin{aligned} f^j(x) &= \int k_0(y) f(x+y) dy + \sum_{l=1}^{j} 2^{ln} \int k(2^l y) f(x+y) dy \\ &= \int k_0(y) f(x + 2^{-j} y) dy \end{aligned} \tag{3}$$

where we omit to indicate the domain of integration. Now we use the explicit form of k_0, see (3.3.2/2). By (3.3.2/15,20) we arrive at

$$\begin{aligned} &f^j(x) - f(x) \\ &= \frac{(-1)^{M+1}}{M!} \sum_{m=1}^{M} (-1)^{M-m} \binom{M}{m} m^{M-n} \int \Phi\left(\frac{y}{m}\right) \sum_{r=0}^{M} (-1)^{M-r} \binom{M}{r} f(x + r2^{-j} y) dy \\ &= \int \tilde{k}(y) \Delta^M_{2^{-j}y} f(x) dy, \end{aligned} \tag{4}$$

where $\tilde{k}$ is a compactly supported C^∞ function with $\operatorname{supp} \tilde{k} \subset \mathbb{R}^n_+$. We can replace f on the right-hand side of (4) by $f - P$ with $P \in \mathbb{P}_{M-1}$. Then it follows

$$|f(x) - f^j(x)| \leqq c' |f(x) - P(x)| + c' \oint_{B(x,c2^{-j})} |f(y) - P(y)| dy \tag{5}$$

for all $P \in \mathbb{P}_{M-1}$ and some $c > 0$ and $c' > 0$. We obtain by integration

$$\left(\oint_{B(x,2^{-j})} |f(y) - f^j(y)|^p dy \right)^{1/p} \leqq c' \operatorname{osc}_p^{M-1} f(x, c2^{-j}) \tag{6}$$

for some $c > 0$ and $c' > 0$. Let $y \in B(x, 2^{-j})$, then we have by Taylor's formula

$$\left| f^j(y) - \sum_{|\alpha| \leqq M-1} \frac{1}{\alpha!} (D^\alpha f^j)(x)(y-x)^\alpha \right| \leqq c 2^{-jM} \sup_{z \in B(x,2^{-j})} \sum_{|\alpha|=M} |D^\alpha f^j(z)|. \quad (7)$$

By (3) it follows

$$D^\alpha f^j(x) = 2^{j|\alpha|} \int D^\alpha k_0(y) f(x + 2^{-j} y) dy, \qquad |\alpha| = M. \quad (8)$$

We can again replace f on the right-hand side of (8) by $f - P$ with $P \in \mathbb{P}_{M-1}$ and obtain by the same arguments as above

$$|D^\alpha f^j(z)| \leqq c' 2^{jM} \operatorname{osc}_p^{M-1} f(x, c2^{-j}), \qquad z \in B(x, 2^{-j}), \ |\alpha| = M, \quad (9)$$

for some $c > 0$ and $c' > 0$. We insert (9) in (7), then the right-hand side of (2) follows from (6) and (7).

3.4.3 Additive inequalities

All notations have the same meaning as in the preceding subsection.

Proposition. Let Ω be either $\mathbb{R}^n$ or $\mathbb{R}^n_+$ or a bounded domain C^∞ in $\mathbb{R}^n$. Let $1 \leqq p \leqq \infty$ and $M \in \mathbb{N}_0$, then there exist two positive numbers c and c' such that

$$\operatorname{osc}_p^{M-1} f(x, 2^{-j}) \leqq c' \sum_{l=0}^{j} 2^{-(j-l)M} \operatorname{osc}_p^M f(x, c2^{-l}) + c' \, 2^{-jM} \int_{B(x,c)} |f(y)| dy \quad (1)$$

for all $f \in L_p(\Omega)$, $j \in \mathbb{N}_0$, and all $x \in \Omega$.

Proof. We may assume that Ω has the properties described at the end of 3.4.1 and that $x \in \Omega$ is a point near the origin. Then we can apply (3.4.2/2) and obtain

$$\operatorname{osc}_p^{M-1} f(x, 2^{-j}) \leqq c' \operatorname{osc}_p^M f(x, c2^{-j}) + c' \, 2^{-jM} \sum_{|\alpha|=M} |D^\alpha f^j(x)|. \quad (2)$$

Next we calculate $D^\alpha f^l(x)$ via (3.4.2/3) and (3.3.2/27,13) and obtain

$$D^\alpha f^l(x) = 2^{lM} \int k'_\alpha(y) \Delta^M_{2^{-l}y} f(x) dy, \qquad |\alpha| = M, \ \ l \in \mathbb{N}_0, \quad (3)$$

where $k'_\alpha(y)$ is a C^∞ function in $\mathbb{R}^n$ with $\operatorname{supp} k'_\alpha \subset \mathbb{R}^n_+$, cf. also with (3.4.2/4), (obvious modification if $M = 0$). Let $l \in \mathbb{N}_0$ then we have

$$D^\alpha f^{l+1}(x) - D^\alpha f^l(x) = 2^{(l+1)M} \int k'_\alpha(y)[\Delta^M_{2^{-l-1}y} f(x) - 2^{-M}\Delta^M_{2^{-l}y} f(x)]dy. \quad (4)$$

Now we use the formula

$$\Delta^M_{2^{-l-1}y} f(x) - 2^{-M}\Delta^M_{2^{-l}y} f(x) = \sum_{r=0}^{M-1} a_r \Delta^{M+1}_{2^{-l-1}y} f(x + r2^{-l-1}y), \quad (5)$$

where a_r are appropriate numbers, see [Triß: (2.5.9/45)]. By (5) it is now clear that one can replace f in (4) by $f - P$ with $P \in \mathbb{P}_M$ (this is also correct if $M = 0$). Because $\int k'_\alpha(y)dy = 0$ if $M \in \mathbb{N}$, see (3.3.2/27), it follows

$$|D^\alpha f^{l+1}(x) - D^\alpha f^l(x)| \leqq c' \, 2^{lM} \fint_{B(x,c2^{-l})} |f(y) - P(y)|dy \quad (6)$$

for some $c > 0$ and $c' > 0$, and for all $P \in \mathbb{P}_M$ (this is also correct if $M = 0$). Hence we have

$$|D^\alpha f^{l+1}(x) - D^\alpha f^l(x)| \leqq c' \, 2^{lM} \operatorname{osc}^M_p f(x, c2^{-l}). \quad (7)$$

We return to (2) and estimate

$$\begin{aligned} 2^{-jM}|D^\alpha f^j(x)| &\leqq 2^{-jM} \sum_{l=0}^{j-1} |D^\alpha f^{l+1}(x) - D^\alpha f^l(x)| + 2^{-jM}|D^\alpha f_0(x)| \quad (8) \\ &\leqq c' \sum_{l=0}^{j-1} 2^{-(j-l)M} \operatorname{osc}^M_p f(x, c2^{-l}) + c' \, 2^{-jM} \int_{B(x,c)} |f(y)|dy, \end{aligned}$$

where $|\alpha| = M$. Now (1) follows from (2) and (8).

3.4.4 Multiplicative inequalities

All notations have the same meaning as in the preceding subsections.

Proposition. Let Ω be either $\mathbb{R}^n$, or $\mathbb{R}^n_+$ or a bounded C^∞ domain in $\mathbb{R}^n$. Let $M \in \mathbb{N}_{-1}$,

$$0 < u < 1 < r \leqq \infty \quad \text{and} \quad 1 = \frac{1-\Theta}{u} + \frac{\Theta}{r}. \quad (1)$$

Then there exists positive numbers c and c' such that

$$\operatorname{osc}_1^M f(x,2^{-j}) \leqq c' \left[\sum_{l=0}^{\infty} \operatorname{osc}_r^M f(x, c2^{-j-l}) \right]^{\Theta} \tag{2}$$

$$\times \sup_{l \in \mathbb{N}} \left[\operatorname{osc}_u^M (x, c2^{-j-l}) + \left(\oint_{B(x,c2^{-j})} \operatorname{osc}_u^M (w, c2^{-j-l})^u dw \right)^{1/u} \right]^{1-\Theta}$$

for all $f \in L_r(\Omega)$, $j \in \mathbb{N}_0$, and $x \in \Omega$.

Proof. We assume that Ω has the properties described at the end of 3.4.1 and that $x \in \Omega$ is a point near the origin. Then (3.4.2/2) is at our disposal, now with M instead of $M-1$. These hypotheses will justify all the calculations below. We may assume $M \in \mathbb{N}_0$ because the case $M=-1$ follows from Hölder's inequality. We use the same technique as in 3.4.2 and estimate

$$\operatorname{osc}_1^M f(x,2^{-j}) \leqq \oint_{B(x,2^{-j})} |f(y) - f^j(y)| dy \tag{3}$$

$$+ \oint_{B(x,2^{-j})} \left| f^j(y) - \sum_{|\alpha| \leq M} \frac{1}{\alpha!} D^\alpha f^j(x)(y-x)^\alpha \right| dy,$$

see (3.4.2/2) with M instead of $M-1$. In order to estimate the second term we use the following integral version of Taylor's formula,

$$f^j(y) - \sum_{|\alpha| \leq M} \frac{1}{\alpha!} D^\alpha f^j(x)(y-x)^\alpha \tag{4}$$

$$= \sum_{|\beta|=M+1} \frac{M+1}{\beta!} (y-x)^\beta \int_0^1 (1-\tau)^M D^\beta f^j(x+\tau(y-x)) d\tau.$$

We may assume $x + \tau(y-x) \in B(x,2^{-j})$ for all $0 \leqq \tau \leqq 1$. Then we have

$$\oint_{B(x,2^{-j})} |D^\beta f^j(x+\tau(y-x))| dy \leqq c \oint_{B(x,\tau 2^{-j})} |D^\beta f^j(y)| dy. \tag{5}$$

We use (3.4.3/3), now with $M+1$ instead of M, and obtain

$$2^{-j(M+1)} |D^\beta f^j(y)| \tag{6}$$

$$\leqq \left(\int |k'_\beta(z)|^u |\Delta^{M+1}_{2^{-j}z} f(y)|^u dz \right)^{(1-\theta)/u} \left(\int |k'_\beta(z)|^r |\Delta^{M+1}_{2^{-j}z} f(y)|^r dz \right)^{\theta/r}$$

$$\leqq c' \left(|f(y) - P_1(y)|^u + \oint_{2^j w \in B} |f(y+w) - P_1(y+w)|^u dw \right)^{(1-\theta)/u}$$

$$\times\Big(|f(y)-P_2(y)|^r+\fint_{B(y,c2^{-j})}|f(z)-P(z)|^r dz\Big)^{\theta/r}$$

where the polynomials $P_1\in\mathbb{P}_M$ and $P_2\in\mathbb{P}_M$ are at our disposal and where B is a suitable fixed ball in $\mathbb{R}^n_+$ which comes from $\operatorname{supp}k'_\beta$. By (1) and Hölder's inequality we arrive at

$$2^{-j(M+1)}\fint_{B(x,\tau 2^{-j})}|D^\beta f^j(y)|dy \tag{7}$$

$$\leqq c\Big[\fint_{B(x,\tau 2^{-j})}|f(y)-P_1(y)|^u dy$$

$$+\fint_{2^j w\in B}\fint_{B(x,\tau 2^{-j})}|f(y+w)-P_1(y+w)|^u dy dw\Big]^{(1-\theta)/u}$$

$$\times\Big[\fint_{B(x,\tau 2^{-j})}|f(y)-P_2(y)|^r dy+\fint_{B(x,\tau 2^{-j})}\fint_{B(y,c2^{-j})}|f(z)-P_2(z)|^r dz dy\Big]^{\theta/r}.$$

The two factors will be treated differently. As for the second factor we choose $P_2(z)$ as an optimal polynomial in the sense of (3.4.2/2) with M instead of $M-1$. Then the second summand in the second factor can be estimated from above by $c'\mathrm{osc}^M_r f(x,c2^{-j})^r$ with a new constant $c>0$. We may assume $\tau=2^{-l}$ with $l\in\mathbb{N}_0$ in the first summand of the second factor. By our choice of P_2 and by (3.4.2/2,1) with $j+l$ and M instead of j and $M-1$, respectively, we have

$$\Big(\fint_{B(x,2^{-j-l})}|f(y)-P_2(y)|^r dy\Big)^{1/r} \tag{8}$$

$$\leqq c'\,\mathrm{osc}^M_r f(x,c2^{-l-j})+c'\sum_{k=1}^{l}\sum_{|\alpha|\leqq M}2^{-(j+l)|\alpha|}|D^\alpha f_{j+k}(x)|.$$

By (3.3.2/10,12) we have

$$D^\alpha f_{j+k}(x)=2^{(j+k)|\alpha|}\sum_{|\beta|=M+1}\int D^\beta k_{\alpha\beta}(y)\Delta^{M+1}_{2^{-j-k}y}f(x)dy \tag{9}$$

if either $|\alpha|=0$ or $|\alpha|=M+1$, where $k_{\alpha\beta}(y)$ are compactly supported C^∞ functions in $\mathbb{R}^n$ with supports in $\mathbb{R}^n_+$. We can replace f in (9) by $f-P$ with $P\in\mathbb{P}_M$. Because the integral over the kernels vanishes we obtain

$$|D^\alpha f_{j+k}(x)|\leqq c'\,2^{(j+k)|\alpha|}\mathrm{osc}^M_r f(x,c2^{-j-k}) \tag{10}$$

if either $|\alpha|=0$ or $|\alpha|=M+1$. Let $0<|\alpha|\leqq M$, then we have

$$2^{-(j+k)|\alpha|}|D^\alpha f_{j+k}(x)| \tag{11}$$

$$\leqq c \sup_{y\in B(x,2^{-j-k})} |f_{j+k}(y)| + c2^{-(j+k)(M+1)} \sum_{|\beta|=M+1} \sup_{y\in B(x,2^{-j-k})} |D^\beta f_{j+k}(y)|$$
$$\leqq c' \operatorname{osc}_r^M f(x, c2^{-j-k}).$$

In other words, (10) holds for all α with $0 \leqq |\alpha| \leqq M+1$. We put (10) in (8) and obtain

$$\left(\fint_{B(x,2^{-j-l})} |f(y) - P_2(y)|^r dy \right)^{1/r} \leqq c' \sum_{k=0}^{\infty} \operatorname{osc}_r^M f(x, c2^{-j-k}). \tag{12}$$

In other words, the second factor in (7) can be estimated from above by

$$c' \left[\sum_{l=0}^{\infty} \operatorname{osc}_r^M f(x, c2^{-j-l}) \right]^\theta . \tag{13}$$

We choose P_1 in the first factor in (7) in an optimal way, then the first factor on the right-hand side of (7) can be estimated from above by the second factor on the right-hand side of (2), independently of τ with $0 < \tau \leqq 1$. Hence (7) can be estimated from above by the right-hand side of (2), independently of τ. Now by (4) and (5) it follows that the second integral in (3) can be estimated from above by the right-hand side of (2). In order to estimate the first integral in (3) we use (3.4.2/4), now with $M+1$ instead of M. But then we have the same estimates as on the right-hand side of (6) and (7), but now with $\tau = 1$. Hence we have again the desired estimate.

3.5 Characterizations via oscillations and differences

3.5.1 Characterizations via oscillations

We wish to characterize some spaces F^s_{pq} and B^s_{pq} on $\mathbb{R}^n$ in terms of oscillations. The spaces F^s_{pq} and B^s_{pq} have the same meaning as in Definition 2.3.1, whereas oscillations had been introduced in Definition 3.4.1, now based on $\Omega = \mathbb{R}^n$. Recall that the numbers

$$\sigma_p = n\left(\frac{1}{p} - 1\right)_+ \quad \text{and} \quad \sigma_{pq} = n\left(\frac{1}{\min(p,q)} - 1\right)_+ \tag{1}$$

play a decisive role in the theory of the spaces B^s_{pq} and F^s_{pq}, see 2.3.3, 2.4.1 and the characterizations of some spaces of the above type in terms of differences, see 2.6. Here we need a slight but natural modification of (1). Recall $a_+ = \max(a, 0)$ if $a \in \mathbb{R}$.

Theorem. (i) Let $0 < p < \infty$, $0 < q \leqq \infty$, $1 \leqq r \leqq \infty$, and

$$s > n\left(\frac{1}{p} - \frac{1}{r}\right)_+ \quad \text{and} \quad s > n\left(\frac{1}{q} - \frac{1}{r}\right)_+ . \tag{2}$$

Let $0 < u \leqq r$ and $M \in \mathbb{N}_0$ with $M > [s]$, then

$$F^s_{pq} = \Big\{ f \in L_{\max(p,r)}: \tag{3}$$

$$\|f \mid L_p\| + \left\| \left(\int_0^1 t^{-sq} \mathrm{osc}_u^M f(\cdot,t)^q \frac{dt}{t} \right)^{1/q} \mid L_p \right\| < \infty \Big\}$$

(modification if $q = \infty$) in the sense of equivalent quasi-norms.

(ii) Let $0 < p \leqq \infty$, $0 < q \leqq \infty$, $1 \leqq r \leqq \infty$ and

$$s > n\left(\frac{1}{p} - \frac{1}{r}\right)_+ . \tag{4}$$

Let $0 < u \leqq r$ and $M \in \mathbb{N}_0$ with $M \geqq [s]$, then

$$B^s_{pq} = \Big\{ f \in L_{\max(p,r)}: \tag{5}$$

$$\|f \mid L_p\| + \left(\int_0^1 t^{-sq} \|\mathrm{osc}_u^M f(\cdot,t) \mid L_p\|^q \frac{dt}{t} \right)^{1/q} < \infty \Big\}$$

(modification if $q = \infty$) in the sense of equivalent quasi-norms.

Proof. Step 1. We concentrate on (i). The proof of (ii) is similar but technically simpler. We begin with some preliminarily remarks. First, by (3.4.1/3) it is clear that the quasi-norm in (3) is equivalent to

$$\|f \mid F^s_{pq}\|_{\mathrm{osc}} = \|f \mid L_p\| + \left\| \left(\sum_{j=1}^{\infty} 2^{jsq} \mathrm{osc}_u^M f(\cdot, 2^{-j})^q \right)^{1/q} \mid L_p \right\| . \tag{6}$$

Secondly, we wish to compare (6) with a modified version of (2.4.6/6), where p, q, r, s have the above meaning. Let k^0 and $k(y) = \Delta^N k^0(y)$ with $2N > s$ be the same functions as in (2.4.6/2–4), then

$$\|f \mid F^s_{pq}\|_k = \|f \mid L_p\| + \left\| \left(\sum_{j=1}^{\infty} 2^{jsq} |k(2^{-j}, f)(\cdot)|^q \right)^{1/q} \mid L_p \right\| \tag{7}$$

is an equivalent quasi-norm in F^s_{pq}. The proof is the same as the proof of Theorem 2.4.6, where we may now choose $k_0(y) = 1$ because $s > \sigma_{pq}$, see Theorem 2.4.1. In particular $f \in L_p$ makes sense if $f \in F^s_{pq}$. Let $r > p$, then

$$F^s_{pq} \subset L_r \tag{8}$$

is a known embedding theorem, an extension of (2.3.2/25) with the same proof and the same references, see also (2.3.2/22). In other words we have $f \in L_{\max(p,r)}$ if $f \in F^s_{pq}$.

Step 2. Let p, q, r, s, and M be the same numbers as in (i) and let u be restricted by $1 \leqq u \leqq r$. Let $f \in L_{\max(p,r)}$ such that (7) is finite. As mentioned above we have $k(y) = \Delta^N k^0(y)$ where we may choose N as large as we want, in particular $2N > M$. Then we have

$$\begin{aligned} |k(2^{-j}, f)(x)| &= \left| \int k(y)(f - P)(x + 2^{-j}y)dy \right| \\ &\leqq c \operatorname{osc}_u^M f(x, 2^{-j}), \quad x \in \mathbb{R}^n, \quad j \in \mathbb{N}. \end{aligned} \tag{9}$$

We used that the equality holds for any $P \in \mathbb{P}_M$. It follows

$$\|f \mid F^s_{pq}\|_k \leqq c\|f \mid F^s_{pq}\|_{\text{osc}}. \tag{10}$$

Step 3. We wish to prove the converse inequality under the same restrictions for the involved parameters, in particular $1 \leqq u \leqq r$. We assume $f \in F^s_{pq}$ (what was not needed in the second step). We use the same techniques and the same notations as in 3.4.2 and obtain

$$\begin{aligned} \operatorname{osc}_u^M f(x, 2^{-j}) &\leqq \left(\oint_{B(x,2^{-j})} |f(y) - f^j(y)|^u dy \right)^{1/u} \\ &+ \left(\oint_{B(x,2^{-j})} \left| f^j(y) - \sum_{|\alpha| \leqq M} \frac{1}{\alpha!}(D^\alpha f^j)(x)(y-x)^\alpha \right|^u dy \right)^{1/u} \\ &\leqq \sum_{l=1}^\infty \left(\oint_{B(x,2^{-j})} |f_{j+l}(y)|^u dy \right)^{1/u} + c \sup \left| \int D^\alpha k_0(y) f(z + 2^{-j}y)dy \right| \end{aligned} \tag{11}$$

where the supremum must be taken over $z \in B(x, c2^{-j})$ and $|\alpha| = M + 1$: see (3.3.2/9), (3.4.2/1) as for the first term and (3.4.2/7,8) (with $M + 1$ instead of M) as far as the second term is concerned. The second term on the right-hand side of (11) will be abbreviated temporarily by $\tilde{f}_j(x)$. Because $M + 1 > s$ we interpet $\tilde{f}_j(x)$ as a maximal function to which Corollary 2.4.1/2 can be applied (for an estimate from above the Tauberian conditions are not necessary). We obtain

$$\left\| \left(\sum_{j=1}^\infty 2^{jsq} \tilde{f}_j(\cdot)^q \right)^{1/q} \mid L_p \right\| \leqq c\|f \mid F^s_{pq}\|. \tag{12}$$

To handle the first term on the right-hand side of (11) we need more maximal functions. Let g^* be the usual Hardy–Littlewood maximal function of g and let

$$f_j^a(x) = \sup_{y \in \mathbb{R}^n} \frac{|f_j(x+y)|}{1+|2^j y|^a} \quad \text{with} \quad a > \frac{n}{\min(p,q)}. \tag{13}$$

Because $\frac{s}{n} > \frac{1}{\min(p,q)} - \frac{1}{r}$ and $1 \leqq u \leqq r$ we find a number $\kappa \in \mathbb{R}$ with $\kappa < 1$ and

$$\frac{1}{u} \min(p,q) > \kappa > 1 - \frac{s}{n} \min(p,q). \tag{14}$$

Then we have

$$\left(⨏_{B(x,2^{-j})} |f_{j+l}(y)|^u dy \right)^{1/u} \leqq c\, 2^{la(1-\kappa)} f_{j+l}^a(x)^{1-\kappa} (|f_{j+l}|^{\kappa u})^{*\frac{1}{u}}(x). \tag{15}$$

Now by (6,11,12,15) we obtain

$$\|f \mid F_{pq}^s\|_{\mathrm{osc}} \leqq c \|f \mid F_{pq}^s\| \tag{16}$$

$$+ c \sum_{l=1}^{\infty} 2^{-(s-\varepsilon)l} 2^{la(1-\kappa)} \left\| \left(\sum_{j=1}^{\infty} 2^{(j+l)sq} f_{j+l}^a(\cdot)^{q(1-\kappa)} (|f_{j+l}|^{\kappa u})^{*\frac{q}{u}}(\cdot) \right)^{1/q} \mid L_p \right\|,$$

where ε is an arbitrary positive number. We may choose a in (13), κ in (14) and $\varepsilon > 0$ in such a way that $s - \varepsilon > a(1-\kappa)$. Then the right-hand side of (16) can be estimated from above by

$$c\|f \mid F_{pq}^s\| + c \left\| \left(\sum_{j=1}^{\infty} 2^{jsq} f_j^a(\cdot)^q \right)^{1/q} \mid L_p \right\|^{1-\kappa} \tag{17}$$

$$\times \left\| \left(\sum_{j=1}^{\infty} 2^{jsq} (|f_j|^{\kappa u})^{*\frac{q}{\kappa u}} \right)^{1/q} \mid L_p \right\|^{\kappa}.$$

By Corollary 2.4.1/2 the first factor of the second summand in (17) can be estimated from above by $c\|f \mid F_{pq}^s\|^{1-\kappa}$. Furthermore by construction we have $p > \kappa u$ and $q > \kappa u$, and hence, by Theorem 2.2.2 the second factor of the second summand in (17) can be estimated from above by $c\|f \mid F_{pq}^s\|^{\kappa}$. We arrive at

$$\|f \mid F_{pq}^s\|_{\mathrm{osc}} \leqq c \|f \mid F_{pq}^s\|. \tag{18}$$

Together with (10) we obtain, that $\|f \mid F_{pq}^s\|_{\mathrm{osc}}$ is an equivalent quasi-norm on F_{pq}^s. Recall that up to now u was restricted by $1 \leqq u \leqq r$.

Step 4. The next task is to extend the just obtained result to $0 < u < 1$. By Hölder's inequality we have

$$\mathrm{osc}_v^M f(x,t) \leqq \mathrm{osc}_w^M f(x,t), \qquad 0 < v < w \leqq \infty, \quad t > 0. \tag{19}$$

Then it follows that (18) can be extended immediately to $0 < u < 1$. We have to prove (10) under the hypothesis $f \in F^s_{pq}$ (which will be removed in the following step). We begin with some preparations. If (2) is satisfied for $r = 1$ then it is also valid for some $r > 1$. Furthermore by (19) it is sufficient to prove (10) for small values of $u > 0$. Hence we may assume without restriction of generality

$$0 < u < 1 < r \leqq \infty \quad \text{and} \quad 0 < u < \min(p,q). \tag{20}$$

By (3.4.4/2) we have

$$\mathrm{osc}_1^M f(x,2^{-j}) \leqq \varepsilon \left(\sum_{l=0}^{\infty} 2^{l\eta q} \mathrm{osc}_r^M f(x, c2^{-j-l})^q \right)^{1/q} \tag{21}$$
$$+ c_\varepsilon \left(\sum_{l=0}^{\infty} |\mathrm{osc}_u^M f(x, 2^{-j-l})^u|^{*\frac{q}{u}} \right)^{1/q}$$

where again the star indicates the Hardy–Littlewood maximal function. The positive numbers ε and η may be chosen arbitrarily small, in particular, $0 < \eta < s$. We multiply (21) with 2^{js} and take the l_q-quasi-norm with respect to j. Then we obtain

$$\left(\sum_{j=1}^{\infty} 2^{jsq} \mathrm{osc}_1^M f(x,2^{-j})^q \right)^{1/q} \leqq \varepsilon \left(\sum_{j=1}^{\infty} 2^{jsq} \mathrm{osc}_r^M f(x, c2^{-j})^q \right)^{1/q} \tag{22}$$
$$+ c'_\varepsilon \left(\sum_{j=1}^{\infty} 2^{jsq} |\mathrm{osc}_u^M f(x, c2^{-j})^u|^{*\frac{q}{u}} \right)^{1/q}.$$

Next we apply the L_p-quasi-norm, add on both sides $\|f \mid L_p\|$, and obtain on the basis of the above results

$$\|f \mid F^s_{pq}\| \leqq \varepsilon \|f \mid F^s_{pq}\| + c\|f \mid L_p\| \tag{23}$$
$$+ c'_\varepsilon \left\| \left(\sum_{j=1}^{\infty} 2^{jsq} |\mathrm{osc}_u^M f(x, c2^{-j})^u|^{*\frac{q}{u}} \right)^{1/q} \mid L_p \right\|$$

where $\varepsilon > 0$ is at our disposal. Recall $u < \min(p,q)$, hence by Theorem 2.2.2 we arrive at

$$\|f \mid F^s_{pq}\| \leqq \varepsilon \|f \mid F^s_{pq}\| + \tilde{c}_\varepsilon \|f \mid F^s_{pq}\|_{\mathrm{osc}} \tag{24}$$

where the latter quasi-norm is given by (6) with the desired value of u. Hence the proof of (10) is complete. In other words, we proved that the quasi-norm in (3) is an equivalent quasi-norm in F^s_{pq} under the full hypotheses of the involved parameters.

Step 5. It remains to prove the following assertion: Let $f \in L_{\max(p,r)}$ and let $\|f \mid F^s_{pq}\|_{\mathrm{osc}} < \infty$, where the latter is given by (6), then $f \in F^s_{pq}$. We begin with two preliminaries. Let $K \in \mathbb{N}$ with $K \geqq [s]$ and let φ be a C^∞ function in $\mathbb{R}^n$ with

$$|D^\gamma \varphi(x)| \leqq 1 \quad \text{if} \quad |\gamma| \leqq K+1 \quad \text{and} \quad x \in \mathbb{R}^n. \tag{25}$$

Then we have

$$\begin{aligned} \mathrm{osc}_u^{K+M}(\varphi f)(x, 2^{-j}) \leqq \Bigg(\fint_{B(x,2^{-j})} \Bigg| \varphi(y) f(y) \\ - \sum_{|\alpha| \leqq K} \frac{1}{\alpha!} D^\alpha \varphi(x)(y-x)^\alpha P(y) \Bigg|^u dy \Bigg)^{1/u} \end{aligned} \tag{26}$$

where $P \in \mathbb{P}_M$ is an arbitrary polynomial. It follows by Taylor's formula and a standard remainder estimate

$$\begin{aligned} \mathrm{osc}_u^{K+M}(\varphi f)(x, 2^{-j}) &\leqq c\, \mathrm{osc}_u^M f(x, 2^{-j}) \\ &\quad + c 2^{-j(K+1)} \left(\fint_{B(x,2^{-j})} |f(y)|^u dy \right)^{1/u}. \end{aligned} \tag{27}$$

The last integral can be estimated from above by $(|f|^u)^{*\frac{1}{u}}(x)$, where the star again indicates the Hardy–Littlewood maximal function. We have

$$\begin{aligned} &\left\| \left(\sum_{j=1}^\infty 2^{jsq} \mathrm{osc}_u^{K+M}(\varphi f)(\cdot, 2^{-j})^q \right)^{1/q} \mid L_p \right\| \\ &\leqq c \|(|f|^u)^{*\frac{1}{u}} \mid L_p\| + c \left\| \left(\sum_{j=1}^\infty 2^{jsq} \mathrm{osc}_u^M f(\cdot, 2^{-j})^q \right)^{1/q} \mid L_p \right\|. \end{aligned} \tag{28}$$

By (19) we may assume without restriction of generality $u < p$. Then we can apply the Hardy–Littlewood maximal inequality to the first term on the right-hand side of (28), which shows that it can be estimated from above by $c\|f \mid L_p\|$. Then we obtain that the left-hand side of (28) can be estimated from above by a constant which is independent of the chosen function φ with (25). We need a second preparation. Let ψ be a compactly supported C^∞ function on $\mathbb{R}^n$ with $\int \psi(y)dy = 1$ and let f be the above function, i.e., $f \in L_{\max(p,r)}$ such that the quasi-norm (6) is finite. Then $\psi(t, f)(x)$ in the sense of (2.4.6/1) is a mollification. We have

$$\psi(t_l, f)(x) \to f(x) \qquad \text{a.e. in } \mathbb{R}^n \tag{29}$$

at least for some sequence $t_l \to 0$. Furthermore

$$\operatorname{osc}_u^L \psi(t,f)(x,2^{-j}) \leqq \left(\oint_{B(x,2^{-j})} |\psi(t,f-P)(y)|^u dy \right)^{1/u} \tag{30}$$

for any $P \in \mathbb{P}_L$ because $\psi(t,P) \in \mathbb{P}_L$. Now we combine these two preparations. Let $f \in L_{\max(p,r)}$ such that the quasi-norm in (6) is finite. Let φ be a cut-off function with (25) and $\varphi(y) = 1$ if $|y| \leqq R$. Then $\psi(t,\varphi f)$ is a compactly supported C^∞ function and belongs to F^s_{pq}. We can apply the result obtained in Step 4 to $\psi(t,\varphi f)$ instead of f and obtain

$$\begin{aligned} \|\psi(t,\varphi f) \mid F^s_{pq}\| \leqq \|\psi(t,\varphi f) \mid L_p\| \\ + \left\| \left(\sum_{j=1}^\infty 2^{jsq} \operatorname{osc}_u^{K+M} \psi(t,\varphi f)(\cdot,2^{-j})^q \right)^{1/q} \mid L_p \right\|. \end{aligned} \tag{31}$$

We use (30) with $L = K+M$, φf instead of f and $P = 0$. By Hölder's inequality it follows

$$\operatorname{osc}_u^{K+M} \psi(t,\varphi f)(x,2^{-j}) \leqq c \left(\oint_{B(x,c2^{-j})} |f(y)|^\rho dy \right)^{1/\rho} \tag{32}$$

where $0 < t \leqq 2^{-j}$ and $\rho = \max(r,p)$. Hence by Lebesgue's bounded convergence theorem and by (29,30) with φf instead of f we obtain for any bounded domain ω and for some sequence $t_l \to 0$

$$\begin{aligned} &\left\| \left(\sum_{j=1}^J 2^{jsq} \operatorname{osc}_u^{K+M} \psi(t_l,\varphi f)(\cdot,2^{-j})^q \right)^{1/q} \mid L_p(\omega) \right\| \\ &\to \left\| \left(\sum_{j=1}^J 2^{jsq} \operatorname{osc}_u^{K+M} (\varphi f)(\cdot,2^{-j})^q \right)^{1/q} \mid L_p(\omega) \right\| \\ &\leqq c\|f \mid L_p\| + c \left\| \left(\sum_{k=1}^\infty 2^{ksq} \operatorname{osc}_u^M f(\cdot,2^{-k})^q \right)^{1/q} \mid L_p \right\|, \end{aligned} \tag{33}$$

where we used (28) and the subsequent arguments, $J \in \mathbb{N}$. By Fatou's lemma we can replace J on the left-hand side of (33) by ∞ and ω by $\mathbb{R}^n$. Hence the F^s_{pq}-quasi-norms of all the functions $\psi(t_l,\varphi f)$ are uniformly bounded. We use the original definition of F^s_{pq}, see (2.3.1/8), and apply again a Fatou argument. Then it follows $f \in F^s_{pq}$. The proof is complete.

3.5.2 Means of differences

We introduce means of differences. This subsection is in some sense the counterpart of 3.4.1, where we introduced oscillations not only on $\mathbb{R}^n$, but also on $\mathbb{R}^n_+$ and on bounded C^∞ domains Ω in $\mathbb{R}^n$. First we recall

$$\Delta_h^M f(x) = \sum_{j=0}^{M} (-1)^{M-j} \binom{M}{j} f(x+jh), \qquad x \in \mathbb{R}^n \text{ and } h \in \mathbb{R}^n. \tag{1}$$

Secondly, let Ω be either $\mathbb{R}^n$, or $\mathbb{R}^n_+$, or a bounded C^∞ domain in $\mathbb{R}^n$. Let $M \in \mathbb{N}$, $t > 0$, and $x \in \Omega$, then

$$V^M(x,t) = \{h \in \mathbb{R}^n \colon |h| < t \text{ and } x + \tau h \in \Omega \text{ for all } 0 \leqq \tau \leqq M\} \tag{2}$$

is the maximal subset of a ball of radius t, centered at the origin, star-shaped with respect to the origin, such that $x + MV^M(x,t) \subset \Omega$. We shall use the abbreviation in (3.4.1/2).

Definition. Let Ω be either $\mathbb{R}^n$, or $\mathbb{R}^N_+$, or a C^∞ domain in $\mathbb{R}^n$. Let $M \in \mathbb{N}$ and $0 < u \leqq \infty$, then

$$d^M_{t,u} f(x) = \left(\fint_{V^M(x,t)} |\Delta_h^M f(x)|^u dh \right)^{1/u}, \qquad x \in \Omega, \quad t > 0 \tag{3}$$

(modification if $u = \infty$).

Remark. This is the counterpart of Definition 3.4.1. Of course, $d^M_{t,u} f$ and also $V^M(x,t)$ depend on Ω. But as in the case of $\mathrm{osc}^M_u f$, see Remark 3.4.1, we shall not indicate this dependence. Usually Ω is a fixed domain. Furthermore it is tacitly assumed that $|\Delta_h^M f|^u$ in (3) is integrable. But this always will be the case.

3.5.3 Characterizations via differences

This chapter deals with spaces on $\mathbb{R}^n$. Spaces on domains will be treated in Chapter 5 with some preparations in Chapter 4. On the other hand we introduced both $\mathrm{osc}^M_u f$ in 3.4.1 and $d^M_{t,u} f$ in 3.5.2 not only for $\mathbb{R}^n$ but also for $\mathbb{R}^n_+$ and bounded C^∞ domains Ω in $\mathbb{R}^n$. The reason will become clear now: The aim of this subsection is to prove characterizations of type (3.5.1/3,5) with means of differences instead of oscillations.

This will be done by reduction of the corresponding claims to Theorem 3.5.1. On the other hand, later on, we shall prove Theorem 3.5.1 with $\mathbb{R}^n_+$ and Ω instead of $\mathbb{R}^n$, where $\mathrm{osc}_u^M f$ is then based on $\mathbb{R}^n_+$ and Ω, respectively, see 4.5.3 and 5.2.1. If one takes for granted that Theorem 3.5.1 holds also for $\mathbb{R}^n_+$ or Ω, then the proof below goes through essentially without any changes and we arrive at corresponding assertions for means of differences with respect to $\mathbb{R}^N_+$ or Ω. The reader is asked to have this point in mind, although we stick at the moment at spaces on $\mathbb{R}^n$, both in the formulation of the theorem and its proof. In other words, the spaces F^s_{pq} and B^s_{pq} in this chapter always have the same meaning as in Definition 2.3.1.

Theorem. (i) Let $0 < p < \infty$, $0 < q \leqq \infty$, $1 \leqq r \leqq \infty$, and

$$s > n\left(\frac{1}{p} - \frac{1}{r}\right)_+ \quad \text{and} \quad s > n\left(\frac{1}{q} - \frac{1}{r}\right)_+ . \tag{1}$$

Let $0 < u \leqq r$, $M \in \mathbb{N}$ and $M > s$. Then

$$F^s_{pq} = \Big\{ f \in L_{\max(p,r)}: \tag{2}$$

$$\|f \mid L_p\| + \left\| \left(\int_0^1 t^{-sq} d^M_{t,u} f(\cdot)^q \frac{dt}{t} \right)^{1/q} \mid L_p \right\| < \infty \Big\}$$

(modification if $q = \infty$) in the sense of equivalent quasi-norms.

(ii) Let $0 < p \leqq \infty$, $0 < q \leqq \infty$, $1 \leqq r \leqq \infty$, and

$$s > n\left(\frac{1}{p} - \frac{1}{r}\right)_+ . \tag{3}$$

Let $0 < u \leqq r$, $M \in \mathbb{N}$, and $M > s$. Then

$$B^s_{pq} = \Big\{ f \in L_{\max(p,r)}: \tag{4}$$

$$\|f \mid L_p\| + \left(\int_0^1 t^{-sq} \|d^M_{t,u} f \mid L_p\|^q \frac{dt}{t} \right)^{1/q} < \infty \Big\}$$

(modification if $q = \infty$) in the sense of equivalent quasi-norms.

Proof. Step 1. We restrict ourselves to the proof of part (i). The proof of (ii) is the same. Furthermore it is obvious by (3.5.2/3) that one can replace the continuous version in (2) by its discrete counterpart

$$\|f \mid F^s_{pq}\|_{\Delta,M} = \|f \mid L_p\| + \left\| \left(\sum_{j=0}^{\infty} 2^{jsq} d^M_{2^{-j},u} f(\cdot)^q \right)^{1/q} \mid L_p \right\|. \tag{5}$$

Let $f \in F^s_{pq}$ with $0 < p < \infty$, $0 < q \leqq \infty$ and s restricted by (1) and let $M \in \mathbb{N}_0$ with $M \geqq [s]$. We wish to estimate (5) with $M+1$ instead of M from above by $c\|f \mid F^s_{pq}\|$. We use the same splitting technique as in (3.5.1/11). By Hölder's inequality we may assume $u \geqq 1$. Then (3.5.2/3) yields

$$\begin{aligned} d^{M+1}_{2^{-j},u} f(x) \leqq & \sum_{l=1}^{\infty} \left(\oint_{V^{M+1}(x,2^{-j})} |\Delta_h^{M+1} f_{j+l}(x)|^u dh \right)^{1/u} \\ & + \left(\oint_{V^{M+1}(x,2^{-j})} |\Delta_h^{M+1} (f^j - P)(x)|^u dh \right)^{1/u} \end{aligned} \tag{6}$$

where we choose the same polynomial as in (3.5.1/11). With the same notations as in (3.5.1/11) we obtain

$$\begin{aligned} d^{M+1}_{2^{-j},u} f(x) \leqq c & \sum_{l=1}^{\infty} \Bigg[|f_{j+l}(x)| \\ & + \left(\oint_{B(x,c2^{-j})} |f_{j+l}(y)|^u dy \right)^{1/u} \Bigg] + c \sup |\cdots| \end{aligned} \tag{7}$$

where $\sup|\cdots|$ is the same supremum as in (3.5.1/11). By the same arguments as after (3.5.1/11) we arrive at

$$\|f \mid F^s_{pq}\|_{\Delta,M+1} \leqq c\|f \mid F^s_{pq}\|. \tag{8}$$

Step 2. Let again $f \in F^s_{pq}$. We wish to prove the converse inequality to (8). For this purpose we may assume

$$0 < u < (\min 1, p, q), \quad r > 1, \quad \text{and} \quad 1 = \frac{1-\theta}{u} + \frac{\theta}{r}. \tag{9}$$

By (3.4.2/4) and (3.4.4/4) with $y \in B(x, 2^{-j})$ we have

$$\begin{aligned} & \left| f(y) - \sum_{|\alpha| \leqq M} \frac{1}{\alpha!} (D^\alpha f^j)(x)(y-x)^\alpha \right|^u \\ & \leqq |f(y) - f^j(y)|^u + \left| f^j(y) - \sum_{|\alpha| \leqq M} \frac{1}{\alpha!} (D^\alpha f^j)(x)(y-x)^\alpha \right|^u \\ & \leqq c' \left(\oint_{V^{M+1}(y,c2^{-j})} |\Delta_h^{M+1} f(y)| dh \right)^u \end{aligned} \tag{10}$$

$$+ c' 2^{-j(M+1)u} \sum_{|\beta|=M+1} \left(\int_0^1 |D^\beta f^j(x+\tau(y-x))|^u d\tau \right)^u (\sup |D^\gamma f^j(z)|)^{(1-u)u},$$

where the last supremum is taken over $|\gamma| = M+1$ and $z \in B(x, c2^{-j})$. By (3.4.2/8) with $M+1$ instead of M, this supremum, multiplied with $2^{-j(M+1)u(1-u)}$, leads us to the same supremum as in (3.5.1/11). By (3.4.2/2), (9), and Hölder's inequality we obtain

$$\begin{aligned} \operatorname{osc}_u^M f(x, 2^{-j}) &\leqq \varepsilon \operatorname{osc}_r^M f(x, c2^{-j}) + c_\varepsilon \left(\oint_{B(x,2^{-j})} d^{M+1}_{c2^{-j},u} f(y)^u dy \right)^{1/u} \\ &\quad + \varepsilon 2^{-j(M+1)} \sup |D^\gamma f^j(z)| + c_\varepsilon \sum_{|\beta|=M+1} 2^{-j(M+1)} \\ &\quad \times \left(\int_0^1 \oint_{B(x,2^{-j})} |D^\beta f^j(x+\tau(y-x))|^u dy d\tau \right)^{1/u}, \end{aligned} \tag{11}$$

where the supremum in (11) is the same as in (10). Furthermore $\varepsilon > 0$ is at our disposal. As for the inner integral in the last summand of the right-hand side we have (3.4.4/5) with $|\cdot|^u$ instead of $|\cdot|$. Then by (3.4.3/3) with $|\beta| = M+1$ instead of $|\alpha| = M$, (9), and Hölder's inequality we obtain

$$\begin{aligned} \oint_{B(x,2^{-j})} |D^\beta f^j(x+\tau(y-x))|^u dy &\leqq c \oint_{B(x,\tau 2^{-j})} |D^\beta f^j(y)|^u dy \\ &\leqq c\, 2^{j(M+1)u} \oint_{B(x,\tau 2^{-j})} d^{M+1}_{c2^{-j},u} f(y)^{(1-\theta)u} \operatorname{osc}_r^M f(x, c' 2^{-j})^{\theta u} dy. \end{aligned} \tag{12}$$

We apply again Hölder's inequality with respect to $1 = (1-\theta) + \theta$, insert (12) in (11) and arrive at

$$\begin{aligned} \operatorname{osc}_u^M f(x, 2^{-j}) &\leqq \varepsilon \operatorname{osc}_r^M f(x, c2^{-j}) + c_\varepsilon (d^{M+1}_{c2^{-j},u} f^u)^{*\frac{1}{u}}(x) \\ &\quad + \varepsilon\, 2^{-j(M+1)} \sup |D^\gamma f^j(z)|, \end{aligned} \tag{13}$$

where the supremum is again the same as in (10,11) and the star indicates the Hardy–Littlewood maximal function. We multiply (13) with 2^{js}, take the l_q-quasi-norm with respect to j and then the L_p-quasi-norm. We obtain by Theorem 3.5.1 and the arguments after (3.5.1/11)

$$\begin{aligned} \| f \mid F^s_{pq} \| &\leqq \varepsilon \| f \mid F^s_{pq} \| + c \| f \mid L_p \| \\ &\quad + c_\varepsilon \left\| \left(\sum_{j=1}^\infty 2^{jsq} (d^{M+1}_{c2^{-j},u} f^u)^{*\frac{1}{u}} (\cdot)^q \right)^{1/q} \mid L_p \right\| \end{aligned} \tag{14}$$

where ε is at our disposal. Because $u < \min(p, q)$ we can apply Theorem 2.2.2 and obtain

$$\|f \mid F^s_{pq}\| \leqq c'\|f \mid L_p\| + c' \left\| \left(\sum_{j=0}^{\infty} 2^{jsq} d^{M+1}_{c2^{-j},u} f(\cdot)^q \right)^{1/q} \mid L_p \right\|. \tag{15}$$

We may choose $c = 1$ in (15) because $d^{M+1}_{c2^{-j},u} f(x)$ with, say $c2^{-j} \geqq \frac{1}{2}$, can be estimated from above by $(\int_{B(x,a)} |f(y)|^p dy)^{1/p}$ for some $a > 0$ which can easily be incorporated in $\|f \mid L_p\|$. Hence we have the converse of (8). In other words, we proved that the quasi-norm in (2) is an equivalent quasi-norm on F^s_{pq}.

Step 3. It remains to prove the following assertion: Let $f \in L_{\max(p,r)}$ and let $\|f \mid F^s_{pq}\|_{\Delta,M+1} < \infty$, where the latter is given by (5) with $M \in \mathbb{N}_0$ and $M \geqq [s]$, then $f \in F^s_{pq}$. We use the same reasoning as in Step 5 of the proof of Theorem 3.5.1. Let $K \in \mathbb{N}$ with $K \geqq [s]$ and let φ be a C^∞ function on $\mathbb{R}^n$ with

$$|D^\gamma \varphi(x)| \leqq 1 \quad \text{if} \quad |\gamma| \leqq K + M + 1 \quad \text{and} \quad x \in \mathbb{R}^n. \tag{16}$$

By mathematical induction it follows

$$\Delta_h^{K+M+1} \varphi f(x) = \sum_{m=0}^{K+M+1} a_m^{K+M+1} \Delta_h^{K+M+1-m} \varphi(x + mh) \Delta_h^m f(x) \tag{17}$$

for some real numbers a_m^{K+M+1}. Then we have

$$|\Delta_h^{K+M+1} \varphi f(x)|^u \leqq c|h|^{(K+1)u} \sum_{l=0}^{M} |f(x + lh)|^u + c \sum_{l=M+1}^{K+M+1} |\Delta_h^l f(x)|^u. \tag{18}$$

We may replace φf on the left-hand side of (18) by $\varphi f - P$ with $P \in \mathbb{P}_{K+M}$. We fix $t = 2^{-j}$, replace x in (18) by $x+z$, integrate over $\{z \in \mathbb{R}^n \colon |z| < 2^{-j}\}$ and afterwards over $\{h \in \mathbb{R}^n \colon |h| < 2^{-j}\}$. Then we arrive at

$$\begin{aligned}(\mathrm{osc}_u^{K+M} \varphi f)(x, 2^{-j}c) \leqq{}& c' 2^{-j(K+1)} \left(\oint_{B(x,c2^{-j})} |f(y)|^u dy \right)^{1/u} \\ &+ c'(d^{M+1}_{c2^{-j},u} f^u)^{*\frac{1}{u}}(x),\end{aligned} \tag{19}$$

where the star again indicates the Hardy–Littlewood maximal function. This is the counterpart of (3.5.1/27) where we may assume $u < \min(p, q)$. Now we can argue in the same way as after that formula. We have again (3.5.1/29). Now one can use a similar reasoning as at the end of Step 5 of the proof of Theorem 3.5.1 and we arrive at $f \in F^s_{pq}$.

Remark 1. We followed essentially [Tri23: 2.3.6 and 3.6] where we restricted ourselves to $r = u = 1$. But the extension to $1 \leqq r \leqq \infty$ and $0 < u \leqq r$ is not only of interest for its own sake, but it seems to be crucial at least in the last step of the above proof, where we used $u < \min(p, q)$ in order to apply Theorem 2.2.2. In this context the corresponding remarks in [Tri23: 3.6.4] look somewhat doubtful, although not wrong. The above version may also be considered as an improvement of these remarks.

Remark 2. Recall our introductory remarks. It is our aim to extend both Theorem 3.5.1 and the above theorem from $\mathbb{R}^n$ to $\mathbb{R}^n_+$ and to bounded C^∞ domains on $\mathbb{R}^n$. This causes a lot of trouble as far as Theorem 3.5.1 is concerned and will be done in detail later on. But afterwards the corresponding extensions of the above theorem to $\mathbb{R}^n_+$ and to the above domain Ω are covered by the above proof. For this purpose one assumes that the bounded C^∞ domain Ω satisfies the hypotheses at the end of 3.4.1 and that $x \in \Omega$ is near to the origin. Then one can follow the above proof line for line, the mollification in Step 3 must be adapted, maybe combined with appropriate translations in direction of the positive x_n-axis.

Remark 3. The above theorem should be compared with the results obtained in 2.6.1 – 2.6.3. Furthermore we refer to 1.7.3 and in particular to 1.10.4. There one can find also the necessary references both for Theorem 3.5.1 and for the above theorem.

Chapter 4

Key Theorems

4.1 Introduction

In Chapters 2 and 3 we developed the technical instruments of the theory of the spaces F^s_{pq} and B^s_{pq} providing the basis for what follows. Although not a major theme of this book, we are looking at function spaces from the point of view of applications, say, to partial differential equations. Whether a function space defined on $\mathbb{R}^n$, $\mathbb{R}^n_+$, or on bounded domains in $\mathbb{R}^n$ is of some use for these purposes depends on its properties. In the last 15 years or so, just those properties of the spaces F^s_{pq} and B^s_{pq} have been studied extensively which are of interest in this context. All crucial problems are solved now and most of them are studied extensively in [Triß]. However the situation at the end of the eighties is quite different of that one around 1980 when [Triß] was written. New techniques came into being or, if existed before in connection with other types of function spaces, proved to be useful for the spaces F^s_{pq} and B^s_{pq}, too, shed new light on these spaces, and simplified and improved older cumbersome (sometimes awful) proofs considerably. It is just the aim of this chapter to present these new proofs based on new techniques such as local means, atoms, and oscillations. But first we give a list of those crucial problems which govern the theory of the spaces F^s_{pq} and B^s_{pq} to a large extent.

Pointwise multipliers. Let φ be a bounded function on $\mathbb{R}^n$. The question is under what conditions the mapping $f \to \varphi f$ (pointwise multiplication) makes sense and generates a bounded operator in a given space F^s_{pq} or B^s_{pq}. For some classical spaces, such as Sobolev spaces, Hölder spaces etc., one gets easily satisfactory answers. But for the general spaces F^s_{pq} and B^s_{pq}, in particular those ones with $p < 1$, the problem turned out to be much harder. The trouble comes from the fact that the original Fourier-analytical Definition 2.3.1 for these spaces and pointwise multiplications do not fit together very well. To circumvent these difficulties, J. Peetre and the author invented independently around 1976 a method which a few years later became widely known as paramultiplication (around 1980 in connection with the work by J.M. Bony and Y. Meyer), see [Pee6, Tri5], and for further details and references about this method [Triß: 2.8]. Far-reaching improvements with final character of the results in

H. Triebel, *Theory of Function Spaces II*, Modern Birkhäuser Classics,
DOI 10.1007/978-3-0346-0419-2_4, © Birkhäuser Verlag 1992

[Triß] had been obtained later on by several authors, mostly based on the method of paramultiplication. A good survey of the state of art has been given by W. Sickel, see [Sic1], including many references. In 4.2 we present a somewhat light-hearted approach which shows how easily some substantial results can be obtained if one uses the new techniques. But it is not our ambition to compete with the almost final results obtained in [Sic1] and the underlying papers.

Diffeomorphisms. Let ψ be a diffeomorphic map of $\mathbb{R}^n$ onto itself. The question is under what conditions the mapping $f \to f \circ \psi$ makes sense and generates a bounded operator in a given space F^s_{pq} or B^s_{pq}. For some classical spaces, such as Sobolev spaces, one gets easily satisfactory answers. But for the general spaces F^s_{pq} and B^s_{pq} the problem seemed to be rather complicated. The first proof covering all spaces is due to the author (1978), see [Tri25] and [Triß: 2.10]. In 4.3 we return to this subject and give a simple proof removing such awkward conditions as $\psi(x) = x$ if $|x|$ is large, needed in the original proof (to adapt the globally defined Fourier transform to the local diffeomorphism).

Traces. If one wishes to apply function spaces of the above type to boundary value problems for, say, elliptic differential operators, then one needs information of traces on surfaces of functions belonging to appropriate function spaces. Let $x = (x', x_n)$ with $x' \in \mathbb{R}^{n-1}$ and $x_n \in \mathbb{R}$. The problem is what can be said about the trace-operator

$$\mathrm{Tr}: \quad f(x) \to f(x', 0). \tag{1}$$

A first satisfactory answer for those spaces F^s_{pq} and B^s_{pq} for which (1) makes sense was given by B. Jawerth [Jaw1], see also [Triß: 2.7.2]. The trace problem for (classical) function spaces of Sobolev–Besov type has been treated by many authors, detailed references may be found in [Triα]. We give in 4.4 a new approach to this subject via atoms and local means. We hesitate a little bit to call this approach simple, even simpler is a matter of taste, but in any case it is new and it shows the interplay of different new techniques developed in this book.

Extensions. The last but not least of the key problems which will be newly discussed is the extension problem. Let $F^s_{pq}(\mathbb{R}^n_+)$ be the restriction of F^s_{pq} on $\mathbb{R}^n_+$, quasi-normed in the usual way. If re stands for the restriction operator, then one is asking for a linear bounded extension operator ext from $F^s_{pq}(\mathbb{R}^n_+)$ into F^s_{pq} such that

$$\text{re} \circ \text{ext} = \text{id} \qquad (\text{identity in } F^s_{pq}(\mathbb{R}^n_+)). \tag{2}$$

Again one faces the situation that the Fourier-analytical Definition 2.3.1 of F^s_{pq} and also of B^s_{pq} does not provide any hint how to cope with this problem. We treated this problem extensively in [Triß: 2.9], based on [Tri25, 9] (1978, 1980). But there remained few gaps for some spaces F^s_{pq} with $p \leqq 1$ which had been filled by J. Franke, [Fra] (1986). But all these proofs are very complicated depending on sophisticated properties especially of F^s_{pq}. (As for the classical Besov–Sobolev spaces one may consult also [Triα: 2.9.3, 2.9.4] where one finds also relevant references). In 4.5 we give new simpler (but not simple) proofs covering all spaces which are based on the distinguished representations derived in 3.3 (and on oscillations in a partial case). A second problem connected with spaces on $\mathbb{R}^n_+$ or on domains is that one of intrinsic characterizations. Here we have also to say something new.

Further problems. The above problems are all of vital importance both for function spaces treated for their own sake and from the standpoint of applications to partial differential equations, pseudodifferential operators, etc. But there are a few other crucial problems, such as embeddings and Fourier multipliers. However here the situation is different, since both embeddings and Fourier multipliers are quite well adapted to the original Fourier-analytical Definition 2.3.1 of B^s_{pq} and F^s_{pq}. We treated both problems extensively in [Triß] and we have essentially nothing new to add. The few new more sophisticated assertions of this type needed in the sequel (in connection with pseudodifferential operators) have already been discussed in 2.4.8, see also 2.2.3 and 2.2.4 as far as the background is concerned. Far-reaching new results about Fourier multipliers in F^s_{pq} spaces and related maximal inequalities have been obtained recently in [See2].

4.2 Pointwise multipliers

4.2.1 An approach via atoms

Let $k \in \mathbb{N}$ then C^k stands for the collection of all complex-valued functions h in $\mathbb{R}^n$ having bounded classical derivatives $D^\alpha h(x)$ for all α with $0 \leqq |\alpha| \leqq k$. This is an immaterial modification of (1.2.1/2).

Proposition. Let $0 < p \leqq 1 < q \leqq \infty$ and $s > n(\frac{1}{p} - 1)$. Let $k \in \mathbb{N}$ and $k > s$ and

$h \in C^k$. Then $f \to hf$ is a linear and bounded operator from F^s_{pq} into itself.

Proof. We use the atomic decomposition (3.2.3/1) with $K = k$ and $L = -1$ in the sense of (3.2.2/1). In particular no moment conditions (3.2.2/5) are necessary. However (3.2.3/1) can be multiplied immediately with $h \in C^k$ and we obtain an atomic decomposition of hf, which completes the proof.

Remark 1. By Theorem 3.2.3 we arrive at

$$\|hf \mid F^s_{pq}\| \leqq \sum_{|\alpha| \leqq k} \sup_{x \in \mathbb{R}^n} |D^\alpha h(x)| \cdot \|f \mid F^s_{pq}\| \tag{1}$$

Remark 2. In 1.9.2 we described the atomic decomposition of the spaces B^s_{pq} due to M. Frazier, B. Jawerth, and H.-Q. Bui. On the basis of Theorem 1.9.2 one obtains by the same arguments as above the following result: Let $0 < p \leqq \infty$, $0 < q \leqq \infty$, $s > n(\frac{1}{p} - 1)_+$ and $k \in \mathbb{N}$ with $k > s$, then $h \in C^k$ is a pointwise multiplier for B^s_{pq} and one has an obvious counterpart of (1) (with B^s_{pq} instead of F^s_{pq}). By duality, see [Trіß: 2.11.2], real and complex interpolation, see [Trіß: 2.4] or 1.6.7, this result can be extended immediately to all spaces B^s_{pq} with $0 < p \leqq \infty$, $0 < q \leqq \infty$, $s \in \mathbb{R}$, and to some spaces F^s_{pq}, provided that the above number k is large enough in dependence on s, p, q. However apparently one does not catch all spaces F^s_{pq} in that way. But this will be done in the following subsection.

Remark 3. Interesting new results on pointwise multipliers for F^s_{pq}-spaces based on atomic representations have been obtained recently in [FrJ3: §13].

4.2.2 An appproach via local means

The shortcoming of Proposition 4.2.1 is not the restriction for s, which can easily be removed as will be seen below, but the restrictions for p and q which can only be partly removed by standard procedures as has been indicated in Remark 4.2.1/2. We present here a second method which covers all cases. It is based on local means in the sense of Theorems 2.4.6 and 2.5.3. Let k_0 and k be the same functions as in (2.4.6/2–4) and let the means $k(t, f)(x)$, and similarly $k_0(1, f)(x)$, be given by

(2.4.6/1). We study again gf with $g \in C^m$, see 4.2.1, and $f \in F^s_{pq}$ or $f \in B^s_{pq}$. We use the Theorems 2.4.6 and 2.5.3 and prefer a straightforward calculation. We add few technical comments how to understand gf afterwards.

Theorem. Let $s \in \mathbb{R}$ and $0 < q \leqq \infty$.

(i) Let $0 < p \leqq \infty$. If $m \in \mathbb{N}$ is sufficiently large then there exists a positive number c_m such that

$$\|gf \mid B^s_{pq}\| \leqq c_m \sum_{|\alpha| \leqq m} \|D^\alpha g \mid L_\infty\| \, \|f \mid B^s_{pq}\| \tag{1}$$

for all $g \in C^m$ and all $f \in B^s_{pq}$.

(ii) Let $0 < p < \infty$. If $m \in \mathbb{N}$ is sufficiently large then there exists a positive number c_m such that

$$\|gf \mid F^s_{pq}\| \leqq c_m \sum_{|\alpha| \leqq m} \|D^\alpha g \mid L_\infty\| \, \|f \mid F^s_{pq}\| \tag{2}$$

for all $g \in C^m$ and all $f \in F^s_{pq}$.

Proof. Step 1. First we prove (2) under the additional assumption $s > n/p$. By (2.4.6/1) we have

$$k(t, gf)(x) = \int_{\mathbb{R}^n} k(y) f(x+ty) g(x+ty) dy \tag{3}$$

$$= \sum_{|\alpha| \leqq m-1} \frac{t^{|\alpha|}}{\alpha!} (D^\alpha g)(x) \int_{\mathbb{R}^n} y^\alpha k(y) f(x+ty) + t^m \int_{\mathbb{R}^n} k(y) r_m(x,t,y) f(x+ty) dy,$$

where $r_m(x,t,y)$ is the remainder term in Taylor's expansion, see e.g., (3.4.4/4). In any case it is uniformly bounded. If we choose N in (2.4.6/4) sufficiently large then each kernel $y^\alpha k(y)$ with $|\alpha| \leqq m-1$ can also be expressed as a finite sum of kernels for which the counterpart of (2.4.6/7) can be estimated from above by $c\|f \mid F^s_{pq}\|$, see the proof of Theorem 2.4.6. Let $m > s$, then we plug (3) in (2.4.6/7) with gf instead of f. By the above remark we obtain

$$\left\| \left(\int_0^1 t^{-sq} |k(t, gf)(\cdot)|^q \frac{dt}{t} \right)^{1/q} \mid L_p \right\| \leqq c \sum_{|\alpha| \leqq m-1} \|D^\alpha g \mid L_\infty\| \, \|f \mid F^s_{pq}\| \tag{4}$$

$$+ c \sum_{|\alpha| \leqq m} \|D^\alpha g \mid L_\infty\| \left\| \sup_{|\cdot - y| \leqq 1} |f(y)| \mid L_p \right\|.$$

By (2.4.7/7), Remark 2.4.7/1 or (2.4.8/4) with $a = 1$ the last factor in (4) can be estimated from above by $c\|f \mid F^s_{pq}\|$. The proof of (2) is complete provided that $s > n/p$.

Step 2. Let $-\infty < s \leqq n/p$ and let $l \in \mathbb{N}$ with $s + 2l > n/p$. Any $f \in F^s_{pq}$ can be represented as $f = (\mathrm{id} + (-\Delta)^l)h$ with

$$h \in F^{s+2l}_{pq} \quad \text{and} \quad \|f \mid F^s_{pq}\| \sim \|h \mid F^{s+2l}_{pq}\|. \tag{5}$$

This is a well-known lift property of the spaces F^s_{pq}, see [Triß: 2.3.8]. But it is also a consequence of the definition (2.3.1/8) and of the arguments in 2.3.2. We have

$$gf = (\mathrm{id} + (-\Delta)^l)gh + \sum_{|\alpha|<2l} D^\alpha(g_\alpha h), \tag{6}$$

where g_α is the sum of terms of type $D^\beta g$ with $|\beta| \leqq 2l$. Then we have

$$\|gf \mid F^s_{pq}\| \leqq c \sum_{|\alpha|\leqq 2l} \|g_\alpha h \mid F^{s+2l}_{pq}\|, \tag{7}$$

see again [Triß: 2.3.8] for an explicit formulation, or 2.3.1 and 2.3.2 from which it can be derived easily. If m is sufficiently large, e.g., $m - 2l > s + 2l$, then we can apply Step 1 to the right-hand side of (7). In other words, the right-hand side of (7) can be estimated from above by $c\|h \mid F^{s+2l}_{pq}\|$. Then (5) proves (2).

Step 3. The proof of (i) is the same. We use Theorem 2.5.3 instead of Theorem 2.4.6.

Remark 1. We have to justify the somewhat formal calculations of Step 1. Let $p < \infty$ and $q < \infty$. Let $f \in S$ be compactly supported and $g \in C^\infty$. Then, of course $gf \in F^s_{pq}$ and we have (2) with $s > n/p$. Recall that the compactly supported $f \in S$ are dense in F^s_{pq}. Furthermore $g \in C^m$ with $m > s > n/p$ can be approximated by C^∞ functions, maybe via Sobolev's mollification method. Now by completion we have (2) for all $f \in F^s_{pq}$ and $g \in C^m$. Let $p < \infty$, $q = \infty$, and $f \in F^s_{p\infty}$ with $s > n/p$. Let f^l be the same functions as in 2.4.2, they belong to any F^σ_{pr} with $\sigma \in \mathbb{R}$, $0 < r \leqq \infty$. Then by the above arguments it follows $gf^l \in F^s_{p\infty}$ and Step 1 can be applied. We arrive at (2) with f^l on the left-hand side (but with f on the right-hand side via $\|f^l \mid F^s_{pq}\| \leqq c\|f \mid F^s_{pq}\|$). We use the same Fatou argument as in 2.4.2 which yields $gf \in F^s_{pq}$ and (2). Similarly for the spaces B^s_{pq}.

Remark 2. The above proof coincides essentially with a corresponding proof in [Tri18]. Furthermore we used the same techniques in a special case in Steps 2 and 3 of the proof of Theorem 2.4.7.

What about the smallest number m in the above theorem. To clarify this point and, even more, to improve the theorem we need the Hölder–Zygmund spaces introduced in 1.2.2. See also (2.6.5/1) and Theorem 1.5.1(ii) which shows that Hölder–Zygmund spaces fit in our scheme, they are special Besov spaces. In particular, let $\rho \in \mathbb{R}$ with

$$0 < \rho = [\rho] + \{\rho\}, \quad [\rho] \in \mathbb{N}_0, \quad 0 < \{\rho\} < 1, \tag{8}$$

then

$$\|g \mid \mathcal{C}^s\| = \sum_{|\alpha| \leqq [\rho]} \sup_{x \in \mathbb{R}^n} |D^\alpha g(x)| + \sum_{|\alpha| = [\rho]} \sup \frac{|D^\alpha f(x) - D^\alpha f(y)|}{|x-y|^{\{\rho\}}}, \tag{9}$$

where the second supremum is taken over all $x \in \mathbb{R}^n$ and $y \in \mathbb{R}^n$ with, say, $0 < |x-y| < 1$, is an equivalent norm in $\mathcal{C}^\rho$, see e.g., Theorem 1.2.2, where (1.2.2/8) holds also for the above spaces C^m.

Corollary. Let $s \in \mathbb{R}$ and $0 < q \leqq \infty$.

(i) Let $0 < p \leqq \infty$ and $\rho > \max(s, n(\frac{1}{p} - 1)_+ - s)$. Then there exists a positive number c such that

$$\|gf \mid B^s_{pq}\| \leqq c\|g \mid \mathcal{C}^\rho\| \, \|f \mid B^s_{pq}\| \tag{10}$$

for all $g \in \mathcal{C}^\rho$ and all $f \in B^s_{pq}$.

(ii) Let $0 < p < \infty$ and $\rho > \max(s, n(\frac{1}{p} - 1)_+ - s)$. Then there exists a positive number c such that

$$\|gf \mid F^s_{pq}\| \leqq c\|g \mid \mathcal{C}^\rho\| \, \|f \mid F^s_{pq}\| \tag{11}$$

for all $g \in \mathcal{C}^\rho$ and all $f \in F^s_{pq}$.

Proof. Step 1. Let $s > n/p$. We use (3) with $m = [\rho]$ where we may assume that ρ is not an element of $\mathbb{N}$. To have the same estimate as in (4) we must look more closely at the second term on the right-hand side of (3), now with $m = [\rho]$, the remainder term $r_{[\rho]}(x,t,y)$ can be calculated by

$$t^{[\rho]} r_{[\rho]}(x,t,y) = \sum_{|\beta| = [\rho]} \frac{[\rho]}{\beta!} t^{[\rho]} y^\beta \int_0^1 (1-\tau)^{[\rho]-1} D^\beta g(x + \tau t y) d\tau, \tag{12}$$

see (3.4.4/4) with $M + 1 = [\rho]$, g instead of f^j and $x + ty$ instead of y (modification if $[\rho] = 0$). We replace $D^\beta g(x + \tau t y)$ in (12) by

$$D^\beta g(x) + [D^\beta g(x+\tau t y) - D^\beta g(x)], \qquad |\beta| = [\rho]. \tag{13}$$

The resulting first term contributes to the first term on the right-hand side of (4) with $|\alpha| = [\rho]$. By (9) the second term can be estimated from above by $c\|g \mid \mathcal{C}^\rho\| t^\rho$ which via (3) gives the desired counterpart of the second term on the right-hand side of (4). The rest is the same as in Step 1 of the proof of the theorem. We arrive at (10) and (11) with $\rho > s$.

Step 2. The extension of this result to the other values of s is a matter of complex interpolation for quasi-Banach spaces in the sense of 1.6.6, 1.6.7, and of duality. The details are the same as in the proof of the Corollary in [Trіß: 2.8.2] and will be omitted here. This covers at least all cases with $q < \infty$. As for $q = \infty$ see [Sic1].

Remark 3. The above corollary improves the corresponding assertions of the Corollary in [Triß: 2.8.2]. However the result itself is not new, it is covered by [Fra], see also the survey [Sic1]. The above restrictions for ρ are sharp in the following sense: Let $s \in \mathbb{R}$, $0 < p < \infty$, $0 < q \leqq \infty$, and $0 < \rho < \max(s, n(\frac{1}{p} - 1)_+ - s)$, then there exists a function $g \in \mathcal{C}^\rho$ which is not a pointwise multiplier for F^s_{pq}. Similarly for B^s_{pq}. This assertion can be proved in the same way as in Step 4 of the proof of the Corollary in [Triß: 2.8.2]. Again we wish to mention that several better results are known, including limiting cases, see [Sic1: III.6] and the references given there.

4.3 Diffeomorphisms

4.3.1 An approach via atoms

Let C be the collection of all complex-valued bounded and continuous functions in $\mathbb{R}^n$. Let $k \in \mathbb{N}$, then we say that the one-to-one mapping $y = \psi(x)$ of $\mathbb{R}^n$ onto $\mathbb{R}^n$ is a k-diffeomorphism if the components $\psi_j(x)$ of $\psi(x) = (\psi_1(x), \ldots, \psi_n(x))$ have classical derivatives up to order k with $D^\alpha \psi_j(x) \in C$ if $0 < |\alpha| \leqq k$, and if $|\det \psi_*(x)| \geqq c > 0$ for some c and all $x \in \mathbb{R}^n$. Here ψ_* stands for the Jacobian matrix. The mapping $\psi = \psi(x)$ is called a diffeomorphism if it is a k-diffeomorphism for any $k \in \mathbb{N}$. Let ψ be a diffeomorphism then

$$f \circ \psi(x) = f(\psi(x)), \qquad f \in S', \tag{1}$$

makes sense (usual interpretation). If ψ is only a k-diffeomorphism, then (1) with $f \in F^s_{pq}$ (or $f \in B^s_{pq}$) has to be understood via limiting procedures in the sense of

Remark 4.2.2/1: first (1) is applied to smooth functions, then one derives the desired inequalities and afterwards one uses approximation procedures, maybe combined with Fatou arguments. So we always assume that (1) makes sense.

Proposition. Let $0 < p \leqq 1 < q \leqq \infty$ and $s > n(\frac{1}{p} - 1)$. Let $k \in \mathbb{N}$ with $k > s$ and let ψ be a k-diffeomorphism. Then $f \rightarrow f \circ \psi$ is a linear and bounded operator from F^s_{pq} onto itself.

Proof. We use the atomic decomposition (3.2.3/1) with $K = k$ and $L = -1$ in the sense of (3.2.2/1). In particular no moment conditions (3.2.2/5) are necessary. If g is a (s, p, q)-atom in the sense of Definition 3.2.2, then $g \circ \psi$ is also a (s, p, q)-atom, based on a new cube, and multiplied with a constant depending on ψ. But this is just what we need and we arrive at the desired assertion.

Remark. By the same arguments as in Remark 4.2.1/2 one can extend this proposition to all spaces B^s_{pq} with $s \in \mathbb{R}$, $0 < p \leqq \infty$, $0 < q \leqq \infty$, and to some spaces F^s_{pq} provided k is chosen large enough. Besides the Frazier–Jawerth atoms for B^s_{pq}, duality, real and complex interpolation, one needs for that purpose Theorem 4.2.2. Instead of discussing details we refer to the following subsection where we extend the above assertion to all spaces B^s_{pq} and F^s_{pq}.

4.3.2 An approach via local means

Let ψ be a diffeomorphism in $\mathbb{R}^n$ in the sense of 4.3.1. We wish to prove that $f \rightarrow f \circ \psi$ is a linear and bounded operator in all spaces F^s_{pq} and in all spaces B^s_{pq} (and hence it is an isomorphic mapping in all these spaces). The proof is based on the local means in the sense of Theorems 2.4.6 and 2.5.3. Let k_0 and k be the same functions as in (2.4.6/2–4) and let the means $k(t, f)(x)$, and similarly $k_0(1, f)(x)$, be given by (2.4.6/1). First we need a slight modification of 2.4.6 and 2.5.3.

Let $a(x)$ be a n–n matrix with real-valued continuous entries $a_{jl}(x)$, where $x \in \mathbb{R}^n$, $j = 1, \ldots, n$ and $l = 1, \ldots, n$. We assume that there exist two positive numbers C and c with

$$|a_{jl}(x)| \leqq C \quad \text{for all} \quad x \in \mathbb{R}^n, \quad j = 1, \ldots, n \text{ and } l = 1, \ldots, n, \tag{1}$$

and

$$|\det a(x)| \geqq c \quad \text{for all} \quad x \in \mathbb{R}^n. \tag{2}$$

Then for every fixed $x \in \mathbb{R}^n$ the affine mapping $y \to a(x)y$ is one-to-one in $\mathbb{R}^n$. We generalize (2.4.6/1) by

$$k(a,t,f)(x) = \int_{\mathbb{R}^n} k(y) f(x + ta(x)y) dy, \qquad x \in \mathbb{R}^n, \tag{3}$$

and similarly $k_0(a,t,f)(x)$.

Proposition. Let $s \in \mathbb{R}$, $0 < q \leqq \infty$, and $0 < p \leqq \infty$ (with $p < \infty$ in the case of F^s_{pq}). Let $a(x)$ be the above matrix with (1), (2), and let k_0, k be the above kernels satisfying (2.4.6/5) or (2.5.3/1). Then

$$\|k_0(a,1,f) \mid L_p\| + \left(\int_0^1 t^{-sq} \|k(a,t,f) \mid L_p\|^q \frac{dt}{t} \right)^{1/q} \leqq \|f \mid B^s_{pq}\| \tag{4}$$

(modification if $q = \infty$) for all $f \in B^s_{pq}$ and

$$\|k_0(a,1,f) \mid L_p\| + \left\| \left(\int_0^1 t^{-sq} |k(a,t,f)(\cdot)|^q \frac{dt}{t} \right)^{1/q} \mid L_p \right\| \leqq \|f \mid F^s_{pq}\| \tag{5}$$

(modification if $q = \infty$) for all $f \in F^s_{pq}$, where $\|f \mid B^s_{pq}\|$ and $\|f \mid F^s_{pq}\|$ are appropriate quasi-norms.

Proof. Let B be the collection of all matrices $b = \{b_{jl}\}^n_{j,l=1}$ with $b_{jl} \in \mathbb{R}$ such that the counterparts of (1), (2) with the same numbers C and c are fulfilled. Then we have

$$k(b,t,f)(x) = k^b(t,f)(x) \quad \text{with} \quad k^b(y) = c' k(b^{-1}y), \tag{6}$$

see (3) and (2.4.6/1). Similarly for $k_0(b,1,f)(x)$. By the same arguments as in the proofs of Theorem 2.4.6 and Theorem 2.4.1 it follows that

$$\left\| \sup_{b \in B} k_0(b,1,f) \mid L_p \right\| + \left\| \left(\int_0^1 t^{-sq} \sup_{b \in B} |k(b,t,f)(\cdot)|^q \right)^{1/q} \mid L_p \right\| \tag{7}$$

is an equivalent quasi-norm in F^s_{pq}. This proves (5). The proof of (4) is the same.

Let ψ be a m-diffeomorphism, then $f \circ \psi$ where f belongs to some spaces B^s_{pq} or F^s_{pq} must be explained, if necessary, via some limiting procedures in the sense of

4.3.1 and Remark 4.2.2/1. We always assume in the calculations below that $f \circ \psi(x)$ makes sense.

Theorem. Let $s \in \mathbb{R}$ and $0 < q \leqq \infty$.

(i) Let $0 < p \leqq \infty$ and let ψ be a m-diffeomorphism where $m \in \mathbb{N}$ is sufficiently large (in dependence on s, p) then $f \to f \circ \psi$ is an isomorphic mapping of B^s_{pq} onto itself.

(ii) Let $0 < p < \infty$ and let ψ be a m-diffeomorphism where $m \in \mathbb{N}$ is sufficiently large (in dependence on s, p) then $f \to f \circ \psi$ is an isomorphic mapping of F^s_{pq} onto itself.

Proof. Step 1. We prove (ii). The proof of (i) is the same. Furthermore it will be sufficient to prove

$$\|f \circ \psi \mid F^s_{pq}\| \leqq c\|f \mid F^s_{pq}\| \quad \text{for all} \quad f \in F^s_{pq} \tag{8}$$

and some constant c which is independent of f, because the reverse inequality follows from the fact that with ψ also the inverse ψ^{-1} is a m-diffeomorphism. By the above remarks we assume that f is smooth.

Step 2. Let $0 < p < \infty$, $0 < q \leqq \infty$, and $s > \frac{2n}{p} + 2$. Then we find a number $K \in \mathbb{N}$ with

$$K + \frac{n}{p} < s < 2K. \tag{9}$$

To prove (8) we use (2.4.6/7) where we may choose N in (2.4.6/5) as large as we want. We estimate the second term in (2.4.6/7), the first term can be treated in the same way. We have

$$\begin{aligned} k(t, f \circ \psi)(x) = \int_{\mathbb{R}^n} k(y) f[\psi(x + ty)]dy = \int_{\mathbb{R}^n} k(y) f\Big[\psi(x) + t\psi_*(x)y \\ + \sum_{2 \leqq |\alpha| \leqq 2K-1} t^{|\alpha|} \frac{1}{\alpha!}(D^\alpha\psi)(x)y^\alpha + t^{2K} R_{2K}(x, t, y)\Big] dy, \end{aligned} \tag{10}$$

where $\psi_*(x)y$ must be interpreted as the multiplication of the Jacobian matrix with $y \in \mathbb{R}^n$, whereas $D^\alpha \psi$ and the remainder term R_{2K} stand for appropriate vectors. We apply Taylor's expansion and obtain

$$f\Big[\psi(x) + t\psi_*(x)y + \sum_{2 \leqq |\alpha| \leqq 2K-1} + t^{2K} R_{2K}\Big] \tag{11}$$

$$= f\left[\psi(x) + t\psi_*(x)y + \sum_{2\leqq|\alpha|\leqq 2K-1}\right] + t^{2K}\tilde{R}_{2K}(\nabla f)(\cdots),$$

where the last term must be understood as a scalar product with immaterially modified remainder terms. We insert (11) in (10) and afterwards the corresponding term in (2.4.6/7). Because $2K > s$ the resulting term which comes from $t^{2K}\tilde{R}_{2K}\nabla f(\cdots)$ can be estimated from above by

$$c\left\| \sup_{|z-\psi(x)|\leqq c'} |\nabla f(z)| \mid L_p \right\| \tag{12}$$

for some positive numbers c and c'. One may replace $\psi(x)$ in (12) by x. Then it follows from (2.4.7/7), Remark 2.4.7/1 and (9) that (12) can be estimated from above by $c\|f \mid F^s_{pq}\|$. Hence, it is sufficient to handle the first term on the right-hand side of (11), which, in turn, can be expanded by

$$\sum_{0\leqq|\beta|\leqq K-1} D^\beta f[\psi(x) + t\psi_*(x)y]\frac{1}{\beta!}\left(\sum_{2\leqq|\alpha|\leqq 2K-1}\right)^\beta \tag{13}$$

$$+ \sum_{|\beta|=K} D^\beta f[\cdots]\left(\sum_{2\leqq|\alpha|\leqq 2K-1}\right)^\beta,$$

where again $\sum_{|\beta|=K}$ is a remainder term. We insert the last term in (13) in (10) and afterwards the resulting term in (2.4.6/7). By

$$\left|\left(\sum_{2\leqq|\alpha|\leqq 2K-1}\right)^\beta\right| \leqq ct^{2K} \quad \text{if} \quad |\beta| = K$$

the resulting term can be estimated from above by

$$c\sum_{|\beta|=K}\left\| \sup_{|z-\psi(x)|\leqq c} |D^\beta f(z)| \mid L_p \right\|, \tag{14}$$

see (12). Again by (2.4.7/7), Remark 2.4.7/1, and (9) all these terms can be estimated from above by $c\|f \mid F^s_{pq}\|$. Hence it is sufficient to handle the first term in (13). We insert these terms in (10) and obtain a sum of terms which can be estimated by

$$ct^b\left|\int_{\mathbb{R}^n} y^\gamma k(y) D^\beta f[\psi(x) + t\psi_*(x)y]dy\right| \tag{15}$$

with $|\beta| \leqq K-1$, $b \geqq 2|\beta|$, and $|\gamma| \leqq (2K-1)|\beta|$. If N in (2.4.6/5) is chosen large enough, then we are in the same position as in Step 1 of the proof of Theorem 4.2.2 as far as the kernels $y^\gamma k(y)$ are concerned. Furthermore we identify $\psi_*(x)$ with $a(x)$ in the above proposition. In other words, we put (15) in (2.4.6/7), use (5), and obtain that the corresponding terms can be estimated from above by $c\|D^\beta f \mid F^{s-b}_{pq}\|$, and hence by $c\|f \mid F^{s-b+|\beta|}_{pq}\|$, and so by $c\|f \mid F^s_{pq}\|$. We obtain (8) under the additional assumption $s > \frac{2n}{p} + 2$.

Step 3. Let $0 < p < \infty$, $0 < q \leqq \infty$, $s \leqq \frac{2n}{p} + 2$ and $l \in \mathbb{N}$ with $s + 2l > \frac{2n}{p} + 2$. As in Step 2 of the proof of Theorem 4.2.2 we represent $f \in F^s_{pq}$ as $f = (\mathrm{id} + (-\Delta)^l)h$ with $h \in F^{s+2l}_{pq}$ and (4.2.2/5). We have

$$f(x) = (\mathrm{id} + (-\Delta)^l)h(\psi^{-1} \circ \psi(x)) = \sum_{|\alpha| \leqq 2l} c_\alpha(x)(D^\alpha h \circ \psi)(\psi^{-1}(x)), \tag{16}$$

with some functions c_α. We may assume that Theorem 4.2.2(ii) can be applied to these functions. Then (16) and (4.2.2/2) with c_α instead of g yield

$$\|f \circ \psi \mid F^s_{pq}\| \leqq c \sum_{|\alpha| \leqq 2l} \|D^\alpha h \circ \psi \mid F^s_{pq}\| \leqq c'\|h \circ \psi \mid F^{s+2l}_{pq}\|. \tag{17}$$

By Step 1 the last term can be estimated from above by $c\|h \mid F^{s+2l}_{pq}\|$. Now (4.2.2/5) yields the desired result.

Remark 1. The above theorem is more or less known, see [Trіß: 2.10]. However the original proof is more complicated, see also the corresponding remarks in 4.1.

Remark 2. One may ask for the best possible number m in the above theorem. Let $0 < p < \infty$, $0 < q \leqq \infty$, and $s > \frac{2n}{p} + 2$. Then m may be chosen as the smallest even number with $m > s$. Such an assertion is covered by Step 2 of the above proof. This is quite satisfactory. Let $s \leqq \frac{2n}{p} + 2$. Then, by Step 3 and Corollary 4.2.2, one can calculate the best possible numbers m which can be obtained in this way. But the result is not so natural. One can improve these results by duality, see [Triß: 2.11], real and complex interpolation. But the outcome is not really satisfactory. So we do not go into detail.

4.4 Traces

4.4.1 Preliminaries

In this section the spaces F^s_{pq} and B^s_{pq} introduced in Definition 2.3.1 will be written as $F^s_{pq}(\mathbb{R}^n)$ and $B^s_{pq}(\mathbb{R}^n)$. The reason is that we deal simultaneously with spaces on $\mathbb{R}^n$ and on $\mathbb{R}^{n-1}$ which, of course, must be distinguished. The trace problem reads as follows. Let $x = (x', x_n) \in \mathbb{R}^n$ with $x' \in \mathbb{R}^{n-1}$ and $x_n \in \mathbb{R}$. We always assume in the sequel $n \in \mathbb{N}$ and $n \geqq 2$. We ask whether

$$\mathrm{Tr}: \quad f(x) \to f(x', 0) \tag{1}$$

makes sense if f belongs to some spaces $F^s_{pq}(\mathbb{R}^n)$ or $B^s_{pq}(\mathbb{R}^n)$. Rather final answers are known:

(i) Let $0 < p \leqq \infty$, $0 < q \leqq \infty$, and $s - \frac{1}{p} > (n-1)(\frac{1}{p} - 1)_+$ then Tr is a linear and bounded operator

$$\text{from } B^s_{pq}(\mathbb{R}^n) \text{ onto } B^{s-(1/p)}_{pq}(\mathbb{R}^{n-1}). \tag{2}$$

(ii) Let $0 < p < \infty$, $0 < q \leqq \infty$, and $s - \frac{1}{p} > (n-1)(\frac{1}{p} - 1)_+$, then Tr is a linear and bounded operator

$$\text{from } F^s_{pq}(\mathbb{R}^n) \text{ onto } F^{s-(1/p)}_{pp}(\mathbb{R}^{n-1}). \tag{3}$$

Remark 1. Recall $F^\sigma_{pp}(\mathbb{R}^{n-1}) = B^\sigma_{pp}(\mathbb{R}^{n-1})$.

Remark 2. One can prove even more: In both cases (2) and (3) there exists a linear and bounded extension operator Ex from the trace space into the original space such that

$$\mathrm{Tr} \circ \mathrm{Ex} = \mathrm{id}, \tag{4}$$

where id stands for the identity in $B^{s-(1/p)}_{pq}(\mathbb{R}^{n-1})$ in case (i) and in $F^{s-(1/p)}_{pp}(\mathbb{R}^{n-1})$ in case (ii).

Remark 3. As we said the above assertions are known. The classical cases, i.e., the Besov spaces $B^s_{pq}(\mathbb{R}^n)$ with $1 < p < \infty$, $1 \leqq q \leqq \infty$, $s > \frac{1}{p}$, and the fractional

Sobolev spaces $H_p^s(\mathbb{R}^n) = F_{p2}^s(\mathbb{R}^n)$ with $1 < p < \infty$, $s > 1/p$, have been studied extensively by many authors. Detailed references may be found in [Triα: 2.9.3, 2.9.4, in particular Remark 2.9.4/2] see also [Nik2]. The general assertions (2–4) have been treated extensively in [Triß: 2.7.2]. We emphasize that the first full proof of (3) is due to B. Jawerth, see [Jaw1] (1977). In [Triß: 2.7.2, in particular Remarks 2.7.2/3,4] one finds further results and references to papers by M.L. Gol'dman, G.A. Kaljabin, and J. Peetre in this connection.

It is the main aim of this section to give a new proof of (3) based on atoms and local means. This will be done in 4.4.2. Minor modifications (and simplifications) yield (2), as will be mentioned shortly. More or less as a by-product we obtain in 4.4.3 a somewhat curious new limiting case. The proofs provide a better understanding of the trace problem and shed new light on the machinery behind the scene. On the other hand, the use of atoms in connection with trace problems is not new. M. Frazier and B. Jawerth proved in [FrJ1] (1985) part (i) of the above assertion via atoms. Furthermore we refer to [FrJ2] and in particular to [FrJ3] where the authors dealt with trace problems of type (3) within the framework of their atomic representations.

Remark 4. We wish to summarize our point of view: We give new proofs of (3), ((2), and some limiting cases) where we follow essentially [Tri22]. A reader who is interested in trace problems should consult [Triß: 2.7.2]. As far as an atomic approach to trace problems is concerned we refer also to the above-mentioned papers by M. Frazier and B. Jawerth. We always assume that the involved functions are smooth enough, for example $f(x)$ in (1). The necessary approximation procedures and Fatou arguments have been discussed in some details in Remark 4.2.2/1 (with a reference to 2.4.2) and will not be repeated here.

4.4.2 The trace theorem

All notations have the meaning explained in the preceding subsection.

Theorem. Let $n \geqq 2$, $0 < p < \infty$, $0 < q \leqq \infty$, and $s - \frac{1}{p} > (n-1)(\frac{1}{p} - 1)_+$, then Tr is a linear and bounded operator from $F_{pq}^s(\mathbb{R}^n)$ onto $F_{pp}^{s-(1/p)}(\mathbb{R}^{n-1})$,

$$\operatorname{Tr} F_{pq}^s(\mathbb{R}^n) = F_{pp}^{s-(1/p)}(\mathbb{R}^{n-1}). \tag{1}$$

Proof. Step 1. Let $0 < p \leqq 1 < q \leqq \infty$ and $s - \frac{1}{p} > (n-1)(\frac{1}{p} - 1)$ with $n \geqq 2$. By Theorem 3.2.3 any $f \in F^s_{pq}(\mathbb{R}^n)$ can be represented as

$$f(x) = \sum_{j=1}^{\infty} (\mu_j a_j(x) + \lambda_j g_j(x)), \qquad x \in \mathbb{R}^n, \tag{2}$$

where the $a_j(x)$'s are s-atoms and the $g_j(x)$'s are (s,p,q)-atoms. We have (3.2.3/2). Since $s > n(\frac{1}{p} - 1)$ we choose $L = -1$ in (3.2.2/1). Hence we have no moment conditions of type (3.2.2/5). Let again $x = (x', x_n) \in \mathbb{R}^n$ with $x' \in \mathbb{R}^{n-1}$ and $x_n \in \mathbb{R}$, then we have

$$f(x',0) = \sum_{j=1}^{\infty} (\mu_j a_j(x',0) + \lambda_j g_j(x',0)), \qquad x' \in \mathbb{R}^{n-1}. \tag{3}$$

In this step we wish to prove

$$f(x',0) \in F^{s-(1/p)}_{pp}(\mathbb{R}^{n-1}), \quad \|f(\cdot,0) \mid F^{s-(1/p)}_{pp}(\mathbb{R}^{n-1})\| \leqq c\|f \mid F^s_{pq}(\mathbb{R}^n)\| \tag{4}$$

for some $c > 0$ which is independent of f. Let $g = g_j$ be one of the (s,p,q)-atoms in (2), then we have

$$g(x',0) = \sum_{(\mu,l)<(\nu,k)} d_{\mu l} a_{\mu l}(x',0) \tag{5}$$

where $a_{\mu l}(x)$ are $(Q_{\mu l}, s, p, q)$-atoms, see (3.2.2/6,7). Only cubes $Q_{\mu l}$ are of interest for which $5Q_{\mu l}$ has a non-empty intersection with the hyperplane $\{x = (x', x_n) \in \mathbb{R}^n\colon x_n = 0\}$. Let $Q = Q_{\mu l}$ be one of these cubes and let Q' be the projection of Q on that hyperplane, now being identified with $\mathbb{R}^{n-1}$. By (3.2.2/4) we have

$$\begin{aligned} |D^\alpha a_{\mu l}(x',0)| &\leqq |Q'|^{\frac{n}{n-1}(-\frac{1}{q}+\frac{s}{n}-\frac{|\alpha|}{n})} \\ &= |Q'|^{-\frac{1}{p}+\frac{1}{n-1}(s-\frac{1}{p})-\frac{|\alpha|}{n-1}} \, |Q'|^{\frac{n}{n-1}(\frac{1}{p}-\frac{1}{q})}, \qquad |\alpha| \leqq K. \end{aligned} \tag{6}$$

Hence,

$$a_{\mu l}(x',0)|Q'|^{\frac{n}{n-1}(\frac{1}{q}-\frac{1}{p})} \tag{7}$$

is a $(n-1)$-dimensional $(Q', s - \frac{1}{p}, p, p)$-atom in the sense of Definition 3.2.2(ii), where we used $s - \frac{1}{p} > (n-1)(\frac{1}{p} - 1)$ which shows that no moment conditions of type (3.2.2/5) are required. Next we prove

$$g(\cdot,0) \in F^{s-(1/p)}_{pp}(\mathbb{R}^{n-1}) \quad \text{and} \quad \|g(\cdot,0) \mid F^{s-(1/p)}_{pp}(\mathbb{R}^{n-1})\| \leqq c \tag{8}$$

where c is independent of the chosen atom $g = g_j$. By $F^{s-(1/p)}_{pp}(\mathbb{R}^{n-1}) = B^{s-(1/p)}_{pp}(\mathbb{R}^{n-1})$, Theorem 1.9.2 with $p = q$, and the above observation (7) we must estimate

$$\sum_{(\mu,l')<(\nu,k)} |d_{\mu l'}|^p 2^{\mu n(\frac{1}{q}-\frac{1}{p})p} = \sum_{\mu=\nu}^{\infty} 2^{\mu n(\frac{1}{q}-\frac{1}{p})p} \sum_{l'} |d_{\mu l'}|^p \tag{9}$$

uniformly with respect to ν and k, where l' indicates that the summation takes place only for those $l \in \mathbb{Z}^n$ for which $5Q_{\mu l'}$ intersects the plane "$x_n = 0$." Then the last sum in (9) has at most $c2^{(\mu-\nu)(n-1)}$ summands. By Hölder's inequality and (3.2.2/7) we have

$$\sum_{l'} |d_{\mu l'}|^p \leqq c \left(\sum_{l'} |d_{\mu l'}|^q \right)^{p/q} 2^{(\mu-\nu)(n-1)(1-\frac{p}{q})} \tag{10}$$

$$\leqq c 2^{-\nu n(\frac{p}{q}-1)} 2^{(\mu-\nu)(n-1)(1-\frac{p}{q})} = c 2^{\nu(1-\frac{p}{q})} 2^{\mu(n-1)(1-\frac{p}{q})}.$$

We put (10) in (9) and sum over μ. Then it follows that the left-hand side of (9) is uniformly bounded with respect to ν and k. Hence we arrive at (8) for any of the (s,p,q)-atoms $g = g_j$ in (2). A corresponding but much simpler assertion holds for the s-atoms a_j in (2). Now (4) is a consequence of (8) with g_j instead of g, its counterpart with a_j, and Theorem 3.2.3.

Step 2. By (4) and the monotonicity of F^s_{pq} with respect to q we have

$$\operatorname{Tr} F^s_{pq}(\mathbb{R}^n) \subset F^{s-(1/p)}_{pp}(\mathbb{R}^{n-1}), \qquad 0 < p \leqq 1,\ 0 < q \leqq \infty, \tag{11}$$

and $s - \frac{1}{p} > (n-1)(\frac{1}{p} - 1)$ in the sense of a linear and bounded mapping. Recall again $B^\sigma_{pp} = F^\sigma_{pp}$ if $\sigma \in \mathbb{R}$ and $0 < p \leqq \infty$ are the Besov spaces. We use the atomic decomposition of these Besov spaces described in Theorem 1.9.2. Then Step 1 can be applied to this case, too, and we obtain

$$\operatorname{Tr} F^s_{pp}(\mathbb{R}^n) \subset F^{s-(1/p)}_{pp}(\mathbb{R}^{n-1}), \qquad 1 \leqq p \leqq \infty, \tag{12}$$

and $s > \frac{1}{p}$ in the sense of a linear and bounded mapping. We use the complex interpolation formula (1.6.7/4) where $F^{s_1}_{p_1 q_1}$ can also be replaced by $F^{s_1}_{\infty\infty} = B^{s_1}_{\infty\infty}$. Then we interpolate the spaces in (11) and the spaces in (12) which are the substitutes of $F^{s_0}_{p_0 q_0}$ and $F^{s_1}_{p_1 q_1}$ in (1.6.7/4), respectively. The trace operator has the interpolation property. Furthermore this interpolation covers all cases of interest. In other words,

$$\operatorname{Tr} F^s_{pq}(\mathbb{R}^n) \subset F^{s-(1/p)}_{pp}(\mathbb{R}^{n-1}), \qquad 0 < p < \infty,\ 0 < q \leqq \infty, \tag{13}$$

and $s - \frac{1}{p} > (n-1)(\frac{1}{p} - 1)_+$ in the sense of a linear and bounded mapping.

Step 3. We prove the extension property: Let $0 < p < \infty$, $0 < q \leqq \infty$, $s - \frac{1}{p} > (n-1)(\frac{1}{p} - 1)_+$ and $g \in F^{s-(1/p)}_{pp}(\mathbb{R}^{n-1}) = B^{s-(1/p)}_{pp}(\mathbb{R}^{n-1})$, then there exists a function $f \in F^s_{pq}(\mathbb{R}^n)$ with

$$f(x', 0) = g(x') \quad \text{and} \quad \|f \mid F^s_{pq}(\mathbb{R}^n)\| \leqq c \|g \mid F^{s-\frac{1}{p}}_{pp}(\mathbb{R}^{n-1})\|, \tag{14}$$

where c is independent of g. By the monotonicity property of F^s_{pq} with respect to q we may assume $0 < q < \min(1, p)$. Let

$$g(x') = \sum_{j=1}^{\infty} \lambda_j a_j(x'), \qquad \sum_{j=1}^{\infty} |\lambda_j|^p \leq c \|g \mid F_{pp}^{s-\frac{1}{p}}(\mathbb{R}^{n-1})\|^p \tag{15}$$

be an atomic decomposition of g in the sense of Theorem 1.9.2. We use again that we may choose $L = -1$ such that no moment conditions of type (1.9.2/6) are necessary. Then (15) covers both s-atoms and $(Q_{\nu k}, s, p)$-atoms. Let $h(t)$ and $H(t)$ be two C^∞ functions on $\mathbb{R}$ with

$$\operatorname{supp} H \subset (-1, 1), \qquad h(t) = H^{(l)}(t) \text{ and } h(0) = 1 \tag{16}$$

for some $l \in \mathbb{N}$. Let $a_j(x')$ be based on a $(n-1)$-dimensional cube with side length d_j. We try to extend $g(x')$ from $\mathbb{R}^{n-1}$ to $\mathbb{R}^n$ by

$$f(x) = \sum_{j=1}^{\infty} \lambda_j \tilde{a}_j(x) \text{ with } \tilde{a}_j(x) = a_j(x') h\left(\frac{x_n}{d_j}\right), \quad x = (x', x_n). \tag{17}$$

The first assertion in (14) is clear. In order to prove the second assertion in (14) we use (2.4.6/6), i.e.,

$$\|f \mid F^s_{pq}(\mathbb{R}^n)\| \sim \|k_0(1, f) \mid L_p(\mathbb{R}^n)\| \tag{18}$$
$$+ \left\| \left(\sum_{m=0}^{\infty} 2^{msq} |k(2^{-m}, f)(\cdot)|^q \right)^{1/q} \mid L_p(\mathbb{R}^n) \right\|$$

where k_0 and k have the same meaning as in Theorem 2.4.6. In particular $N \in \mathbb{N}$ in (2.4.6/4) is at our disposal. Let φ and ψ be the $(n-1)$-dimensional and the 1-dimensional counterparts of the above function k, respectively. In particular ψ is a C^∞ function on $\mathbb{R}$ with

$$\operatorname{supp} \Psi \subset (-1, 1) \quad \text{and} \quad \psi(t) = \Psi^{(L)}(t) \tag{19}$$

for some C^∞ function Ψ, where $L \in \mathbb{N}$ is at our disposal. By the proof of Theorem 2.4.6, which in turn is based on Theorem 2.4.1, we may replace the above function k by a finite sum

$$k(x) = k^1(x) + k^2(x) + \cdots \quad \text{with} \quad k^1(x) = \varphi(x')\psi(x_n) \tag{20}$$

and $k^r(x) = k^1(U^r x)$, where U^r are some rotation matrices in $\mathbb{R}^n$. In this step we deal with $k^1(x)$ and in the next step we indicate the necessary modifications. By (2.4.6/1) we have

$$k^1(2^{-m}, f)(x) = \sum_{j=1}^{\infty} \lambda_j \varphi(2^{-m}, a_j)(x') \psi(2^{-m}, h(d_j^{-1} \cdot))(x_n). \tag{21}$$

We estimate the last factor. By the assumptions (16) and (19) we have

$$\psi(2^{-m}, h(d_j^{-1}\cdot))(x_n) = \int_{\mathbb{R}} \psi(t) h(d_j^{-1}x_n + 2^{-m}d_j^{-1}t)dt \tag{22}$$

$$= \int_{\mathbb{R}} \Psi^{(L)}(t) H^{(l)}(d_j^{-1}x_n + 2^{-m}d_j^{-1}t)dt.$$

We use $H^{(l)}(\cdots) = (2^m d_j)^l \frac{d^l}{dt^l} H(\cdots)$ and shift in (22) by partial integration all derivatives either to Ψ or to H. We arrive at

$$|\psi(2^{-m}, h(d_j^{-1}\cdot))(x_n)| \leqq c(2^m d_j)^K (1 + |d_j^{-1}x_n|)^{-\kappa} \tag{23}$$

where both the integer K and the number $\kappa > 0$ are at our disposal. The constant c in (23) may depend on K and κ. Now we replace a_j in (21) by a_{jr} with $j \in \mathbb{N}_0$ and $r \in \mathbb{Z}^{n-1}$, simply to indicate that a_{jr} is based on the $(n-1)$-dimensional cube Q_{jr} in the sense of Theorem 1.9.2 with side length $d_j = 2^{-j}$. Let j be fixed then the number of overlappings of the supports of $\varphi(2^{-m}, a_{jr})(x')$ can be estimated from above by $c2^{|j-m|N}$ for some positive numbers c and N, which are independent of j and m. Then (21), now with λ_{jr} instead of λ_j, yields

$$|k^1(2^{-m}, f)(x)|^q \leqq \sum_{j=0}^{\infty} \sum_{r\in\mathbb{Z}^{n-1}} |\lambda_{jr}|^q |\varphi(2^{-m}, a_{jr})(x')|^q \frac{2^{-K|m-j|}}{(1 + |2^j x_n|)^{\kappa}} \tag{24}$$

$$\leqq \sum_{j=0}^{\infty} \left(\sum_{r\in\mathbb{Z}^{n-1}} |\lambda_{jr}|^p |\varphi(2^{-m}, a_{jr})(x')|^p \right)^{q/p} \frac{2^{-L|m-j|}}{(1 + |2^j x_n|)^{\kappa}}$$

where the positive numbers L and κ are at our disposal. By construction the integration over $\mathbb{R}^n$ on the right-hand side of (18) can be restricted to $\mathbb{R}^{n-1} \times (0, c)$. We may assume $c = 1$ and decompose the integration over x_n in the dyadic intervals $(2^{-u}, 2^{-u+1})$ with $u \in \mathbb{N}_0$. Then we have

$$\sum_{u=0}^{\infty} \int_{2^{-u}}^{2^{-u+1}} \int_{\mathbb{R}^{n-1}} \left(\sum_{m=0}^{\infty} 2^{smq} |k^1(2^{-m}, f)(\cdot)|^q \right)^{p/q} dx'dx_n \tag{25}$$

$$\leqq \sum_{u=0}^{\infty} 2^{-u} \int_{\mathbb{R}^{n-1}} \left[\sum_{m=0}^{\infty} \sum_{j=0}^{\infty} \left(\sum_{r\in\mathbb{Z}^{n-1}} |\lambda_{jr}|^p 2^{smp} |\varphi(2^{-m}, a_{jr})(x')|^p \right)^{q/p} \frac{2^{-L|m-j|}}{(1 + 2^{j-u})^{\kappa}} \right]^{p/q} dx'.$$

We split $\sum_{j=0}^{\infty} = \sum_{j=0}^{u} + \sum_{j=u+1}^{\infty}$. In the first sum we omit the factor $(1 + 2^{j-u})^{\kappa}$, observe that

$$2^{-u} 2^{smp} = 2^{(s-\frac{1}{p})mp} 2^{m-j} 2^{j-u}, \tag{26}$$

and arrive for some $\rho > 0$ at

$$\sum_{u=0}^{\infty} 2^{-u} \int_{\mathbb{R}^{n-1}} \left[\sum_{j=0}^{u} \sum_{m=0}^{\infty} (\cdots)^{q/p} 2^{-L|m-j|} \right]^{p/q} dx' \tag{27}$$

$$\leqq c \sum_{u=0}^{\infty} \int_{\mathbb{R}^{n-1}} \sum_{j=0}^{u} \sum_{m=0}^{\infty} \sum_{r \in \mathbb{Z}^{n-1}} 2^{\rho(j-u)} |\lambda_{jr}|^p 2^{(s-\frac{1}{p})mp} |\varphi(2^{-m}, a_{jr})(x')|^p dx$$

$$\leqq c' \sum_{j=0}^{\infty} \sum_{r \in \mathbb{Z}^{n-1}} |\lambda_{jr}|^p \|a_{jr} \mid F_{pp}^{s-\frac{1}{p}}(\mathbb{R}^{n-1})\|^p \leqq c'' \|g \mid F_{pp}^{s-\frac{1}{p}}(\mathbb{R}^{n-1})\|^p,$$

see (15). As for the sum $\sum_{j=u+1}^{\infty}$ we have in addition the factors $2^{-\kappa(j-u)}$ where $\kappa > 0$ is at our disposal. We choose $\kappa > 1$. Then we have a corresponding estimate for the terms with $\sum_{j=u+1}^{\infty}$. We put these two estimates together and obtain

$$\left\| \left(\sum_{m=0}^{\infty} 2^{msq} |k^1(2^{-m}, f)(\cdot)|^q \right)^{1/q} \mid L_p(\mathbb{R}^n) \right\| \leqq c \|g \mid F_{pp}^{s-\frac{1}{p}}(\mathbb{R}^{n-1})\|^p. \tag{28}$$

This corresponds to the second term on the right-hand side of (18). Next we estimate the first term on the right-hand side of (18) with k_0^1 instead of k_0. We assume $k_0(x) = \varphi(x')\psi(x_n)$, where φ and ψ are appropriate functions. We have again (17) with (16). By the above reasoning we have (23) with $m = 0$, where now $K > 0$ and $\kappa > 0$ are at our disposal. But this is sufficient to get the counterpart of (24) with $m = 0$. Then one obtains (28) with $m = 0$ and k_0^1 instead of $\sum_{m=0}^{\infty}$ and k^1.

Step 4. We indicate the necessary changes if one replaces k^1 in the above considerations by $k^1(Ux)$ where U is a rotation matrix in $\mathbb{R}^n$, see (20). First we remark that one can shift the rotation from $k^1(Uy)$ to $\tilde{a}_j(U^{-1}y)$, see (17) and also (2.4.6/1), where the x-variable must be replaced by $U^{-1}x$. Hence we have again $k^1(x) = \varphi(x')\psi(x_n)$, see (20), however $\tilde{a}_j(U^{-1}y)$ has not the desired product structure. But we can expand $\tilde{a}_j(U^{-1}y)$ in the respective cube in a trigonometric series and after multiplication with appropriate functions we have

$$\tilde{a}_j(U^{-1}x) = \sum_{k \in \mathbb{Z}^n} a_{jk}^U(x') h_k^U \left(\frac{x_n}{d_j} \right) d_{jk} \tag{29}$$

where $a_{jk}^U(x')$ and $h_k^U(t)$ have the same properties as $a_j(x')$ and $h(t)$, respectively, and

$$|d_{jk}| \leqq c(1 + |k|)^{-\sigma}, \qquad k \in \mathbb{Z}^n, \tag{30}$$

where $\sigma > 0$ is at our disposal and c is independent of j. To prove the last claim one begins with atoms in the unit cube, the rest is a matter of translation and dilation. If $k \in \mathbb{Z}^n$ is fixed then we can apply the above considerations to each term $a_{jk}^U(x')h_k^U(\frac{x_n}{d_j})$. Afterwards one sums over k. If one chooses σ in (30) sufficiently large, then the corresponding sum converges. Now the proof of (14) is complete.

Remark 1. References and comments may be found in 4.4.1. The most difficult part of the present proof is to show that f in (17) belongs to any space $F^s_{pq}(\mathbb{R}^n)$ if g in (15) is an element of $F^{s-\frac{1}{p}}_{pp}(\mathbb{R}^{n-1}) = B^{s-\frac{1}{p}}_{pp}(\mathbb{R}^{n-1})$. It is the main aim of the following corollary to demonstrate that the corresponding counterpart for the B^s_{pq}-spaces is much simpler.

Corollary. Let $n \geqq 2$, $0 < p \leqq \infty$, $0 < q \leqq \infty$, and $s - \frac{1}{p} > (n-1)(\frac{1}{p} - 1)_+$ then Tr is a linear and bounded operator from $B^s_{pq}(\mathbb{R}^n)$ onto $B^{s-\frac{1}{p}}_{pq}(\mathbb{R}^{n-1})$,

$$\operatorname{Tr} B^s_{pq}(\mathbb{R}^n) = B^{s-\frac{1}{p}}_{pq}(\mathbb{R}^{n-1}). \tag{31}$$

Proof. Step 1. Let $p < \infty$ and $0 < q \leqq \infty$ then (1), the interpolation property, and (1.6.7/7) prove

$$\operatorname{Tr} B^s_{pq}(\mathbb{R}^n) \subset B^{s-\frac{1}{p}}_{pq}(\mathbb{R}^{n-1}), \qquad s - \frac{1}{p} > (n-1)\left(\frac{1}{p} - 1\right)_+ . \tag{32}$$

Step 2. Let $0 < p \leqq \infty$, $0 < q \leqq \infty$, and $s - \frac{1}{p} > (n-1)(\frac{1}{p} - 1)_+$. Then any $g \in B^{s-\frac{1}{p}}_{pq}(\mathbb{R}^{n-1})$ can be represented by Theorem 1.9.2, where we may assume $L = -1$, i.e., no moment conditions of type (1.9.2/6) are required. However any $(n-1)$-dimensional $(Q, s - \frac{1}{p}, p)$-atom in the sense of Definition 1.9.2(iii) can easily be extended to a n-dimensional $(\tilde{Q}, s, p)$-atom, where Q and $\tilde{Q}$ have the same side length. Similarly for s-atoms. But the resulting function $f(x', x_n)$ is an element of $B^s_{pq}(\mathbb{R}^n)$ and we may assume $f(x', 0) = g(x')$. This observation and (32) prove the corollary provided that $p < \infty$.

Step 3. It remains to prove (32) with $p = \infty$. Let $k \in \mathbb{N}_0$, then $C^k(\mathbb{R}^n)$ stands now for the collection of all functions f having uniformly continuous classical derivatives up to order k such that

$$\| f \mid C^k(\mathbb{R}^n) \| = \sum_{|\alpha| \leqq k} \sup_{x \in \mathbb{R}^n} |D^\alpha f(x)| < \infty. \tag{33}$$

Let $k \in \mathbb{N}$, then we have

$$(C^k(\mathbb{R}^n), C^0(\mathbb{R}^n))_{\theta,q} = B^{k(1-\theta)}_{\infty q}(\mathbb{R}^n), \qquad 0 < \theta < 1,\ 0 < q \leqq \infty, \tag{34}$$

see [Triß: 2.2.2, (2.5.7/13), (2.4.2/1)]. Then

$$\operatorname{Tr} C^k(\mathbb{R}^n) = C^k(\mathbb{R}^{n-1}), \qquad k \in \mathbb{N}_0, \tag{35}$$

and (34) prove (32) with $p = \infty$.

Remark 2. The main aim of this subsection (and also of the following one) is to demonstrate that atoms and local means are useful tools in the theory of function spaces. On the other hand, as we said, both the theorem and the corollary are not new, references have been given in 4.4.1. Furthermore the simple striking argument of Step 2 of the proof of the above corollary may be found in [FrJ1]. In contrast to the corresponding proofs in [Trië: 2.7.2] we have not proved the existence of a linear and bounded extension operator Ex in the sense of (4.4.1/4). Maybe one can overcome this shortcoming if one uses in (15) standardized atoms, such as wavelets or splines.

4.4.3 Curiosities

All notations have the same meaning as in the preceding subsection. We discuss some limiting cases which are not covered by Theorem 4.4.2 and Corollary 4.4.2. Let $1 \leqq p < \infty$ then

$$\mathrm{Tr}\ B^{1/p}_{p1}(\mathbb{R}^n) = L_p(\mathbb{R}^{n-1}) \tag{1}$$

has been proved independently by J. Peetre, M.L. Gol'dman and V.I. Burenkov, see [Pee7, Gol3, BuG]. This result was extended by M. Frazier and B. Jawerth to

$$\mathrm{Tr}\ B^{1/p}_{pq}(\mathbb{R}^n) = L_p(\mathbb{R}^{n-1}), \qquad 0 < p < \infty,\ q \leqq \min(1,p), \tag{2}$$

see [FrJ1]. It is possible to prove that (2) cannot be extended to $q > \min(1,p)$. We do not go into detail, but it follows from the technique developed in [Nik2: Note section, Notes to 8.9].

Proposition. Let $n \geqq 2$, $0 < p \leqq 1$, and $0 < q \leqq \infty$, then

$$\mathrm{Tr}\ F^{1/p}_{pq}(\mathbb{R}^n) = L_p(\mathbb{R}^{n-1}). \tag{3}$$

Proof. Step 1. We wish to prove

$$\mathrm{Tr}\ F^{1/p}_{pq}(\mathbb{R}^n) \subset L_p(\mathbb{R}^{n-1}). \tag{4}$$

For this purpose we may assume $0 < p \leqq 1 < q \leqq \infty$. We modify Step 1 of the proof of Theorem 4.4.2. We represent $f \in F^{1/p}_{pq}(\mathbb{R}^n)$ by (4.4.2/2), where $a_j(x)$ are $\frac{1}{p}$-atoms and $g_j(x)$ are $(\frac{1}{p}, p, q)$-atoms. We have (4.4.2/3,5) and the counterpart

$$|a_{\mu l}(x',0)| \leqq c|Q'|^{-1/p} 2^{\mu n(\frac{1}{q}-\frac{1}{p})}, \quad \operatorname{supp} a_{\mu l}(x',0) \subset 5Q' \tag{5}$$

of (4.4.2/6). By the same arguments as in (4.4.2/9,10) we arrive at

$$\|g(\cdot,0) \mid L_p(\mathbb{R}^{n-1})\|^p \leqq \sum_{(\mu,l')<(\nu,k)} |d_{\mu l'}|^p 2^{\mu n(\frac{1}{q}-\frac{1}{p})p} \leqq c, \tag{6}$$

where c is independent of the chosen atom $g = g_j$. Similarly for the atom a_j (4.4.2/2). Now (6) with g_j instead of g, its counterpart with a_j, and Theorem 3.2.3 yield

$$f(x',0) \in L_p(\mathbb{R}^{n-1}), \quad \|f(x',0) \mid L_p(\mathbb{R}^{n-1})\| \leqq c\|f \mid F^{1/p}_{pq}(\mathbb{R}^n)\| \tag{7}$$

and consequently (4).

Step 2. By monotonicity we have (4) for all $0 < q \leqq \infty$. To prove the converse assertion we recall the elementary embedding

$$B^\sigma_{p,\min(p,q)}(\mathbb{R}^n) \subset F^\sigma_{pq}(\mathbb{R}^n) \subset B^\sigma_{p,\max(p,q)}(\mathbb{R}^n) \tag{8}$$

which follows from Definition 2.3.1, see also [Triß: (2.3.2/9)]. Now (2) with $0 < q \leqq p \leqq 1$, (4), and (8) prove (3).

Remark 1. The assertion (3) cannot be extended to $1 < p < \infty$ and $q > 1$. This follows from (8) and the remarks in front of the above proposition.

Remark 2. Both 4.4.2 and this subsection are based on [Tri22]. We remark that (3) has been proved independently by M. Frazier and B. Jawerth, see [FrJ3: §11], where one also finds other results in this direction.

4.5 Spaces on $\mathbb{R}^n_+$, extensions

4.5.1 Definitions and preliminaries

In this section we write $F^s_{pq}(\mathbb{R}^n)$ instead of F^s_{pq} and $B^s_{pq}(\mathbb{R}^n)$ instead of B^s_{pq}, where the spaces F^s_{pq} and B^s_{pq} have the same meaning as in Definition 2.3.1. We put

$$\mathbb{R}^n_+ = \{x = (x', x_n) : x' \in \mathbb{R}^{n-1}, x_n > 0\} \tag{1}$$

with an obvious modification if $n = 1$.

Definition. (i) Let $0 < p < \infty$, $0 < q \leqq \infty$ and $s \in \mathbb{R}$, then $F^s_{pq}(\mathbb{R}^n_+)$ is the restriction of $F^s_{pq}(\mathbb{R}^n)$ on $\mathbb{R}^n_+$, quasi-normed by

$$\|f \mid F^s_{pq}(\mathbb{R}^n_+)\| = \inf \|g \mid F^s_{pq}(\mathbb{R}^n)\|, \tag{2}$$

where the infimum is taken over all $g \in F^s_{pq}(\mathbb{R}^n)$ with $g \mid \mathbb{R}^n_+ = f$ in the sense of $D'(\mathbb{R}^n_+)$.

(ii) Let $0 < p \leqq \infty$, $0 < q \leqq \infty$, and $s \in \mathbb{R}$, then $B^s_{pq}(\mathbb{R}^n_+)$ is the restriction of $B^s_{pq}(\mathbb{R}^n)$ on $\mathbb{R}^n_+$, quasi-normed by

$$\|f \mid B^s_{pq}(\mathbb{R}^n_+)\| = \inf \|g \mid B^s_{pq}(\mathbb{R}^n)\| \tag{3}$$

where the infimum is taken over all $g \in B^s_{pq}(\mathbb{R}^n)$ with $g \mid \mathbb{R}^n_+ = f$ in the sense of $D'(\mathbb{R}^n_+)$.

Remark. It follows by standard arguments that both $F^s_{pq}(\mathbb{R}^n_+)$ and $B^s_{pq}(\mathbb{R}^n_+)$ are quasi-Banach spaces.

Extension problem. Let re be the restriction operator from $F^s_{pq}(\mathbb{R}^n)$ onto $F^s_{pq}(\mathbb{R}^n_+)$ or from $B^s_{pq}(\mathbb{R}^n)$ onto $B^s_{pq}(\mathbb{R}^n_+)$. Obviously, re is a linear and bounded operator. We are looking for a linear and bounded operator ext from $F^s_{pq}(\mathbb{R}^n_+)$ into $F^s_{pq}(\mathbb{R}^n)$ or from $B^s_{pq}(\mathbb{R}^n_+)$ into $B^s_{pq}(\mathbb{R}^n)$ with

$$\mathrm{re} \circ \mathrm{ext} = \mathrm{id} \quad (\text{identity in } F^s_{pq}(\mathbb{R}^n_+) \text{ or } B^s_{pq}(\mathbb{R}^n_+)). \tag{4}$$

This extension property is well known for the classical spaces (Sobolev, fractional Sobolev, Besov, Hölder–Zygmund). For general spaces F^s_{pq} and B^s_{pq} including in particular smoothness parameters $s \leqq 0$ and values $p < 1$ this property has been studied in [Triß: 2.9] (based on [Tri25,9] (1978, 1980)), see also [Triα: 2.9.3, 2.9.4] where one finds also relevant references. There remained few gaps for some spaces F^s_{pq} with $p < 1$ which had been filled by J. Franke in [Fra] (1986). It is the main aim of this section to present two new proofs. The first one is based on oscillations. If p, q, s are restricted in the same way as in Theorem 3.5.1, then one obtains the desired assertion by strikingly simple arguments, see 4.5.2. This result can be extended to all values p, q, s by standard arguments: interpolation and duality. However these two devices, interpolation and duality, have not been studied in detail in this book. So we formulate this way only as a corollary. In 4.5.5 we give a second proof, covering all values p, q, s with one stroke, and completely within the framework of this book. It is based on distinguished representations in the sense of 3.3.3.

Intrinsic descriptions. A second major problem of spaces on $\mathbb{R}^n_+$ is the search for intrinsic descriptions, i.e., for descriptions of, say, $F^s_{pq}(\mathbb{R}^n_+)$ where one needs only a knowledge of $f \in F^s_{pq}(\mathbb{R}^n_+)$ on $\mathbb{R}^n_+$. We obtain some interesting assertions of this type as a by-product of the proof of Theorem 4.5.2: Characterizations via oscillations and differences on $\mathbb{R}^n_+$. We formulate these results separately in 4.5.3 and 4.5.4, respectively.

4.5.2 An approach via oscillations

First we recall that the numbers

$$\sigma_p = n\left(\frac{1}{p} - 1\right)_+ \quad \text{and} \quad \sigma_{pq} = n\left(\frac{1}{\min(p,q)} - 1\right)_+ \tag{1}$$

play a decisive role in the theory of the spaces B^s_{pq} and F^s_{pq}, see 2.3.3, 2.4.1, 2.6, and 3.5.1. Furthermore we always assume $n \in \mathbb{N}$ with $n \geqq 2$.

Theorem. (i) Let

$$0 < p < \infty, \quad 0 < q \leqq \infty, \quad \text{and} \quad s > \sigma_{pq}. \tag{2}$$

Let $L \in \mathbb{N}_0$ with $L \geqq [s]$, let $0 < \lambda_0 < \lambda_1 < \cdots < \lambda_L$ and $a_0, \ldots, a_L$ be real numbers with

$$\sum_{k=0}^{L} a_k(-\lambda_k)^l = 1 \quad \text{where} \quad l = 0, \ldots, L, \tag{3}$$

then ext_L, given by

$$\text{ext}_L f(x) = \begin{cases} f(x) & \text{if } x \in \mathbb{R}^N_+ \\ \sum_{k=0}^{L} a_k f(x', -\lambda_k x_n) & \text{if } x = (x', x_n) \text{ with } x_n \leqq 0, \end{cases} \tag{4}$$

is an extension operator from $F^s_{pq}(\mathbb{R}^n_+)$ into $F^s_{pq}(\mathbb{R}^n)$.

(ii) Let

$$0 < p \leqq \infty, \quad 0 < q \leqq \infty, \quad \text{and} \quad s > \sigma_p. \tag{5}$$

Let $L \in \mathbb{N}_0$ with $L \geqq [s]$, and let both $\lambda_0, \ldots, \lambda_L$ and $a_0, \ldots, a_L$ be the same numbers as in (i), then ext_L, given by (4), is also an extension operator from $B^s_{pq}(\mathbb{R}^n_+)$ into $B^s_{pq}(\mathbb{R}^n)$.

Proof. Step 1. We prove (i). Let $\lambda_0, \ldots, \lambda_L$ be given then (3) has a unique solution $a_0, \ldots a_L$ (Vandermonde's determinant). Let p, q, s be given by (2). By (3.5.1/3) and Definition 4.5.1 we have $F^s_{pq}(\mathbb{R}^n_+) \subset L_{\max(p,r)}(\mathbb{R}^n_+)$ for $r \geqq 1$ with (3.5.1/2). In particular, $\mathrm{ext}_L f(x)$ makes sense if $f \in F^s_{pq}(\mathbb{R}^n_+)$. In order to use Theorem 3.5.1 we have to distinguish between oscillations based on $\mathbb{R}^n$, denoted by $\mathrm{osc}^M_u f$, and oscillations based on $\mathbb{R}^n_+$ in the sense of Definition 3.4.1, which will be temporarily denoted by $\mathrm{Osc}^M_u f$. Recall that $\mathbb{P}_L$ stands for the collection of all polynomials of degree less than or equal to L. Let $P \in \mathbb{P}_L$ then we have by (3)

$$\mathrm{ext}_L P(x) = P(x), \qquad x \in \mathbb{R}^n. \tag{6}$$

In particular $\mathrm{ext}_L f(x) - P(x) = \mathrm{ext}_L(f - P)(x)$ for any $f \in F^s_{pq}(\mathbb{R}^n_+)$ and $P \in \mathbb{P}_L$. Let $0 < u \leqq r$, $x = (x', x_n)$ with $x' \in \mathbb{R}^{n-1}$ and $0 < t \leqq 1$, then (3.4.1/3) with respect to $\mathbb{R}^n$ and $\mathbb{R}^n_+$ yield

$$\mathrm{osc}^L_u(\mathrm{ext}_L f)(x,t) = \mathrm{Osc}^L_u f(x,t) \ \text{ if } \ x \in \mathbb{R}^n, \ x_n > t \tag{7}$$

and

$$\mathrm{osc}^L_u(\mathrm{ext}_L f)(x,t) \leqq c \sum_{k=0}^{L} \mathrm{Osc}^L_u f((x', -\lambda_k x_n), ct) \tag{8}$$

if $x \in \mathbb{R}^n$, $x_n < -t$, for some $c > 0$. Finally if $x = (x', x_n)$ with $|x_n| \leqq t$, then it follows

$$\mathrm{osc}^L_u f(\mathrm{ext}_L f)(x,t) \leqq c\, \mathrm{Osc}^L_u f((x', |x_n|), ct) \tag{9}$$

for some $c > 0$. Recall $L \geqq [s]$, then we claim

$$\begin{aligned} \|f \mid F^s_{pq}(\mathbb{R}^n_+)\| &\leqq \|\mathrm{ext}_L f \mid F^s_{pq}(\mathbb{R}^n)\| \\ &\leqq c'\|f \mid L_p(\mathbb{R}^n_+)\| + c' \left\| \left(\int_0^1 t^{-sq} \mathrm{Osc}^L_u f(\cdot, ct)^q \frac{dt}{t} \right)^{1/q} \mid L_p(\mathbb{R}^n_+) \right\| \\ &\leqq c''\|f \mid F^s_{pq}(\mathbb{R}^n_+)\| \end{aligned} \tag{10}$$

for some positive numbers c, c', and c'' which are independent of $f \in F^s_{pq}(\mathbb{R}^n_+)$. Furthermore we assume that p, q, s and also r and u are the same numbers as in Theorem 3.5.1(i). By (7–9) and Theorem 3.5.1(i) follow $\mathrm{ext}_L f \in F^s_{pq}(\mathbb{R}^n)$ and the middle inequality in (10). Then the first inequality in (10) is obvious and the last one is again a consequence of Theorem 3.5.1(i). The proof of part (i) is complete including the fact that ext_L is an extension operator.

Step 2. The proof of (ii) is the same. Instead of part (i) of Theorem 3.5.1 we have to use part (ii).

Remark 1. Extension operators of type (4) are well known in literature. They go back to Fichtengol'z. We used them several times, see [Triα: 2.9.1, 2.9.3], [Triß: 2.9], and [Tri24] in an extended form which covers also values $s \leqq 0$. We return to this point in 4.5.5. The above proof is based on [Tri23].

Corollary. Let $\varepsilon > 0$ and $|s| < \varepsilon^{-1}$ then there exists a common extension operator ext^ε

$$\text{from } F^s_{pq}(\mathbb{R}^n_+) \text{ into } F^s_{pq}(\mathbb{R}^n) \text{ with } \varepsilon < p < \infty,\ \varepsilon < q \leqq \infty, \tag{11}$$

and

$$\text{from } B^s_{pq}(\mathbb{R}^n_+) \text{ into } B^s_{pq}(\mathbb{R}^n) \text{ with } \varepsilon < p \leqq \infty,\ 0 < q \leqq \infty. \tag{12}$$

Proof. (Outline) By part (i) of the above theorem and duality arguments one can prove the extension property for the spaces $F^s_{pq}(\mathbb{R}^n_+)$ with $s \in \mathbb{R}$, $1 < p < \infty$, $1 < q < \infty$, see [Triß: 2.9.2, 2.11.2] for details. Afterwards one can use complex interpolation which proves the extension property for all spaces $F^s_{pq}(\mathbb{R}^n_+)$ with $s \in \mathbb{R}$, $0 < p < \infty$, $0 < q < \infty$, see [Triß: 2.9.2, 2.9.4] for details and (1.6.7/4) as far as complex interpolation is concerned (the latter can be extended to spaces on $\mathbb{R}^n_+$). Similarly one can prove the extension property for all spaces $B^s_{pq}(\mathbb{R}^n_+)$ with $s \in \mathbb{R}$, $0 < p \leqq \infty$, $0 < q \leqq \infty$, where one can use duality and real interpolation, see (1.6.7/7) and [Triß: 2.11.2]. Finally complex interpolation of $F^{s_0}_{p\infty}$ with $s_0 > n(\frac{1}{p} - 1)_+$, covered by the above theorem, and $B^{s_1}_{\infty\infty} = F^{s_1}_{\infty\infty}$ covered by the preceding considerations, yields the desired assertion for the spaces $F^s_{p\infty}$.

Remark 2. The main assertion of this subsection is the above theorem, not the corollary. The outlined proof of the latter uses interpolation and duality, which have not been treated in detail in this book. In particular, the duality arguments are not obvious: one needs the modification of (4) which will be introduced in 4.5.5. What is meant by the a little sloppy notation of a "common restriction operator" is quite clear: ext^ε is defined on the union of all spaces in question and its restriction to a particular space has the desired property.

4.5.3 Characterizations of spaces on $\mathbb{R}^n_+$ via oscillations

More or less as a by-product of the preceding subsection we obtain an intrinsic characterization of those spaces $F^s_{pq}(\mathbb{R}^n_+)$ and $B^s_{pq}(\mathbb{R}^n_+)$ which are covered by Theorem 4.5.2. In 4.5.2 we denoted temporarily oscillations based on $\mathbb{R}^n_+$ by Osc^M_u. Now we return to the notations introduced in Definition 3.4.1 and Remark 3.4.1 and denote in the following theorem oscillations based on $\mathbb{R}^n_+$ by osc^M_u.

Theorem. Let $\Omega = \mathbb{R}^n_+$ in Definition 3.4.1.

(i) Let $0 < p < \infty$, $0 < q \leqq \infty$, $1 \leqq r \leqq \infty$, and

$$s > n\left(\frac{1}{p}-\frac{1}{r}\right)_+ \quad \text{and} \quad s > n\left(\frac{1}{q}-\frac{1}{r}\right)_+ . \tag{1}$$

Let $0 < u \leqq r$ and $M \in \mathbb{N}_0$ with $M \geqq [s]$, then

$$F^s_{pq}(\mathbb{R}^n_+) = \Big\{ f \in L_{\max(p,r)}(\mathbb{R}^n_+): \tag{2}$$

$$\|f \mid L_p(\mathbb{R}^n_+)\| + \left\| \left(\int_0^1 t^{-sq} \mathrm{osc}^M_u f(\cdot,t)^q \frac{dt}{t} \right)^{1/q} \mid L_p(\mathbb{R}^n_+) \right\| < \infty \Big\}$$

(modification if $q=\infty$) in the sense of equivalent quasi-norms.

(ii) Let $0 < p \leqq \infty$, $0 < q \leqq \infty$, $1 \leqq r \leqq \infty$, and

$$s > n\left(\frac{1}{p}-\frac{1}{r}\right)_+ . \tag{3}$$

Let $0 < u \leqq r$ and $M \in \mathbb{N}_0$ with $M \geqq [s]$, then

$$B^s_{pq}(\mathbb{R}^n_+) = \Big\{ f \in L_{\max(p,r)}(\mathbb{R}^n_+): \tag{4}$$

$$\|f \mid L_p(\mathbb{R}^n_+)\| + \left(\int_0^1 t^{-sq} \|\mathrm{osc}_u^M f(\cdot,t) \mid L_p(\mathbb{R}^n_+)\|^q \frac{dt}{t} \right)^{1/q} < \infty \Big\}$$

(modification if $q = \infty$) in the sense of equivalent quasi-norms.

Proof. If $f \in F^s_{pq}(\mathbb{R}^n_+)$ then we have $f \in L_{\max(p,r)}(\mathbb{R}^n_+)$, see the beginning of Step 1 of the proof of Theorem 4.5.2. Furthermore the last inequality in (4.5.2/10) holds where Osc_u^M must be identified now with the above osc_u^M. On the other hand let f be an element of the right-hand side of (2). Then (4.5.2/7–9) and (3.5.1/3) prove $\mathrm{ext}_L f \in F^s_{pq}(\mathbb{R}^n)$. Hence $f \in F^s_{pq}(\mathbb{R}^n_+)$ and we have (4.5.2/10) with the above interpretation of osc_u^M instead of Osc_u^M. The proof of (i) is complete. The proof of (ii) is the same.

Remark. References and comments may be found in Remarks 1.7.3/1 and 1.10.4/1. The above theorem is the $\mathbb{R}^n_+$-counterpart of Theorem 3.5.1.

4.5.4 Characterizations of spaces on $\mathbb{R}^n_+$ via differences

On the basis of Theorem 4.5.3 it is easy to extend Theorem 3.5.3 from $\mathbb{R}^n$ to $\mathbb{R}^n_+$.

Theorem. Let $\Omega = \mathbb{R}^n_+$ in Definition 3.5.2.

(i) Let $0 < p < \infty$, $0 < q \leqq \infty$, $1 \leqq r \leqq \infty$, and

$$s > n\left(\frac{1}{p} - \frac{1}{r}\right)_+ \quad \text{and} \quad s > n\left(\frac{1}{q} - \frac{1}{r}\right)_+ . \tag{1}$$

Let $0 < u \leqq r$, $M \in \mathbb{N}$, and $M > s$. Then

$$F^s_{pq}(\mathbb{R}^n_+) = \Big\{ f \in L_{\max(p,r)}(\mathbb{R}^n_+): \tag{2}$$

$$\|f \mid L_p(\mathbb{R}^n_+)\| + \left\| \left(\int_0^1 t^{-sq} d^M_{t,u} f(\cdot)^q \frac{dt}{t} \right)^{1/q} \mid L_p(\mathbb{R}^n_+) \right\| < \infty \Big\}$$

(modification if $q = \infty$) in the sense of equivalent quasi-norms.

(ii) Let $0 < p \leqq \infty$, $0 < q \leqq \infty$, $1 \leqq r \leqq \infty$, and

$$s > n\left(\frac{1}{p} - \frac{1}{r}\right)_+ . \tag{3}$$

Let $0 < u \leqq r$, $M \in \mathbb{N}$ and $M > s$. Then

$$B^s_{pq}(\mathbb{R}^n_+) = \Big\{ f \in L_{\max(p,r)}(\mathbb{R}^n_+): \tag{4}$$

$$\|f \mid L_p(\mathbb{R}^n_+)\| + \left(\int_0^1 t^{-sq} \|d^M_{t,u} f \mid L_p(\mathbb{R}^n_+)\|^q \frac{dt}{t}\right)^{1/q} < \infty \Big\}$$

(modification if $q = \infty$) in the sense of equivalent quasi-norms.

Proof. The above theorem follows from Theorem 4.5.3 and the considerations in 3.5.3, see in particular the beginning of 3.5.3 and Remark 3.5.3/2.

Remark. References may be found in 1.10.4.

4.5.5 An approach via distinguished representations

In this subsection we modify ext_L in (4.5.2/4) such that the modified operators can be used to construct explicitly extension operators for all spaces $F^s_{pq}(\mathbb{R}^n)$ and $B^s_{pq}(\mathbb{R}^n)$. The considerations are based on the distinguished representations in the sense of 3.3. Let $L \in \mathbb{N}$ and let

$$1 < \lambda_1 < \lambda_2 < \cdots < \lambda_{2L+1}. \tag{1}$$

Let $a_1, \ldots, a_{2L+1}$ be the uniquely determined real numbers such that

$$\sum_{l=1}^{2L+1} a_l (-\lambda_l)^{j-1} = 1; \qquad j = -L, -L+1, \ldots, 0, \ldots L-1, L, \tag{2}$$

(Vandermonde's determinant). If $x \in \mathbb{R}^n$ then we write $x = (x', x_n)$ with $x' \in \mathbb{R}^{n-1}$ and $x_n \in \mathbb{R}$. Let $f \in S'$ be represented in the sense of Proposition 3.3.2(ii) as

$$f = \sum_{j=0}^{\infty} f_j \tag{3}$$

where the C^∞ functions f_j are given by (3.3.2/4) if $j \in \mathbb{N}$ and by (3.3.2/5) if $j = 0$. In (3.3.2/2) the number M is at our disposal. In any case

$$\operatorname{ext}^L f_j(x) = \begin{cases} f_j(x) & \text{if } x \in \mathbb{R}^n_+ \\ \sum_{k=1}^{2L+1} a_k f_j(x', -\lambda_k x_n) & \text{if } x' \in \mathbb{R}^{n-1},\ x_n \leqq 0, \end{cases} \tag{4}$$

makes sense. Of course, ext^L depends also on the chosen numbers $\lambda_1, \ldots, \lambda_{2L+1}$. In order to construct $\operatorname{ext}^L f_j(x)$ one needs only a knowledge of f in $\mathbb{R}^n_+$, see (3.3.2/4,5,7).

Theorem. (i) Let $0 < p < \infty$, $0 < q \leqq \infty$, $s \in \mathbb{R}$, let $L \in \mathbb{N}$ with

$$L > |s| + \frac{n}{\min(p,q)}. \tag{5}$$

Let $f \in F^s_{pq}(\mathbb{R}^n_+)$, then

$$\operatorname{ext}^L f = \sum_{j=0}^{\infty} \operatorname{ext}^L f_j \tag{6}$$

converges in $F^s_{pq}(\mathbb{R}^n)$, and ext^L is an extension operator from $F^s_{pq}(\mathbb{R}^n_+)$ into $F^s_{pq}(\mathbb{R}^n)$.

(ii) Let $0 < p \leqq \infty$, $0 < q \leqq \infty$, $s \in \mathbb{R}$, let $l \in \mathbb{N}$ with

$$L > |s| + \frac{n}{p}. \tag{7}$$

Let $f \in B^s_{pq}(\mathbb{R}^n_+)$, then (6) converges in $B^s_{pq}(\mathbb{R}^n)$ and ext^L is an extension operator from $B^s_{pq}(\mathbb{R}^n_+)$ into $B^s_{pq}(\mathbb{R}^n)$.

Proof. Step 1. We begin with a preparation. Let $0 < p < \infty$, $0 < q \leqq \infty$, $s \in \mathbb{R}$, and let $L \in \mathbb{N}$ with $L > |s| + \frac{n}{p}$, then

$$\left| \int_{\mathbb{R}^n} g(x)\varphi(x)dx \right| \leqq \|g \mid F^s_{pq}(\mathbb{R}^n)\| \sup(1+|x|)^{n+1}|D^\beta \varphi(x)| \tag{8}$$

for all $g \in F^s_{pq}(\mathbb{R}^n)$ and all functions φ having classical derivatives in $\mathbb{R}^n$ up to order L, where the supremum is taken over all $x \in \mathbb{R}^n$ and all $|\beta| \leqq L$. By simple arguments based on Definition 2.3.1 and inequalities for entire analytic functions it follows

$$\left| \int_{\mathbb{R}^n} g(x)\varphi(x)dx \right| \leqq \|g \mid B^s_{p\infty}(\mathbb{R}^n)\| \, \|\varphi \mid B^{-s+\frac{n}{p}}_{11}(\mathbb{R}^n)\|, \tag{9}$$

see [Triß: Step 3 of the proof of Theorem 2.3.3] for details. Now (2.3.2/22), say, Theorem 2.6.1 applied to $B^{|s|+\varepsilon}_{11}(\mathbb{R}^n)$, $\varepsilon > 0$, and $L > |s| + \frac{n}{p}$, show that (8) is a consequence of (9). Of course (8) remains valid if one replaces F^s_{pq} by B^s_{pq}.

Step 2. We prove that (6) converges at least in S'. Let $f \in F^s_{pq}(\mathbb{R}^n)$ or $f \in B^s_{pq}(\mathbb{R}^n)$ decomposed by (3). Then we have for any $\varphi \in S$

$$\int_{\mathbb{R}^n} \varphi(y)\mathrm{ext}^L f_j(y)dy \tag{10}$$

$$= \int_{\mathbb{R}^n_+} \left[\varphi(y) + \sum_{l=1}^{2L+1} \frac{a_l}{\lambda_l}\varphi\left(y', -\frac{y_n}{\lambda_l}\right)\right] f_j(y)dy = \int_{\mathbb{R}^n_+} \psi(y) f_j(y)dy.$$

Let $r = 0, \ldots, L$, then (2) yields

$$\frac{\partial^r}{\partial y_n^r}\psi(x',0) = \left[1 - \sum_{l=1}^{2L+1} a_l(-\lambda_l)^{-r-1}\right] \frac{\partial^r \varphi}{\partial y_n^r}(y',0) = 0. \tag{11}$$

We extend $\psi(y)$ from $\mathbb{R}^n_+$ to $\mathbb{R}^n \backslash \mathbb{R}^n_+$ by zero. Then the extended function has classical derivatives in $\mathbb{R}^n$ up to order L. We may apply (8) and arrive at

$$\left|\int_{\mathbb{R}^n} \varphi(y)\mathrm{ext}^L f_j(y)dy\right| \leqq \|f_j \mid F^s_{pq}(\mathbb{R}^n_+)\| \sup(1+|x|)^{n+1}|D^\beta\varphi(x)|, \tag{12}$$

where the supremum is taken over $x \in \mathbb{R}^n$ and $|\beta| \leqq L$. Furthermore, $\{f^j\}_{j=0}^\infty$ with $f^j = \sum_{l=0}^j f_l$ is a fundamental sequence, say, in $F^{s-\varepsilon}_{pp}(\mathbb{R}^n_+)$ with $\varepsilon > 0$, see also Proposition 3.3.3. Then (12) with $F^{s-\varepsilon}_{pp}$ instead of F^s_{pq} shows that $\mathrm{ext}^L f^j$ converges at least in S'. Its limit will be denoted by $\mathrm{ext}^L f$, see (6).

Step 3. In Step 4 we shall prove $\mathrm{ext}^L f \in F^s_{pq}(\mathbb{R}^n)$ if $f \in F^s_{pq}(\mathbb{R}^n_+)$ on the basis of Theorem 3.3.3. For that purpose we have to calculate

$$k(2^{-l}, \mathrm{ext}^L f)(x) = \sum_{j=0}^\infty k(2^{-l}, \mathrm{ext}^L f_j)(x) \quad \text{if } l \in \mathbb{N} \tag{13}$$

and

$$k_0(2^{-l}, \mathrm{ext}^L f)(x) = \sum_{j=0}^\infty k_0(2^{-l}, \mathrm{ext}^L f_j)(x) \quad \text{if } l = 0, -1, \ldots, -K, \tag{14}$$

where k and k_0 are the same functions as in (3.3.2/4,5). In this step we derive the necessary estimates which prove to be the heart of the matter. By (3.3.2/4) we have

$$k(2^{-l}, \mathrm{ext}^L f_j)(x) = \int_{\mathbb{R}^n} k(y)\mathrm{ext}^L f_j(x + 2^{-l}y)dy \tag{15}$$

$$= 2^{ln}\int_{\mathbb{R}^n} k(2^l y - 2^l x)\mathrm{ext}^L f_j(y)dy$$

$$= 2^{ln}\int_{\mathbb{R}^n_+} \left[k(2^l y - 2^l x) + \sum_{r=1}^{2L+1} \frac{a_r}{\lambda_r} k\left(2^l y' - 2^l x', -2^l\frac{y_n}{\lambda_r} - 2^l x_n\right)\right] f_j(y)dy,$$

where we used (4). Let $j \in \mathbb{N}$, then (3.3.2/4,7) yield

$$f_j(y) = 2^{jn}\int_{\mathbb{R}^n_+} k(2^j z - 2^j y) f(z)dz \quad \text{if } y \in \mathbb{R}^n_+. \tag{16}$$

Now we choose M in (3.2.2/2) sufficiently large, then we have by (3.2.2/2,3,17)

$$k(2^j z - 2^j y) = \sum_u \sum_{|\beta|=L} D^\beta_y[\rho_{u\beta}(2^j z - 2^j y)]2^{-jL} \tag{17}$$

where $\sum_u$ is a finite sum, $\rho_{u\beta}$ are C^∞ functions in $\mathbb{R}^n$ with compact supports in $\mathbb{R}^n_+$ and

$$\hat{\rho}_{u\beta}(\xi) = O(|\xi|^G) \qquad \text{near the origin,} \tag{18}$$

where $G \in \mathbb{N}$ is at our disposal. Let $x_n \geqq 0$, then

$$\begin{aligned} k(2^{-l}, \mathrm{ext}^L f_j)(x) &= 2^{ln} \int_{\mathbb{R}^n_+} k(2^l y - 2^l x) f_j(y) dy \\ &= c 2^{(l+j)n} 2^{-L(l+j)} \sum_{u,v} \int_{\mathbb{R}^n_+} f(z) \\ \times \int_{\mathbb{R}^n_+} &\sum_{|\alpha|=|\beta|=L} D^\beta_y[\rho_{u\beta}(2^j z - 2^j y)] D^\alpha_y[\rho_{v\alpha}(2^l y - 2^l x)] dy dz. \end{aligned} \tag{19}$$

For fixed u, v, α, and β with $|\alpha| = |\beta| = L$ and $\gamma + \delta = \alpha + \beta$ we have

$$\begin{aligned} &2^{(l+j)(n-L)} \int_{\mathbb{R}^n_+} D^\beta_y[\rho_{u\beta}(2^j z - 2^j y)] D^\alpha_y[\rho_{v\alpha}(2^l y - 2^l x)] dy \\ &= 2^{(l+j)n} 2^{(L-|\gamma|)(l-j)} \int_{\mathbb{R}^n_+} (D^\gamma \rho_{u\beta})(2^j z - 2^j y)(D^\delta \rho_{v\alpha})(2^l y - 2^l x) dy. \end{aligned} \tag{20}$$

Let $l \leqq j$ then we have $\gamma = 0$ in (20), plug the result in (19) and obtain a sum of terms

$$\begin{aligned} I_{j,l}(x) &= 2^{-L(j-l)} 2^{(l+j)n} \int_{\mathbb{R}^n_+} (D^\delta \rho_{v\alpha})(2^l y - 2^l x) \\ &\quad \times \int_{\mathbb{R}^n_+} \rho_{u\beta}(2^j z - 2^j y) f(z) dz dy \\ &= 2^{-L(j-l)} 2^{ln} \int_{\mathbb{R}^n_+} (D^\delta \rho_{v\alpha})(2^l y - 2^l x) \int_{\mathbb{R}^n_+} \rho_{u\beta}(z) f(y + 2^{-j} z) dz dy \\ &= 2^{-L(j-l)} 2^{ln} \int_{\mathbb{R}^n_+} (D^\delta \rho_{v\alpha})(2^l y - 2^l x) \rho_{u\beta}(2^{-j}, f)(y) dy, \end{aligned} \tag{21}$$

see (3.3.2/4), where we may assume that G in (18) is chosen sufficiently large. Let

$$\rho^+_{u\beta}(2^{-j}, f)(x) = \sup_{x+w \in \mathbb{R}^n_+} \frac{|\rho_{u\beta}(2^{-j}, f)(x+w)|}{1 + |2^j w|^a}, \qquad x \in \mathbb{R}^n, \tag{22}$$

$a > 0$, be a maximal function. Then (21) yields

$$|I_{j,l}(x)| \leqq c\,2^{-L(j-l)+a(j-l)} \rho^+_{u\beta}(2^{-j}, f)(x) \tag{23}$$

and

$$|k(2^{-l}, \text{ext}^L f_j)(x)| \leqq c \sum_u \sum_{|\beta|=L} \rho^+_{u\beta}(2^{-j}, f)(x) 2^{-(L-a)|j-l|}, \quad x \in \mathbb{R}^n_+,\ j \geqq l. \tag{24}$$

Let again $x_n \geqq 0$ and $j < l$. Then we choose $\delta = 0$ in (20). In particular $L - |\gamma| = -L$. We arrive at (24) with some function $D^\gamma \rho_{u\beta}$ instead of $\rho_{u\beta}$ but this modification is immaterial. By the same arguments one can extend (24) to $j = 0$ or to $l = -1, \ldots, -K$, where K has the same meaning as in Theorem 3.3.3. In these exceptional cases one has to replace k by k_0, for example, if $j = 0$ and $l = -1, \ldots, -K$, then one has to replace both k and $\rho_{u\beta}$ in (24) by k_0. In other words, with the indicated modifications, (24) with $x_n \geqq 0$ holds for all numbers j and l of interest. Next we discuss the case $x_n < 0$, where we return to the general case, i.e., $l \in \mathbb{N}$ and $j \in \mathbb{N}$. We restrict x_n by $x_n \leqq -c2^{-l}$, where c is an appropriate positive number. Then we have $k(2^l y - 2^l x) = 0$ if $y \in \mathbb{R}^n_+$ and each of the remaining $2L+1$ terms in (15) can be treated separately in the same way as above. The support of $k(2^l y' - 2^l x', -2^l \frac{y_n}{\lambda_r} - 2^l x_n)$ is concentrated around $(x', -\lambda_r x_n)$. Otherwise one can follow the above reasoning and arrives at

$$|k(2^{-l}, \text{ext}^L f_j)(x)| \leqq c \sum_u \sum_{r=1}^{2L+1} \rho^+_{u\beta}(2^{-j}, f)(x', -\lambda_r x_n) 2^{-(L-a)|j-l|}, \tag{25}$$

$x \in \mathbb{R}^n$, $x_n \leqq -c2^{-l}$, $j \geqq l$. As above we have corresponding estimates for $l > j$ and if either $j = 0$ and/or $l \leqq 0$, provided that $x \in \mathbb{R}^n$ and $x_n \leqq -c2^{-l}$. Finally let $x \in \mathbb{R}^n$ with $-c2^{-l} \leqq x_n \leqq 0$, where we again assume $l \in \mathbb{N}$ and $j \in \mathbb{N}$. We have again (15). We must study the kernels in (15) more closely. For this purpose we begin with a preparation. Let $x \in \mathbb{R}^n$ with $-c \leqq x_n < 0$ be fixed, then we consider

$$K_x(y) = \begin{cases} k(y-x) + \sum_{r=1}^{2L+1} \frac{a_r}{\lambda_r} k\left(y' - x', -\frac{y_n}{\lambda_r} - x_n\right) & \text{if } y_n \geqq 0 \\ 0 & \text{if } y_n < 0. \end{cases} \tag{26}$$

In other words, $K_x(y)$ is just the kernel in (15) with $l = 0$. Similarly as in (11) it follows that $K_x(y)$ has classical derivatives up to order L in $\mathbb{R}^n$ (with respect to y). We claim that $K_x(y)$ can be represented as a finite sum

$$K_x(y) = \sum_v \sum_{|\alpha|=L} D^\alpha_y \kappa_{v\alpha}(x, y), \tag{27}$$

where $\kappa_{v\alpha}(x, y)$ has the necessary classical derivatives, $D^\alpha_y \kappa_{v\alpha}(x, y)$ has classical derivatives up to order L with respect to y and

$$\operatorname{supp} \kappa_{v\alpha}(x,\cdot) \subset \mathbb{R}^n_+ \cap \{y\colon |x-y| \leqq c\}, \qquad x \text{ fixed}, \tag{28}$$

for some $c > 0$. By (3.2.2/2,3,17) (the latter with $M = L$) the problem can be reduced to

$$\Phi^\beta_x(y) = \begin{cases} D^\beta \varphi_\beta(y-x) + \sum\limits_{r=1}^{2L+1} \dfrac{a_r}{\lambda_r} (D^\beta \varphi_\beta)\left(y'-x', -\dfrac{y_n}{\lambda_r} - x_n\right), & \text{if } y_n \geqq 0 \\ 0, & \text{if } y_n < 0 \end{cases} \tag{29}$$

can be represented in the desired way. Here φ_β has the same meaning as in (3.3.2/17) with $|\beta| = L$. We have

$$\int_{\mathbb{R}} y_n^l \Phi^\beta_x(y', y_n) dy_n = \int_0^\infty y_n^l (D^\beta \varphi_\beta)(y'-x', y_n - x_n) dy_n \tag{30}$$

$$+ \sum_{r=1}^{2L+1} a_r (-\lambda_r)^l \int_{-\infty}^0 y_n^l (D^\beta \varphi_\beta)(y'-x', y_n - x_n) dy_n$$

$$= \int_{\mathbb{R}} y_n^l (D^\beta \varphi_\beta)(y'-x', y_n - x_n) dy_n; \qquad l = 0, \ldots, L-1,$$

where we used (2). We obtain

$$\int_{\mathbb{R}} y_n^l \Phi^\beta_x(y', y_n) dy_n = 0 \quad \text{if } \; l = 0, \ldots, \beta_n - 1, \tag{31}$$

where $\beta = (\beta_1, \ldots, \beta_n)$. Now we apply (3.3.1/4) with $\Phi^\beta_x(y', y_n)$ and β_n instead of $\varphi(\tau) - \frac{1}{2}\varphi(\frac{\tau}{2})$ and M, respectively. The resulting counterpart of the function ψ in (3.3.1/4) has the desired property. Furthermore $D^\alpha_y \kappa_{v\alpha}(x,y)$ has the desired differentiability properties. After these preparations we return to the remaining case $l \in \mathbb{N}$, $j \in \mathbb{N}$, and $-c2^{-l} \leqq x_n \leqq 0$. The kernel in (15) is given by

$$K_{2^l x}(2^l y) = \sum_v \sum_{|\alpha|=L} D^\alpha_y [\kappa_{v\alpha}(2^l x, 2^l y)] 2^{-lL} \tag{32}$$

with

$$\operatorname{supp} \kappa_{v\alpha}(2^l x, 2^l \cdot) \subset \mathbb{R}^n_+ \cap \{y\colon |x-y| \leqq c2^{-l}\} \tag{33}$$

where one can replace x by $(x', -x_n)$. But now we are in the same position as in (19) with $\kappa_{v\alpha}(2^l x, 2^l y)$ instead of $\rho_{v\alpha}(2^l y - 2^l x)$, but with the unchanged functions $\rho_{u\beta}(2^j z - 2^j y)$. Then the counterpart of (24) is given by

$$|k(2^{-l}, \operatorname{ext}^L f_j)(x)| \leqq c \sum_u \sum_{|\beta|=L} \rho^+_{u\beta}(2^{-j}, f)(x', -x_n) 2^{-(L-a)|j-l|}, \tag{34}$$

$x \in \mathbb{R}^n$, $-c2^{-l} \leqq x_n \leqq 0$, $j \geqq l$. We have also corresponding estimates for $l > j$ and also for the exceptional cases where either $j = 0$ or $l = 0, -1, \ldots, -K$.

Step 4. We prove (i). Let $0 < p < \infty$, $0 < q \leqq \infty$, $s \in \mathbb{R}$, let $L \in \mathbb{N}$ with

$$L > |s| + a \quad \text{for some} \quad a > \frac{n}{\min(p,q)}, \tag{35}$$

where a has the same meaning as in (22). We claim that there exists a positive number c such that

$$\left\| \left(\sum_{j=0}^{\infty} 2^{jsq} \rho_{u\beta}^{+q}(2^{-j}, f)(\cdot) \right)^{1/q} \mid L_p(\mathbb{R}^n_+) \right\| \leqq c \|f \mid F^s_{pq}(\mathbb{R}^n_+)\| \tag{36}$$

for all $f \in F^s_{pq}(\mathbb{R}^n)$, where $\rho_{u\beta}$ has the above meaning, see (17,22): Let $g \in F^s_{pq}(\mathbb{R}^n)$ with $g \mid \mathbb{R}^n_+ = f$. Furthermore by (2.4.1/55) we have

$$\rho^+_{u\beta}(2^{-j}, g)(x) \leqq ((\rho_{u\beta})^*_j g)_a(x). \tag{37}$$

Recall $\rho^+_{u\beta}(2^{-j}, g)(x) = \rho^+_{u\beta}(2^{-j}, f)(x)$ if $x \in \mathbb{R}^n_+$. Then (35,37) and Corollary 2.4.1/2 prove that the left-hand side of (36) can be estimated from above by $c\|g \mid F^s_{pq}(\mathbb{R}^n_+)\|$. Taking the infimum over all admissible g we arrive at (36). Let again $x \in \mathbb{R}^n_+$ and $l \in \mathbb{N}$, then (24) yields

$$2^{ls}|k(2^{-l}, \mathrm{ext}^L f)(x)| \leqq 2^{ls} \sum_{j=0}^{\infty} |k(2^{-l}, \mathrm{ext}^L f_j)(x)| \tag{38}$$

$$\leqq c \sum_u \sum_{|\beta|=L} \sum_{m=-\infty}^{\infty} 2^{(l+m)s} \rho^+_{u\beta}(2^{-l-m}, f)(x) 2^{-(L-a-|s|)|m|} b_{ml} + +$$

with $b_{ml} = 1$ if $l + m \geqq 0$ and $b_{ml} = 0$ if $l + m < 0$, where $++$ stands for similar terms in the sense of our remarks after (24), including exceptional cases. Let $\lambda = \min(1, p, q)$. By (35), (36), and their counterparts related to $++$ in (38) we obtain

$$\left\| \left(\sum_{l=1}^{\infty} 2^{lsq} |k(2^{-l}, \mathrm{ext}^L f)(\cdot)|^q \right)^{1/q} \mid L_p(\mathbb{R}^N_+) \right\|^\lambda \tag{39}$$

$$\leqq c \sum_u \sum_{|\beta|=L} \sum_{m=-\infty}^{\infty} 2^{-(L-a-|s|)|m|\lambda} \left\| \left(\sum_{j=0}^{\infty} 2^{jsq} \rho^{+q}_{u\beta}(2^{-j}, f)(\cdot) \right)^{1/q} \mid L_p(\mathbb{R}^n_+) \right\|^\lambda + +$$

$$\leqq c' \|f \mid F^s_{pq}(\mathbb{R}^n_+)\|^\lambda.$$

By the above mentioned modification we have a corresponding estimate with $l = 0$ and k_0 instead of k. Let $x \in \mathbb{R}^n_- = \{y \in \mathbb{R}^n \colon y_n < 0\}$ then we use (25) and (34) instead of (24) and the corresponding modifications if $j < l$ and arrive at

$$\left\| \left(\sum_{l=1}^{\infty} 2^{lsq} |k(2^{-l}, \mathrm{ext}^L f)(\cdot)|^q \right)^{1/q} \mid L_p(\mathbb{R}^n_-) \right\|^{\lambda} \leqq c \|f \mid F^s_{pq}(\mathbb{R}^n_+)\|^{\lambda} \tag{40}$$

with the indicated modifications. Let us assume temporarily $\mathrm{ext}^L f \in F^s_{pq}(\mathbb{R}^n)$, then (39,40), the indicated exceptional cases and Theorem 3.3.3 yield

$$\|\mathrm{ext}^L f \mid F^s_{pq}(\mathbb{R}^n)\| \leqq c \|f \mid F^s_{pq}(\mathbb{R}^n_+)\|, \tag{41}$$

which completes the proof of (i). It remains to justify $\mathrm{ext}^L f \in F^s_{pq}(\mathbb{R}^n)$. By Step 2 we have

$$\mathrm{ext}^L f = \lim_{j \to \infty} \mathrm{ext}^L f^j \quad \text{in} \quad S'. \tag{42}$$

Furthermore it can be seen easily $f^j \in F^\sigma_{pq}(\mathbb{R}^n_+)$ for any $\sigma \in \mathbb{R}$. For large values of σ we apply Theorem 4.5.2(i) with ext^L instead of ext_L and obtain $\mathrm{ext}^L f^j \in F^\sigma_{pq}(\mathbb{R}^n)$. In particular, by the above reasoning we have

$$\|\mathrm{ext}^L f^j \mid F^s_{pq}(\mathbb{R}^n)\| \leqq c \|f \mid F^s_{pq}(\mathbb{R}^n_+)\|, \tag{43}$$

where c is independent of j. Now (43,42), (2.3.1/8) and a Fatou argument prove $\mathrm{ext}^L f \in F^s_{pq}(\mathbb{R}^n)$. The proof of (i) is complete.

Step 5. The proof of (ii) is the same. Instead of Corollary 2.4.1/2 we use Corollary 2.5.1/2.

Remark 1. Corollary 4.5.2 is an immediate consequence of the above theorem.

Remark 2. We complement the references given in 4.5.1 and Remark 4.5.2/1. The typical construction (4) which proved to be useful also for spaces F^s_{pq} and B^s_{pq} with $s < 0$ had been used first in [Triα: 2.9.3] and later in [Triß: 2.9] and [Fra]. The main result of [Triß: 2.9] and [Fra] reads as follows: Let $a > 0$ and $b > 0$ be given and let $L = L(a, b) \in \mathbb{N}$ be chosen sufficiently large, then ext^L is a common extension operator from $F^s_{pq}(\mathbb{R}^n_+)$ into $F^s_{pq}(\mathbb{R}^n)$ and from $B^s_{pq}(\mathbb{R}^n_+)$ into $B^s_{pq}(\mathbb{R}^n)$ provided that $|s| \leqq a$, $b \leqq p \leqq \infty$ ($p < \infty$ in the case of the F-spaces), $0 < q \leqq \infty$. This result is slightly different of part (i) of the above theorem, because the assertion is independent of q. The proof in [Triß: 2.9] and [Fra] is rather complicated (which does not mean that we claim that the above proof is simple) and one needs several deep properties of the spaces F^s_{pq}: characteristic functions as pointwise multipliers, Fubini properties, duality, and complex interpolation (extended to values $p < 1$).

Chapter 5

Spaces on Domains

5.1 Preliminaries

5.1.1 An introductory remark

Up to now we dealt with spaces $F^s_{pq} = F^s_{pq}(\mathbb{R}^n)$, $B^s_{pq} = B^s_{pq}(\mathbb{R}^n)$ on the one hand and their restrictions $F^s_{pq}(\mathbb{R}^n_+)$, $B^s_{pq}(\mathbb{R}^n_+)$ on $\mathbb{R}^n_+$ on the other hand. As for the latter spaces see Definition 4.5.1. Now we introduce in precisely the same way spaces $F^s_{pq}(\Omega)$ and $B^s_{pq}(\Omega)$, where Ω always stands for a bounded C^∞ domain in $\mathbb{R}^n$. We are not interested in this book to deal with non-smooth domains, avoiding serious technical difficulties which may be especially hard for spaces with $p < 1$ and/or $s < 0$. The two outstanding problems treated here are the same as we described in 4.5.1: the extension problem and intrinsic descriptions. Based on Theorem 4.5.5 the extension problem for spaces on Ω can be solved very easily, see 5.1.3. As for intrinsic characterizations we prove in 5.2.1 the counterpart of Theorem 4.5.3 with Ω instead of $\mathbb{R}^n_+$. This may be considered as one of the main results of this chapter. Afterwards it will be easy to find corresponding characterizations in terms of differences: practically we have already done this in 3.5.3, but we return to this subject in 5.2.2. Section 5.3 may be considered as a digression from our main line. Oscillations had been used since a long time to introduce and study spaces of Morrey–Campanato type, spaces of bounded mean oscillations, and, more recently, spaces of type C^s_p which are closely connected with so-called sharp maximal functions. We introduce these spaces in 5.3.1. Our technique developed so far is good enough to prove some basic properties of these spaces, to identify some of these spaces with spaces of type $C^s = B^s_{\infty\infty}$ and $F^s_{p\infty}$, and to characterize the Sobolev spaces W^k_p in terms of sharp maximal functions. All this will be done in 5.3. In 5.4 we list some further problems.

5.1.2 Definition

In this chapter we again write $F^s_{pq}(\mathbb{R}^n)$ and $B^s_{pq}(\mathbb{R}^n)$ instead of F^s_{pq} and B^s_{pq}, respectively, where the latter spaces have the same meaning as in Definition 2.3.1. Recall

H. Triebel, *Theory of Function Spaces II*, Modern Birkhäuser Classics,
DOI 10.1007/978-3-0346-0419-2_5, © Birkhäuser Verlag 1992

that we introduced in 4.5.1 the spaces $F^s_{pq}(\mathbb{R}^n_+)$ and $B^s_{pq}(\mathbb{R}^n_+)$.

Definition. Let Ω be a bounded C^∞ domain in $\mathbb{R}^n$.

(i) Let $0 < p < \infty$, $0 < q \leqq \infty$, and $s \in \mathbb{R}$, then $F^s_{pq}(\Omega)$ is the restriction of $F^s_{pq}(\mathbb{R}^n)$ on Ω, quasi-normed by

$$\|f \mid F^s_{pq}(\Omega)\| = \inf \|g \mid F^s_{pq}(\mathbb{R}^n)\|, \tag{1}$$

where the infimum is taken over all $g \in F^s_{pq}(\mathbb{R}^n)$ with $g \mid \Omega = f$ in the sense of $D'(\Omega)$.

(ii) Let $0 < p \leqq \infty$, $0 < q \leqq \infty$, and $s \in \mathbb{R}$, then $B^s_{pq}(\Omega)$ is the restriction of $B^s_{pq}(\mathbb{R}^n)$ on Ω, quasi-normed by

$$\|f \mid B^s_{pq}(\Omega)\| = \inf \|g \mid B^s_{pq}(\mathbb{R}^n)\|, \tag{2}$$

where the infimum is taken over all $g \in B^s_{pq}(\mathbb{R}^n)$ with $g \mid \Omega = f$ in the sense of $D'(\Omega)$.

Remark 1. What is meant by a bounded C^∞ domain is quite clear, we refer for details, e.g., to [Triß: 3.2.1]. The above definition is the direct counterpart of Definition 4.5.1. It follows by standard arguments that both $F^s_{pq}(\Omega)$ and $B^s_{pq}(\Omega)$ are quasi-Banach spaces, see [Triß: 3.2.3] for details.

Remark 2. We are interested in the extension property and in intrinsic characterizations, see 4.5.1 for the $\mathbb{R}^n_+$-counterpart. Furthermore we refer to [Triß: Chapter 3] where we treated these spaces in greater detail.

5.1.3 Extension property

Let again Ω be a bounded C^∞ domain in $\mathbb{R}^n$. Let re be the restriction operator from $F^s_{pq}(\mathbb{R}^n)$ onto $F^s_{pq}(\Omega)$ or from $B^s_{pq}(\mathbb{R}^n)$ onto $B^s_{pq}(\Omega)$. Obviously, re is a linear and bounded operator. We are looking for a linear and bounded operator ext from $F^s_{pq}(\Omega)$ into $F^s_{pq}(\mathbb{R}^n)$ or from $B^s_{pq}(\Omega)$ into $B^s_{pq}(\mathbb{R}^n)$ with

$$\text{re} \circ \text{ext} = \text{id} \qquad (\text{identity in } F^s_{pq}(\Omega) \text{ or } B^s_{pq}(\Omega)). \tag{1}$$

See also 4.5.1 where we formulated the corresponding problem with $\mathbb{R}^n_+$ instead of Ω. The references given there may also be applied to the present situation. In the theorem below we use the notation of a common extension operator which means that the operator in question is defined on the union of all the considered spaces, and its restriction to each concrete space is an extension operator in the above sense.

Theorem. Let Ω be a bounded C^∞ domain in $\mathbb{R}^n$. Let $\varepsilon > 0$ and $|s| < \varepsilon^{-1}$ then there exists a common extension operator ext^ε

$$\text{from } F^s_{pq}(\Omega) \text{ into } F^s_{pq}(\mathbb{R}^n) \text{ with } \varepsilon < p < \infty,\ \varepsilon < q \leqq \infty \tag{2}$$

and

$$\text{from } B^s_{pq}(\Omega) \text{ into } B^s_{pq}(\mathbb{R}^n) \text{ with } \varepsilon < p \leqq \infty,\ 0 < q \leqq \infty. \tag{3}$$

Proof. Let $\varphi_0, \varphi_1, \ldots, \varphi_N$ be a C^∞ resolution of unity with respect to Ω. In other words: $\varphi_0, \ldots, \varphi_N$ are C^∞ functions in $\mathbb{R}^n$ with

$$\operatorname{supp} \varphi_0 \subset \Omega, \quad \operatorname{supp} \varphi_j \subset B_r(x^j) \text{ if } j = 1, \ldots, N, \tag{4}$$

where $B_r(x^j)$ is a ball of radius $r > 0$, centered at, say, $x^j \in \partial\Omega$, and

$$\sum_{j=0}^{N} \varphi_j(x) = 1 \quad \text{in a neighbourhood of } \Omega. \tag{5}$$

We have

$$\|f \mid F^s_{pq}(\Omega)\| \sim \sum_{j=0}^{N} \|\varphi_j f \mid F^s_{pq}(\Omega)\| \tag{6}$$

(equivalent quasi-norms) for all admissible (s, p, q), and a corresponding assertion with B^s_{pq} instead of F^s_{pq}. The equivalence (6) follows immediately from Theorem 4.2.2 (pointwise multipliers) and Definition 5.1.2. If $r > 0$ in (4) is chosen sufficiently small then there exists diffeomorphisms $\psi_1, \ldots, \psi_N$ of $\mathbb{R}^n$ onto itself in the sense of 4.3.1 such that

$$\psi_j[B_r(x^j) \cap \Omega] = \psi_j[B_r(x^j)] \cap \mathbb{R}^n_+, \qquad j = 1, \ldots, N, \tag{7}$$

or, in other words, ψ_j sends $B_r(x^j) \cap \partial\Omega$ into $\mathbb{R}^{n-1} = \{x \in \mathbb{R}^n\colon x_n = 0\}$. By Theorem 4.3.2 and Definition 5.1.2 it follows

$$\|\varphi_j f \mid F^s_{pq}(\Omega)\| \sim \|(\varphi_j f) \circ \psi_j^{-1} \mid F^s_{pq}(\mathbb{R}^n_+)\|, \qquad j = 1, \ldots, N. \tag{8}$$

Now we extend each of the functions $(\varphi_j f)\circ\psi_j^{-1}$ from $F^s_{pq}(\mathbb{R}^n_+)$ to $F^s_{pq}(\mathbb{R}^n)$ by Theorem 4.5.5, where we may assume

$$\operatorname{supp}[\operatorname{ext}^L(\varphi_j f)\circ\psi_j^{-1}] \subset \psi_j[B_r(x^j)], \tag{9}$$

re-transform the result, sum up the terms with $j = 1, \ldots, N$, extend $\varphi_0 f$ outside Ω by zero and arrive at the desired extension operator. Of course we can replace F^s_{pq} by B^s_{pq} which completes the proof.

Remark. The proof is based on three major ingredients: pointwise multipliers (Theorem 4.2.2), diffeomorphism properties (Theorem 4.3.2) and extensions from $\mathbb{R}^n_+$ into $\mathbb{R}^n$ (Theorem 4.5.5). The reduction of the above theorem to these three key theorems is more or less obvious, see also [Triß: 3.3.4]. This principle can also be applied to many other assertions for spaces on Ω. For example, there is an immediate counterpart of the pointwise multiplier assertions in Theorem 4.2.2 and Corollary 4.2.2. Furthermore, via local charts one can introduce spaces of type F^s_{pq} and B^s_{pq} on $\partial\Omega$. Afterwards one has obvious counterparts of the trace assertions in Theorem 4.4.2 and Corollary 4.4.2, and even of the curiosities mentioned in 4.4.3. We do not go into detail but we refer to [Triß: Chapter 3] where some of these assertions have been treated in great detail.

5.2 Intrinsic characterizations

5.2.1 Oscillations

It is one of the main aims of this chapter to extend Theorems 3.5.1 and 4.5.3 from $\mathbb{R}^n$ and $\mathbb{R}^n_+$, respectively, to bounded C^∞ domains Ω in $\mathbb{R}^n$.

Theorem. Let Ω be a bounded C^∞ domain in $\mathbb{R}^n$ and let $\operatorname{osc}^M_u f$ be given by Definition 3.4.1.

(i) Let $0 < p < \infty$, $0 < q \leqq \infty$, $1 \leqq r \leqq \infty$, and

$$s > n\left(\frac{1}{p} - \frac{1}{r}\right)_+ \quad \text{and} \quad s > n\left(\frac{1}{q} - \frac{1}{r}\right)_+ . \tag{1}$$

Let $0 < u \leqq r$ and $M \in \mathbb{N}_0$ with $M > [s]$, then

$$F^s_{pq}(\Omega) = \Big\{ f \in L_{\max(p,r)}(\Omega): \tag{2}$$

$$\|f \mid L_p(\Omega)\| + \left\| \left(\int_0^1 t^{-sq} \mathrm{osc}_u^M f(\cdot, t)^q \frac{dt}{t} \right)^{1/q} \mid L_p(\Omega) \right\| < \infty \Big\}$$

(modification if $q = \infty$) in the sense of equivalent quasi-norms.

(ii) Let $0 < p \leqq \infty$, $0 < q \leqq \infty$, $1 \leqq r \leqq \infty$, and

$$s > n \left(\frac{1}{p} - \frac{1}{r} \right)_+ . \tag{3}$$

Let $0 < u \leqq r$ and $M \in \mathbb{N}_0$ with $M \geqq [s]$ then

$$B^s_{pq}(\Omega) = \Big\{ f \in L_{\max(p,r)}(\Omega): \tag{4}$$

$$\|f \mid L_p(\Omega)\| + \left(\int_0^1 t^{-sq} \|\mathrm{osc}_u^M f(\cdot, t) \mid L_p(\Omega)\|^q \frac{dt}{t} \right)^{1/q} < \infty \Big\}$$

(modification if $q = \infty$) in the sense of equivalent quasi-norms.

Proof. Step 1. We prove (i), the proof of (ii) is the same. We have to distinguish between oscillations based on $\mathbb{R}^n$, which will be temporarily denoted by $\mathrm{Osc}_u^M f$ and oscillation based on Ω, which will be denoted as above by $\mathrm{osc}_u^M f$. Let $\mathrm{ext} = \mathrm{ext}^\varepsilon$ be the extension operator constructed in 5.1.3, where we may use ext_L from Theorem 4.5.2 instead of ext^L from Theorem 4.5.5. We begin with a local consideration and assume that Ω satisfies the hypotheses of 3.4.1 with

$$\frac{\partial \psi}{\partial x_j}(0) = 0 \quad \text{if } j = 1, \ldots, n-1$$

in (3.4.1/4). Let $f \in F^s_{pq}(\Omega)$ with a support near the origin. Let χ be a diffeomorphism of, say, the unit ball B in $\mathbb{R}^n$ into itself such that $\chi(0) = 0$ and $\chi(B \cap \Omega) \subset B \cap \mathbb{R}^n_+$. By the construction in 5.1.3 we can describe ext locally by

$$\mathrm{ext}\, f(x) = (\mathrm{ext}_L f \circ \chi^{-1})(\chi(x)), \qquad x \in B, \tag{5}$$

where ext_L has the same meaning as in Theorem 4.5.2, see the proof of Theorem 5.1.3 for details. In this step we restrict the above number u by $1 \leqq u \leqq r$. Then Proposition 3.4.2 is at our disposal, both for $\mathrm{Osc}_u^L \mathrm{ext}\, f$ and $\mathrm{osc}_u^L f$. We need the counterparts of (4.5.2/7–9) now with Ω instead of $\mathbb{R}^n_+$. We restrict ourselves to the most complicated case, this is the counterpart (4.5.2/9). The transformation $\lambda(x) = (x', |x_n|)$ in (4.5.2/9) must be replaced by $\mu(x) = \chi^{-1} \circ \lambda \circ \chi(x)$. We have

$(x', |x_n|)$ in (4.5.2/9) must be replaced by $\mu(x) = \chi^{-1} \circ \lambda \circ \chi(x)$. We have

$$\mathrm{Osc}_u^L \mathrm{ext}\, f(\chi^{-1}(x),t) \leqq \left(\oint_{\{y:|\chi^{-1}(x)-y|<t\}} \Big| \mathrm{ext}_L f \circ \chi^{-1}(\chi(y)) - \sum_{|\alpha|\leqq L} a_\alpha(x)(y-\chi^{-1}(x))^\alpha \Big|^u dy \right)^{1/u}, \tag{6}$$

where we try to determine $a_\alpha(x)$ in an optimal way, such that we can apply (3.4.2/2) to osc at the point $\mu \circ \chi^{-1}(x)$. In this sense we assume that $a_\alpha(x)$ is a linear combination of $D^\gamma f^j(\mu \circ \chi^{-1}(x))$ with $|\gamma| \leqq L$ and $t = 2^{-j}$ multiplied with harmless C^∞ functions (sums and products of some derivatives of the components of χ and χ^{-1}). We substitute $y = \chi^{-1}(z)$ in (6). We may assume

$$(\chi^{-1}(z) - \chi^{-1}(x))^\alpha = c_\alpha(x)(z-x)^\alpha + \cdots + O(|z-x|^{L+1}) \tag{7}$$

with $c_\alpha(x) \neq 0$, where $+\cdots$ indicates terms of order between $|\alpha|+1$ and L, where the coefficients can be calculated explicitly, see (3.4.1/4). We have

$$\sum_{|\alpha|\leqq L} a_\alpha(x)(\chi^{-1}(z)-\chi^{-1}(x))^\alpha = \sum_{|\beta|\leqq L} b_\beta(x)(z-x)^\beta + O(|z-x|^{L+1}), \tag{8}$$

where the coefficients $b_\beta(x)$ are also linear combinations of $D^\gamma f^j(\mu \circ \chi^{-1}(x))$ in the above sense, what also applies to the remainder term. On the other hand, if the coefficients $b_\beta(x)$ are given, then the coefficients $a_\alpha(x)$ can be calculated uniquely by induction with respect to $|\alpha|$. This will be the crucial point. But first we handle the remainder term in (8). For this purpose we choose L large, for example $L = 2M$ with $M \in \mathbb{N}$ and $M \geqq [s]$. Then we have $L+1-M > M$ and by (3.4.2/8) (with $|\beta| \leqq M$ instead of $|\alpha| = M$)

$$\begin{aligned} 2^{-j(L+1)}|D^\beta f^j(\mu \circ \chi^{-1}(x))| &= 2^{-j(M+1)}2^{-Mj}|D^\beta f^j(\mu \circ \chi^{-1}(x))| \\ &\leqq c\, 2^{-j(M+1)} f^*(\mu \circ \chi^{-1}(x)), \qquad |\beta| \leqq M, \quad 2^{-j} \sim t, \end{aligned} \tag{9}$$

where f^* indicates the Hardy–Littlewood maximal function restricted to $\mathbb{R}^n_+$. If $M < |\beta| \leqq L$, then we estimate $D^\beta f^j(\mu \circ \chi^{-1}(x))$ in the same way as in (3.4.2/9), where osc is based on Ω. Recall that we are looking for the counterpart of (4.5.2/9). Now we insert the first term on the right-hand side of (8) into (6) where we substituted $y = \chi^{-1}(z)$. We use (4.5.2/6) in the same way as in the proof of Theorem 4.5.2. By the same arguments as in 4.5.2, besides constants the integration over $\{z\colon |\chi^{-1}(x) - \chi^{-1}(z)| < t\}$ can be replaced by

$$\{z\colon |\mu \circ \chi^{-1}(x) - \chi^{-1}(z)| < ct, \quad z \in \mathbb{R}^n_+\}.$$

Re-transform $y = \chi^{-1}(z)$, the choice

$$\sum_{|\alpha|\leqq L} a_\alpha(x)(y-\chi^{-1}(x))^\alpha = \sum_{|\alpha|\leqq L} \frac{1}{\alpha!}(D^\alpha f^j)(\mu \circ \chi^{-1}(x))(y - \mu \circ \chi^{-1}(x))^\alpha, \tag{10}$$

Proposition 3.4.2, and the above remainder estimates yields

$$\operatorname{Osc}_u^L \operatorname{ext} f(\chi^{-1}(x), t) \tag{11}$$

$$\leqq c \sum_{k=M}^{L} \operatorname{osc}_u^k f(\mu \circ \chi^{-1}(x), c't) + ct^{M+1} f^*(\mu \circ \chi^{-1}(x)).$$

Similar but somewhat simpler one obtains the counterpart of (4.5.2/7,8).

Step 2. The rest of the proof is more or less based on the crucial estimate (11) and the just-mentioned counterparts. We replace $\chi^{-1}(x)$ in (11) by x, multiply this estimate and their two counterparts with t^{-s}, take the l_q-quasi-norm and then the L_p-quasi-norm and arrive at

$$\left\| \left(\int_0^1 t^{-sq} \operatorname{Osc}_u^L \operatorname{ext} f(\cdot, t)^q \frac{dt}{t} \right)^{1/q} \mid L_p(\mathbb{R}^n) \right\| \tag{12}$$

$$\leqq c' \sum_{k=M}^{L} \left\| \left(\int_0^1 t^{-sq} \operatorname{osc}_u^k f(\cdot, ct)^q \frac{dt}{t} \right)^{1/q} \mid L_p(\Omega) \right\| + c' \| f^* \mid L_p(\Omega) \|,$$

where now f^* stands for the Hardy–Littlewood maximal function with respect to Ω. We used $M + 1 > s$. We may assume $r > 1$ in the theorem. Let $v = \max(p, r)$. By Hölder's inequality we can replace $L_p(\Omega)$ in the last term of (12) by $L_v(\Omega)$. Afterwards we apply the Hardy–Littlewood maximal inequality which shows that the last term in (12) can be estimated from above by $c\|f \mid L_v(\Omega)\|$. Next we apply Proposition 3.4.3 with $k = M, \ldots, L-1$ instead of $M - 1$, with u instead of p and $t \sim 2^{-j}$. Because $k + 1 > s$ it follows

$$\left\| \left(\int_0^1 t^{-sq} \operatorname{Osc}_u^L \operatorname{ext} f(\cdot, t)^q \frac{dt}{t} \right)^{1/q} \mid L_p(\mathbb{R}^n) \right\| \tag{13}$$

$$\leqq c' \left\| \left(\int_0^1 t^{-sq} \operatorname{osc}_u^L f(\cdot, ct)^q \frac{dt}{t} \right)^{1/q} \mid L_p(\Omega) \right\| + c' \| f \mid L_v(\Omega) \|.$$

By the embedding $F^s_{pq}(\mathbb{R}^n) \subset L_v(\mathbb{R}^n)$, see (3.5.1/8) or [Trip: 2.7.1], and the homogeneity assertions in Remark 2.3.3 it follows easily

$$\|\operatorname{ext} f \mid L_v(\mathbb{R}^n)\| \leqq \varepsilon \|\operatorname{ext} f \mid F^s_{pq}(\mathbb{R}^n)\| + c_\varepsilon \|f \mid L_p(\Omega)\| \tag{14}$$

where $\varepsilon > 0$ is at our disposal. Now by Theorem 3.5.1 we arrive at

$$\|f \mid F^s_{pq}(\Omega)\| \leqq \|\operatorname{ext} f \mid F^s_{pq}(\mathbb{R}^n)\| \leqq \|f \mid L_p(\Omega)\| \tag{15}$$

$$+ \left\| \left(\int_0^1 t^{-sq} \mathrm{osc}_u^L f(\cdot, ct)^q \frac{dt}{t} \right)^{1/q} \mid L_p(\Omega) \right\| \leqq c' \| f \mid F^s_{pq}(\Omega) \|.$$

In other words, the main term in (15) is an equivalent quasi-norm in $F^s_{pq}(\Omega)$ provided that $f \in F^s_{pq}(\Omega)$ has a support near the origin, $L = 2M \in \mathbb{N}$ with $M \geqq [s]$ and $1 \leqq u \leqq r$. The extension of this assertion to all $L \in \mathbb{N}_0$ with $L \geqq [s]$ follows from Proposition 3.4.3. Finally let $f \in L_v(\Omega)$ with a support near the origin such that the right-hand side of (2) is finite. Then we have (13) and hence by Theorem 3.5.1 it follows that $\mathrm{ext}\, f \in F^s_{pq}(\mathbb{R}^n)$ and $f \in F^s_{pq}(\Omega)$. In other words, we proved part (i) of the theorem provided $\mathrm{supp}\, f$ is sufficiently small and $1 \leqq u \leqq r$.

Step 3. We wish to remove the assumption about $\mathrm{supp}\, f$. But we stick at $u \geqq 1$. Then we may assume $1 \leqq u \leqq r$. Let $f \in F^s_{pq}(\Omega)$ and let $1 = \sum_{k=1}^K \varphi^k(x)$ if $x \in \Omega$ be an appropriate resolution of unity. If $\mathrm{supp}\, \varphi^k \cap \partial\Omega \neq \emptyset$ then we may assume that the middle part of (15) can be applied to $\varphi^k f$ instead of f. We choose L large, use (3.5.1/27) and estimate

$$\| (|f|^u)^{*\frac{1}{u}} \mid L_p(\Omega) \| \leqq c \| (|f|^u)^{*\frac{1}{u}} \mid L_r(\Omega) \| \leqq c' \| f \mid L_r(\Omega) \| \tag{16}$$

by the Hardy–Littlewood maximal inequality provided that $1 \leqq u < r \leqq \infty$. But the case $u = r = \infty$ fits also in the scheme. We apply (14) with L_r instead of L_v. Then we obtain the desired estimate. Let $\mathrm{supp}\, \varphi^k \cap \partial\Omega = \emptyset$ then we have

$$\mathrm{Osc}_u^L \varphi^k f(x, c2^{-j}) = \mathrm{osc}_u^L \varphi^k f(x, c2^{-j})$$

at least for $j \geqq J$. The terms with $1 \leqq j < J$ can be treated in the above way. This proves the middle part of (15) provided that $f \in F^s_{pq}(\Omega)$, $L \in \mathbb{N}_0$ with $L \geqq [s]$ and $1 \leqq u \leqq r$. The first inequality in (15) is obvious and the last one follows from Theorem 3.5.1. Hence, the main term in (15) is an equivalent quasi-norm in $F^s_{pq}(\Omega)$ for all $L \in \mathbb{N}_0$ with $L \geqq [s]$ and $1 \leqq u \leqq r$. If $f \in L_v(\Omega)$ with $v = \max(p, r)$ such that the quasi-norm in (2) is finite then we apply (3.5.1/27) to $\varphi^k f$ and obtain by the above arguments $\varphi^k f \in F^s_{pq}(\Omega)$ and hence $f \in F^s_{pq}(\Omega)$. This completes the proof of (i) provided that $1 \leqq u \leqq r$.

Step 4. Let $0 < u < 1$. By the same arguments as in Step 4 of the proof of Theorem 3.5.1 it follows that the main term in (15) is an equivalent quasi-norm in $F^s_{pq}(\Omega)$, now with $0 < u \leqq r$. Now let $f \in L_r(\Omega)$ such that the quasi-norm in (2) is finite. By the same arguments as at the beginning of Step 5 of the proof of Theorem 3.5.1 we may assume that Ω satisfies the hypotheses of 3.4.1 and that f has a support near the origin. Now we combine the mollification in Step 5 of the proof of Theorem 3.5.1 with an additional translation $f(x) \to f(x', x_n + \varepsilon_l)$ for some $\varepsilon_l > 0$ and $\varepsilon_l \to 0$. The rest is the same as in Step 5 of the proof of Theorem 3.5.1. The proof is complete.

Remark. We again followed [Tri23: Theorem 2.2.2]. On the other hand we formulated corresponding results in the Theorems 1.7.3 and 1.10.4. In Remarks 1.7.3/1 and 1.10.4/1 we gave some references.

5.2.2 Differences

On the basis of Theorem 5.2.1 it is now easy to extend Theorems 3.5.3 and 4.5.4 to Ω.

Theorem. Let Ω be a bounded C^∞ domain in $\mathbb{R}^n$ and let $d^M_{t,u}f(x)$ be given by Definition 3.5.2.

(i) Let $0 < p < \infty$, $0 < q \leqq \infty$, $1 \leqq r \leqq \infty$, and

$$s > n\left(\frac{1}{p} - \frac{1}{r}\right)_+ \quad \text{and} \quad s > n\left(\frac{1}{q} - \frac{1}{r}\right)_+ . \tag{1}$$

Let $0 < u \leqq r$, $M \in \mathbb{N}$ and $M > s$. Then

$$F^s_{pq}(\Omega) = \Big\{ f \in L_{\max(p,r)}(\Omega): \tag{2}$$

$$\|f \mid L_p(\Omega)\| + \left\| \left(\int_0^1 t^{-sq} d^M_{t,u} f(\cdot)^q \frac{dt}{t} \right)^{1/q} \mid L_p(\Omega) \right\| < \infty \Big\}$$

(modification if $q = \infty$) in the sense of equivalent quasi-norms.

(ii) Let $0 < p \leqq \infty$, $0 < q \leqq \infty$, $1 \leqq r \leqq \infty$, and

$$s > n\left(\frac{1}{p} - \frac{1}{r}\right)_+ . \tag{3}$$

Let $0 < u \leqq r$, $M \in \mathbb{N}$, and $M > s$. Then

$$B^s_{pq}(\Omega) = \Big\{ f \in L_{\max(p,r)}(\Omega): \tag{4}$$

$$\|f \mid L_p(\Omega)\| + \left(\int_0^1 t^{-sq} \|d^M_{t,u} f \mid L_p(\Omega)\|^q \frac{dt}{t} \right)^{1/q} < \infty \Big\}$$

(modification if $q = \infty$) in the sense of equivalent quasi-norms.

Proof. The above theorem follows from Theorem 5.2.1 and the consideration in 3.5.3, see in particular the beginning of 3.5.3 and Remark 3.5.3/2.

Remark. Thus subsection is the direct counterpart of 4.5.4. References may be found in 1.10.4.

5.3 Morrey–Campanato spaces, the spaces C_p^s

5.3.1 Definitions

We always assume that Ω is a bounded C^∞ domain in $\mathbb{R}^n$. We use the notations introduced in 3.4.1, in particular $\mathrm{osc}_u^M f$ is given by (3.4.1/3) with respect to the above domain Ω. We introduce the sharp maximal function

$$f_u^{M,s}(x) = \sup_{0<t<1} t^{-s} \mathrm{osc}_u^M f(x,t), \qquad x \in \Omega, \tag{1}$$

where $s \in \mathbb{R}$, $M \in \mathbb{N}_{-1}$ and $0 < u \leqq \infty$. The notation comes from the fact that $f_u^{-1,0}(x)$ can be interpreted as the Hardy–Littlewood maximal function $(|f|^u)^{*\frac{1}{u}}(x)$ with respect to Ω, see also (1.7.2/4).

Definition. Let Ω be a bounded C^∞ domain in $\mathbb{R}^n$.

(i) Let $1 \leqq p < \infty$, $s \geqq -\frac{n}{p}$ and $M = \max(-1, [s])$. Then

$$L_p^s(\Omega) = \Big\{ f \in L_p(\Omega): \tag{2}$$

$$\|f \mid L_p^s(\Omega)\| = \|f \mid L_p(\Omega)\| + \sup_{x\in\Omega} f_p^{M,s}(x) < \infty \Big\}.$$

(ii) Let $s > 0$, $0 < p \leqq \infty$, and $\overline{p} = \max(1,p)$. Then

$$C_p^s(\Omega) = \Big\{ f \in L_{\overline{p}}(\Omega): \tag{3}$$

$$\|f \mid C_p^s(\Omega)\| = \|f \mid L_p(\Omega)\| + \|f_p^{[s],s} \mid L_p(\Omega)\| < \infty \Big\}.$$

Remark. The spaces $L_p^s(\Omega)$ are the Morrey–Campanato spaces, the spaces $C_p^s(\Omega)$ have been introduced by R.A. DeVore, R.C. Sharpley, and M. Christ. Detailed references, further explanations, and a short history of these spaces may be found in Remark 1.7.2/1.

5.3.2 Hölder–Zygmund and Sobolev spaces

We wish to identify some of the above spaces $L_p^s(\Omega)$ and $C_p^s(\Omega)$ with appropriate spaces $B^s_{\infty\infty}(\Omega)$ and $F^s_{p\infty}(\Omega)$ in the sense of Definition 5.1.2, where again Ω stands for a bounded C^∞ domain in $\mathbb{R}^n$. As usual we put

$$\mathcal{C}^s(\Omega) = B^s_{\infty\infty}(\Omega) \quad \text{if} \quad s > 0. \tag{1}$$

By Theorem 5.2.2(ii) with $p = q = r = u = \infty$, $M \in \mathbb{N}$ with $M > s$, and (3.5.2/3) with $u = \infty$ we have

$$\mathcal{C}^s(\Omega) = \Big\{ f \in L_\infty(\Omega): \tag{2}$$

$$\|f \mid \mathcal{C}^s(\Omega)\| = \|f \mid L_\infty(\Omega)\| + \sup |h|^{-s} |\Delta_h^M f(x)| < \infty \Big\}$$

where the supremum is taken over all $x \in \Omega$, $0 < t < 1$, and $h \in V^M(x,t)$, see (3.5.2/2).

Besides $\mathcal{C}^s(\Omega)$ with $s > 0$ we need the classical spaces $W_p^k(\Omega)$ with $k \in \mathbb{N}_0$ and $1 < p < \infty$. In extension of (2.6.5/2) we may define

$$W_p^k(\Omega) = F^k_{p2}(\Omega), \qquad k \in \mathbb{N}_0, \quad 1 < p < \infty. \tag{3}$$

Recall that $W_p^k(\mathbb{R}^n)$ has been introduced in (1.2.3/1,2). By the above extension method, see, e.g., (5.2.1/5) and 4.5.2 it follows via (3), or better directly,

$$W_p^k(\Omega) = \Big\{ f \in L_p(\Omega): \|f \mid W_p^k(\Omega)\| = \sum_{|\alpha| \le k} \|D^\alpha f \mid L_p(\Omega)\| < \infty \Big\}. \tag{4}$$

This coincides with the original definition of Sobolev spaces on domains.

5.3.3 The main theorem

All notations have the same meaning as in the two preceding subsections. We put $F^s_{\infty\infty}(\Omega) = B^s_{\infty\infty}(\Omega) = \mathcal{C}^s(\Omega)$. Recall $\mathbb{N}_0 = \{0, 1, 2, \ldots\}$ and $\mathbb{N}_{-1} = \mathbb{N}_0 \cup \{-1\}$.

Theorem. Let Ω be a bounded C^∞ domain in $\mathbb{R}^n$.

(i) Let $1 \leqq p < \infty$ and $s > 0$, then

$$L_p^{-\frac{n}{p}}(\Omega) = L_p(\Omega) \quad \text{and} \quad L_p^s(\Omega) = \mathcal{C}^s(\Omega). \tag{1}$$

(ii) Let $1 \leqq p < \infty$, $M \in \mathbb{N}_{-1}$, and $-\frac{n}{p} \leqq s < M+1$, then

$$L_p^s(\Omega) = \Big\{ f \in L_p(\Omega): \qquad (2)$$

$$\|f \mid L_p^s(\Omega)\|^M = \|f \mid L_p(\Omega)\| + \sup_{x \in \Omega} f_p^{M,s}(x) < \infty \Big\}$$

(equivalent norms).

(iii) Let $0 < p \leqq \infty$ and $s > n(\frac{1}{p} - 1)_+$, then

$$C_p^s(\Omega) = F_{p\infty}^s(\Omega) \qquad (3)$$

and, for $M \in \mathbb{N}_0$ with $M \geqq [s]$,

$$C_p^s(\Omega) = \Big\{ f \in L_{\bar{p}}(\Omega): \qquad (4)$$

$$\|f \mid C_p^s(\Omega)\|^M = \|f \mid L_p(\Omega)\| + \|f_p^{M,s} \mid L_p(\Omega)\| < \infty \Big\}$$

(equivalent quasi-norms).

(iv) Let $k \in \mathbb{N}_0$ and $1 \leqq u < p < \infty$, then

$$W_p^k(\Omega) = \{ f \in L_p(\Omega): \|f \mid L_p(\Omega)\| + \|f_u^{k-1,k} \mid L_p(\Omega)\| < \infty \} \qquad (5)$$

(equivalent norms).

Proof. Step 1. Part (iii) follows from (5.3.1/3) and Theorem 5.2.1(i) with $r = \max(1, p)$ and $u = p$, where the case $p = \infty$ is covered by Theorem 5.2.1(ii).

Step 2. To prove (ii) it is sufficient to show that there exists a constant $c > 0$ such that

$$\|f \mid L_p^s(\Omega)\|^M \leqq c \|f \mid L_p^s(\Omega)\|^{M+1}, \qquad f \in L_p(\Omega), \qquad (6)$$

provided that $M \in \mathbb{N}_{-1}$ and $-\frac{n}{p} \leqq s < M+1$. However (6) follows from Proposition 3.4.3 (with $M+1$ instead of M).

Step 3. We prove the first part of (1). Let $s = -\frac{n}{p}$, then we have

$$t^{-s} \operatorname{osc}_p^{-1} f(x,t) = t^{n/p} \left(⨍_{B(x,t)} |f(y)|^p dy \right)^{1/p} \leqq c \left(\int_{B(x,t)} |f(y)|^p dy \right)^{1/p}, \qquad (7)$$

see 3.4.1 for definitions. By (5.3.1/2) follows $L_p(\Omega) = L_p^{-s}(\Omega)$.

Step 4. In the next three steps we prove the second part of (1). Let $s > 0$ and $f \in C^s(\Omega)$. We calculate $\operatorname{osc}_p^M f(x,t)$ with $1 \leqq p < \infty$ and $M \geqq [s]$. We may assume that Ω satisfies the hypotheses of 3.4.1 and that all calculations take place near the origin. In particular we can use the same technique as in 3.4.2. By (3.4.2/2,4,7) and (3.4.3/3) (with $M+1$ instead of M and $t \sim 2^{-j}$) we obtain

$$\operatorname{osc}_p^M f(x,t) \leqq c' \sup_{z\in B(x,t)} \fint_{V^{M+1}(z,ct)} |\Delta_y^{M+1} f(z)| dy. \tag{8}$$

By (5.3.1/2) it follows

$$\|f \mid L_p^s(\Omega)\| \leqq c\|f \mid L_\infty(\Omega)\| + c \sup |h|^{-s} |\Delta_h^{M+1} f(x)|, \tag{9}$$

where the supremum is taken over all $x \in \Omega$, $0 < t < 1$, and $h \in V^{M+1}(x,t)$. Then we have by (5.3.2/2)

$$\|f \mid L_p^s(\Omega)\| \leqq c\|f \mid \mathcal{C}^s(\Omega)\|. \tag{10}$$

In particular $f \in L_p^s(\Omega)$.

Step 5. Next we prove the converse of (10) under the assumption $f \in \mathcal{C}^s(\Omega)$. By (3.3.2/9,4,5) and the properties of the kernels $k(y)$ we have

$$|f(x)| = \left|\sum_{j=0}^{\infty} f_j(x)\right| \tag{11}$$

$$\leqq \left|\int k_0(y) f(x+y) dy\right| + \sum_{j=1}^{\infty} \left|\int k(y)(f - P_j)(x + 2^{-j}y) dy\right|$$

with $P_j \in \mathbb{P}_M$ (recall that the latter stands for the collection of all polynomials of degree less than or equal to M). In particular

$$\|f \mid L_\infty(\Omega)\| \leqq c'\|f \mid L_p(\Omega)\| + c' \sup_{x\in\Omega, j\in\mathbb{N}} 2^{j\varepsilon} \operatorname{osc}_p^M f(x, c2^{-j}) \tag{12}$$

$$\leqq c''\|f \mid L_p^s(\Omega)\|,$$

where we used that ε with $0 < \varepsilon < s$ is at our disposal. Next we estimate $\Delta_h^{M+1} f(x)$ with $M \geqq [s]$ and $|h| \sim 2^{-j}$. We use (3.3.2/9) and (3.4.2/1) and obtain

$$\Delta_h^{M+1} f(x) = \Delta_h^{M+1} f^{j+R}(x) + \sum_{r=1}^{\infty} \Delta_h^{M+1} f_{j+R+r}(x), \tag{13}$$

where $R \in \mathbb{N}_0$ will be determined later on. By (3.4.2/3) we have

$$\Delta_h^{M+1} f^{j+R}(x) = 2^{(j+R)n} \Delta_h^{M+1} \int k_0(2^{j+R}y - 2^{j+R}x) f(y) dy \tag{14}$$

$$= \int \Delta_{-h2^{j+R}}^{M+1} k_0(y) f(x + 2^{-j-R}y) dy.$$

We can replace f on the right-hand side of (14) by $f - P$ with $P \in \mathbb{P}_M$. Recall $|h| \sim 2^{-j}$, then it follows

$$|\Delta_h^{M+1} f^{j+R}(x)| \leqq c' \operatorname{osc}_p^M f(x, c2^{-j}). \tag{15}$$

Similar as in (3.5.1/13) we introduce the maximal function

$$f_k^a(x) = \sup \frac{|f_k(x+y)|}{1+|2^k y|^a} \quad \text{with} \quad a>0, \tag{16}$$

where $k = j+R+r$ with $r \in \mathbb{N}$ and where the supremum is taken over y with $x+y \in B(x, c2^{-j})$ for some $c>0$, see (3.4.1/1). Recall again $|h| \sim 2^{-j}$, then we have

$$|\Delta_h^{M+1} f_{j+R+r}(x)| \leqq c\, 2^{(R+r)a} f^a_{j+R+r}(x) \tag{17}$$

and

$$|h|^{-s} |\Delta_h^{M+1} f_{j+R+r}(x)| \leqq c\, 2^{-(R+r)(s-a)} 2^{(j+R+r)s} f^a_{j+R+r}(x) \tag{18}$$

and finally

$$\sum_{r=1}^{\infty} |h|^{-s} |\Delta_h^{M+1} f_{j+R+r}(x)| \leqq \varepsilon \sup_{k\in\mathbb{N}} 2^{ks} f_k^a(x), \tag{19}$$

where $\varepsilon > 0$ is at our disposal. One has to choose $0 < a < s$ and R large. By (12,13,15,19) we arrive at

$$\|f \mid \mathcal{C}^s(\Omega)\| \leqq c \|f \mid L_p^s(\Omega)\| + \varepsilon \sup_{k\in\mathbb{N}, x\in\Omega} 2^{ks} f_k^a(x), \tag{20}$$

where $\varepsilon > 0$ is at our disposal. Recall that $\mathcal{C}^s(\Omega)$ is the restriction of $\mathcal{C}^s(\mathbb{R}^n) = B^s_{\infty\infty}(\mathbb{R}^n)$ to Ω. Hence we may apply Corollary 2.5.1/2 with the maximal function (2.4.1/55), which is the $\mathbb{R}^n$-counterpart of (16). Then it follows that the last term in (20) can be estimated from above by $c\varepsilon \|f \mid \mathcal{C}^s(\Omega)\|$. We choose ε small and obtain

$$\|f \mid \mathcal{C}^s(\Omega)\| \leqq c \|f \mid L_p^s(\Omega)\| \tag{21}$$

under the assumption $f \in \mathcal{C}^s(\Omega)$.

Step 6. Let $f \in L_p^s(\Omega)$. Then we use the same approximation procedures as in Step 5 of the proof of Theorem 3.5.1 and Step 4 of the proof of Theorem 5.2.1: translations and mollifications. For the approximating function we have (21). Now it follows by the previous arguments that (21) can be extended to all $f \in L_p^s(\Omega)$. The proof of the second part of (1) is complete.

Step 7. In the remaining two steps we prove (iv). The case $k=0$ follows from the Hardy–Littlewood maximal inequality, where we have to use $u<p$. Let $f \in W_p^k(\Omega)$ with $k \in \mathbb{N}$ and $1<p<\infty$. Let $1 \leqq u < p$. We use again the above technique, in particular

$$\operatorname{osc}_u^{k-1} f(x, 2^{-j}) \leqq \sum_{r=1}^{\infty} \left(⨍_{B(x,2^{-j})} |f_{j+r}(y)|^u dy \right)^{1/u} \tag{22}$$

$$+\,c2^{-jk}\sup_{z\in B(x,2^{-j})}\sum_{|\alpha|=k}|D^\alpha f^j(z)|,$$

see Proposition 3.4.2 and (3.4.2/7). By (3.4.2/3) we have

$$|D^\alpha f^j(z)| = \left|\int k_0(y)D^\alpha f(z+2^{-j}y)dy\right| \leqq c(D^\alpha f)^*(x), \tag{23}$$

where the star indicates a Hardy–Littlewood maximal function restricted to Ω, furthermore $z\in B(x,2^{-j})$. Then (22) yields

$$2^{jk}\mathrm{osc}_u^{k-1}f(x,2^{-j}) \leqq c\sum_{|\alpha|=k}(D^\alpha f)^*(x) + c\Big(\sup_r 2^{rku}|f_r|^u\Big)^{*\frac{1}{u}}(x). \tag{24}$$

We take the supremum with respect to j on the left-hand side of (24) and obtain

$$\|f_u^{k-1,k}\mid L_p(\Omega)\| \leqq c\sum_{|\alpha|=k}\|(D^\alpha f)^*\mid L_p(\Omega)\| \tag{25}$$

$$+c\left\|\Big(\sup_r 2^{rku}|f_r|^u\Big)^{*\frac{1}{u}}\mid L_p(\Omega)\right\|.$$

Recall $1\leqq u<p<\infty$. By the Hardy–Littlewood maximal inequality we arrive at

$$\|f_u^{k-1,k}\mid L_p(\Omega)\| \leqq c\sum_{|\alpha|=k}\|D^\alpha f\mid L_p(\Omega)\| + c\left\|\sup_r 2^{rk}|f_r|\mid L_p(\Omega)\right\|. \tag{26}$$

By Theorem 3.3.3 and Definition 5.1.2 it follows that the last term in (26) can be estimated from above by $c\|f\mid F^k_{p\infty}(\Omega)\|$, which, in turn, can be estimated from above by $c''\|f\mid F^k_{p2}(\Omega)\|\sim\|f\mid W^k_p(\Omega)\|$. In other words we have

$$\|f\mid L_p(\Omega)\| + \|f_u^{k-1,k}\mid L_p(\Omega)\| \leqq c\|f\mid W_p^k(\Omega)\|. \tag{27}$$

Step 8. We prove the converse inequality. Let $f\in L_p(\Omega)$ such that the left-hand side of (27) is finite. We assume temporarily that (3.3.2/9) with $D^\alpha f$ instead of f holds pointwise almost everywhere, $|\alpha|=k$. Then we have by (3.4.2/1,3)

$$D^\alpha f(x) = \lim_{j\to\infty}D^\alpha f^j(x) = \lim_{j\to\infty}\int k_0(y)D^\alpha f(x+2^{-j}y)dy \tag{28}$$

$$=\lim_{j\to\infty}2^{jk}\int D^\alpha k_0(y)(f-P_j)(x+2^{-j}y)dy$$

with $|\alpha|=k$ and $P_j\in\mathbb{P}_{k-1}$. We obtain

$$|D^\alpha f(x)| \leqq c\,f_u^{k-1,k}(x), \qquad |\alpha|=k \tag{29}$$

and consequently $f \in W_p^k(\Omega)$, including the converse inequality to (27). Finally, we have to remove the additional assumptions made at the beginning of this step. Let $f \in L_p(\Omega)$ such that the left-hand side of (27) is finite. Then we argue in the same way as in Step 6 and arrive at $f \in W_p^k(\Omega)$.

Remark. We followed essentially [Tri23], but almost all results are known. The above theorem coincides essentially with Theorem 1.7.2. Detailed references may be found in Remark 1.7.2/3, see also Remarks 1.7.3/1,2.

5.4 Further problems

We discuss very briefly some further problems in connection with spaces on domains. We always assume that Ω is a bounded C^∞ domain in $\mathbb{R}^n$. The spaces $F^s_{pq}(\Omega)$ and $B^s_{pq}(\Omega)$ have the same meaning as in Definition 5.1.2. First we refer to [Triß: Chapter 3] where we dealt with spaces of this type in detail.

Pointwise multipliers. The pointwise multiplier assertions of Theorem 4.2.2 and Corollary 4.2.2 for the spaces $B^s_{pq}(\mathbb{R}^n)$ and $F^s_{pq}(\mathbb{R}^n)$ can be extended to the spaces $B^s_{pq}(\Omega)$ and $F^s_{pq}(\Omega)$. This follows immediately from Definition 5.1.2 and the just quoted result.

Traces. There are no problems to introduce spaces $F^s_{pq}(\partial\Omega)$ and $B^s_{pq}(\partial\Omega)$ on the compact manifold $\partial\Omega$ via finitely many local charts and a related resolution of unity. The independence of these spaces of the chosen local charts and accompanying resolutions of unity follows from the Theorems 4.2.2 (pointwise multipliers) and 4.3.2 (preservation under diffeomorphisms), see [Triß: 3.2.2]. Now the trace assertions of Theorem 4.4.2 and Corollary 4.4.2 have counterparts with Ω instead of $\mathbb{R}^n$ and $\partial\Omega$ instead of $\mathbb{R}^{n-1}$. Similarly for the curiosities in 4.4.3. Again one has to use the quoted results combined with the Theorems 4.2.2 and 4.3.2. See also [Triß: 3.3.3] for greater details.

Periodic spaces. One can replace $\mathbb{R}^n$ in $F^s_{pq}(\mathbb{R}^n)$ and $B^s_{pq}(\mathbb{R}^n)$ by the n-torus T^n. The theory of the periodic spaces $F^s_{pq}(T^n)$ and $B^s_{pq}(T^n)$ can be developed to a large extent parallel to the theory of the corresponding spaces on $\mathbb{R}^n$. One begins with a more or less obvious counterpart of Definition 2.3.1. There are some pecularities coming from the compactness of T^n (in contrast to $\mathbb{R}^n$) and the close connection with trigonometric series. In [ScT] we gave a detailed account of the above periodic spaces, mostly based on Fourier-analytic techniques.

Spaces on compact manifolds. Chapter 7 of this book deals with spaces of type F^s_{pq} and B^s_{pq} on manifolds Ω which are, in general, non-compact and equipped with a Riemannian or Lie structure. If the manifold Ω is compact, then in principle the situation is simpler, at least there is only one reasonable way to introduce the spaces $B^s_{pq}(\Omega)$ and $F^s_{pq}(\Omega)$ via finitely many local charts. On the other hand one would like to have intrinsic descriptions in natural terms. For example, descriptions of type (2.3.1/7,8) in terms of appropriate pseudo-differential operators are highly desirable. In the periodic case, i.e., $\Omega = T^n$, one has a completely satisfactory solution of this problem. In the case of a general smooth compact manifold Ω we refer to [SeS] where one finds interesting characterizations of type (2.3.1/7,8) in terms of pseudodifferential operators. Intrinsic characterizations for Morrey–Campanato spaces on compact manifolds may be found in [Gei2], see also [Gei1] as far as spaces of the above type on compact Lie groups are concerned.

Chapter 6

Mapping Properties of Pseudodifferential Operators

6.1 Introduction

This chapter is not a self-contained introduction to the theory of pseudodifferential operators (abbreviated as ψDO in the sequel). Just on the contrary, we restrict ourselves to a special aspect, global and local mapping properties of ψDO's and Fourier integral operators in spaces of type F^s_{pq} and B^s_{pq}, see Definition 2.3.1. In this field of research much has been done in the framework of a L_2-theory. Our intension is quite different, we are especially interested in spaces of the above type with $p \leqq 1$. Even worse, in the first line we are eager to demonstrate that the, as we think, powerful techniques for function spaces developed in this book so far, can be highly efficiently applied to ψDO's. In this sense we shall not hesitate to quote without proofs well-known general results about ψDO's which can be found nowadays in relevant books. Section 2 deals with ψDO's of type $S^0_{1\delta}$ with $0 \leqq \delta < 1$. We show that the local means treated in 2.4.6 can be used to prove global mapping properties of corresponding ψDO's in F^s_{pq}-spaces. Unfortunately the interesting exotic case $\delta = 1$ is excluded. But rescue comes from the second approach, the use of atomic representations in the sense of 3.2, which we present in Section 3. This method is especially well adapted to the exotic case. The price to pay is the occasional restriction to $p \leqq 1$. We admitted that the method is more interesting for us than the result. But nevertheless we hope that the assertions, in particular those ones in connection with exotic ψDO's, are of interest for their own sake. Some of them are rather sharp.

H. Triebel, *Theory of Function Spaces II*, Modern Birkhäuser Classics,
DOI 10.1007/978-3-0346-0419-2_6, © Birkhäuser Verlag 1992

6.2 An approach via local means

6.2.1 Definitions and preliminaries

This section deals with global mapping properties of ψDO's acting in F^s_{pq} or B^s_{pq}, where the latter spaces had been introduced in Definition 2.3.1. In particular, all spaces of interest in this chapter are defined on $\mathbb{R}^n$, which justifies to omit "$\mathbb{R}^n$" as a suffix. In this section it is reasonable to introduce the corresponding symbol classes globally. Local versions will be considered in 6.3. Let $a(x,\xi)$ with $x \in \mathbb{R}^n$ and $\xi \in \mathbb{R}^n$ be a C^∞ function in $\mathbb{R}^{2n}$. We use the abbreviation

$$D^\alpha_x a(x,\xi) = \frac{\partial^{|\alpha|} a(x,\xi)}{\partial x_1^{\alpha_1} \cdots \partial x_n^{\alpha_n}} \quad \text{and} \quad D^\beta_\xi a(x,\xi) = \frac{\partial^{|\beta|} a(x,\xi)}{\partial \xi_1^{\beta_1} \cdots \partial \xi_n^{\beta_n}}, \tag{1}$$

where $\alpha = (\alpha_1, \ldots, \alpha_n)$ and $\beta = (\beta_1, \ldots, \beta_n)$ are multi-indices in the usual sense.

Definition. Let $0 \leqq \delta \leqq 1$, then S_δ is the collection of all complex-valued C^∞ functions $a(x,\xi)$ in $\mathbb{R}^{2n}$, where $x \in \mathbb{R}^n$ and $\xi \in \mathbb{R}^n$, such that for any multi-indices $\alpha = (\alpha_1, \ldots, \alpha_n)$ and $\beta = (\beta_1, \ldots, \beta_n)$ there exist a positive number $c_{\alpha\beta}$ with

$$|D^\alpha_x D^\xi_\beta a(x,\xi)| \leqq c_{\alpha\beta}(1+|\xi|)^{-|\beta|+\delta|\alpha|}, \qquad x \in \mathbb{R}^n, \quad \xi \in \mathbb{R}^n. \tag{2}$$

Remark 1. This is a special case of the Hörmander class $S^\mu_{\rho\delta}$ of symbols: Let $\mu \in \mathbb{R}$, $0 \leqq \rho \leqq 1$, and $0 \leqq \delta \leqq 1$, then $S^\mu_{\rho\delta}$ is the collection of all complex-valued C^∞ functions $a(x,\xi)$ in $\mathbb{R}^{2n}$, where $x \in \mathbb{R}^n$ and $\xi \in \mathbb{R}^n$, such that (2) is modified by

$$|D^\alpha_x D^\beta_\xi a(x,\xi)| \leqq c_{\alpha\beta}(1+|\xi|)^{\mu+\delta|\alpha|-\rho|\beta|}, \qquad x \in \mathbb{R}^n, \quad \xi \in \mathbb{R}^n. \tag{3}$$

Two further generalizations of (3) should be mentioned. First, (3) is required only for some derivatives, e.g., $|\alpha| \leqq N$ and $|\beta| \leqq N$. Secondly, instead of $x \in \mathbb{R}^n$ in (3) it is assumed that (3) holds for $x \in K$, where K is any compact set in $\mathbb{R}^n$, and $c_{\alpha\beta}$ may depend on K. This is the local version of (3) or (2), maybe restricted to finitely many α's and β's. In 6.3 we return to some of these more general symbol classes.

Pseudodifferential operators. Let $a(x,\xi) \in S_\delta$ or, more generally, $a(x,\xi) \in S^\mu_{\rho\delta}$, then

$$a(x,D)f(x) = \int_{\mathbb{R}^n} e^{ix\xi} a(x,\xi) \hat{f}(\xi) d\xi, \qquad x \in \mathbb{R}^n, \tag{4}$$

is the corresponding pseudodifferential operator. Here $\hat{f}$ is the Fourier transform of f, see 2.2.1, whereas $x\xi = \sum_{j=1}^{n} x_j \xi_j$ stands for the scalar product of $x \in \mathbb{R}^n$ and $\xi \in \mathbb{R}^n$ in $\mathbb{R}^n$. As we said in 6.1 we are not going to discuss the basic properties of the ψDO's. A good introduction may be found in [Tay: Chapter 2], see also [Tre; Hör 3: III].

Remark 2. It can be seen easily that $a(x, D)$ maps S into S, where S and also S' have the same meaning as in 2.2.1. If $\delta < 1$, then one can apply a duality argument which shows that $a(x, D)$ also maps S' into S', see [Tay, II, §4] for details. However there is no need for us to discuss these general global mapping properties in detail. Our method, both in 6.2 and 6.3, is strictly local, and it will be completely sufficient to discuss (4) locally for smooth functions, the rest will be done afterwards via some limiting arguments.

Remark 3. Recall that

$$I_\sigma : f \longrightarrow ((1 + |\xi|^2)^{\sigma/2} \hat{f})^\vee, \qquad \sigma \in \mathbb{R}, \tag{5}$$

maps F^s_{pq} isomorphically onto $F^{s-\sigma}_{pq}$. This follows easily from the arguments in the proof of Theorem 2.3.2 and is stated explicitly in [Triß: 2.3.8]. If $a(x, \xi) \in S^\mu_{\rho\delta}$ then $a(x, \xi)(1 + |\xi|^2)^{\sigma/2} \in S^{\mu+\sigma}_{\rho\delta}$ and the corresponding ψDO's are given by $a(x, D)$ and $a(x, D) \circ I_\sigma$, respectively. The mentioned lifting property of I_σ shows that for our purposes, i.e., mapping properties of $a(x, D)$ in F^s_{pq} and B^s_{pq}, the value of μ is completely immaterial. This is the reason why we restricted the above definition of S_δ to $\mu = 0$, compared with $S^\mu_{\rho\delta}$. The other restriction, i.e., $\rho = 1$ in, say, $S^0_{\rho\delta}$ is also quite natural in our context. We are looking for mappings of $a(x, D)$ from F^s_{pq} into itself, in particular from L_p with $1 < p < \infty$ into itself. It is a known fact that $\rho = 1$ is natural in this context, see [Tay: Chapter XI]. This justifies our definition of S_δ compared with the indicated more general possibilities.

Remark 4. Symbols $a(x, \xi) \in S_1$ are called exotic. In recent times they attracted much attention. However in this section we are forced to restrict δ in S_δ to $0 \leqq \delta < 1$. Based on another method exotic ψDO's will be treated extensively in 6.3.

6.2.2 Mapping properties

All notations have the same meaning as in the preceding subsection. In the proof of the following theorem we prefer a straightforward calculation. We add a technical remark afterwards.

Theorem. Let $0 \leqq \delta < 1$ and $a(x,\xi) \in S_\delta$. Let $s \in \mathbb{R}$ and $0 < q \leqq \infty$.

(i) Let $0 < p < \infty$. Then $a(x,D)$ is a linear and bounded map from F^s_{pq} into itself.

(ii) Let $0 < p \leqq \infty$. Then $a(x,D)$ is a linear and bounded map from B^s_{pq} into itself.

Proof. Step 1. We prove part (i) under the additional assumption that s is a large positive number. We wish to use Theorem 2.4.6. For this purpose we have to calculate the local means (2.4.6/1) with $a(x,D)f(x)$ instead of f. First we expand $a(x+ty,D)f(x+ty)$ with $x \in \mathbb{R}^n$, $|y| \leqq 1$, and $0 < t < 1$ in a Taylor polynomial with remainder term,

$$
\begin{aligned}
a(x+ty,D)f(x+ty) &= \int_{\mathbb{R}^n} e^{i(x+ty)\xi} a(x+ty,\xi)\hat{f}(\xi)d\xi \qquad (1)\\
&= \sum_{|\gamma| \leqq L-1} \frac{t^{|\gamma|}}{\gamma!} y^\gamma \int_{\mathbb{R}^n} e^{i(x+ty)\xi} D^\gamma_x a(x,\xi)\hat{f}(\xi)d\xi \\
&+ \sum_{|\gamma|=L} \frac{Lt^L}{\gamma!} y^\gamma \int_{\mathbb{R}^n} e^{i(x+ty)\xi} \hat{f}(\xi) \int_0^1 (1-\tau)^{L-1} D^\gamma_x a(x+t\tau y,\xi)d\tau d\xi,
\end{aligned}
$$

see (3.4.4/4), where L with $L > s$ is a natural number. Let k be the same function as in (2.4.6/4) and let $k_\gamma(y) = y^\gamma k(y)$ with $|\gamma| \leqq L-1$. Then $k_\gamma(t,f)(x)$ is given by (2.4.6/1) with k_γ instead of k. In particular

$$k_\gamma(t,f)^\wedge(\xi) = k^\vee_\gamma(t\xi)\hat{f}(\xi). \qquad (2)$$

Now we insert (1) in (2.4.6/1) with $a(x,D)f(x)$ instead of f and obtain

$$
\begin{aligned}
k(t,a(x,D)f)(x) &= \int_{\mathbb{R}^n} k(y)a(x+ty,D)f(x+ty)dy \qquad (3)\\
&= \sum_{|\gamma| \leqq L-1} c_\gamma t^{|\gamma|} \int_{\mathbb{R}^n} e^{ix\xi} D^\gamma_x a(x,\xi) k_\gamma(t,f)^\wedge(\xi)d\xi + R(x,t),
\end{aligned}
$$

where we used (2). In order to estimate the remainder term $R(x,t)$ we modify (6.2.1/4) by

$$D_z^\gamma a(z,D)f(x) = \int_{\mathbb{R}^n} e^{ix\xi} D_z^\gamma a(x,\xi)\hat{f}(\xi)d\xi, \qquad z\in\mathbb{R}^n,\ x\in\mathbb{R}^n. \tag{4}$$

Then we arrive at

$$|R(x,t)| \leqq c\,t^L \sum_{|\gamma|=L} \sup_{|x-y|\leqq 1} \sup_{z\in\mathbb{R}^n} |D_z^\gamma a(z,D)f(y)|. \tag{5}$$

By (6.2.1/2) we have

$$a_\gamma(x,\xi) = D_x^\gamma a(x,\xi)(1+|\xi|^2)^{-\frac{\delta}{2}|\gamma|} \in S_\delta. \tag{6}$$

By (6.2.1/5) and (2,3) we have

$$\begin{aligned} &|k(t,a(x,D)f)(x)| \\ &\quad\leqq c \sum_{|\gamma|\leqq L-1} t^{|\gamma|} \sup_{z\in\mathbb{R}^n} |a_\gamma(z,D)k_\gamma(t,f)I_{\delta|\gamma|}f(x)| + |R(x,t)|. \end{aligned} \tag{7}$$

We wish to apply (2.4.8/3). Let s and L be fixed, then we choose N large enough such that Proposition 2.4.8(i) with a_γ, k_γ, and $s-|\gamma|$ instead of a, k, and s can be applied, cf. (2.4.8/1) and (6), furthermore $|\hat{k}_\gamma(\xi)| \leqq c|\xi|^M$ with M large and ξ near the origin, is sufficient. By (7,5) we arrive at

$$\begin{aligned} &\left\| \left(\int_0^1 t^{-sq} |k(t,a(x,D)f(x)|^q \frac{dt}{t} \right)^{1/q} \mid L_p \right\| \\ &\leqq c \sum_{|\gamma|\leqq L-1} \|I_{\delta|\gamma|}f \mid F_{pq}^{s-|\gamma|}\| + c\sum_{|\gamma|=L} \left\| \sup_{|x-y|\leqq 1} \sup_{z\in\mathbb{R}^n} |a_\gamma(z,D)I_{\delta|\gamma|}f| \mid L_p \right\|, \end{aligned} \tag{8}$$

where we used (6) with $L > s$. By Remark 6.2.1/3 the terms in the first sum on the right-hand side of (8) can be estimated from above by $c\|f \mid F_{pq}^{s+\delta|\gamma|-|\gamma|}\|$ and hence by $c'\|f \mid F_{pq}^s\|$. Now we apply Proposition 2.4.8(ii) to the terms in the second sum on the right-hand side of (8). Hence these terms can be estimated from above by $c\|I_{\delta L}f \mid F_{pq}^\sigma\|$ with $\sigma > \frac{n}{p}$ and consequently by $c'\|f \mid F_{pq}^{\sigma+\delta L}\|$. Now we fix the assumptions for s and L by

$$\frac{n}{p} + \delta L < s < L, \qquad L\in\mathbb{N}. \tag{9}$$

If p (and q) are fixed then (9) covers all sufficiently large values of s (here one needs for the first time $\delta < 1$). Under these assumptions we arrive at

$$\left\| \left(\int_0^1 t^{-sq} |k(t,a(x,D)f)(x)|^q \frac{dt}{t} \right)^{1/q} \mid L_p \right\| \leqq c\|f \mid F_{pq}^s\|. \tag{10}$$

We use Theorem 2.4.6, where (10) is the modified second term in (2.4.6/7). However the corresponding first term in (2.4.6/7) can be treated in the same way, see the first term in (2.4.8/3). In other words, we have

$$\|a(x, D)f \mid F^s_{pq}\| \leqq c\|f \mid F^s_{pq}\|, \qquad f \in F^s_{pq}, \tag{11}$$

under the restriction that s is large, see (9).

Step 2. We prove part (i) for arbitrary values of s. By the composition theorem for ψDO's which may be found in [Hör3, III, pp. 71/94] we have $a^d(x,\xi) \in S_\delta$ where the corresponding ψDO is given by

$$a^d(x, D) = I_{-d} \circ a(x, D) \circ I_d, \qquad d \in \mathbb{R}, \tag{12}$$

see (6.2.1/5). Here we used again $\delta < 1$. Consequently

$$a(x, D) = I_d \circ a^d(x, D) \circ I_{-d}. \tag{13}$$

We choose d such that we have (11) with $d + s$ instead of s. By Remark 6.2.1/3 the lift I_{-d} maps F^s_{pq} onto F^{s+d}_{pq}, then we use (11) with $s + d$ instead of s and a^d instead of a, and map finally F^{s+d}_{pq} by $I_{-\alpha}$ onto F^s_{pq}. We arrive at (11) without any restriction for s.

Step 3. The proof of part (ii) is the same. Instead of Theorem 2.4.6 we have to use Theorem 2.5.3.

Remark 1. Theorem 4.2.2 and its proof may be considered as a simple case of the above theorem and its proof. The proofs are based on the same idea. Roughly speaking, in 4.2.2 we applied local means to pointwise multipliers, whereas the above proof extends the use of local means to pseudodifferential operators, which may be considered as a complicated mixture of pointwise and Fourier multipliers.

Remark 2. One can extend the technical Remark 4.2.2/1 to the above situation: If $f \in S$ then the above arguments cause no problems and we arrive at (11). If $p < \infty$ and $q < \infty$ then the rest is a matter of completion. In the limiting cases, i.e., $p = \infty$ or $q = \infty$, one can use the same Fatou argument as in Remark 4.2.2/1 with a reference to 2.4.2.

Remark 3. Beside some limiting cases, the above theorem is due to L. Päivärinta, see [Päi3]. The presented proof is a slightly modified version of a corresponding proof in [Tri26].

Remark 4. As we mentioned in the introduction, ψDO's have been studied preferably in a L_2-setting, see the books referred to in 6.2.1. But there has also been done a lot to study diverse types of ψDO's in L_p with $1 < p < \infty$, in Hölder and Hardy spaces. Besides [Tay: XI] we refer to [Bea1,2, Ill, KuN2, Nag1–4, NaS1,2, Yam1,4, Bou4, Miy1–3, MiY1,2, PäS, MuN, RoC, Sug1]. Some of these papers deal with the problem to weaken the smoothness assumptions for the symbol $a(x,\xi) \in S^{\mu}_{\rho\delta}$ in particular with respect to the x-variable. Mapping properties of ψDO in B^s_{pq} and F^s_{pq} spaces have been studied in [Bou1, Gib, Päi3, Bui2, Yam2,3, Mar1–3, Yab1,2, Sug2, FHJW, AlH]. We say nothing about boundary value problems for ψDO in L_p-spaces and, more general, in B^s_{pq} and F^s_{pq} spaces. A good reference for the general background and a L_2-theory is [ReS]. An extension of this theory to B^s_{pq} and F^s_{pq} spaces has been given by J. Franke, see [Fra2,3] and also [Gru1,2]. Further references, in particular in connection with exotic ψDO, will be given in Remark 6.3.4/3.

6.3 An approach via atoms

6.3.1 Local mappings

Definition 6.2.1, its generalization (6.2.1/3), and Theorem 6.2.2 are global by nature. But this is not the typical point of view in the theory of ψDO's and Fourier integral operators. Just on the contrary, often the behaviour at infinity or, in a more general context where ψDO's are studied in domains, near the boundary of that domain, is out of interest. One confines the investigations to local questions, which we shall again express in terms of function spaces.

Definition. Let $s \in \mathbb{R}$, $0 < p < \infty$, $0 < q \leqq \infty$.

(i) $F^{s\,\mathrm{com}}_{pq}$ is the collection of all $f \in F^s_{pq}$ having a compact support.

(ii) $F^{s\,\mathrm{loc}}_{pq}$ is the collection of all $f \in S'$ such that $\psi f \in F^s_{pq}$ for any compactly supported C^∞ function ψ in $\mathbb{R}^n$.

Remark 1. It is clear what is meant by $B^{s\,\mathrm{com}}_{pq}$ and $B^{s\,\mathrm{loc}}_{pq}$, where $s \in \mathbb{R}$, $0 < p \leqq \infty$, $0 < q \leqq \infty$.

Local mappings. Let $a(\cdot, D)$ be the ψDO defined in (6.2.1/4). Then

$$a(\cdot, D):\quad F^{s\,\mathrm{com}}_{pq} \longrightarrow F^{s\,\mathrm{loc}}_{pq} \tag{1}$$

should be interpreted as follows: For any bounded domain Ω in $\mathbb{R}^n$ and for any compactly supported C^∞ function ψ in $\mathbb{R}^n$

$$\psi(x)a(x, D):\quad \{f \in F^s_{pq}:\ \operatorname{supp} f \subset \overline{\Omega}\} \to F^s_{pq} \tag{2}$$

is a (linear) and continuous operator where the first space may be considered as a naturally quasi-normed subspaces of F^s_{pq}. Again it is clear what is meant by

$$a(\cdot, D):\quad B^{s\,\mathrm{com}}_{pq} \longrightarrow B^{s\,\mathrm{loc}}_{pq}.$$

Remark 2. The following reformulation of (1) is of some interest. Let Ω be a bounded C^∞ domain in $\mathbb{R}^n$ and let $F^s_{pq}(\Omega)$ be given by Definition 5.1.2. By Theorem 5.1.3 there exists an extension operator ext from $F^s_{pq}(\Omega)$ into F^s_{pq} with $\operatorname{supp} \operatorname{ext} f \subset \omega$, where ω is a given bounded domain in $\mathbb{R}^n$ with $\overline{\Omega} \subset \omega$. Let re be the restriction operator from F^s_{pq} onto $F^s_{pq}(\Omega)$, see (5.1.3/1). Then (1) can be reformulated as

$$\mathrm{re} \circ a(\cdot, D) \circ \mathrm{ext}:\quad F^s_{pq}(\Omega) \longrightarrow F^s_{pq}(\Omega) \tag{3}$$

for any bounded C^∞ domain Ω in $\mathbb{R}^n$. The two versions, (1) and (3) are equivalent. But we stick at (1) in the sequel.

6.3.2 The main theorem

We prove a mapping property of type (6.3.1/1) for a rather general class of exotic ψDO's. The proof is based on the atomic representations of $f \in F^s_{pq}$ described in 3.2.3 and 3.2.4. This explains the restrictions for the parameters s, p, q. Furthermore we postpone to 6.3.4 comments, references, and a detailed discussion what is meant by the conditions forced upon the symbol $a(x,\xi)$, in particular why $a(x,\xi)$ is rightly called exotic. L^{loc}_1 stands for the locally integrable functions on $\mathbb{R}^n$.

Theorem. Let $0 < p \leqq 1 < q \leqq \infty$ and $s > n(\frac{1}{p} - 1)$. Let $N \in \mathbb{N}$ with $N > \frac{n}{p} + 1$. Let $a(x,\xi)$ be a complex-valued function, defined in $\mathbb{R}^n \times (\mathbb{R}^n \setminus \{0\})$, having there classical derivatives $D^\alpha_x D^\beta_\xi a(x,\xi) = a_{\alpha\beta}(x,\xi)$ where $|\alpha| \leqq 1 + [s]$ and $|\beta| \leqq N$. Let the following two hypotheses be satisfied:

(i) For any $R > 0$ and any $|\alpha| \leqq 1 + [s]$ holds

$$\sup_{|x|\leqq R} |D_x^\alpha a(x,\xi)| \in L_1^{\rm loc}. \tag{1}$$

(ii) There exist two positive numbers ρ and λ such that

$$\sup_{|x|\leqq R} |\xi|^\rho |a_{\alpha\beta}(x,2^j\xi)| \leqq c_R(1+|\xi|)^\lambda 2^{j|\alpha|-j|\beta|} \tag{2}$$

holds for any $R>0$, $|\alpha|\leqq 1+[s]$, $|\beta|\leqq N$, $j\in\mathbb{N}_0$ and $\xi\in\mathbb{R}^n$ with $|2^j\xi|\geqq 1$, where c_R is independent of ξ and j. Then

$$a(\cdot,D)\colon\quad F_{pq}^{s\,{\rm com}} \longrightarrow F_{pq}^{s\,{\rm loc}} \tag{3}$$

in the sense of (6.3.1/1), where $a(\cdot,D)$ is given by (6.2.1/4).

Proof. Step 1. We begin with two preparations. Let φ and ψ be two compactly supported C^∞ functions in $\mathbb{R}^n$ with $\varphi(\xi)=1$ near the origin and $\psi(x)=1$ if $x\in\operatorname{supp}\varphi$. Let $f\in F_{pq}^s$ and $g=(\psi\hat{f})^\vee$ then $g\in L_p$ with $p\leqq 1$. By (1.5.3/3) or [Triß: Theorem 1.4.1] we have $g\in L_1$ and hence $\hat{g}\in L_\infty$. Then (1) and standard arguments show that

$$h(x)=\int_{\mathbb{R}^n} e^{ix\xi}a(x,\xi)\varphi(\xi)\hat{f}(\xi)d\xi = \int_{\mathbb{R}^n} e^{ix\xi}a(x,\xi)\varphi(\xi)\hat{g}(\xi)d\xi \tag{4}$$

has classical derivatives in $\mathbb{R}^n$ up to order $1+[s]$. Then we have $h\in F_{pq}^{s\,{\rm loc}}$, see, e.g., Proposition 4.2.1. In other words we may assume in the following steps that the symbol $a(x,\xi)$ vanishes in $\mathbb{R}^n\times\Omega$, where Ω is a given bounded domain in $\mathbb{R}^n$ with $0\in\Omega$.

Step 2. We need a second preparation. Let $f\in F_{pq}^{s\,{\rm com}}$ be given by

$$f=\sum_{j=1}^\infty(\mu_j b_j+\lambda_j g_j), \qquad \sum_{j=1}^\infty(|\mu_j|^p+|\lambda_j|^p)\sim \|f\mid F_{pq}^s\|^p, \tag{5}$$

where b_j are s-atoms, g_j are (s,p,q)-atoms, μ_j and λ_j are complex numbers, see Theorem 3.2.3 and Proposition 3.2.4(ii). To avoid confusions with the symbol $a(x,\xi)$ we replaced a_j in (3.2.3/1) by b_j. Let b be one of the s-atoms b_j where K in (3.2.3/3) is at our disposal. In particular

$$|\hat{b}(\xi)|\leqq c(1+|\xi|)^{-K}, \tag{6}$$

where c is independent of $b(x)$. By (2) with $j=0$ and $K>\lambda-\rho+n+1+[s]$ it follows that

$$a(x,D)b(x)=\int_{\mathbb{R}^n} e^{ix\xi}a(x,\xi)\hat{b}(\xi)d\xi \tag{7}$$

has classical derivatives up to order $1+[s]$ in $\mathbb{R}^n$ and the constant c in

$$\sup_{|x|\leqq R} |D^\gamma a(x,D)b(x)| \leqq c, \qquad |\gamma| \leqq 1+[s], \tag{8}$$

is independent of the chosen s-atom b (but may depend on R). Let χ_1 and χ_2 be two compactly supported C^∞ functions in $\mathbb{R}^n$ with $\chi_2(x)=1$ if $x \in \operatorname{supp}\chi_1$. Then (8) and Remark 4.2.1/1 yield

$$\begin{aligned} &\|\chi_1(x)a(x,D)b(x) \mid F^s_{pq}\| \\ &\qquad \sum_{0\leqq|\gamma|\leqq 1+[s]} \sup_{x\in\mathbb{R}^n} |D^\gamma \chi_1(x)a(x,D)b(x)| \, \|\chi_2 \mid F^s_{pq}\| \leqq c', \end{aligned} \tag{9}$$

where c' is independent of the chosen s-atom b. By (5) and (9) we arrive at

$$\left\| \chi_1(x) \sum_{j=1}^\infty \mu_j a(x,D)b_j(x) \mid F^s_{pq} \right\|^p \leqq c \sum_{j=1}^\infty |\mu_j|^p \leqq c' \|f \mid F^s_{pq}\|^p. \tag{10}$$

Step 3. After these preparations we come to the crucial point of the proof which reads as follows: Let g be an (s,p,q)-atom based on the cube $Q = Q_{00}$ with side length 1 and centered at the origin. Then there exist two constants C^R and C_R which depend only on the constants R, ρ, λ, c_R (and n) in (2) such that

$$a(x,D)g(x) = C^R G(x) + H(x), \qquad |x| \leqq R, \tag{11}$$

where $G(x)$ is an (s,p,q)-atom based on $4Q$ and $H(x)$ has classical derivatives in $\mathbb{R}^n$ up to order $1+[s]$ with

$$|D^\gamma H(x)| \leqq C_R(1+|x|)^{-N} \text{ for all } |\gamma| \leqq 1+[s] \text{ and } |x| \leqq R, \tag{12}$$

where $N > 1+\frac{n}{p}$ has the same meaning as in the theorem. By Definition 3.2.2(iii) we have

$$g = \sum_{(\mu,l)<(0,0)} d_{\mu l} b_{\mu l}(x), \tag{13}$$

where $b_{\mu l}(x)$ is a $(Q_{\mu l}, s, p, q)$-atom and $d_{\mu l}$ are complex numbers with

$$\left(\sum_{(\mu,l)<(0,0)} |d_{\mu l}|^q \right)^{1/q} \leqq 1. \tag{14}$$

Let $b(x) = b_{00}(x)$, where the numbers K and L in (3.2.2/4,5) are at our disposal. In particular we have

$$|D^\gamma \hat{b}(\xi)| \leqq c|\xi|^\kappa (1+|\xi|)^{-\delta}, \qquad \xi \in \mathbb{R}^n, \ |\gamma| \leqq N, \tag{15}$$

where $\kappa > 0$ and $\delta > 0$ may be chosen as large as one wants, and c is independent of b. By Step 1 we may assume that (2) holds for all $\xi \in \mathbb{R}^n$. Then (2) with $j = 0$ and (7), where b has now the present meaning, yield

$$|D^\alpha a(x,D)b(x)| \leqq c(1+|x|)^{-N}, \quad |\alpha| \leqq 1+[s], \quad |x| \leqq R, \tag{16}$$

where c depends only on R, ρ, λ, and c_R in (2). Let $b_\mu(x) = b_{\mu 0}(x)$ with $\mu \in \mathbb{N}_0$, see (3.2.2/4,5). Then

$$b(x) = 2^{\mu(s-\frac{n}{q})} b_\mu(2^{-\mu}x), \qquad x \in \mathbb{R}^n, \tag{17}$$

is an (Q_{00}, s, p, q)-atom of the above type. Then we have

$$\begin{aligned} a(x,D)b_\mu(x) &= 2^{\mu(\frac{n}{q}-s)} \int_{\mathbb{R}^n} e^{ix\xi} a(x,\xi) 2^{-\mu n} \hat{b}(2^{-\mu}\xi) d\xi \\ &= 2^{\mu(\frac{n}{q}-s)} a_\mu(2^\mu x, D) b(2^\mu x) \end{aligned} \tag{18}$$

with the symbol

$$a_\mu(x,\xi) = a(2^{-\mu}x, 2^\mu \xi). \tag{19}$$

However all the symbols $a_\mu(x,\xi)$ with $\mu \in \mathbb{N}_0$ satisfy (2) with a_μ instead of a uniformly with respect to μ. Recall that we may assume that (2) holds for all $\xi \in \mathbb{R}^n$. Hence we can apply (16) with a_μ instead of a and with (17). Then (18) yields

$$|D^\alpha a(x,D)b_\mu(x)| \leqq c 2^{\mu(\frac{n}{q}-s+|\alpha|)} (1+|2^\mu x|)^{-N}, \quad |\alpha| \leqq 1+[s], \; |x| \leqq R, \tag{20}$$

where c is the same constant as in (16). Finally let $b_{\mu l}(x)$ be an arbitrary $(Q_{\mu l}, s, p, q)$-atom in the sense of the representation (13). By a translation argument we arrive at

$$|D^\alpha a(x,D)b_{\mu l}(x)| \leqq c 2^{\mu(\frac{n}{q}-s+|\alpha|)} (1+|2^\mu x - l|)^{-N}, \quad |\alpha| \leqq 1+[s], \; |x| \leqq R, \tag{21}$$

where c is independent of the chosen atom $b_{\mu l}$. Armed with (21) we return to (13) and arrive at

$$a(x,D)g = \sum_{\mu=0}^{\infty} \sum_{l}{}^{\mu} d_{\mu l} a(x,D) b_{\mu l}(x), \tag{22}$$

where $\sum_l^\mu$ indicates the summation over those $l \in \mathbb{Z}^n$ such that $(\mu, l) < (0,0)$. Let

$$1 = \sum_{k \in \mathbb{Z}^n} \chi_{\mu k}(x), \qquad x \in \mathbb{R}^n, \tag{23}$$

be a smooth resolution of unity where $\chi_{\mu k}(x)$ are C^∞ functions in $\mathbb{R}^n$ with $\operatorname{supp} \chi_{\mu k} \subset 2Q_{\mu k}$ and, say, $\chi_{\mu k}(x) = \chi_{\mu 0}(x - 2^{-\mu}k)$. Then we have

$$a(x,D)g = \sum_{\mu=0}^{\infty} \sum_{l}{}^{\mu} \sum_{k}{}_{1}^{\mu} d_{\mu l} \chi_{\mu k}(x) a(x,D) b_{\mu l}(x) \tag{24}$$

$$+ \sum_{\mu=0}^{\infty} \sum_{l}{}^{\mu} \sum_{k}{}_{2}^{\mu} d_{\mu l} \chi_{\mu k}(x) a(x,D) b_{\mu l}(x) = G_0(x) + H(x),$$

where $\sum_k{}_1^{\mu}$ is the sum over those $k \in \mathbb{Z}^n$ such that $Q_{\mu k} \subset 4Q$ and $\sum_k{}_2^{\mu}$ is the sum over the remaining lattice-points of $\mathbb{Z}^n$. We claim that (24) coincides with (11). Let μ and k be fixed with $Q_{\mu k} \subset 4Q$ and let $x \in 2Q_{\mu k}$. By (21) we have for $|\alpha| \leqq 1 + [s]$

$$\left| D^{\alpha} \sum_{l}{}^{\mu} d_{\mu l} a(x,D) b_{\mu l}(x) \right| \leqq c\, 2^{\mu(\frac{n}{q} - s + |\alpha|)} \tag{25}$$

$$\times \left(\sum_{l}{}^{\mu} |d_{\mu l}|^q (1 + |l-k|)^{-\kappa q} \right)^{1/q} \left(\sum_{l}{}^{\mu} (1 + |l-k|)^{\kappa q' - N q'} \right)^{1/q'},$$

where κ is a positive number with $(N-\kappa)q' > n$. Hence

$$G_{\mu k} = c \chi_{\mu k}(x) \left(\sum_{l}{}^{\mu} |d_{\mu l}|^q (1 + |l-k|)^{-\kappa q} \right)^{-1/q} \sum_{l}{}^{\mu} d_{\mu l} a(x,D) b_{\mu l}(x) \tag{26}$$

is an $(Q_{\nu k}, s, p, q)$-atom in the sense of Definition 3.2.2(ii) (recall that $s > n(\frac{1}{p} - 1)$, such that no moment conditions (3.2.2/5) are required) where c is a suitable constant depending on R in (21). By (24) we have

$$G_0(x) = \sum_{(\mu,k)} D_{\mu k} G_{\mu k}(x), \tag{27}$$

where (μ, k) is restricted by $Q_{\mu k} \subset 4Q$, and

$$D_{\mu k} = c \left(\sum_{l}{}^{\mu} |d_{\mu l}|^q (1 + |l-k|)^{-\kappa q} \right)^{1/q} \tag{28}$$

with some $c > 0$ depending on R. Let $\kappa q > n$ which can be materialized simultaneously with the above assumption $\kappa + \frac{n}{q'} < N$. Then (28) and (14) yield

$$\left(\sum_{(\mu,k)} |D_{\mu k}|^q \right)^{1/q} \leqq c, \tag{29}$$

which shows that $G_0(x) = C^R G(x)$, where $G(x)$ is an (s,p,q)-atom based on $4Q$ in the sense of Definition 3.2.2(iii), see also (11). Next we prove that $H(x)$ in (24), and hence in (11), has the desired properties (12). First we remark that

$$H(x) = 0 \quad \text{if} \quad x \in 2Q. \tag{30}$$

Let $x \notin 2Q$, then (21) yields

$$|D^\alpha a(x,D)b_{\mu l}(x)| \leqq c\, 2^{\mu(\frac{n}{q}-s+|\alpha|-N)}|x|^{-N}. \tag{31}$$

Then we have by (24)

$$|D^\alpha H(x)| \leqq c \sum_{\mu=0}^{\infty}\sum_{l}{}^{\mu}\, |d_{\mu l}| 2^{\mu(\frac{n}{q}-s+|\alpha|-N)}|x|^{-N} \tag{32}$$

$$\leqq c \sum_{\mu=0}^{\infty}\left(\sum_{l}{}^{\mu}\, |d_{\mu l}|^q\right)^{1/q} 2^{\mu\frac{n}{q'}+\mu(\frac{n}{q}-s+|\alpha|-N)}|x|^{-N}$$

with $|\alpha| \leqq 1+[s]$, where c may depend on R. By (14) the sum over $d_{\mu l}$ can be estimated by 1, and the exponent of 2 can be estimated by $\mu\delta$ with $\delta = -s+|\alpha|+n-N \leqq -1$, because $N \geqq n+2$. Now by (30) and (32) we have (12). The proof of (11) is complete.

Step 4. Let g be the same (s,p,q)-atom as in the previous step and let $H(x)$ be the same function as in (11) and (12). Let χ be a C^∞ function in $\mathbb{R}^n$ with $\operatorname{supp}\chi \subset \{y \in \mathbb{R}^n\colon\ |y|<1\}$ and let $\chi_R(x) = \chi(\frac{x}{R})$. We claim $\chi_R H \in F^s_{pq}$ and

$$\|\chi_R H \mid F^s_{pq}\| \leqq \tilde{C}_R, \tag{33}$$

where $\tilde{C}_R$ depends only on R, ρ, λ, and c_R in (2). For our later purposes it is convenient to extend this claim slightly: Let $h(x)$ be a function in $\mathbb{R}^n$, having there classical derivatives up to order $1+[s]$ such that

$$|D^\gamma h(x)| \leqq \frac{c}{(1+|x|)^N} \quad \text{for } |\gamma| \leqq 1+[s], \quad x \in \mathbb{R}^n, \tag{34}$$

for some $N > 1+\frac{n}{p}$ and $c>0$, then $h \in F^s_{pq}$ and

$$\|h \mid F^s_{pq}\| \leqq c' \tag{35}$$

where c' depends only on c in (34). By Theorem 2.4.7 and (4.2.1/1) we have

$$\|h \mid F^s_{pq}\|^p \sim \sum_{k\in\mathbb{Z}^n} \|\psi_k h \mid F^s_{pq}\|^p \leqq c_1 \sum_{k\in\mathbb{Z}^n} |k|^{-Np}\|\psi_k \mid F^s_{pq}\|^p \leqq c_2. \tag{36}$$

Obviously, (33) is a consequence of (35). We return to (11). We have $G(x) \in F^s_{pq}$ and

$$\|G \mid F^s_{pq}\| \leqq c, \tag{37}$$

where c is independent of the chosen atom and also of R, see (3.2.3/36). Now (11), (33), and (37) yield

$$\|\chi_R(x)a(x,D)g(x) \mid F^s_{pq}\| \leqq c^R, \tag{38}$$

where c^R depends only on R, ρ, λ, and c_R in (2).

Step 5. Next we assume that g stands for an (s,p,q)-atom based on the cube $Q_{\nu 0}$ with the side-length $2^{-\nu}$ and centered at the origin, $\nu \in \mathbb{N}_0$. Let g be given by

$$g = \sum_{(\mu,l)<(\nu,0)} d_{\mu l} b_{\mu l}(x), \tag{39}$$

where $b_{\mu l}(x)$ are $(Q_{\mu l}, s, p, q)$-atoms and $d_{\mu l}$ are complex numbers with

$$\left(\sum_{(\mu,l)} |d_{\mu l}|^q \right)^{1/q} \leqq 2^{-\nu n(\frac{1}{q}-\frac{1}{p})}, \tag{40}$$

see Definition 3.2.2(iii). We normalize g and put

$$h(x) = 2^{\nu(s-\frac{n}{p})} g(2^{-\nu}x) = 2^{\nu(s-\frac{n}{p})} \sum_{(\mu,l)<(\nu,0)} d_{\mu l} 2^{-\nu(s-\frac{n}{q})} c_{\mu l}(x) \tag{41}$$

with

$$c_{\mu l}(x) = 2^{\nu(s-\frac{n}{q})} b_{\mu l}(2^{-\nu}x). \tag{42}$$

Then $h(x)$ is an (s,p,q)-atom based on the unit cube Q_{00}, see again Definition 3.2.2(iii). By (17,18) we have

$$\chi_R(x) a(x,D) g(x) = 2^{\nu(\frac{n}{p}-s)} \chi_{2^\nu R}(2^\nu x) a_\nu(2^\nu x, D) h(2^\nu x), \tag{43}$$

where χ_R is the same function as in the previous step and $a_\nu(x,\xi)$ is given by (19). As we remarked after (19), the symbols $a_\nu(x,\xi)$ with $\nu \in \mathbb{N}_0$ satisfy (2) uniformly. We apply (38) to $\chi_{2^\nu R}(x) a_\nu(x,D)$. By (19) it follows that this can be done with the same constant c_R, independently of ν, hence

$$\|\chi_{2^\nu R}(x) a_\nu(x,D) h(x) \mid F^s_{pq}\| \leqq c^R, \qquad \nu \in \mathbb{N}_0. \tag{44}$$

Recall $0 < p \leqq 1$ and $s > n(\frac{1}{p} - 1)$, then we can use the homogeneity argument at the end of 2.3.3 which yields

$$\begin{aligned} &\|\chi_{2^\nu R}(2^\nu x) a_\nu(2^\nu x, D) h(2^\nu x) \mid F^s_{pq}\| \\ &\qquad \leqq c\, 2^{\nu(s-\frac{n}{p})} \|\chi_{2^\nu R}(x) a_\nu(x,D) h(x) \mid F^s_{pq}\| \leqq \tilde{c}^R 2^{\nu(s-\frac{n}{p})}. \end{aligned} \tag{45}$$

By (43) and (45) we arrive at (38), where g is now an arbitrary (s,p,q)-atom based on $Q_{\nu 0}$.

Step 6. Let $f \in F^{s\,\mathrm{com}}_{pq}$ be given by (5) with $\operatorname{supp} f \subset Q$ for some large cube Q. By Proposition 3.2.4(ii) we may assume $\operatorname{supp} g_j \subset 10\,Q$. By a translation argument, depending on Q, we may extend (38) to g_j, where now c^R depends on R, ρ, λ, c_R in (2), and Q. Then (5), (9,10), and (38) yield

$$\|\chi_R(x) a(x,D) f(x) \mid F^s_{pq}\|^p \leqq c \sum_{j=1}^{\infty} (|\mu_j|^p + |\lambda_j|^p) \leqq c' \|f \mid F^s_{pq}\|^p. \tag{46}$$

The proof is complete.

6.3.3 Complements

Before discussing in the next subsection the meaning of the conditions for the symbol $a(x,\xi)$ in Theorem 6.3.2 we add two corollaries. The spaces $B^{s\,\mathrm{com}}_{pq}$ and $B^{s\,\mathrm{loc}}_{pq}$ are the obvious counterparts of the spaces $F^{s\,\mathrm{com}}_{pq}$ and $F^{s\,\mathrm{loc}}_{pq}$ introduced in Definition 6.3.1.

Corollary 1. Let $0 < p \leqq 1$ and $s > n(\frac{1}{p} - 1)$. Let $a(x,\xi)$ be the same symbol as in Theorem 6.3.2.

(i) Let $p \leqq q \leqq \infty$. Then

$$a(\cdot, D)\colon\ F^{s\,\mathrm{com}}_{pq} \longrightarrow F^{s\,\mathrm{loc}}_{pq} \tag{1}$$

in the sense of (6.3.1/1), where $a(\cdot, D)$ is given by (6.2.1/4).

(ii) Let $0 < q \leqq \infty$. Then

$$a(\cdot, D)\colon\ B^{s\,\mathrm{com}}_{pq} \longrightarrow B^{s\,\mathrm{loc}}_{pq}. \tag{2}$$

Proof. (6.3.1/1) can be reformulated as

$$\psi_1(x)a(x,D) \circ \psi_2(x)\colon\ F^s_{pq} \longrightarrow F^s_{pq} \tag{3}$$

for any two compactly supported C^∞ functions in $\mathbb{R}^n$. By Theorem 6.3.2 we have (3) if $0 < p \leqq 1 < q \leqq \infty$ and $s > n(\frac{1}{p} - 1)$. We apply the real interpolation formula (1.6.7/7) and obtain (2). Now complex interpolation between $F^s_{pp} = B^s_{pp}$ and F^s_{pr} with $1 < r \leqq \infty$ in the sense of (1.6.7/4) yields (1).

Corollary 2. Let $0 < p \leqq 1$ and $s > n(\frac{1}{p} - 1)$. Let $N \in \mathbb{N}$ with $N > \frac{n}{p} + 1$. Let $a(x,\xi)$ be a complex-valued function, defined on $\mathbb{R}^{2n}$, having there classical derivatives $D^\alpha_x D^\beta_\xi a(x,\xi) = a_{\alpha\beta}(x,\xi)$ where $|\alpha| \leqq 1 + [s]$ and $|\beta| \leqq N$. Let the following two hypotheses be satisfied:

(i) For some $\varepsilon > 0$ it holds $a(x,\xi) = 0$ if $x \in \mathbb{R}^n$ and $|\xi| \leqq \varepsilon$.

(ii) There exists three positive numbers ρ, λ, and c such that

$$|\xi|^\rho |a_{\alpha\beta}(x, 2^j\xi)| \leqq c(1 + |\xi|)^\lambda\, 2^{j|\alpha| - j|\beta|} \tag{4}$$

holds for $|\alpha| \leqq 1 + [s]$, $|\beta| \leqq N$, $j \in \mathbb{N}_0$, $x \in \mathbb{R}^n$, and $\xi \in \mathbb{R}^n$. Then

$$a(\cdot, D)\colon\ F^s_{pq} \longrightarrow F^s_{pq} \ \text{ if } \ p \leqq q \leqq \infty \tag{5}$$

and

$$a(\cdot, D):\ B^s_{pq} \longrightarrow B^s_{pq} \ \text{ if } \ 0 < q \leqq \infty. \tag{6}$$

Proof. Step 1. We prove (5) with $1 < q \leqq \infty$. We may assume, say, $\varepsilon = 2$. Then it will be sufficient to show

$$\|a(\cdot, D)f \mid F^s_{pq}\| \leqq c\|f \mid F^s_{pq}\| \tag{7}$$

for all $f \in F^s_{pq}$ with $\hat{f}(\xi) = 0$ if $|\xi| \leqq 1$. This is a simple Fourier multiplier assertion. However if f has this property then (6.3.2/5) may be replaced by

$$f = \sum_{j=1}^{\infty} \lambda_j g_j, \qquad \sum_{j=1}^{\infty} |\lambda_j|^p \sim \|f \mid F^s_{pq}\|^p, \tag{8}$$

where g_j are the same (s,p,q)-atoms as in (6.3.2/5). In other words, we have an atomic representation of f without s-atoms. Now we follow the proof of Theorem 6.3.2. Steps 1 and 2 are not necessary now. We have (6.3.2/11,12) now for all $x \in \mathbb{R}^n$ and with constants independent of R. We apply (35) with (34) to H instead of h. We have (37) and arrive at

$$\|a(x, D)g(x) \mid F^s_{pq}\| \leqq c. \tag{9}$$

Now Steps 5 and 6 prove (7).

Step 2. The proof of (6) as well as the extension of (5) from $1 < q \leqq \infty$ to $p \leqq q \leqq \infty$ can be done in the same way as in the proof of Corollary 1 via interpolation.

6.3.4 Exotic symbol classes

The aim of this subsection is twofold. First we wish to discuss the hypotheses for the symbol $a(x, \xi)$ in Theorem 6.3.2. Secondly we introduce some symbol classes complementing the classes S_δ and $S^\mu_{\rho\delta}$ from Definition 6.2.1 and Remark 6.2.1/1.

A discussion. Let $a(x, \xi)$ be the symbol introduced in Theorem 6.3.2. The weak assumption (6.3.2/1) reflects our restriction to local mappings of type (6.3.2/3). If one wishes to have global mappings of type (6.3.3/5) or so then even global versions of type (6.3.2/1) are not sufficient. A good way to find out which conditions are sufficient in order to get global mapping theorems is to decompose a given symbol $a(x,\xi)$ into a part supported near the origin $\xi = 0$ (maybe uniformly with respect to $x \in \mathbb{R}^n$) and into a rest with a ξ-support off the origin, maybe of the same type as in Corollary 6.3.3/2. Then one can ask under which conditions the first part yields the desired mapping property. But we do not go into detail and return to $a(x,\xi)$ referred to in Theorem 6.3.2. Condition (6.3.2/2) is crucial. The factor $|\xi|^\rho$ again indicates, besides (6.3.2/1), that $a_{\alpha\beta}(x,\xi)$ may have some singularities with respect to ξ near the origin. On the other hand, the factor $(1+|\xi|)^\lambda$ is unimportant, it simply ensures that all the ψDO's discussed in this chapter can be treated within the realm of tempered distributions. More interesting is the following consequence of (6.3.2/2). Let $\eta = 2^j\xi$ with $|\xi| \sim 1$, then we have

$$\sup_{|x|\leqq R} |a_{\alpha\beta}(x,\eta)| \leqq c_R |\eta|^{|\alpha|-|\beta|}, \qquad |\eta| \geqq 1. \tag{1}$$

The global version of (1) with respect to $x \in \mathbb{R}^n$ and without singularities near $\eta = 0$ reads as follows,

$$|a_{\alpha\beta}(x,\eta)| \leqq c(1+|\eta|)^{|\alpha|-|\beta|}, \qquad x \in \mathbb{R}^n,\ \eta \in \mathbb{R}^n,\ |\alpha| \leqq M,\ |\beta| \leqq N. \tag{2}$$

But this is the exotic class S_1 discussed in Definition 6.2.1 and Remark 6.2.1/1 with restricted differentability requirements. If singularities of $a(x,\eta)$ near $\eta = 0$ are admitted, then the global homogeneous counterpart of (1) and (2) is given by

$$|a_{\alpha\beta}(x,\eta)| \leqq c|\eta|^{|\alpha|-|\beta|}, \qquad x \in \mathbb{R}^n,\ 0 \neq \eta \in \mathbb{R}^n,\ |\alpha| \leqq M,\ |\beta| \leqq N. \tag{3}$$

We may replace η in (3) by $2^j\eta$ with $j \in \mathbb{N}_0$. This shows that (3) comes near to (6.3.3/4). In any case, this discussion makes clear, that the crucial assumption (6.3.2/2) is well reflected by exotic symbol classes of type (1–3) and that it is justified to turn the tide and to start with symbol classes of this type.

Definition. (i) Let $\mu \in \mathbb{R}$, $M \in \mathbb{N}$ and $N \in \mathbb{N}$. Then $S_1^{\mu\,\mathrm{loc}}(M,N)$ is the collection of all complex-valued functions $a(x,\xi)$, defined in $\mathbb{R}^n \times (\mathbb{R}^n \setminus \{0\})$, having there classical derivatives $D_x^\alpha D_\xi^\beta a(x,\xi) = a_{\alpha\beta}(x,\xi)$, $|\alpha| \leqq M$, $|\beta| \leqq N$, such that for any $R > 0$ there exists a constant c_R with

$$\sup_{|x|\leqq R} |a_{\alpha\beta}(x,\xi)| \leqq c_R |\xi|^{\mu+|\alpha|-|\beta|}, \qquad |\xi| \geqq 1,\ |\alpha| \leqq M,\ |\beta| \leqq N. \tag{4}$$

(ii) Let $\mu \in \mathbb{R}$, then $S_1^{\mu\,\mathrm{loc}} = \bigcap S_1^{\mu\,\mathrm{loc}}(M,N)$, where the intersection is taken with respect to all $M \in \mathbb{N}$ and $N \in \mathbb{N}$.

Remark 1. Assumption (4) with $\mu = 0$ coincides with (1). Furthermore, $S_1^{\mu\,\mathrm{loc}}(M,N)$ and $S_1^{\mu\,\mathrm{loc}}$ are the local versions of S_{11}^{μ} from Remark 6.2.1/1. Finally $S_1^{\mu\,\mathrm{loc}}$ is the C^∞-counterpart of $S_1^{\mu\,\mathrm{loc}}(M,N)$, and the counterpart of (4) is given by

$$\sup_{|x|\leqq R} |a_{\alpha\beta}(x,\xi)| \leqq c_R^{\alpha\beta} |\xi|^{\mu+|\alpha|-|\beta|}, \qquad |\xi| \geqq 1, \tag{5}$$

for all α and β.

Remark 2. The restriction of (4) and (5) to $|\xi| \geqq 1$ is natural and justified by the above discussion. The behaviour of $a(x,\xi)$ near $\xi = 0$ is not so important and should be added (or omitted) from case to case.

Remark 3. We complement the references given in Remark 6.2.2/4, now concentrating on exotic ψDO's of the above type. It was observed by Ching (1972), see [Chi], that exotic ψDO's do not necessarily map L_2 into L_2. But at the beginning of the eighties it was proved that they map $H_p^s = F_{pq}^s$, $s > 0$, $1 < p < \infty$, into itself, see [Mey3]. Afterwards exotic ψDO's attracted more and more attention, in particular in connection with Bony's applications of exotic ψDO's to non-linear problems. Corresponding investigations covering general spaces of B_{pq}^s and F_{pq}^s type may be found in [Run1], see also [Bou3, Hör4,5, Tor] for other recent aspects of exotic ψDO's. Section 6.3 is based on [Tri22], the results had been announced before in [Tri19].

Remark 4. There is a striking difference between mapping properties for non-exotic ψDO's of type S_δ with $\delta < 1$, see Theorem 6.2.2, on the one hand, and mapping properties for exotic ψDO's, see Theorem 6.3.2 and Corollaries 6.3.3/1,2, on the other hand: For exotic ψDO's there do not exist composition formulas of type (6.2.2/12) and an extension of Theorem 6.3.2 and Corollaries 6.3.3/1,2 from $s > n(\frac{1}{p} - 1)$ to all $s \in \mathbb{R}$ cannot be expected, in sharp contrast to the situation in Theorem 6.2.2, cf. also the preceding remark.

6.3.5 Exotic Fourier integral operators

The Fourier integral operators we have in mind are given by

$$Af(x) = \int_{\mathbb{R}^n} e^{ix\xi + i\Phi(x,\xi)} a(x,\xi) \hat{f}(\xi) d\xi \tag{1}$$

where $\Phi(x,\xi)$ is the real-valued phase function. In other words, Fourier integral operators are ψDO's in the sense of (6.2.1/4) with the symbol $e^{i\Phi(x,\xi)}a(x,\xi)$. Such a splitting of the symbol has to be justified. Of course, the assumption that the phase function $\Phi(x,\xi)$ is real, is a severe restriction (however sometimes also Fourier integral operators with a complex phase function are considered). Furthermore, usually it is assumed that $\Phi(x,\xi)$ is homogeneous of degree 1 (or of another degree) with respect to ξ. These restrictions, and also the name "phase function," hint to the physical origin of operators of this type. In mathematics they attracted some attention since the early seventies. Mapping properties are mostly studied in an L_2-setting under further restrictions for the real phase function. We refer the interested reader to [Tre: II, Ch. 8, 6.2] and [Hör3: IV, 2.5.3]. Operators of type (1) in L_p-spaces, $1 < p < \infty$, have been treated in [Per, Bea3]. Extensions of these considerations to Besov spaces and also to more general symbol classes may be found in [Qin1,2]. Our point of view is the following: We wish to apply Theorem 6.3.2 with $e^{i\Phi(x,\xi)}a(x,\xi)$ instead of $a(x,\xi)$, where $\Phi(x,\xi)$ and $a(x,\xi)$ belong to the exotic symbol classes introduced in Definition 6.3.4(i).

Theorem. Let $0 < p \leqq 1$ and $s > n(\frac{1}{p}-1)$. Let $N \in \mathbb{N}$ with $N > 1 + \frac{n}{p}$. Let $\nu \geqq 0$, $\mu \geqq \nu(1+[s]+N)$ and

$$\Phi(x,\xi) \in S_1^{\nu\,\mathrm{loc}}(1+[s],N), \qquad a(x,\xi) \in S_1^{-\mu\,\mathrm{loc}}(1+[s],N). \tag{2}$$

Let $\Phi(x,\xi)$ be real and

$$\sup_{|x|\leqq R} |D_x^\alpha(e^{i\Phi(x,\xi)}a(x,\xi))| \in L_1^{\mathrm{loc}} \tag{3}$$

for all $|\alpha| \leqq 1+[s]$ and all $R > 0$. Let A be given by (1).

(i) Let $p \leqq q \leqq \infty$, then

$$A\colon\; F_{pq}^{s\,\mathrm{com}} \longrightarrow F_{pq}^{s\,\mathrm{loc}}. \tag{4}$$

(ii) Let $0 < q \leqq \infty$, then

$$A\colon\; B_{pq}^{s\,\mathrm{com}} \longrightarrow B_{pq}^{s\,\mathrm{loc}}. \tag{5}$$

Proof. Step 1. It is sufficient to prove (4) with $0 < p \leqq 1 < q \leqq \infty$. The extension of (4) to $p \leqq q \leqq \infty$ and the proof of (5) follow by interpolation in the same way as in the proof of Corollary 6.3.3/1.

Step 2. Let $0 < p \leqq 1 < q \leqq \infty$ and $s > n(\frac{1}{p}-1)$. We apply Theorem 6.3.2 with $e^{i\Phi(x,\xi)}a(x,\xi)$ instead of $a(x,\xi)$. Of course, (3) is the counterpart of (6.3.2/1). We have to check (6.3.2/2) with $e^{i\Phi(x,\xi)}a(x,\xi)$ instead of $a(x,\xi)$, where ρ and λ are at our disposal. However this can easily be done by straightforward calculations where $\mu \geqq \nu(1+[s]+N)$ relects the worst case. Then the counterpart of (6.3.2/3) proves (4).

Remark 1. Let $\nu < 0$ in (2) then $\mu \geqq \nu(1+[s]+N)$ must be replaced by $\mu \geqq 0$, in other words there are no improvements compared with $\nu = 0$.

Remark 2. We used essentially that $\Phi(x,\xi)$ is real. But otherwise our assumptions for $\Phi(x,\xi)$ are very general, more general than usual. If $\Phi(x,\xi)$ satisfies some homogeneity conditions with respect to ξ, e.g., of degree 1, or if it is independent of x, then further improvements can be expected. This will be done in the following two subsections where we obtain optimal results (besides limiting cases).

Remark 3. Let $\Phi(x,\xi)$ in the above theorem be independent of $x \in \mathbb{R}^n$. Then the above indicated straightforward calculation shows that $\mu \geqq \nu N$ is sufficient, instead of $\mu \geqq \nu(1+[s]+N)$.

6.3.6 Exotic pseudodifferential operators

We specialize (6.3.5/1) by assuming that the phase function is independent of $x \in \mathbb{R}^n$, i.e.

$$Af(x) = \int_{\mathbb{R}^n} e^{ix\xi + i\Phi(\xi)} a(x,\xi) \hat{f}(\xi) d\xi, \tag{1}$$

where $\Phi(\xi)$ is real-valued. We use the symbol class $S_1^{\mu\,\mathrm{loc}}$ introduced in Definition 6.3.4(ii) and Remark 6.3.4/1. Applied to $\Phi(\xi)$ which is independent of $x \in \mathbb{R}^n$, we have $\Phi(\xi) \in S_1^{\nu\,\mathrm{loc}}$ if and only if for any multi-index β there exists a constant c^β such that

$$|D_\xi^\beta \Phi(\xi)| \leqq c^\beta |\xi|^{\nu-|\beta|}, \qquad |\xi| \geqq 1. \tag{2}$$

Although the superscript "loc" looks now a little bit suspicious we stick at this notation.

Theorem. Let $0 < p \leqq 2$ and $s > n(\frac{1}{p} - \frac{1}{2})$. Let either $\nu > 0$, $\mu > \nu n(\frac{1}{p} - \frac{1}{2})$ or $\mu = \nu = 0$. Let

$$\Phi(\xi) \in S_1^{\nu\,\mathrm{loc}} \quad \text{and} \quad a(x,\xi) \in S_1^{-\mu\,\mathrm{loc}}. \tag{3}$$

Let $\Phi(\xi)$ be real and

$$\sup_{|x| \leqq R} |D_x^\alpha a(x,\xi)| \in L_1^{\mathrm{loc}} \tag{4}$$

for any multi-index α and any $R > 0$. Let A be given by (1).

(i) Let $p \leqq q \leqq 2$, then

$$A\colon\ F_{pq}^{s\,\mathrm{com}} \longrightarrow F_{pq}^{s\,\mathrm{loc}}. \tag{5}$$

(ii) Let $0 < q \leqq \infty$, then

$$A\colon\ B_{pq}^{s\,\mathrm{com}} \longrightarrow B_{pq}^{s\,\mathrm{loc}}. \tag{6}$$

Proof. Step 1. First we remark that (6) follows from (5) by real interpolation, see the proof of Corollary 6.3.3/1.

Step 2. By Theorem 6.3.5 and Remark 6.3.5/3 we have

$$A\colon\ F_{pq}^{s\,\mathrm{com}} \longrightarrow F_{pq}^{s\,\mathrm{loc}}, \quad 0 < p \leqq 1,\ p \leqq q \leqq \infty, \tag{7}$$

$$s > n\left(\frac{1}{p} - 1\right), \quad \mu \geqq \nu\left(\frac{n}{p} + 2\right).$$

Furthermore, A can be written as

$$Af(x) = a(x,D)g(x) \quad \text{with} \quad g(x) = (e^{i\Phi(\xi)}\hat{f}(\xi))^\vee(x), \tag{8}$$

where $a(x,D)$ has the same meaning as in (6.2.1/4). Obviously, the mapping $f \to g$ is an isometric mapping of L_2 onto itself, and hence also an isometric mapping of $H_2^\sigma = F_{22}^\sigma$ onto itself for any $\sigma \in \mathbb{R}$. Furthermore we need the following result about exotic ψDO's. Let

$$b(x,\xi) \in S_1^{0\,\mathrm{loc}} \ \text{and} \ b(x,\xi) = 0 \ \text{if} \ |\xi| \leqq \varepsilon \ \text{and} \ x \in \mathbb{R}^n \tag{9}$$

for some $\varepsilon > 0$. Let $\chi(x)$ be a compactly supported C^∞ function in $\mathbb{R}^n$. Then $\chi(x)b(x,\xi) \in S_1$ in the sense of Definition 6.2.1, and we have

$$\chi(x)b(x,D)\colon\ F_{22}^\sigma \longrightarrow F_{22}^\sigma, \quad \sigma > 0, \tag{10}$$

see [Mey3, Run1]. Next we wish to use the complex interpolation method for quasi-Banach spaces described in 1.6.6 specified to the function spaces in 1.6.7. We apply the interpolation property to an analytic family of ψDO's given by the analytic family of symbols

$$e^{i\Phi(\xi)}a(x,\xi)(1+|\xi|^2)^{\frac{\mu}{2}(1-z)}, \qquad z \in \mathbb{C}, \quad 0 \leqq \operatorname{Re} z \leqq 1, \tag{11}$$

where $\mathbb{C}$ stands for the complex plane. In order to prove (5) we may assume

$$a(x,\xi) = 0 \text{ if either } |x| \geqq R \text{ or } |\xi| \leqq \varepsilon \tag{12}$$

for some positive numbers R and ε. The first restriction comes from the possibility to multiply with the above function χ, whereas the second restriction is based on (4) and Step 1 of the proof of Theorem 6.3.2. Now by (3) we have

$$a(x,\xi)(1+|\xi|^2)^{\frac{\mu}{2}+i\tau} \in S_1, \qquad a(x,\xi)(1+|\xi|^2)^{i\tau} \in S_1^{-\mu\,\mathrm{loc}} \tag{13}$$

uniformly with respect to $\tau \in \mathbb{R}$. Then

$$A_z f(x) = \int_{\mathbb{R}^n} e^{ix\xi+i\Phi(\xi)}a(x,\xi)(1+|\xi|^2)^{\frac{\mu}{2}(1-z)}\hat{f}(\xi)d\xi \tag{14}$$

with $0 \leqq \operatorname{Re} z \leqq 1$ is the desired analytic family of ψDO's we are looking for. By (7) and (10) applied to the above operators we have

$$A_{i\tau}\colon \; F^{s_0}_{22} \longrightarrow F^{s_0}_{22}, \qquad s_0 > 0, \tag{15}$$

$$A_{1+i\tau}\colon \; F^{s_1}_{p_1q_1} \longrightarrow F^{s_1}_{p_1q_1}, \qquad 0 < p_1 \leqq 1, \quad p_1 \leqq q_1 \leqq \infty, \tag{16}$$

$$s_1 > n\left(\frac{1}{p_1} - 1\right), \quad \mu \geqq \nu\left(\frac{n}{p_1} + 2\right)$$

uniformly with respect to $\tau \in \mathbb{R}$. By complex interpolation we arrive at

$$A_\theta\colon \; F^s_{pq} \longrightarrow F^s_{pq}, \tag{17}$$

with $0 < \theta < 1$,

$$a_\theta(x,\xi) = a(x,\xi)(1+|\xi|^2)^{\frac{\mu}{2}(1-\theta)} \in S_1^{-\mu_\theta\,\mathrm{loc}} \tag{18}$$

and

$$s = s_0(1-\theta) + s_1\theta, \quad \frac{1}{p} = \frac{1-\theta}{2} + \frac{\theta}{p_1}, \tag{19}$$

$$\frac{1}{q} = \frac{1-\theta}{2} + \frac{\theta}{q_1}, \quad \mu_\theta = \theta\mu,$$

see (1.6.7/4). Let p and s with $0 < p \leqq 2$ and $s > n(\frac{1}{p} - \frac{1}{2})$ be given. Let $\varepsilon = s - n(\frac{1}{p} - \frac{1}{2})$, $s_0 = \varepsilon$, and $s_1 = \varepsilon + n(\frac{1}{p_1} - \frac{1}{2})$ with $p_1 < \min(1,p)$. Then we have

$$s = s_0(1-\theta) + s_1\theta, \quad \frac{1}{p} = \frac{1-\theta}{2} + \frac{\theta}{p_1}, \quad \mu_\theta \geqq \nu\left(\frac{\theta n}{p_1} + 2\theta\right). \tag{20}$$

If $p_1 \to 0$ then $\theta \to 0$ and $\frac{\theta}{p_1} \to \frac{1}{p} - \frac{1}{2}$. Now (18) and (20) cover all desired cases (and a few more) if θ is chosen appropriately. The proof is complete.

Remark 1. The restriction $\mu > \nu n(\frac{1}{p} - \frac{1}{2})$ is sharp and natural, some references will be given in the following subsection. The interpolation of (15) and (16) covers few more cases than explicitly stated in the theorem, but the last limiting argument cannot be applied to these additional cases.

Remark 2. It is one of the advantages of complex interpolation methods compared with the real ones, that they can be applied not only to a single operator but also to analytic families of operators as we did in the above proof. Further information and more elaborated examples may be found, e.g., in [Triß: 2.4.4, 2.4.7, 2.4.8].

6.3.7 Strongly singular integral operators

Now we assume that both the phase function Φ and the symbol a in (6.3.5/1) (and (6.3.6/1)) are independent of $x \in \mathbb{R}^n$. In other words

$$Af(x) = \int_{\mathbb{R}^n} e^{ix\xi + i\Phi(\xi)} a(\xi)\hat{f}(\xi)d\xi = (2\pi)^n (e^{i\Phi} a\hat{f})^\vee(x), \tag{1}$$

where again $\wedge$ and $\vee$ indicate the Fourier transform and its inverse. Then mapping properties of type (6.3.6/5,6) mean that $e^{i\Phi(\xi)}a(\xi)$ is a (local) Fourier multiplier in the considered space. To simplify the formulations we restrict ourselves in the following theorem to the spaces B^s_{pq} and

$$H^s_p = F^s_{p2}, \qquad 0 < p < \infty, \quad s \in \mathbb{R}, \tag{2}$$

where the latter spaces cover both fractional Sobolev spaces and spaces of Hardy type, see 2.6.5 or Theorem 1.5.1. It is clear what is meant by $H^{s\,\mathrm{com}}_p$ and $H^{s\,\mathrm{loc}}_p$.

Theorem. (i) Let $0 < p < \infty$, and $s \in \mathbb{R}$. Let

$$\text{either} \quad \nu > 0, \quad \mu > \nu n \left|\frac{1}{p} - \frac{1}{2}\right| \quad \text{or} \quad \nu = \mu = 0 \tag{3}$$

and

$$\Phi(\xi) \in S_1^{\nu \,\mathrm{loc}} \text{ real}, \qquad a(\xi) \in S_1^{-\mu \,\mathrm{loc}} \cap L_1^{\mathrm{loc}}. \tag{4}$$

Let A be given by (1), then

$$A\colon \; H_p^{s \,\mathrm{com}} \longrightarrow H_p^{s \,\mathrm{loc}}. \tag{5}$$

(ii) Let $0 < p \leqq \infty$, $0 < q \leqq \infty$, and $s \in \mathbb{R}$. Let (3,4) be satisfied and let A be given by (1), then

$$A\colon \; B_{pq}^{s \,\mathrm{com}} \longrightarrow B_{pq}^{s \,\mathrm{loc}}. \tag{6}$$

Proof. Step 1. By the same argument as in Step 1 of the proof of Theorem 6.3.2 we may assume $\Phi(\xi) = a(\xi) = 0$ if $|\xi| \leqq 1$. Then (4), explicated in (6.3.6/2), and the global assertion in Corollary 6.3.3/2, extended globally in the sense of Theorem 6.3.6, yield

$$A\colon \; H_p^s \longrightarrow H_p^s, \qquad 0 < p \leqq 2, \quad s > n\left(\frac{1}{p} - \frac{1}{2}\right). \tag{7}$$

We are looking for Fourier multipliers and then it is well-known that (7) holds for all $s \in \mathbb{R}$, see [Triß: 2.6.6, 2.3.8]. Recall the duality formula

$$(H_p^s)' = H_{p'}^{-s}, \qquad 1 < p < \infty, \quad \frac{1}{p} + \frac{1}{p'} = 1, \tag{8}$$

see [Triß: 2.11.2]. Fourier multipliers generate formally self-adjoint operators. Hence (7) extended to $s \in \mathbb{R}$, and (8) prove (5).

Step 2. In the same way one proves (6), now based on the global version of (6.3.6/6) and the duality formula

$$(B_{pq}^s)' = B_{p'q'}^{-s}, \quad s \in \mathbb{R}, \quad 1 \leqq p < \infty, \quad 1 \leqq q \leqq \infty, \tag{9}$$

$$\frac{1}{p} + \frac{1}{p'} = \frac{1}{q} + \frac{1}{q'} = 1.$$

The extension to all values of q with $0 < q \leqq \infty$ is again a matter of real interpolation.

Example. We describe a typical example. Let $\Phi(\xi)$ and $a(\xi)$ be C^∞ functions in $\mathbb{R}^n \setminus \{0\}$ with

$$a(\xi) = |\xi|^{-\mu} a\left(\frac{\xi}{|\xi|}\right) \quad \text{and} \quad \Phi(\xi) = |\xi|^{\nu} \Phi\left(\frac{\xi}{|\xi|}\right) \quad \text{if } |\xi| \geqq 1. \tag{10}$$

Let $a(\xi) \in L_1^{\rm loc}$ and $\Phi(\xi)$ be real. Then we have (4), see also (6.3.6/2). Hence the theorem can be applied. Fourier multipliers of this type have been extensively discussed, in particular the radial symmetric case, e.g.

$$e^{i\Phi(\xi)}a(\xi) = e^{i|\xi|^\nu}(1+|\xi|^2)^{-\frac{\mu}{2}}, \qquad \xi \in \mathbb{R}^n, \tag{11}$$

which is an immaterial modification of the homogeneity assumption (10).

Remark. In other words, the above theorem may be considered as an extension of some known Fourier multiplier assertions for functions of type (11), see [Ste1: IV, 7.4] for further references. The restriction $\mu > \nu n|\frac{1}{p} - \frac{1}{2}|$ with $\nu > 0$ is sharp (with exception of limiting cases), at least if $0 < \nu < 1$, see [FeS2: p. 160] and [Sjö, Per].

Chapter 7

Spaces on Riemannian Manifolds and Lie Groups

7.1 Introduction

In 1.11 we gave a historically minded description of function spaces of type F^s_{pq} and B^s_{pq} on Riemannian manifolds and Lie groups. Now we return to this subject in detail. But we shall not rely on what had been said in 1.11: as always in this book we use only the material presented from Chapter 2 onwards. On the other hand we shall not repeat those references which are not directly needed. This chapter has six sections, the first five deal with spaces on Riemannian manifolds, and only the last one with spaces on Lie groups. The reason will be clear: We convert a Lie group in an analytic Riemannian manifold and reduce via that way spaces on Lie groups to spaces on manifolds.

The reader of this chapter is expected to have a working knowledge of basic facts of Riemannian geometry and Lie groups. But we shall explain briefly the needed notations, in particular the more specific ones. Our general references are [Hel, Kli] and, as far as the connection between Lie groups and Riemannian manifolds is concerned, [Pri]. This chapter is mostly based on the author's results [Tri13,14] (Riemannian manifolds) and [Tri15,16] (Lie groups). Further references will be given in due course.

7.2 The spaces F^s_{pq} on manifolds

7.2.1 Basic notations

We collect those basic facts about Riemannian manifolds which we need in the sequel. Besides [Hel, Kli] we refer also to [Aub2].

H. Triebel, *Theory of Function Spaces II*, Modern Birkhäuser Classics,
DOI 10.1007/978-3-0346-0419-2_7, © Birkhäuser Verlag 1992

Complete manifolds. Let M be always a connected C^∞ manifold of dimension n furnished with a smooth Riemannian metric g. Recall that a smooth Riemannian metric is given by a real twice-covariant C^∞ tensor field g such that g_P at each point $P \in M$ is a positive definite bilinear symmetric form

$$g_P(X,Y) = g_P(Y,X) \quad \text{and} \quad g_P(X,X) > 0 \text{ if } T_PM \ni X \neq 0,\ Y \in T_PM. \quad (1)$$

Here T_PM stands for the tangent space at $P \in M$. Covariant (or Levi–Civita) derivatives are taken with respect to the (unique) Riemannian connection. In particular, in a local chart (Ω, φ), where Ω is an open set in M and φ is a homeomorphism from Ω onto an open set U in $\mathbb{R}^n$, the Christoffel symbols are given by

$$\Gamma^l_{ij} = \frac{1}{2} g^{kl} [\partial_i g_{kj} + \partial_j g_{ki} - \partial_k g_{ij}], \qquad (2)$$

where we used the summation convention, furthermore $\partial_j = \frac{\partial}{\partial x_j}$, g_{kj} are the components of the covariant tensor g (more precisely $\varphi^{-1^*} g$) and g^{kl} are the components of its contravariant version, i.e., it is the inverse matrix, $g^{kl} g_{lm} = \delta^k_m$. Let $c(t) = c(P,X,t)$ be the geodesic with $c(P,X,0) = P \in M$ and $\frac{dc}{dt}(P,X,0) = X \in T_PM$ where $X \neq 0$ and $t \in \mathbb{R}$ (usually only $t \geqq 0$ will be of interest). We always assume that the Riemannian manifold is complete, i.e., all geodesics are infinitely extendable with respect to the arc length. By the Hopf–Rinow-theorem this is equivalent to the assumption that M is complete as a metric space.

Geodesics. We recall two fundamental properties of geodesics. Let $C(t) = \varphi(c(t))$ be the geodesic $c(P,X,t)$ in the above local chart, then $C(t)$ obeys the system of ordinary differential equations

$$\frac{d^2 C^j(t)}{dt^2} + \Gamma^j_{ik}(C(t)) \frac{dC^i(t)}{dt} \frac{dC^k(t)}{dt} = 0; \qquad j = 1, \ldots, n; \qquad (3)$$

with $C(0) = \varphi(P)$ and $\frac{dC}{dt}(0) = \varphi_* X \in T_{\varphi(P)} U$, where $C^j(t)$ are the components of $C(t)$. Furthermore

$$g_{ij}(C(t)) \frac{dC^i(t)}{dt} \frac{dC^j(t)}{dt} \qquad (4)$$

is constant along the geodesic $C(t)$. In particular we have $\sigma = \|X\| t$ where σ stands for the arc length, $t \geqq 0$, $\sigma = 0$ corresponds to $\varphi(P)$, and $\|X\| = \sqrt{g(X,X)}$.

The exponential map. Of special interest for us is the exponential map $\exp_P$ given by

$$\exp_P(X) = c(P, X, 1), \qquad P \in M, \quad X \in T_P M, \tag{5}$$

where $c(P, X, t)$ is the above geodesic. We put $\exp_P(0) = P$. It is convenient and usual, although a slight abuse of notation, to identify $T_P M$ with $\mathbb{R}^n$. This can be justified by identifying $T_P M$ with $T_{\varphi(P)} U$ where the latter has a natural $\mathbb{R}^n$-structure via $\varphi_* X = a^i \partial_i$. If $r > 0$ is small then $\exp_P$ is a diffeomorphism from

$$B(r) = \{X \in \mathbb{R}^n \colon |X| < r\} \quad \text{onto} \quad \Omega_P(r) = \exp_P B(r), \tag{6}$$

in particular, $(\Omega_P(r), \exp_P^{-1})$ is a local chart, where we used the above identification of $T_P M$ with $\mathbb{R}^n$. The corresponding local coordinates are denoted as normal geodesic coordinates, where "normal" refers to

$$g_{ij}(0) = \delta_{ij} \quad \text{and} \quad \partial_k g_{ij}(0) = 0, \tag{7}$$

where i, j, k are natural numbers between 1 and n. In particular we have $\Gamma^l_{ij}(0) = 0$. Again we do not distinguish between g and $\exp_P^* g$.

Injectivity radius. Let r_P be the supremum of all numbers r such that $\exp_P$ yields a diffeomorphism in the sense of (6). Then $r_0 = \inf r_P$ is called the injectivity radius of M, where the infimum is taken over all $P \in M$. We always assume $r_0 > 0$, in other words, only manifolds with positive injectivity radius are of interest for us. Compact C^∞ manaifolds have always a positive injectivity radius, see [Kli: p. 131]. On the other hand, non-compact manifolds need not to have a positive injectivity radius But there are distinguished large classes of manifolds having a positive injectivity radius. We refer especially to manifolds with negative curvature (more precisely: non-positive sectional curvatures, see [Hel: p. 72]). In this case we have $r_0 = \infty$. In particular if M is a simply connected complete Riemannian manifold with negative curvature then $\exp_P$ is for any $P \in M$ a diffeomorphism from $\mathbb{R}^n \sim T_P M$ onto M, see [Hel: p. 74]. A detailed discussion of these so-called Cartan–Hadamard manifolds may be found in [GrW]. As a special case, $\exp_P$ is a diffeomorphism from $\mathbb{R}^n$ onto M if M is a symmetric Riemannian manifold of non-compact type, see [Kli: p. 152]. An example of a manifold with $r_0 = 0$ may be found in [Kli: p. 131].

Bounded geometry. Besides the assumption that M has a positive injectivity radius we supose that M has a bounded geometry. Let $0 < r < r_0$ where again r_0 stands for the (positive) injectivity radius. Then M is said to have a bounded geometry if there exists a positive number c and, for every multi-index α, positive numbers c_α with

$$\det g \geqq c, \qquad |D^\alpha g_{ij}| \leqq c_\alpha, \tag{8}$$

in normal geodesic coordinates of every local chart $(\Omega_P(r), \exp_P^{-1})$ with $P \in M$ in the sense of (6) where $g = (g_{jk})_{j,k}^{1,\dots,n}$ is the metric tensor interpreted as a matrix. (Again we identified g with $\exp_P^* g$.) To discuss this notation somewhat we assume that M is a complete Riemannian C^∞ manifold with positive injectivity radius. Then M is of bounded geometry in the above sense if and only if all covariant derivatives of the curvature tensor are bounded, i.e., all the scalars

$$g^{\alpha_1\beta_1} \cdots g^{\alpha_k\beta_k} \nabla_{\beta_1} \cdots \nabla_{\beta_k} R^{rstu} \nabla_{\alpha_1} \cdots \nabla_{\alpha_k} R_{rstu}; \qquad k \in \mathbb{N}_0, \tag{9}$$

are bounded C^∞ functions on M (recall that we use extensively the summation convention). Here R_{rstu} stands for the components of the curvature tensor and ∇_α indicates the covariant derivatives. One direction is obvious: If (8) is fulfilled then the C^∞ functions in (9) are bounded. As far as the non-trivial opposite direction is concerned we refer to [CGT: p. 33], [Aub1: in particular Lemma 3], [Loh: Theorem 3], [Mic3], and [Pet: 1, §7, p. 35]. In differential geometry one often prefers to express geometric quantities in terms of sectional curvature. Of course the boundedness of the C^∞ functions in (9) includes a corresponding assertion with respect to the sectional curvature instead of the curvature. However the opposite assertion is also correct. This follows from the fact that the Riemannian curvature can be recovered from the sectional curvature, see [BiG: p. 254]. We summarize our assumptions about the manifold.

Hypotheses (about the manifold). M is a connected n-dimensional complete Riemannian manifold with positive injectivity radius and bounded geometry: If $0 < r < r_0$, where r_0 stands for the injectivity radius, then (8) holds in the distinguished local charts $(\Omega_P(r), \exp_P^{-1})$ of normal geodesic coordinates for every $P \in M$ and every multi-index α.

Remark 1. In 7.2 – 7.5 we always assume that M satisfies these Hypotheses with the exception of 7.4.7 where we suppose that the above manifold is not only C^∞ but analytic.

Proposition. Let M be a manifold satisfying the above Hypotheses. If $\delta > 0$ is small then there exist a uniformly locally finite covering of M by a sequence of open balls $\Omega_{P_j}(\delta)$ and a corresponding smooth resolution of unity $\{\psi_j\}$ where the ψ_j's are C^∞ functions on M with $\operatorname{supp} \psi_j \subset \Omega_{P_j}(\delta)$.

Explanation and Proof. $\Omega_{P_j}(\delta)$ stands for the geodesic ball centered at $P_j \in M$ and of radius δ, see (6). "Uniformly locally finite" means that there exists a $L \in \mathbb{N}$ such that any fixed ball $\Omega_{P_j}(\delta)$ has a nonempty intersection with at most L of the remaining other balls. This part of the proposition follows essentially from the above Hypotheses and the Lemmas 2.25 and 2.26 in [Aub2], which are due to E. Calabi and T. Aubin. We assume that $\delta > 0$ is small. Then the construction of a smooth resolution of unity can be done in normal geodesic coordinates, see (6) with P_j instead of P and $0 < \delta < r < r_0$. The proof in [Aub2: Lemma 2.26] shows that one can proceed similar as in the euclidean case. In other words, there exists functions ψ_j with

$$\psi_j \in C^\infty(M), \qquad 0 \leqq \psi_j \leqq 1, \quad \sum_j \psi_j = 1 \quad \text{on } M, \tag{10}$$

$$\operatorname{supp} \psi_j \subset \Omega_{P_j}(\delta), \tag{11}$$

and for every multi-index α there exists a positive number b_α with

$$|D^\alpha(\psi_j \circ \exp_{P_j})(x)| \leqq b_\alpha, \qquad x \in \mathbb{R}^n \quad \text{with} \quad |x| \leqq r, \tag{12}$$

and all admissible j, see (6). In other words, (10–12) give the necessary explanation what is meant in the proposition by a smooth resolution of unity.

Remark 2. In the sequel we assume $\{\psi_j\} = \{\psi_j\}_{j=1}^\infty$ which refers to a noncompact manifold. But, of course, our approach covers also the case of a finite resolution of unity $\{\psi_j\} = \{\psi_j\}_{j=1}^J$ which refers to a compact manifold. Furthermore we shall always assume that $\delta > 0$ is small enough, in any case smaller than $\frac{r_0}{8}$. This gives the possibility to handle for every fixed j all functions ψ_k with $\operatorname{dist}(\operatorname{supp} \psi_j, \operatorname{supp} \psi_k) \leqq \delta$ within the same local chart, e.g., $(\Omega_{P_j}(r), \exp_{P_j}^{-1})$ with r near r_0.

7.2.2 Definition

We define the spaces $F^s_{pq}(M)$ where M is the above manifold satisfying the Hypotheses 7.2.1. We shift the definition of $B^s_{pq}(M)$ to 7.3.1. The deeper reason for this splitting lies in the rather different nature of the spaces $F^s_{pq}(M)$ on the one hand and $B^s_{pq}(M)$ on the other hand. In sharp contrast to the situation on $\mathbb{R}^n$, the spaces $F^s_{pq}(M)$ are now simpler in many respects than the spaces $B^s_{pq}(M)$ and they have priority compared with $B^s_{pq}(M)$. We return to this claim in Remark 2. As usual, $D'(M)$ stands for the collection of all complex-valued distributions on M. Furthermore, let F^s_{pq} be the spaces on $\mathbb{R}^n$ introduced in Definition 2.3.1, complemented by $F^s_{\infty\infty} = B^s_{\infty\infty}$.

Definition. Let M be a manifold satisfying Hypotheses 7.2.1 and let $\psi = \{\psi_j\}_{j=1}^\infty$ be the resolution of unity described in Proposition 7.2.1 and Remark 7.2.1/2. Let $s \in \mathbb{R}$ and let either $0 < p < \infty$, $0 < q \leqq \infty$ or $p = q = \infty$. Then

$$F^s_{pq}(M) = \Big\{ f \in D'(M) \colon \ \|f \mid F^s_{pq}(M)\|^\psi \tag{1}$$

$$= \left(\sum_{j=1}^\infty \|\psi_j f \circ \exp_{P_j} \mid F^s_{pq}\|^p \right)^{1/p} < \infty \Big\}$$

(usual modification if $p = \infty$).

Remark 1. In (1) we assume that $\psi_j f \circ \exp_{P_j}$ is extended from $|x| \leqq r$ to $\mathbb{R}^n$ by zero, see (7.2.1/12). In particular, $\psi_j f \circ \exp_{P_j} \in S'$ in any case. Hence (1) makes sense, where we have to justify that $F^s_{pq}(M)$ is independent of the chosen resolution of unity ψ. This will be done in the following subsection. Furthermore, intrinsic characterizations are desirable.

Proposition. Let $s \in \mathbb{R}$ and let either $0 < p < \infty$, $0 < q \leqq \infty$ or $p = q = \infty$. Then

$$F^s_{pq} = F^s_{pq}(M) \quad \text{with} \quad M = \mathbb{R}^n, \tag{2}$$

where F^s_{pq} must be understood in the sense of Definition 2.3.1 and the right-hand side of (2) is given by (1) (equivalent quasi-norms).

Proof. Step 1. We note that $\exp_{P_j}$ is a translation in $\mathbb{R}^n$ which maps 0 in $P_j \in \mathbb{R}^n$. Let $f \in F^s_{pq}$. Then $f \in F^s_{pq}(M)$ in the sense of (1) follows from an obvious modification of Theorem 2.4.7(i) and its proof, see also Remark 2.4.7/2.

Step 2. Let $f \in F^s_{pq}(M)$ in the sense of (1), where again $\exp_{P_j}$ is the above translation. Let φ be a C^∞ function in $\mathbb{R}^n$ with $\varphi(x) = 1$ if $|x| \leqq 1$ and $\varphi(y) = 0$ if $|y| \geqq 2$. By Theorems 4.2.2 and 2.4.7 we have $\varphi(2^{-j}\cdot)f \in F^s_{pq}$ for any $j \in \mathbb{N}$ and

$$\sup_j \|\varphi(2^{-j}\cdot)f \mid F^s_{pq}\| \leqq c\|f \mid F^s_{pq}\|^\psi \tag{3}$$

for some $c > 0$. Then it follows easily $\varphi(2^{-j}\cdot)f \to f$ in S' if $j \to \infty$, convergence in S', see Remark 2.3.2/2. Finally a Fatou argument applied to (2.3.1/7,8) yields $f \in F^s_{pq}$, including the desired estimate for the involved quasi-norms.

Remark 2. The Proposition justifies the Definition. A similar definition of corresponding spaces $B^s_{pq}(M)$ in the sense of (1) is not possible, see Theorem 2.4.7(ii). We shall introduce the spaces $B^s_{pq}(M)$ in 7.3 via real interpolation of the F-spaces. In other words, it is natural to give priority to $F^s_{pq}(M)$ in comparison with $B^s_{pq}(M)$.

7.2.3 Basic properties

All notations have the same meaning as in the preceding subsection.

Theorem. Let $s \in \mathbb{R}$ and let either $0 < p < \infty$, $0 < q \leqq \infty$ or $p = q = \infty$. Then $F^s_{pq}(M)$ from Definition 7.2.2 is a quasi-Banach space (Banach space if $p \geqq 1$, $q \geqq 1$). It is independent of the chosen local charts $\{(\Omega_{P_j}(r), \exp^{-1}_{P_j})\}$ and the corresponding resolution of unity $\psi = \{\psi_j\}$.

Proof. Step 1. The proof of the independence is based on pointwise multiplier assertions and diffeomorphism properties for the spaces F^s_{pq}, see Theorem 4.2.2 and 4.3.2, respectively. Let $\psi = \{\psi_j\}^\infty_{j=1}$ and $\varphi = \{\varphi_k\}^\infty_{k=1}$ be two admissible resolutions of unity in the sense of Proposition 7.2.1 and Remark 7.2.1/2, and let $(\Omega_{P_j}(r), \exp^{-1}_{P_j})$ and $(\Omega_{Q_k}(t), \exp^{-1}_{Q_k})$ be the respective local charts. If k is given then $K(k)$ collects all j's for which $\operatorname{supp}\psi_j$ and $\operatorname{supp}\varphi_k$ have a non-empty intersection. Here card $K(k)$ can be estimated from above by a natural number which is independent of k. By our assumptions ψ_j and φ_k with $j \in K(k)$ can be treated within the same local chart, e.g., $(\Omega_{Q_k}(t), \exp^{-1}_{Q_k})$. We assume $f \in F^s_{pq}(M)$ in the sense of Definition 7.2.2, i.e., with respect to $(\Omega_{P_j}(r), \exp^{-1}_{P_j})$ and ψ_j. By the mentioned diffeomorphism property of the spaces F^s_{pq} we have

$$\|\psi_j f \circ \exp_{Q_k} \mid F^s_{pq}\| \leqq c\|\psi_j f \circ \exp_{P_j} \mid F^s_{pq}\|, \qquad j \in K(k). \tag{1}$$

By our assumptions about the resolutions of unity $\{\psi_j\}$ and $\{\varphi_k\}$, and by the properties of normal geodesic coordinates, see (7.2.1/3,8), it follows that c in (1) can be chosen independently of k and $j \in K(k)$, the diffeomorphic map in question is given by $\exp^{-1}_{P_j} \circ \exp_{Q_k}$ in a small neighbourhood of the origin. In particular $\psi_j f \circ \exp_{Q_k} \in F^s_{pq}$ for all $k \in \mathbb{N}$ and $j \in K(k)$. Then we have by the indicated pointwise multiplier property

$$\|\varphi_k f \circ \exp_{Q_k} \mid F^s_{pq}\| \leqq c \sum_{j\in K(k)} \|\psi_j f \circ \exp_{Q_k} \mid F^s_{pq}\|, \tag{2}$$

where c is independent of k. However (1) and (2) prove

$$\left(\sum_{k=1}^{\infty} \|\varphi_k f \circ \exp_{Q_k} \mid F^s_{pq}\|^p\right)^{1/p} \leqq c \left(\sum_{j=1}^{\infty} \|\psi_j f \circ \exp_{P_j} \mid F^s_{pq}\|^p\right)^{1/p}. \tag{3}$$

In other words, the definition of $F^s_{pq}(M)$ in (7.2.2/1) is independent of the chosen resolution of unity.

Step 2. We prove the completeness of the spaces $F^s_{pq}(M)$. Let $\{\psi_j\}$ be a fixed resolution of unity in the above sense. Let $J(j)$ be the collection of all k such that $\operatorname{supp} \psi_k$ and $\operatorname{supp} \psi_j$ have a non-empty intersection. Let $\lambda_j = (\sum_{k\in J(j)} \psi_k) \circ \exp_{P_j}$. Furthermore $l_p(F^s_{pq})$ has the usual meaning, in particular it is a complete quasi-Banach space with respect to the quasi-norm

$$\|\{f_l\} \mid l_p(F^s_{pq})\| = \left(\sum_{l=1}^{\infty} \|f_l \mid F^s_{pq}\|^p\right)^{1/p} \tag{4}$$

(usual modification if $p = \infty$). By the pointwise multiplier property and the diffeomorphism property it follows that the operator Λ, given by

$$\Lambda\{f_l\} = \sum_{j=1}^{\infty} \lambda_j f_j \circ \exp^{-1}_{P_j}, \tag{5}$$

is a linear and bounded operator from $l_p(F^s_{pq})$ into $F^s_{pq}(M)$. (Of course, $\lambda_j f_j \circ \exp^{-1}_{P_j}$ is extended outside of $\Omega_{P_j}(r)$ by zero). Furthermore, Ψ given by

$$\Psi f = \{\psi_j f \circ \exp_{P_j}\}_{j=1}^{\infty}, \tag{6}$$

is a linear and bounded operator from $F^s_{pq}(M)$ into $l_p(F^s_{pq})$. We have

$$\Lambda\Psi = \mathrm{id} \qquad (\text{identity in } F^s_{pq}(M)). \tag{7}$$

Let $\{f^k\}$ be a fundamental sequence in $F^s_{pq}(M)$. Then $\{\Psi f^k\}$ is a fundamental sequence in $l_p(F^s_{pq})$. Since $l_p(F^s_{pq})$ is a quasi-Banach space this fundamental sequence converges to a limit element denoted by f. We have $\Lambda f \in F^s_{pq}(M)$ and

$$\|\Lambda f - f^k \mid F^s_{pq}(M)\| = \|\Lambda f - \Lambda\Psi f^k \mid F^s_{pq}\| \leqq c\|f - \Psi f^k \mid l_p(F^s_{pq})\|. \tag{8}$$

Hence, Λf is the desired limit element and $F^s_{pq}(M)$ is complete.

7.2.4 Intrinsic local means and the main theorem

Definition 7.2.2 reduces spaces on M to spaces on $\mathbb{R}^n$. This is convenient for proofs, but, of course, invariant intrinsic descriptions of F^s_{pq} are desirable. For this purpose we introduce invariant local means, the Riemannian generalizations of the euclidean local means (2.4.6/1). One of the main goals of this chapter is the proof of the Riemannian counterpart of Theorem 2.4.6. We formulate a corresponding theorem for $F^s_{pq}(M)$ in this subsection and give a proof for large values of s. In order to extend the proof to all values $s \in \mathbb{R}$ we need some mapping properties of the Laplace–Beltrami operator on M. This will be done in the following subsection. Armed with these results we complete the proof of our main theorem in 7.2.6.

Basic functions. First we recall some notations introduced in 2.4.6 which we modify slightly. Let again $B = \{y \in \mathbb{R}^n \colon |y| < 1\}$ be the unit ball in $\mathbb{R}^n$, and let K_0 and K^0 be two C^∞ functions in $\mathbb{R}^n$ with

$$\operatorname{supp} K_0 \subset B, \qquad \operatorname{supp} K^0 \subset B, \tag{1}$$

and

$$K_0^\wedge(y) \neq 0 \quad \text{for all} \quad y \in \mathbb{R}^n, \quad K^{0\wedge}(0) \neq 0. \tag{2}$$

We put again

$$K(y) = \left(\sum_{j=1}^n \frac{\partial^2}{\partial y_j^2} \right)^N K^0(y), \qquad y \in \mathbb{R}^n, \tag{3}$$

where $N \in \mathbb{N}$ is at our disposal. (1–3) is the counterpart of (2.4.6/2–4), where it is convenient for us to strengthen (2.4.6/3) somewhat. Furthermore it will be useful to assume that K_0 and K^0, and hence also K, are rotational invariant, in other words

$$K_0(x) = k_0(|x|), \quad K^0(x) = k^0(|x|), \qquad x \in \mathbb{R}^n, \tag{4}$$

and hence

$$K(x) = k(|x|) \quad \text{with} \quad k(\rho) = \left(\frac{d^2}{d\rho^2} + \frac{n-1}{\rho}\frac{d}{d\rho} \right)^N k^0(\rho) \tag{5}$$

with $x \in \mathbb{R}^n$ and $\rho \in \mathbb{R}$. There exist functions with the properties (1–5). Let, for instance, κ be a non-trivial rotational invariant real-valued C^∞ function in $\mathbb{R}^n$ with a support near the origin, then $K_0(x) = e^{-|x|^2}(\kappa * \kappa)(x)$ has the desired properties.

Local means. Now we introduce the counterpart of (2.4.6/1). Let (Ω, φ) be a local chart, $P \in \Omega$, and $U = \varphi(\Omega)$, then $T_{\varphi(P)}U$ has a natural $\mathbb{R}^n$-structure via the representation $\varphi_* X = a^i \partial_i$ with $X \in T_P M$. Of course, M is the same manifold as in Hypotheses 7.2.1. Let $d\varphi_* X$ be the usual euclidean volume element, taken with respect to the components a^i interpreted as cartesian coordinates. By standard arguments, see, e.g., [Aub2: 1.74] it follows that $\sqrt{\det g_{\varphi(P)}}\, d\varphi_* X$ has an invariant meaning which allows us to introduce on that way the invariant volume element dX on $T_P M$. The geodesic $c(P, X, t)$ has the same meaning as in 7.2.1, recall $C(\varphi(P), \varphi_* X, t) = \varphi(c(P, X, t))$. Then the counterpart of (2.4.6/1) is given by

$$k(t, f)(P) = \int_{T_P M} k(\|X\|) f(c(P, X, t)) dX \tag{6}$$

$$= \int_{T_{\varphi(P)}U} k(\|\varphi_* X\|)(f \circ \varphi^{-1})(C(\varphi(P), \varphi_* X, t)) \sqrt{|\det g_{\varphi(P)}|}\, d\varphi_* X,$$

where the latter expression is the definition of the former one, $t > 0$ small, and the integrand is extended outside of the unit ball in $T_{\varphi(P)}U$ by zero. We have to check that these expressions have an invariant meaning and that they make sense for any $f \in D'(M)$. Let $(\tilde{\Omega}, \tilde{\varphi})$ be a second local chart with $P \in \Omega \cap \tilde{\Omega}$. Recall $\|X\| = \|\varphi_* X\| = \|\tilde{\varphi}_* X\|$ if $X \in T_P M$ and $\sigma = t\|X\|$ where σ is the arc length of the geodesic $c(P, X, t)$. Then we have

$$(f \circ \varphi^{-1})(C(\varphi(P), \varphi_* X, t)) = f \circ \tilde{\varphi}^{-1}(C(\tilde{\varphi}(P), \tilde{\varphi}_* X, t)) = f(c(P, X, t)) \tag{7}$$

with $t > 0$ small. Together with the above remarks about the invariant volume element dX it follows that the last expression (6) is independent of the chosen local chart (Ω, φ). In particular in normal geodesic coordinates with the local chart $(\Omega_P(r), \exp_P^{-1})$ we have

$$k(t, f)(P) = \int_{T_0 U} k(\|\exp_{P*}^{-1} X\|)(f \circ \exp_P)(t \exp_{P*}^{-1} X) d \exp_{P*}^{-1} X. \tag{8}$$

This shows that $k(t, f)(P)$ makes sense for every t with $0 < t < r_0$, where r_0 again stands for the injectivity radius of M (as above we may assume that $f \circ \exp_P$ is extended outside of the unit ball in $T_0 U$ by zero). We again identify $T_0 U$ with $\mathbb{R}^n$, put $\exp_{P*}^{-1} X = Y$, and arrive at

$$k(t, f)(P) = \int_{\mathbb{R}^n} k(|Y|)(f \circ \exp_P)(tY) dY. \tag{9}$$

This version makes clear that (9), and hence also (6), make sense for every $f \in D'(M)$: We have $f \circ \exp_P(t, \cdot) \in D'(\mathbb{R}^n)$ after extending $f \circ \exp_P(t \cdot)$ outside of the unit ball in an appropriate way. Of course $k_0(t, f)$ is defined by (6) with k_0 instead of k. If $0 < p \leqq \infty$ then $L_p(M)$ has the usual meaning with respect to the invariant Riemannian volume element, which in local cartesian coordinates is given by $\sqrt{|\det g|}dx$, see, e.g., [Aub2: 1.74].

Theorem. Let M be a Riemannian manifold in the sense of Hypotheses 7.2.1 with the injectivity radius $r_0 > 0$. Let $s \in \mathbb{R}$ and let either $0 < p < \infty$, $0 < q \leqq \infty$ or $p = q = \infty$.

(i) Let $0 < r < r_0$, let $\varepsilon > 0$ be sufficiently small and $N \in \mathbb{N}$ in (3,5,6) be sufficiently large (in dependence on s, p, q), then

$$\|k_0(\varepsilon, f) \mid L_p(M)\| + \left\| \left(\int_0^r t^{-sq} |k(t,f)(\cdot)|^q \frac{dt}{t} \right)^{1/q} \mid L_p(M) \right\| \tag{10}$$

(modification if $q = \infty$) is an equivalent quasi-norm in $F^s_{pq}(M)$.

(ii) Let $\psi = \{\psi_j\}_{j=1}^\infty$ be the resolution of unity referred to in Definition 7.2.2 and let $\|f \mid F^s_{pq}(M)\|$ an (equivalent) quasi-norm in $F^s_{pq}(M)$, then

$$\left(\sum_{j=1}^\infty \|\psi_j f \mid F^s_{pq}(M)\|^p \right)^{1/p} \tag{11}$$

(modification if $p = \infty$) is also an equivalent quasi-norm in $F^s_{pq}(M)$.

Proof (for large values of s). Step 1 (Preliminaries). We prove the theorem for $0 < p < \infty$, $0 < q \leqq \infty$, and $s > \frac{2n}{p} + 4$. By slight technical modifications the proof can be extended to $p = q = \infty$ and $s > 4$. The proof for the remaining values of s will be shifted to 7.2.6. Let $s > \frac{2n}{p} + 4$ and $L \in \mathbb{N}$ with $L = s - \frac{n}{p} - \kappa$, $1 < \kappa \leqq 2$, then

$$L + \frac{n}{p} < s - 1 < s < 2s - \frac{2n}{p} - 4 \leqq 2L. \tag{12}$$

This number L will play a crucial role in the sequel. Besides this preparatory step we break the proof of the theorem in three steps. For the time being we abbreviate (10) by $\|f \mid F^s_{pq}(M)\|$. In Step 2 we prove

$$\|f \mid F^s_{pq}(M)\|^p \leqq c \sum_{j=1}^\infty \|\psi_j f \mid F^s_{pq}(M)\|^p, \qquad f \in F^s_{pq}(M). \tag{13}$$

The heart of the proof is to check

$$\|\psi_j f \mid F^s_{pq}(M)\| \sim \|(\psi_j f) \circ \exp_{P_j} \mid F^s_{pq}\|, \qquad f \in F^s_{pq}(M), \tag{14}$$

see (7.2.2/1), where this equivalence must be understood independently of $j \in \mathbb{N}$. This will be done in Step 3. Finally in Step 4 we prove

$$\sum_{j=1}^{\infty} \|\psi_j f \mid F^s_{pq}(M)\|^p \leqq c\|f \mid F^s_{pq}(M)\|^p, \qquad f \in F^s_{pq}(M). \tag{15}$$

All assertions of the theorem follows from (13–15) and Definition 7.2.2.

Step 2 (Proof of (13)). Recall that $\psi = \{\psi_j\}_{j=1}^{\infty}$ has the same meaning as in Proposition 7.2.1 and Remark 7.2.1/2, in particular we always assume that $\delta > 0$ is small,

$$\delta + r < r' < r_0. \tag{16}$$

Then (13) follows easily from the local character of (6).

Step 3 (Proof of (14)). Let $h = \psi_j f \circ \exp_{P_j}$ and $U = B(r')$, see (7.2.1/6), with (16). In order to calculate $k_0(\varepsilon, \psi_j f)(P)$ and $k(t, \psi_j f)(P)$ with $\varepsilon < r$ and $t < r$ we may assume $P \in \Omega_{P_j}(r') = \exp_{P_j} U$, see (7.2.1/6), because these means are zero for all other points $P \in M$. Let $P \in \Omega_{P_j}(r')$, then only those geodesics $c(P, X, t)$ with $0 < t < r$, $X \in T_P M$, $\|X\| \leqq 1$, are of interest which have a non-empty intersection with $\operatorname{supp} \psi_j$, and which consequently lie completely in $\Omega_{P_j}(r')$, see (16). Hence the proof of (14) can be reduced to the proof of

$$\|h \mid F^s_{pq}\| \sim \|k_0(\varepsilon, h) \mid L_p(U)\| + \left\| \left(\int_0^r t^{-sq} |k(t,h)(\cdot)|^q \frac{dt}{t} \right)^{1/q} \mid L_p(U) \right\| \tag{17}$$

with $h(x) = 0$ if $|x| > \delta$. In the underlying local chart $(\Omega, \varphi) = (\Omega_{P_j}(r'), \exp^{-1}_{P_j})$ we may assume that $k(t,h)$ is given by

$$k(t,h)(x) = \int_{\mathbb{R}^n} k(|X|) h(C(x,X,t)) dX, \qquad x \in U, \tag{18}$$

$|X| \leqq 1$, $0 < t < r$, and $C(x, X, t) \subset U$. We eliminated an uninteresting factor $c(x)$ with $0 < c_1 \leqq c(x) \leqq c_2$ coming from $\sqrt{|\det g_{\varphi(P)}|}$ in (6). Similar for $k_0(\varepsilon, h)(x)$. For the components C^a of C we have

$$C^a(x,X,t) = x^a + tX^a + \sum_{l=2}^{2L-1} \frac{t^l}{l!} \frac{d^l C^a}{dt^l}(x,X,0) + \frac{t^{2L}}{(2L)!} \frac{d^{2L} C^a}{dt^{2L}}(x,X,\vartheta^a t) \tag{19}$$

with $0 \leqq \vartheta^a \leqq 1$, where $L \in \mathbb{N}$ has the same meaning as in Step 1. By (7.2.1/2,3,8) we have

$$C^a(x,X,t) = x^a + tX^a + \sum_{2 \leqq |\alpha| \leqq 2L-1} t^{|\alpha|} b^a(x) X^\alpha + t^{2L} R^a_{2L}(x,X,t), \tag{20}$$

$a = 1, \ldots, n$ and $0 < t < r$. Here the $b^a_\alpha(x)$'s are uniformly bounded C^∞-functions on U (with respect to j). Furthermore because (7.2.1/4) is constant along a geodesic we have for some $c > 0$

$$c \sum_{k=1}^{n} \left| \frac{dC^k}{dt} \right|^2 \leqq g_{ij} \frac{dC^i}{dt} \frac{dC^j}{dt} = |X|^2 \leqq 1 \tag{21}$$

along the geodesic $C(x, X, t)$ and hence $|R^a_{2L}(x, X, t)| \leqq c$ for all $x \in U$, $0 < t < r$, $|X| \leqq 1$. In particular with $b_\alpha(x) = (b^a_\alpha(x))$ we arrive at

$$\begin{aligned} \Big| h(C(x, X, t)) - h\Big(x + tX + \sum_{2 \leqq |\alpha| \leqq 2L-1} t^{|\alpha|} b_\alpha(x) X^\alpha\Big)\Big| \qquad \\ \leqq c t^{2L} \sum_{a=1}^{n} \left| \frac{\partial h}{\partial x^a}(\cdots) \right|. \end{aligned} \tag{22}$$

We insert (20) in (18) and use (22). The term which comes from the right-hand side of (22) can be estimated from above by

$$c t^{2L} \sum_{a=1}^{n} \sup_{|x-y| \leqq c'} \left| \frac{\partial h}{\partial y^a}(y) \right|, \tag{23}$$

where c and c' are independent of t, X, and x. Recall $2L > s$, see (12). We insert (18) in the second term on the right-hand side of (17) and use (22,23). Then the term which results from (23) can be estimated from above by

$$c \left\| \sup_{|x-y| \leqq c'} \left| \frac{\partial h}{\partial y^a}(y) \right| \mid L_p(U) \right\| \leqq c'' \|h \mid F^\sigma_{pq}\|, \tag{24}$$

with $\sigma > 1 + \frac{n}{p}$, where the latter is a special case of (2.4.8/4) with $a(\eta, \xi)$ identically 1. Next we expand

$$\begin{aligned} & h\left(x + tX + \sum_{2 \leqq |\alpha| \leqq 2L-1} t^{|\alpha|} b_\alpha(x) X^\alpha\right) \\ & \quad = \sum_{0 \leqq |\beta| \leqq L-1} D^\beta h(x + tX) \Big(\sum_{2 \leqq |\alpha| \leqq 2L-1} t^{|\alpha|} b_\alpha(x) X^\alpha \Big) \frac{1}{\beta!} \\ & \qquad + \sum_{|\beta| = L} D^\beta h(\cdots) \left(\sum_{2 \leqq |\alpha| \leqq 2L-1} \cdots \right)^\beta \frac{1}{\beta!}. \end{aligned} \tag{25}$$

By $s > \frac{n}{p} + L$ and $h \in F^s_{pq}$ it follows that h has classical derivatives up to order L, see [Triß: 2.7.1] for details. Hence, (25), and also (24), make sense. We return to (18), now with the left-hand side of (25) instead of $h(C(x, X, t))$. We replace the left-hand side of (25) by its right-hand side and put the result in the second term on the right-hand side of (17). The resulting term which comes from the last expression in (25) can be estimated from above by

$$c \sum_{|\beta|=L} \left\| \sup_{|x-y| \leqq c'} |(D^\beta h)(y)| \mid L_p(U) \right\| \leqq c \|h \mid F^\kappa_{pq}\| \tag{26}$$

with $\kappa > L + \frac{n}{p}$, where we used again $2L > s$ and the above-mentioned special case of (2.4.8/4). In other words after estimating remainder terms we may concentrate ourselves on (18) with the first sum on the right-hand side of (25) instead of $h(C(x, X, t))$. The term with $|\beta| = 0$, inserted in the second term on the right-hand side of (17) via (18) is just what we want to have, see (14) and (2.4.6/1). Then there remain the terms on the right-hand side of (25) with $1 \leqq |\beta| \leqq L - 1$, which yield means of the type

$$\int_{\mathbb{R}^n} k(|X|) t^{|\alpha|} X^\alpha (D^\beta h)(x + tX) dX, \tag{27}$$

$$1 \leqq |\beta| \leqq L - 1, \quad 2|\beta| \leqq |\alpha| \leqq (2L-1)|\beta|,$$

multiplied with smooth bounded functions depending only on x. There are two cases. First if $|\alpha| > s$ in (27) then the corresponding expressions in the second term on the right-hand side of (17) can again be estimated as in (26), now with $|\beta| < L$, but with the same κ. Secondly, let $|\alpha| \leqq s$. The kernels $k(|X|)$ are given by (3,5). Now (27) are $\mathbb{R}^n$-means in the sense of (2.4.6/1) and the rotational symmetry is no longer of interest. As the proof of Theorem 2.4.6 shows the only property we really need in the $\mathbb{R}^n$-context is $|\hat{k}(X)| \leqq c|X|^{2N}$ near the origin. Then (2.4.6/7) can be estimated from above by $\|f \mid F^s_{pq}\|$. We apply this observation to the kernels $k(|X|)X^\alpha$ and $F^{s-|\alpha|}_{pq}$. Recall $s > \frac{2n}{p} + 4$ and choose $N > s + n(\frac{1}{p} - 1)_+$. By these remarks and

$$N - |\alpha| > s - |\alpha| + n(\frac{1}{p} - 1)_+ \geqq \max(s - |\alpha|, n(\frac{1}{p} - 1)_+)$$

it follows from Theorem 2.4.6 that the terms in (27), inserted in the second term on the right-hand side of (17), can be estimated from above by

$$c\|D^\beta h \mid F^{s-|\alpha|}_{pq}\| \leqq c'\|h \mid F^{s-1}_{pq}\|. \tag{28}$$

To distinguish the Riemannian means (6) and their euclidean counterparts (2.4.6/1) we denote the latter ones temporarily by $k^{euc}(t, f)(x)$. By (22, 24, 25, 26 extended to $|\beta| \leqq L$, 28) we arrive at

$$\left\| \left(\int_0^r t^{-sq} |k(t, h)(\cdot) - k^{euc}(t, h)(\cdot)|^q \frac{dt}{t} \right)^{1/q} \mid L_p(U) \right\| \leqq c\|h \mid F^{s-1}_{pq}\|, \tag{29}$$

where we used $s - 1 > L + \frac{n}{p}$, see (12), which gives us the possibility to choose $\kappa = s - 1$ in (26). The number c in (29) can be estimated by $c'L$ and hence by $c''s$, where c'' is independent of s. In the same way one obtains

$$\|k_0(\varepsilon,h) - k_0^{euc}(\varepsilon,h) \mid L_p(U)\| \leqq c\,\varepsilon \|h \mid F^\sigma_{pq}\| \tag{30}$$

with $0 < \varepsilon < 1$, where c is independent of ε, and $\sigma > 1 + \frac{n}{p}$: For this purpose it is sufficient to use (22) with $L = 1$ and $t = \varepsilon$, and (24). By (29,30) and Theorem 2.4.6 we arrive at

$$\|\psi_j f \mid F^s_{pq}(M)\| \leqq c\|h \mid F^s_{pq}\|. \tag{31}$$

Furthermore by (29,30) we have

$$\|h \mid F^s_{pq}\| \leqq c\|\psi_j f \mid F^s_{pq}(M)\| + c'\|h \mid F^{s-1}_{pq}\| \tag{32}$$

$$\leqq c\|\psi_j f \mid F^s_{pq}(M)\| + \frac{1}{2}\|h \mid F^s_{pq}\| + c'' \sum_{b=0}^{B} 2^{bs}\|(\varphi_b \hat{h})^\vee \mid L_p\|,$$

see 2.3.1 where B is an appropriate natural number with $2^B \sim s$. Furthemore $c'' \sim s$. By our assumptions about k_0, see (2,4), it is possible to prove

$$\|\varphi_b \hat{h})^\vee \mid L_p\| \leqq \eta\|h \mid F^s_{pq}\| + c_\eta \|k_0^{euc}(\varepsilon,h) \mid L_p\|, \tag{33}$$

where $\eta > 0$ is at our disposal and c_η is independent of s and ε. We prove this claim in Remark 3 below. Now (30, 32, 33) prove

$$\|h \mid F^s_{pq}\| \leqq c\|\psi_j f \mid F^s_{pq}(M)\|, \tag{34}$$

provided that $\varepsilon > 0$ is chosen sufficiently small. But (31, 34) complete the proof of (14).

Step 4. (Proof of (15)). We fix j. As in Step 3 all calculations can be done within the local chart $(\Omega_{P_j}(r'), \exp^{-1}_{P_j})$. Again we can identify $\|\psi_j f \mid F^s_{pq}(M)\|$ with the right-hand side of (17) where $k(t,h)$ is given by (18) and

$$h(C(x,X,t)) = \psi_j \circ \exp_{P_j}(C(x,X,t)) f \circ \exp_{P_j}(C(x,X,t)). \tag{35}$$

We use (20), but in contrast to Step 2 restricted to the factor $\psi(C(x,X,t))$ with $\psi = \psi_j \circ \exp_{P_j}$. We have (22) with ψ instead of h, where we now incorporate tX in the sum $\sum_{2\leqq|\alpha|\leqq 2L-1}$, which we write as

$$\sum_{1\leqq|\alpha|\leqq 2L-1} t^{|\alpha|} b_\alpha(x) X^\alpha.$$

Similarly we replace $x+tX$ on the right-hand side of (25) (with ψ instead of h) by x. The term of interest is again that one with $\beta = 0$ which in this modified formula is simply $\psi(x)$. Now we put (35) with $\psi_j \circ \exp_{P_j}(x)$ instead of $\psi_j \circ \exp_{P_j}(C(x,X,t))$ in the second term on the right-hand side of (17). The resulting term can be estimated from above by

$$c\left\|\left(\int_0^r t^{-sq}|k(t,f\circ\exp_{P_j})(\cdot)|^q\frac{dt}{t}\right)^{1/q} \mid L_p(\operatorname{supp}(\psi_j\circ\exp_{P_j}))\right\|, \tag{36}$$

see (6). Summation of the pth power of (36) with respect to j and the properties of the resolution of unity $\{\psi_j\}$ yield the right-hand side of (15). The remaining terms come from the above-mentioned counterpart of (22,25) with $|\beta|>0$. In (35) with

$$\psi_j\circ\exp_{P_j}(C(x,X,t))-\psi_j\circ\exp_{P_j}(x) \quad \text{instead of } \psi_j\circ\exp_{P_j}(C(x,X,t))$$

we may replace $f\circ\exp_{P_j}(C(x,X,t))$ by $\tilde{\psi}_j f\circ\exp_{P_j}(C(x,X,t))$ with $\tilde{\psi}_j=\sum\psi_l$ where the sum is taken over all l with $\Omega_{P_j}(r')\cap\operatorname{supp}\psi_l\neq\emptyset$. This replacement is justified by $\psi_j=\psi_j\tilde{\psi}_j$ and

$$\psi_j\circ\exp_{P_j}(x)=\psi_j\circ\exp_{P_j}(x)\tilde{\psi}_j(C(x,X,t))$$

for all x, X, t of interest. Furthermore the number of elements in $\sum\psi_l$ can be estimated from above by a natural number which is independent of j. Now we treat $(\tilde{\psi}_j f)\circ\exp_{P_j}$ in the same way as in the previous step. Together with the remaining expressions and factors from the modified expansion (25) with ψ instead of h it follows that all the other terms can be estimated from above by

$$\eta\|\tilde{\psi}_j f \mid F^s_{pq}(M)\| + c_\eta\|k_0(\varepsilon,\tilde{\psi}_j f\circ\exp_{P_j}) \mid L_p(U)\|, \tag{37}$$

where $\eta>0$ is at our disposal, $\varepsilon>0$ is small and c_η is independent of ε. We use (20) with $t=\varepsilon$ in order to expand $\tilde{\psi}_j(C(x,X,\varepsilon))$ in the second summand in (37). We obtain

$$\|k_0(\varepsilon,f\circ\exp_{P_j}) \mid L_P(\operatorname{supp}\tilde{\psi}_j\circ\exp_{P_j})\| \tag{38}$$

and remainder terms with a factor ε in front which can be treated as above. We choose $\varepsilon>0$ and $\eta>0$ in (37) small. Summation over j yields

$$\sum_{j=1}^\infty\|\psi_j f \mid F^s_{pq}(M)\|^p \leqq c\|f \mid F^s_{pq}(M)\|^p + \frac{1}{2}\sum_{j=1}^\infty\|\psi_j f \mid F^s_{pq}(M)\|^p. \tag{39}$$

The proof of (15) is complete.

Remark 1. We proved the theorem under the restriction $s>\frac{2n}{p}+4$ and $N>s+\sigma_p$ with $\sigma_p=n(\frac{1}{p}-1)_+$, see the arguments between (27) and (28). In the euclidean case we have the natural restriction $2N>s$, see (2.4.6/5). One can ask how small ε must be chosen. We are not going to discuss this point and refer to the underlying paper [Tri13: Remark 20].

Corollary. Let M be a Riemannian manifold in the sense of Hypotheses 7.2.1 with the injectivity radius $r_0 > 0$. Let either $0 < p < \infty$, $0 < q \leqq \infty$, or $p = q = \infty$. Let $s > \frac{2n}{p} + 4$, $0 < r < r_0$, and $N \in \mathbb{N}$ with $N > s + \sigma_p$ in (3,5,6) then

$$\|f \mid L_p(M)\| + \left\| \left(\int_0^r t^{-sq} |k(t,f)(\cdot)|^q \frac{dt}{t} \right)^{1/q} \mid L_p(M) \right\| \tag{40}$$

(modification if $q = \infty$) is an equivalent quasi-norm in $F_{pq}^s(M)$.

Proof. Step 1. Let $\kappa > \sigma_p = n(\frac{1}{p} - 1)_+$. We prove that

$$\|f \mid L_p\| + \left\| \left(\int_0^\lambda t^{-\kappa q} |k^{euc}(t,f)(\cdot)|^q \frac{dt}{t} \right)^{1/q} \mid L_p \right\| \tag{41}$$

is an equivalent quasi-norm in F_{pq}^κ, where the numbers p and q and the kernel k have the same meaning as in the corollary, $k^{euc}(t,f)$ stands for the euclidean means (2.4.6/1) with the same k as above and $\lambda > 0$. By the arguments in the proofs of Theorem 2.3.3 and Theorem 2.4.6 it follows that (41) can be estimated from above by $\|f \mid F_{pq}^\kappa\|$. To prove the converse assertion we replace in the proof of Theorem 2.4.6 the function $k_0^\vee$ by a compactly supported C^∞ function φ_0 with $\varphi_0(0) \neq 0$. Then we have

$$\|f \mid F_{pq}^\kappa\| \sim \|(\varphi_0 \hat{f})^\vee \mid L_p\| + \left\| \left(\int_0^\lambda t^{-\kappa q} |k^{euc}(t,f)(\cdot)|^q \frac{dt}{t} \right)^{1/q} \mid L_p \right\|. \tag{42}$$

Based on Fourier multiplier assertions for L_p-spaces of entire analytic functions one can prove

$$\|(\varphi_0 \hat{f})^\vee \mid L_p\| \leqq \varepsilon \|f \mid F_{pq}^\kappa\| + c_\varepsilon \|f \mid L_p\|, \tag{43}$$

where $\varepsilon > 0$ is at our disposal, see [Triß: 2.5.9, formula (37)] for details. Now (42) and (43) yield the desired estimate.

Step 2. Now the proof of the corollary can be done in the same way as the proof of the theorem. The counterpart of (13) is obvious. The proof of the crucial equivalence (14) is the same as in Step 3 of the proof of the theorem, now based on the equivalent quasi-norm (41) with $\kappa = s$ and on (33) with $\|h \mid L_p\|$ instead of $\|k_0^{euc}(\varepsilon, h) \mid L_p\|$. However the latter inequality is covered by (43) with $\kappa = s$. Finally, (15) can be proved in the same way as in Step 4 of the proof of the theorem.

Remark 2. Let $M = \mathbb{R}^n$, then we proved in Step 1 that (40), now coinciding with (41), is an equivalent quasi-norm in F_{pq}^s under the natural restriction $s > \sigma_p$, see also

2.3.3 in this context. The question is whether in the above corollary the unnatural restriction $s > \frac{2n}{p} + 4$ can be replaced by the natural restriction $s > \sigma_p$. We return to this problem in 7.4.6 where we give an affirmative answer.

Remark 3. As we shall see at the end of this remark inequality (33) is essentially covered by the following assertion: Let $\varphi \in S$ and $\psi \in S$, where φ has a compact support and $\psi(y) \neq 0$ for all $y \in \mathbb{R}^n$. Let either $0 < p < \infty$, $0 < q \leqq \infty$, or $p = q = \infty$ and let $s \in \mathbb{R}$, then

$$\|(\varphi\hat{f})^\vee \mid L_p\| \leqq \eta \|f \mid F^s_{pq}\| + c_\eta \|(\psi\hat{f})^\vee \mid L_p\|, \qquad f \in F^s_{pq}, \tag{44}$$

where $\eta > 0$ is at our disposal. We have

$$(\varphi\hat{f})^\vee(x) = \left(\frac{\varphi}{\psi}\psi\hat{f}\right)^\vee(x) = \int_{\mathbb{R}^n} \left(\frac{\varphi}{\psi}\right)^\vee(y)(\psi\hat{f})^\vee(x-y)dy. \tag{45}$$

Let $1 \leqq p \leqq \infty$, then (44) even with $\eta = 0$, follows from the triangle inequality. Let $0 < p < 1$, then we have

$$|(\varphi\hat{f})^\vee(x)| \leqq \left(\sup_{z\in\mathbb{R}^n} \frac{|(\psi\hat{f})^\vee(x-z)|}{1+|z|^a}\right)^{1-p} \tag{46}$$

$$\times \int_{\mathbb{R}^n} (1+|y|^a)^{1-p} \left|\left(\frac{\varphi}{\psi}\right)^\vee(y)\right| |(\psi\hat{f})^\vee(x-y)|^p dy$$

where we choose $a > \frac{n}{p}$. The first factor on the right-hand side is a maximal function in the sense of the Corollaries 2.4.1/2 and 2.5.1/2 with ψ instead of φ_0. We apply the L_p-quasi-norm to (46), use Hölder's inequality based on $p + (1-p) = 1$, and arrive at

$$\|(\varphi\hat{f})^\vee \mid L_p\| \leqq c \left\| \sup_{z\in\mathbb{R}^n} \frac{|(\psi\hat{f})^\vee(x-z)|}{1+|z|^a} \mid L_p \right\|^{1-p} \|(\psi\hat{f})^\vee \mid L_p\|^p. \tag{47}$$

Now the above-mentioned Corollaries 2.5.1/2 and 2.4.1/2 yield (44). In order to prove (33) we need (44) with $\psi(\varepsilon\cdot)$ instead of ψ. We follow the above arguments and arrive at (44) with $\psi(\varepsilon\cdot)$ instead of ψ, where c_η is independent of ε.

7.2.5 The Laplace–Beltrami operator

It is one of the main goals of this chapter to prove Theorem 7.2.4. This was done in 7.2.4 under the restriction $s > \frac{2n}{p} + 4$. The extension of this proof from $s > \frac{2n}{p} + 4$

to arbitrary values of $s \in \mathbb{R}$ will be based on lifting and mapping properties of the Laplace–Beltrami operator which we are going to discuss now. The proof of Theorem 7.2.4 will be completed afterwards in 7.2.6.

Laplace–Beltrami operator. We always assume that M is a manifold satisfying the Hypotheses 7.2.1. Recall that the Laplace–Beltrami operator Δ on M in local coordinates is given by

$$\Delta f = \frac{1}{\sqrt{\det g}}\,\partial_j\left(\sqrt{\det g}\,g^{jk}\partial_k f\right) \tag{1}$$

with $\partial_j = \frac{\partial}{\partial x_j}$. Again we do not distinguish between g on M and $\varphi^{-1*}g$ on a local chart (Ω,φ). This outstanding operator on M attracted much attention and some deeper results have been obtained only rather recently. Recall that $D(M)$ always stands for the collection of all complex-valued C^∞ functions on M with compact support. Then $-\Delta$ with $D(M)$ as its domain of definition is a positive essentially self-adjoint operator with respect to the Hilbert space $L_2(M)$, see [Str2: Sect. 2] and the references mentioned there. Let $G \in \mathbb{N}$, $\rho > 0$, and let id the identity, then

$$(\rho\,\mathrm{id} - \Delta)^G f = h, \qquad h \in D(M), \tag{2}$$

has a uniquely determined $L_2(M)$-solution f. By the well-known smoothness theory for elliptic differential operators, which is completely local and hence an $\mathbb{R}^n$-matter, it follows that f is also a C^∞ function on M. Of interest are the decay properties of f.

Proposition 1. Let M be a manifold in the sense of Hypotheses 7.2.1. Let h and f be given by (2). Let $s \in \mathbb{R}$, and let either $0 < p < \infty$, $0 < q \leqq \infty$, or $p = q = \infty$. Then there exists a non-negative number $\rho(M,p)$, which may depend on M and p, with the following property: If $\rho > \rho(M,p)$ then f belongs to $F^s_{pq}(M)$.

Proof. Step 1. We prove $f \in F^s_{22}(M)$ by mathematical induction. Let $f \in F^{s-1}_{22}(M)$ which is ensured if $s = 1$. We use (7.2.2/1) and have by well-known a priori estimates for elliptic differential operators in $\mathbb{R}^n$

$$\begin{aligned} &\|\psi_j f \circ \exp_{P_j} \mid F^s_{22}\| \qquad\qquad (3)\\ &\leqq c\left\|(\rho\,\mathrm{id} - \Delta)^G \psi_j f \circ \exp_{P_j} \mid F^{s-2G}_{22}\right\| + c\left\|\psi_j f \circ \exp_{P_j} \mid F^{s-2G}_{22}\right\| \end{aligned}$$

$$\leqq c' \left\| \psi_j h \circ \exp_{P_j} \mid F_{22}^{s-2G} \right\| + c' \sum \left\| \psi_l f \circ \exp_{P_j} \mid F_{22}^{s-1} \right\|$$

where we may assume that the sum $\sum$ is restricted to those l's with $\operatorname{supp} \psi_l \cap \operatorname{supp} \psi_j \neq \emptyset$. Hence by (7.2.2/1) we arrive at $f \in F_{22}^s$. The rest is a matter of induction beginning with $f \in L_2(M)$.

Step 2. Let $2 < p < \infty$. Recall the well-known embedding

$$F_{22}^{s_0} \subset F_{pq}^{s_1}, \qquad 0 < q \leqq \infty, \quad s_0 - \frac{n}{2} > s_1 - \frac{n}{p}, \tag{4}$$

and a corresponding assertion with $p = q = \infty$, see [Triß: 2.7.1] and $l_2 \subset l_p$. Then (7.2.2/1) and Step 1 prove $f \in F_{pq}^s(M)$ for any $s \in \mathbb{R}$, where either $2 \leqq p < \infty$, $0 < q \leqq \infty$ or $p = q = \infty$.

Step 3. To deal with the remaining cases we need some preparations. By [Str2: (4.2) – (4.4), Theorem 3.5] (with ρ instead of 1) we have

$$f(P) = (\rho \operatorname{id} - \Delta)^{-G} h(P) = \int_M J_{2G,\rho}(P,Q) h(Q) d\mu(Q), \tag{5}$$

where μ stands for the invariant volume element on M, and $J_{2G,\rho}(P,Q) \geqq 0$ on $M \times M$ with

$$\int_M J_{2G,\rho}(P,Q) d\mu(Q) \leqq c < \infty \quad \text{for all} \quad P \in M. \tag{6}$$

By Step 2 we have $f \in L_\infty(M)$. It is crucial for us that this estimate can be essentially improved. By [CGT: Corollary 3.2] we have

$$J_{2G,\rho}(P,Q) \leqq c\, e^{-\kappa R}, \qquad R = \operatorname{dist}(P,Q), \tag{7}$$

for any κ with $0 < \kappa < \sqrt{\rho}$ and $R \geqq R_0 > 0$. Since h has a compact support we arrive at

$$|f(P)| \leqq c\, e^{-\kappa R}, \qquad R = \operatorname{dist}(P, P_0), \tag{8}$$

where $P_0 \in M$ may be considered as a fixed point whereas P varies on M. Because M is a manifold with bounded geometry we have

$$\operatorname{vol}\{P \in M \colon \operatorname{dist}(P, P_0) \leqq R\} \leqq c\, e^{KR} \tag{9}$$

for some positive numbers c and K. This can be proved quite elementary, see also [CGT: (3.51)]. Let $0 < p < \infty$ and $\kappa p > K$, i.e., $\rho > \frac{K^2}{p^2}$, then (8,9) yield $f(P) \in L_p(M)$, and by (7.2.2/1) and, say, Theorem 2.4.6,

$$f \in F_{pq}^s(M), \qquad s < 0, \ \ 0 < p < \infty, \ \ 0 < q \leqq \infty. \tag{10}$$

Step 4. Now we combine (10) with Step 2, i.e.,

$$f \in F_{pq}^s(M), \quad s \in \mathbb{R}, \quad 2 \leqq p < \infty, \quad 0 < q \leqq \infty, \tag{11}$$

for any $\rho > 0$. Let $\Psi^l = \sum_{j=1}^l \psi_j$ where $\{\psi_j\}_{j=1}^\infty$ is the same resolution of unity as in Definition 7.2.2. Since f is a C^∞ function we have $\Psi^l f \in D(M)$. Let

$$s_0 < 0, \quad 0 < p_0 < \infty, \quad 0 < q_0 \leqq \infty; \tag{12}$$

$$s_1 \in \mathbb{R}, \quad 2 \leqq p_1 < \infty, \quad 0 < q_1 \leqq \infty; \tag{13}$$

and $0 < \theta < 1$,

$$s = (1-\theta)s_0 + \theta s_1, \quad \frac{1}{p} = \frac{1-\theta}{p_0} + \frac{\theta}{p_1}, \quad \frac{1}{q} = \frac{1-\theta}{q_0} + \frac{\theta}{q_1}. \tag{14}$$

Then (7.2.2/1), Theorem 2.4.6, and Hölder's inequality yield

$$\begin{aligned} \|\Psi^l f \mid F^s_{pq}(M)\| &\leqq c\|\Psi^l f \mid F^{s_0}_{p_0q_0}(M)\|^{1-\theta}\|\Psi^l f \mid F^{s_1}_{p_1q_1}(M)\|^\theta \\ &\leqq c'\|f \mid F^{s_0}_{p_0q_0}(M)\|^{1-\theta}\|f \mid F^{s_1}_{p_1q_1}(M)\|^\theta, \end{aligned} \tag{15}$$

where the second inequality comes from (10,11) and the pointwise multiplier assertion in Theorem 4.2.2. Now (7.2.2/1) and $l \to \infty$ prove $f \in F^s_{pq}(M)$. Finally we remark that (12–14) covers all possible values of s, p, q.

Remark 1. We are not interested to find out best possible values of $\rho(M,p)$. But we collect what we needed in the course of the above proof. First we remark that $\rho(M,p) = 0$ is sufficient if $2 \leqq p \leqq \infty$, see Steps 1 and 2. Secondly by Steps 3 and 4 we may choose $\rho(M,p) = \frac{K^2}{p^2}$, where K has the same meaning as in (9). If the volume in (9) is restricted by polynomial growth then it follows immediately that $\rho(M,p) = 0$ is an admissible value for all $0 < p \leqq \infty$.

Remark 2. Nowadays there exist many proofs that Δ with $D(M)$ as domain of definition is essentially self-adjoint, see [Str2], also for further references, and [Dav: Theorem 5.2.3]. The deepest result we used is hidden in the apparently innocent formula (7), the exponential decay of the resolvent kernels. There is a well-known close connection between heat kernels and resolvent kernels, see [CGT] or [Dav: 3.4], and a decay of type (7) follows from a Gaussian decay of heat kernels. This problem attracted much attention since the beginning of the eighties. We used [CGT] (1982) which is one of the earliest papers containing sufficiently strong results. Afterwards several mathematicians contributed to this field of research, see [LiY] (1986) and [Dav: in particular Chapter 5] (1989) and the references given there.

Recall that $k(t,f)$ stands for the means (7.2.4/6) with the kernels k given by (7.2.4/3,5), similarly $k_0(t,f)$.

Proposition 2. Let M be a manifold in the sense of Hypotheses 7.2.1. Let $s \in \mathbb{R}$ and let either $0 < p < \infty$, $0 < q \leqq \infty$ or $p = q = \infty$. Let $G \in \mathbb{N}$ with $s + 2G > \frac{2n}{p} + 4$, $N \in \mathbb{N}$ with $N > s + 2G + \sigma_p$, $\varepsilon > 0$ be sufficiently small, and $\rho > 0$ be sufficiently large. Then holds

$$\|k_0(\varepsilon, f) \mid L_p(M)\| + \left\| \left(\int_0^r t^{-(s+2G)q} |k(t,f)(\cdot)|^q \frac{dt}{t} \right)^{1/q} \mid L_p(M) \right\| \tag{16}$$

$$\sim \|k_0(\varepsilon, (\rho\,\mathrm{id} - \Delta)^G f) \mid L_p(M)\| + \left\| \left(\int_0^r t^{-sq} |k(t, (\rho\,\mathrm{id} - \Delta)^G f)(\cdot)|^q \frac{dt}{t} \right)^{1/q} \mid L_p(M) \right\|$$

(modification if $q = \infty$) in the sense of equivalent quasi-norms in $F^{s+2G}_{pq}(M)$.

Proof. Step 1. The conditions for G, N, and, as we suppose, also for ε are chosen in such a way that the left-hand side of (16) is an equivalent quasi-norm in $F^{s+2G}_{pq}(M)$ by the proven part of Theorem 7.2.4, see Step 1 of the proof of that theorem and Remark 7.2.4/1. Furthermore for the time being it will be convenient to indicate the dependence of the means in (7.2.4/6) and of the underlying kernels on N, i.e.

$$k_N(t,f)(P) = \int_{T_P M} k_N(\|X\|) f(c(P,X,t)) dX, \tag{17}$$

when $k = k_N$ is given by (7.2.4/3,5). We prefer a representation of $k_N(t,f)$ in normal geodesic coordinates, see (7.2.4/9), where we write f instead of $f \circ \exp_P$, i.e.,

$$k_N(t,f)(P) = \int_{\mathbb{R}^n} k_N(|X|) f(tX) dX. \tag{18}$$

Step 2. We prove (16) with $G = 1$. Besides the identification of $f \circ \exp_P$ with f, we always write g instead of $\exp_P^* g$, etc. By (1,18) and (7.2.1/7) we have

$$k_N(t, \Delta f)(P) = \int_{\mathbb{R}^n} k_N(|X|)(\Delta f)(tX) dX \tag{19}$$

$$= t^{-2} \int_{\mathbb{R}^n} k_N(|X|) \frac{1}{\sqrt{\det g(tX)}} \frac{\partial}{\partial X^j} \left(\sqrt{\det g(tX)}\, g^{ij}(tX) \frac{\partial f(tX)}{\partial X^i} \right) dX$$

$$= t^{-2}\int_{\mathbb{R}^n} \frac{\partial}{\partial X^i}\left(g^{ij}(tX)\sqrt{\det g(tX)}\,\frac{\partial}{\partial X^j}\frac{k_N(|X|)}{\sqrt{\det g(tX)}}\right) f(tX)dX$$

$$= t^{-2}\int_{\mathbb{R}^n} k_{N+1}(|X|)f(tX)dX + \cdots$$

(summation convention) where we used the Taylor expansion of $g^{ij}(tX)$ and $\det g(tX)$ with respect to tX based on (7.2.1/7). As in (18) we can give the explicit term on the right-hand side of (19) an invariant meaning, and hence also the term $+\cdots$ has an invariant meaning. This term $+\cdots$ can be represented as a finite sum of terms with factors $t^{|\alpha|-2}X^\alpha$, $2 \leqq |\alpha| \leqq L$, or $t^{|\beta|-1}X^\beta$, $1 \leqq |\beta| \leqq L$, or $t^{|\gamma|}X^\gamma$, $0 \leqq |\gamma| \leqq L$, multiplied with appropriate kernels, and a remainder term with, say, t^{L-2} in front. Recall that $s+2 > \frac{2n}{p}+4$. We choose L large in the sense of Step 1 of the proof in 7.2.4 and obtain by the same technique as in that proof

$$\left\| \left(\int_0^r t^{-sq}|(+\cdots)(\cdot)|^q \frac{dt}{t}\right)^{1/q} \mid L_p(M)\right\| \leqq c\|f \mid F^{s+2-\delta}_{pq}(M)\| \tag{20}$$

for some $\delta > 0$. The term $k_0(\varepsilon, \Delta f)(P)$ can be treated in the same way. Hence the right-hand side of (16) with k_N instead of k can be estimated from above by the left-hand side of (16) with k_{N+1} instead of k and an additional term $c\|f \mid F^{s+2-\delta}_{pq}(M)\|$. We may assume that the proof of 7.2.4 can be applied both to $F^{s+2}_{pq}(M)$ and $F^{s+2-\delta}_{pq}(M)$. In particular the left-hand side of (16) is an equivalent quasi-norm in $F^{s+2}_{pq}(M)$ where it is unimportant whether we choose k_{N+1} or k_N instead of k on the left-hand side of (16). Furthermore $\|f \mid F^{s+2-\delta}_{pq}(M)\|$ can be estimated from above by the left-hand side of (16). In other words, the right-hand side of (16) can be estimated from above by the left-hand side.

Step 3. We prove the converse estimate, again under the assumption $G = 1$. For this purpose we replace $k_0(|X|)$ by $k_0'(|X|) = k_0(|X|) - k_1(|X|)$ where k_1 is given by (7.2.4/3,5) with k_0 instead of k^0 and $N = 1$. The counterparts of (7.2.4/1,2) are satisfied. Because the left-hand side of (16) is an equivalent quasi-norm in $F^{s+2}_{pq}(M)$ we can replace k_0 by k_0'. By (19) and a remainder estimate of type (7.2.4/30) we arrive at

$$\|k_0'(\varepsilon, f) \mid L_p(M)\| \leqq c\|k_0(\varepsilon, (\mathrm{id} - \varepsilon^2\Delta)f) \mid L_p(M)\| + c'\varepsilon\|f \mid F^{s+2-\delta}_{pq}(M)\|, \tag{21}$$

$\delta > 0$ small, where c and c' are independent of ε. Now by (19,20) we have

$$\|f \mid F^{s+2}_{pq}(M)\| \leqq c\|k_0(\varepsilon, (\mathrm{id} - \varepsilon^2\Delta)f) \mid L_p(M)\| \tag{22}$$
$$+ \left\| \left(\int_0^r t^{-sq}|k_N(t, (\varepsilon^{-2}\mathrm{id} - \Delta)f)(\cdot)|^q \frac{dt}{t}\right)^{1/q} \mid L_p(M)\right\| + \frac{c'}{\varepsilon^2}\|f \mid F^{s+2-\delta}_{pq}(M)\|,$$

where c and c' are independent of ε. In Remark 3 we outline a proof of

$$\|k_0(\varepsilon, (\mathrm{id} - \varepsilon^2\Delta)f) \mid L_p(M)\| \leqq \mu\|f \mid F^{s+2}_{pq}(M)\| \qquad (23)$$
$$+ c_\mu\|k_0(\lambda, (\mathrm{id} - \varepsilon^2\Delta)f) \mid L_p(M)\|,$$

where $\mu > 0$ is at our disposal, λ with $0 < \lambda \leqq \varepsilon$ is sufficiently small, and c_μ is independent of λ (and ε). As for the last term in (22) we use

$$\|f \mid F^{s+2-\delta}_{pq}(M)\| \leqq \mu\|f \mid F^{s+2}_{pq}(M)\| + c'_\mu\|k_0(\lambda, (\mathrm{id} - \varepsilon^2\Delta)f) \mid L_p(M)\| \qquad (24)$$

where $\mu > 0$ is at our disposal, c'_μ is independent of ε, and λ with $0 < \lambda < \varepsilon$ is sufficiently small. (24) follows from the technique used in 7.2.4: the second estimate in (7.2.4/32), (7.2.4/33) with $k_0^{euc}(\lambda, (\mathrm{id} - \varepsilon^2\Delta)f)$ where the respective constants in (7.2.4/33) are independent of ε and λ as it follows from Remark 7.2.4/3, and the counterpart of (7.2.4/30), where one now needs that $\lambda > 0$ is small. Now we insert (23) and (24) in (22), choose μ in dependence on ε small and arrive at

$$\|f \mid F^{s+2}_{pq}(M)\| \leqq c\|k_0(\lambda, (\varepsilon^{-2}\mathrm{id} - \Delta)f) \mid L_p(M)\| \qquad (25)$$
$$+ \left\| \left(\int_0^r t^{-sq} |k_N(t, (\varepsilon^{-2}\mathrm{id} - \Delta)f)(\cdot)|^q \frac{dt}{t} \right)^{1/q} \mid L_p(M) \right\|.$$

This is the desired estimate with λ instead of ε and $\rho = \varepsilon^{-2}$. There is no problem to replace afterwards ε on the left-hand side of (16) by λ with $0 < \lambda < \varepsilon$ (equivalent quasi-norms in $F^{s+2}_{pq}(M)$).

Step 4. The proof of (16) for arbitrary $G \in \mathbb{N}$ is essentially the same as for $G = 1$. First one must iterate (19). For the remainder term one has an obvious counterpart of (20). Then it follows that the right-hand side of (16) can be estimated from above by the left-hand side. Furthermore one has corresponding counterparts of (21–25), where k'_0 must be chosen in an appropriate way. This yields the desired estimate.

Remark 3. The proof of (23) can be based on the same technique as in (7.2.4/30–33): Reduction to the euclidean case in the sense of (7.2.4/30), use of an estimate in the sense of the second inequality in (7.2.4/32) which reduces the problem to an inequality of type (7.2.4/33) with $k_0^{euc}(\lambda, (\mathrm{id} - \varepsilon^2\Delta)f)$ instead of $k_0^{euc}(\varepsilon, h)$. However by the proof in Remark 7.2.4/3 such an inequality is correct with constants which are independent of λ and ε.

Remark 4. Let the hypotheses of Proposition 2 for s, p, q, G, N, and ρ be satisfied and let in addition $s > \sigma_p = n(\frac{1}{p} - 1)_+$. Then (16) can be replaced by

$$\|f \mid L_p(M)\| + \left\| \left(\int_0^r t^{-(s+2G)q} |k(t,f)(\cdot)|^q \frac{dt}{t} \right)^{1/q} \mid L_p(M) \right\| \tag{26}$$

$$\sim \|(\rho\,\mathrm{id} - \Delta)^G f \mid L_p(M)\| + \left\| \left(\int_0^r t^{-sq} |k(t,(\rho\,\mathrm{id} - \Delta)^G f)(\cdot)|^q \frac{dt}{t} \right)^{1/q} \mid L_p(M) \right\|$$

(modification if $q = \infty$). We prove this claim. By $s+2G > \frac{2n}{p}+4$ and Corollary 7.2.4 we know that the left-hand side of (26) is an equivalent quasi-norm in $F^{s+2G}_{pq}(M)$. Because now $s + 2G > \sigma_p + 2G$ one can even replace $\|f \mid L_p(M)\|$ on the left-hand side of (26) by $\|(\rho\,\mathrm{id} - \Delta)^G f \mid L_p(M)\|$ in the sense of equivalent quasi-norms. This follows in the same way as in the proof of Corollary 7.2.4 based on the euclidean counterpart (7.2.4/41) with $\|(\rho\,\mathrm{id} - \Delta)^G f \mid L_p\|$ instead of $\|f \mid L_p\|$. The latter assertion can be proved in the same way as in Step 1 of the proof of Corollary 7.2.4 again with a reference to [Triß: 2.5.9, formula (37)]. Now one can follow the proof of Proposition 2 with the just indicated modified quasi-norms. The counterpart of Step 2 causes no problem. As for Step 3 we need the counterpart of (24),

$$\|f \mid F^{s+2G-\delta}_{pq}(M)\| \leqq \mu \|f \mid F^{s+2G}_{pq}(M)\| + c_\mu \|(\rho\,\mathrm{id} - \Delta)^G f \mid L_p(M)\|. \tag{27}$$

In order to prove (27) we assume $s - \delta > \sigma_p$, then we have the $\mathbb{R}^n$-assertion

$$\|f \mid F^{s-\delta}_{pq}\| \leqq \eta \|f \mid F^s_{pq}\| + c_\eta \|f \mid L_p\|, \tag{28}$$

where $\eta > 0$ is at our disposal, see again [Triß: 2.5.9, formula (37)] and the technique used in the second estimate in (7.2.4/32). By (7.2.2/1) we have a corresponding assertion with M instead of $\mathbb{R}^n$. In 7.4.3 we prove that $(\rho\,\mathrm{id} - \Delta)^G$ is a lift in the $F^s_{pq}(M)$ scale, provided $\rho > 0$ is large. Then (27) is a consequence of this lifting property and the M-counterpart of (28). Of course, the proof of the lifting property is independent of this remark. Now (27) completes the proof of the above claim.

Remark 5. We proved Theorem 7.2.4 under the assumption $s > \frac{2n}{p} + 4$ and $N > s + \sigma_p$, see Remark 7.2.4/1. Our main goal in 7.2.6 is to extend this proof from $s > \frac{2n}{p} + 4$ to arbitrary values of s. This extension is based on the two propositions in this subsection. Let s with $s \leqq \frac{2n}{p} + 4$ be given. Then we may choose $G \in \mathbb{N}$ such that

$$\frac{2n}{p} + 4 < s + 2G \leqq \frac{2n}{p} + 6. \tag{29}$$

Furthermore $N > \frac{2n}{p} + 6 + \sigma_p$ is sufficient to apply Proposition 2. Hence

$$N > \max\left(s, \frac{2n}{p} + 6\right) + \sigma_p \tag{30}$$

will be sufficient in any case.

7.2.6 Proof of Theorem 7.2.4

Recall that we proved Theorem 7.2.4 under the restriction $s > \frac{2n}{p} + 4$ and $N \in \mathbb{N}$ with $N > s + \sigma_p$, see Step 1 of the proof in 7.2.4 and Remark 7.2.4/1. Let $s \leqq \frac{2n}{p} + 4$, then we choose $G \in \mathbb{N}$ and $N \in \mathbb{N}$ such that (7.2.5/29,30) are satisfied. This gives us the possibility to apply the two propositions in 7.2.5. Finally we recall that the proof of Theorem 7.2.4 is reduced to the proof of (7.2.4/13–15) which will be done now in several steps. Again we restrict ourselves to $p < \infty$ without restriction of generality.

Step 1. As in 7.2.4 we abbreviate (7.2.4/10) for the time being by $\|f \mid F^s_{pq}(M)\|$, not to be mixed with $\|f \mid F^s_{pq}(M)\|^\psi$ from (7.2.2/1). The proof of (7.2.4/13), i.e.,

$$\|f \mid F^s_{pq}(M)\|^p \leqq c \sum_{j=1}^{\infty} \|\psi_j f \mid F^s_{pq}(M)\|^p, \qquad f \in F^s_{pq}(M), \tag{1}$$

remains unchanged, see Step 2 in 7.2.4. Next we assume that $f \in F^s_{pq}(M)$ can be represented as $f = (\rho\,\mathrm{id} - \Delta)^G h$ with $h \in F^{s+2G}_{pq}(M)$. Then one obtains as in the proof of Proposition 7.2.5/2

$$\sum_{j=1}^{\infty} \|\psi_j f \mid F^s_{pq}(M)\|^p \leqq c\|h \mid F^{s+2G}_{pq}(M)\|^p \leqq c'\|f \mid F^s_{pq}(M)\|^p, \tag{2}$$

where the last inequality comes from (7.2.5/16) (of course we assume that ρ is chosen sufficiently large). Hence (1,2) prove

$$\|f \mid F^s_{pq}(M)\|^p \sim \sum_{j=1}^{\infty} \|\psi_j f \mid F^s_{pq}(M)\|^p \tag{3}$$

under the assumption $f = (\rho\,\mathrm{id} - \Delta)^G h$ with $h \in F^{s+2G}_{pq}(M)$.

Step 2. Let $\|f \mid F^s_{pq}(M)\|^\psi$ be the quasi-norm in (7.2.2/1) where it is convenient to assume that $\|\cdot \mid F^s_{pq}\|$ is given by the euclidean counterpart of (7.2.4/10), see Theorem 2.4.6, complemented by Theorem 2.5.3 if $p = q = \infty$. Let again $f = (\rho\,\mathrm{id} - \Delta)^G h$ with $h \in F^{s+2G}_{pq}(M)$, then we can replace $\|\psi_j f \mid F^s_{pq}(M)\|$ on the left-hand side of (2) by $\|\psi_j f \circ \exp_{P_j} \mid F^s_{pq}\|$. This follows from the above technique of estimates. Hence we have

$$\|f \mid F^s_{pq}(M)\|^\psi \leqq c\|f \mid F^s_{pq}(M)\|. \tag{4}$$

We prove the converse assertion of (4), again under the assumption $f = (\rho\,\mathrm{id} - \Delta)^G h$ with $h \in F^{s+2G}_{pq}(M)$. We have

$$\|h \mid F^{s+2G}_{pq}(M)\|^p \leqq c \sum_{j=1}^{\infty} \|\psi_j h \circ \exp_{P_j} \mid F^{s+2G}_{pq}\|^p \tag{5}$$

$$\leqq c' \sum_{j=1}^{\infty} \|\psi_j f \circ \exp_{P_j} \mid F^s_{pq}\|^p + c'' \|h \mid F^{s+2G-\delta}_{pq}(M)\|^p$$

for some $\delta > 0$, where we used $\psi_j f = (\rho\,\mathrm{id} - \Delta)^G \psi_j h + \cdots$, the euclidean version of Proposition 7.2.5/2 and the above technique of estimate as far as the remainder term is concerned (modification if $p = \infty$). We estimate the last term in (5) by (7.2.5/24) with $2G$ instead of 2. The resulting Riemannian term $\|k_0(\lambda, f) \mid L_p(M)\|$ can be compared with its euclidean counterpart by (7.2.4/30). All together we arrive at

$$\|f \mid F^s_{pq}(M)\| \sim \|h \mid F^{s+2G}_{pq}(M)\| \leqq c\|f \mid F^s_{pq}(M)\|^{\psi} \tag{6}$$

where the first equivalence comes again from Proposition 7.2.5/2. Now (4,6) and (3) prove Theorem 7.2.4 under the assumption $f = (\rho\,\mathrm{id} - \Delta)^G h$ with $h \in F^{s+2G}_{pq}(M)$.

Step 3. We have to discuss the additional assumption that $f \in F^s_{pq}(M)$ can be represented by $f = (\rho\,\mathrm{id} - \Delta)^G h$ with $h \in F^{s+2G}_{pq}(M)$. For this purpose we have now carefully to distinguish between $p < \infty$, $q < \infty$, and the limiting cases $p < \infty$, $q = \infty$, and $p = q = \infty$. Let $p < \infty$, $q < \infty$. By Definition 7.2.2 and the proof of Theorem 7.2.3 (pointwise multiplier property of F^s_{pq}) it follows that the distributions $f \in F^s_{pq}(M)$ with compact support are dense in $F^s_{pq}(M)$. Recall that the compactly supported C^∞ functions in $\mathbb{R}^n$ are dense in F^s_{pq}, see [Triß: 2.3.3]. Both facts combined with pointwise multipliers result in the assertion that the compactly supported C^∞ functions on M are dense in $F^s_{pq}(M)$. By Proposition 7.2.5/1 and the above considerations we have

$$\|f \mid F^s_{pq}(M)\|^{\psi} \sim \|h \mid F^{s+2G}_{pq}(M)\|^{\psi} \sim \|f \mid F^s_{pq}(M)\| \tag{7}$$

with

$$(\rho\,\mathrm{id} - \Delta)^G h = f \in D(M), \qquad h \in F^{s+2G}_{pq}(M). \tag{8}$$

By completion (7,8) can be extended to $f \in F^s_{pq}(M)$. Hence what has been said in Steps 1, 2 can be applied to any $f \in F^s_{pq}(M)$. The proof of Theorem 7.2.4 is complete if $p < \infty$, $q < \infty$.

Step 4. Let $p < \infty$, $q = \infty$. Again the distributions $f \in F^s_{p\infty}(M)$ with compact support are dense in $F^s_{p\infty}(M)$. We approximate $f \in F^s_{p\infty}(M)$ with compact support, in $D'(M)$ by functions $f^l \in D(M)$ where we may assume $\operatorname{supp} f^l \subset V$, where V is a compact subset of M. This is essentially an euclidean procedure. Note that $k(t, f^l) \to k(t, f)$ holds pointwise. Let

$$(\rho\,\mathrm{id} - \Delta)^G h^l = f^l \quad \text{and} \quad (\rho\,\mathrm{id} - \Delta)^G h = f, \tag{9}$$

which makes sense by the above results and the elementary embedding $F^s_{p\infty}(M) \subset F^{s-\varepsilon}_{pq}(M)$, $\varepsilon > 0$. By the above considerations we have $h^l \to h$ in $D'(M)$ and hence

$k(t, h^l) \to k(t, h)$ pointwise. Furthermore (7.2.5/16) holds with h^l and f^l instead of f and $(\rho\,\mathrm{id} - \Delta)^G f$. Under our compactness assumptions the right-hand side of this modified equivalence (7.2.5/16) may be reduced to the euclidean case, say, with the above operator $(\rho\,\mathrm{id} - \Delta)^G$. Then we may assume that the right-hand side of (7.2.5/16) is uniformly bounded and hence $\|h^l \mid F^{s+2G}_{p\infty}(M)\| \leqq c < \infty$ where c is independent of l. By Fatou's lemma, see (7.2.2/1) and (2.3.1/8), it follows $h \in F^{s+2G}_{p\infty}(M)$. Now the rest is the same as above.

Step 5. We complete the proof in the case $p = q = \infty$. Because $D(M)$ is dense in $F^s_{11}(M)$, the dual space $(F^s_{11}(M))'$ of $F^s_{11}(M)$ can be interpreted in the usual way in the sense of the dual pairing $(D(M), D'(M))$. We have

$$(F^s_{11}(M))' = F^{-s}_{\infty\infty}(M), \qquad s \in \mathbb{R}. \tag{10}$$

The $\mathbb{R}^n$- counterpart of (10) may be found in [Triß: 2.11.2]. The extension of this assertion from $\mathbb{R}^n$ to M via Definition 7.2.2 is quite standard. Details may be found in the underlying paper [Tri13: Remark 24]. Furthermore in 7.4.3 we prove that $(\rho\,\mathrm{id} - \Delta)^G$ yields an isomorphic map from $F^{s+2G}_{11}(M)$ onto $F^s_{11}(M)$. Recall that $(\rho\,\mathrm{id} - \Delta)^G$ is formally selfadjoint. Then (10) proves that $(\rho\,\mathrm{id} - \Delta)^G$ is an isomorphic map from $F^{-s}_{\infty\infty}(M)$ onto $F^{-s-2G}_{\infty\infty}(M)$ for any $s \in \mathbb{R}$ and $G \in \mathbb{N}$. Hence Steps 1, 2 can also be applied to this case. The proof is complete.

Remark. We fix the following assertion referred to in Step 3: Let $0 < p < \infty$, $0 < q < \infty$, and $s \in \mathbb{R}$, then $D(M)$ is dense in $F^s_{pq}(M)$.

7.3 The spaces B^s_{pq} on manifolds

7.3.1 Definition and basic properties

The spaces B^s_{pq} and F^s_{pq} on $\mathbb{R}^n$ are usually introduced independently of each other, see Definition 2.3.1, and as a rule of thumb the spaces B^s_{pq} are in many respects the simpler ones. On manifolds M of the above type the situation is quite different. Motivated by the localization principle for F^s_{pq}, see Theorem 2.4.7(i), we introduced the spaces $F^s_{pq}(M)$ in Definition 7.2.2 by reducing them to the corresponding spaces on $\mathbb{R}^n$, which is justified by Proposition 7.2.2. The situation for spaces of B^s_{pq} type is completely different. There is no localization principle which can be taken as a starting point, see Theorem 2.4.7(ii). So we are forced to look for other properties of the spaces B^s_{pq} and F^s_{pq} on $\mathbb{R}^n$ which can be employed in order to introduce spaces of type B^s_{pq} on manifolds. We rely on interpolation theory. In 1.6 we gave a brief description of interpolation methods accompanied by the necessary references. In particular we shall use the real interpolation method for quasi-Banach spaces described in 1.6.2. Recall

$$(F^{s_0}_{pp}, F^{s_1}_{pp})_{\theta,q} = B^s_{pq}, \tag{1}$$

where $0 < p \leqq \infty$, $0 < q \leqq \infty$, $-\infty < s_0 < s_1 < \infty$, $0 < \theta < 1$, and $s = (1-\theta)s_0 + \theta s_1$, see (1.6.7/7) complemented by the limiting case $F^\sigma_{\infty\infty} = B^\sigma_{\infty\infty}$. Historical references may be found in 1.6.7. Furthermore a detailed proof of (1), including the just-mentioned limiting case has been given in [Triß: 2.4.2]. We take the extension of (1) from $\mathbb{R}^n$ to M as our starting point.

Definition. Let M be a manifold satisfying Hypotheses 7.2.1. Let $0 < p \leqq \infty$, $0 < q \leqq \infty$, and $-\infty < s_0 < s < s_1 < \infty$. Then

$$B^s_{pq}(M) = (F^{s_0}_{pp}(M), F^{s_1}_{pp}(M))_{\theta,q} \tag{2}$$

with $s = (1-\theta)s_0 + \theta s_1$, where $F^{s_0}_{pp}(M)$ and $F^{s_1}_{pp}(M)$ must be understood in the sense of Definition 7.2.2.

Remark. In order to justify this definition of $B^s_{pq}(M)$, where s, p, q are given, we have to show that (2) is independent of the chosen numbers s_0 and s_1.

Theorem. (i) Let $0 < p \leqq \infty$, $0 < q \leqq \infty$, and $s \in \mathbb{R}$. Then $B^s_{pq}(M)$ from (2) is a quasi-Banach space (Banach space if $p \geqq 1$, $q \geqq 1$). It is independent of the numbers s_0 and s_1.

(ii) Let $0 < p \leqq \infty$, then

$$B^s_{pp}(M) = F^s_{pp}(M). \tag{3}$$

Proof. Step 1. We prove (3). Let Λ and Ψ be the same operators as in (7.2.3/5,6) with the property (7.2.3/7). This is a standard situation in interpolation theory, the so-called method of retraction-corectraction. Let $0 < p \leqq \infty$, $-\infty < s_0 < s < s_1 < \infty$, $0 < \theta < 1$, and $s = (1-\theta)s_0 + \theta s_1$ be given. Let $l_p(F^s_{pp})$ be the same spaces as in Step 2 of the proof of Theorem 7.2.3. We have

$$(l_p(F^{s_0}_{pp}), l_p(F^{s_1}_{pp}))_{\theta,p} = l_p((F^{s_0}_{pp}, F^{s_1}_{pp})_{\theta,p}) = l_p(F^s_{pp}). \tag{4}$$

The first equality comes from [Triα: Theorem 1.18.1, Remark 1.18.1/4, extension to $p=\infty$]. The second equality is covered by (1). Now by (2, 4), and the interpolation property we have

$$\|\Psi f \mid l_p(F^s_{pp})\| \leqq c\|f \mid B^s_{pp}(M)\| \quad \text{if} \quad f \in B^s_{pp}(M) \tag{5}$$

and

$$\|\Lambda\{f_l\} \mid B^s_{pp}(M)\| \leqq c\|\{f_l\} \mid l_p(F^s_{pp})\| \quad \text{if} \quad \{f_l\} \in l_p(F^s_{pp}). \tag{6}$$

We insert $\{f_l\} = \Psi f$ with $f \in F^s_{pp}(M)$ in (6) and use $\Lambda\Psi f = f$, see (7.2.3/7), then (6) is the reversion of (5), i.e.,

$$\|f \mid B^s_{pp}(M)\| \sim \|\Psi f \mid l_p(F^s_{pp})\|, \tag{7}$$

and $B^s_{pp}(M) = F^s_{pp}(M)$ by Definition 7.2.2.

Step 2. We prove (i). By (3) it follows that $B^s_{pp}(M)$ is independent of s_0 and s_1. Next we use the reiteration theorem of interpolation theory for the spaces $(\cdot,\cdot)_{\theta,q}$ with $0<\theta<1$ and $0<q\leqq\infty$, see [Triα: 1.10], which holds also for quasi-Banach spaces. This property and (3) prove that $B^s_{pq}(M)$ in (2) is independent of s_0 and s_1. Furthermore as an interpolation space of complete spaces, $B^s_{pq}(M)$ is complete, too.

7.3.2 The main theorem

We prove the counterpart of Theorem 7.2.4 for the spaces $B^s_{pq}(M)$. Let $k(t,f)(P)$ and $k_0(\varepsilon,f)(P)$ be the same means as in 7.2.4, see in particular (7.2.4/1–6). Again we omit to indicate the value of $N \in \mathbb{N}$ in $k(t,f)(P)$.

Theorem. Let M be a Riemannian manifold in the sense of Hypotheses 7.2.1 with the injectivity radius $r_0>0$. Let $s\in\mathbb{R}$, $0<p\leqq\infty$ and $0<q\leqq\infty$. Let $0<r<r_0$, let $\varepsilon>0$ be sufficiently small and $N\in\mathbb{N}$ in (7.2.4/3,5,6) be large enough ($N>\max(s,\frac{2n}{p}+6)+\sigma_p$ where $\sigma_p = n(\frac{1}{p}-1)_+$ is sufficient), then

$$\|k_0(\varepsilon,f) \mid L_p(M)\| + \left(\int_0^r t^{-sq}\|k(t,f) \mid L_p(M)\|^q \frac{dt}{t}\right)^{1/q} \tag{1}$$

(modification if $q=\infty$) is an equivalent quasi-norm in $B^s_{pq}(M)$.

Proof. Step 1. First we deal with large values of s as in Step 1 of the proof in Theorem 7.2.4, i.e., $0<p\leqq\infty$, $0<q\leqq\infty$, $s>\frac{2n}{p}+4$. Let $L\in\mathbb{N}$ be given by

(7.2.4/12). For the time being we abbreviate (1) by $\|f \mid B^s_{pq}(M)\|$ and the quasi-norm coming from (7.3.1/2) with $\|f \mid B^s_{pq}(M)\|^\psi$, where we may assume that the quasi-norms of the underlying spaces $F^{s_0}_{pp}(M)$ and $F^{s_1}_{pp}(M)$ are given by (7.2.2/1). In other words we know that $\|f \mid B^s_{pq}(M)\|^\psi$ is an equivalent quasi-norm in $B^s_{pq}(M)$ and we have to prove that $\|f \mid B^s_{pq}(M)\|$ is also an equivalent quasi-norm. In this step we show that there exists a constant $c > 0$ such that

$$\|f \mid B^s_{pq}(M)\|^\psi \leqq c\|f \mid B^s_{pq}(M)\|, \qquad f \in B^s_{pq}(M). \tag{2}$$

The basic idea is to reduce the interpolation (7.3.1/2) to its euclidean counterpart and to argue otherwise as in 7.2.4. Let $\psi = \{\psi_j\}_{j=1}^\infty$ be the same resolution of unity as in Definition 7.2.2, Proposition 7.2.1, and Remark 7.2.1/2. We decompose ψ in a finite number of families $\psi^a = \{\psi^a_j\}_{j=1}^\infty$ with $a = 1, \ldots, A$ such that

$$\operatorname{dist}(\Omega_{P^a_j}(\delta), \Omega_{P^a_k}(\delta)) \geqq 4\delta \quad \text{if} \quad j \neq k, \tag{3}$$

see (7.2.1/6) and Proposition 7.2.1 with the necessary modifications in the notations, $\psi = \bigcup_{a=1}^A \psi^a$. Let $\Psi^a = \sum_{j=1}^\infty \psi^a_j$ and $f^a = \Psi^a f$ with $f \in B^s_{pq}(M)$. Then we have

$$\operatorname{supp} f^a \subset \bigcup_{j=1}^\infty \Omega_{P^a_j}(\delta) \subset \bigcup_{j=1}^\infty \Omega_{P^a_j}(2\delta) = M^a. \tag{4}$$

Of course $M = \bigcup_{a=1}^A M^a$. In order to make the next conclusions more transparent we combine the map $\exp_{P_j}$ in (7.2.2/1) with a translation T_j: $x \to x - x^j$ in $\mathbb{R}^n$. In other words we replace $\psi_j f \circ \exp_{P_j}$ in (7.2.2/1) by $\psi_j f \circ \exp_{P_j} \circ T_j$. This is completely immaterial in (7.2.2/1) but now of great service for us. Similarly $\exp_{P^a_j} \circ T_j$ maps a ball B_j in $\mathbb{R}^n$ centered at x^j and of radius 2δ onto $\Omega_{P^a_j}(2\delta)$ and we assume in addition

$$\operatorname{dist}(B_j, B_k) \geqq 4\delta + r_0 \quad \text{if} \quad j \neq k, \tag{5}$$

which is the $\mathbb{R}^n$-counterpart of (3). Let $B = \bigcup_{j=1}^\infty B_j$. Let

$$F^s_{pq}(M^a) = \{f \in F^s_{pq}(M)\colon \ \operatorname{supp} f \subset \overline{M^a}\}. \tag{6}$$

Then it follows that $\{\exp_{P^a_j} \circ T_j\}_{j=1}^\infty$ generates an one-to-one mapping from

$$F^s_{pq}(B) = \{g \in F^s_{pq}\colon \ \operatorname{supp} g \subset \overline{B}\} \tag{7}$$

onto $F^s_{pq}(M^a)$. We interpolate this assertion based on (7.3.1/1,2). By (4), in particular the little play with δ and 2δ, and pointwise multiplier properties it follows

$$\|f^a \mid B^s_{pq}(M)\|^\psi \sim \left\| \sum_{j=1}^\infty \psi^a_j f \circ \exp_{P^a_j} \circ T_j \mid B^s_{pq} \right\|, \tag{8}$$

where we may assume that B^s_{pq} is quasi-normed (2.5.3/3) with ε and r instead of 1, respectively. We apply the technique developed in 7.2.4 to the right-hand side of (8) and arrive at

$$\|f^a \mid B^s_{pq}(M)\|^{\psi} \leqq c\|f \mid B^s_{pq}(M)\| + \cdots \tag{9}$$

where the remainder terms can be estimated from above by terms of the type

$$\left(\int_0^r t^{-sq+\kappa q}\|\cdots \mid L_p(M)\|^q \frac{dt}{t}\right)^{1/q} \leqq c\left\|\sup_{0<t<r} t^{-s+\kappa'}|\cdots| \mid L_p(M)\right\| \tag{10}$$

with $0<\kappa'<\kappa$ and for some $\kappa>0$. Here $\cdots$ indicates terms of the same type as in the Steps 3 and 4 in the proof in Theorem 7.2.4. However the right-hand side of (10) can be estimated from above by $c\|f \mid F^{s-\kappa'}_{p\infty}(M)\|$. We use an inequality of type (7.2.5/24) where $\|\cdot \mid F^{s+2}_{pq}(M)\|$ can be replaced by $\|\cdot \mid B^{s+2}_{pq}(M)\|^{\psi}$ (as a consequence of interpolation properties). Hence

$$\|f \mid F^{s-\kappa'}_{pq}(M)\| \leqq \eta\|f \mid B^s_{pq}(M)\|^{\psi} + c_\eta\|f \mid B^s_{pq}(M)\| \tag{11}$$

where $\eta>0$ is at our disposal. The last term in (11) can be incorporated in the right-hand side of (9). Then summation over a yields (2).

Step 2. We prove

$$\|f \mid B^s_{pq}(M)\| \leqq c\|f \mid B^s_{pq}(M)\|^{\psi}, \qquad f \in B^s_{pq}(M), \tag{12}$$

where c is independent of $f \in B^s_{pq}(M)$. By (8) and the arguments in Step 1 we have

$$\|f^a \mid B^s_{pq}(M)\| \leqq c\|f^a \mid B^s_{pq}(M)\|^{\psi} + \cdots \tag{13}$$

where the remainder terms can be estimated in the same way as in (10). Recall $f^a = \Psi^a f$ and

$$\|\Psi^a f \mid F^s_{pq}(M)\|^{\psi} \leqq c\|f \mid F^s_{pq}(M)\|^{\psi} \tag{14}$$

by pointwise multiplier properties. By the interpolation property we have a corresponding inequality with $B^s_{pq}(M)$ instead of $F^s_{pq}(M)$, which we apply to the explicit term on the right-hand side of (13). Furthermore the remainder term in (13) can be estimated from above by $c\|f \mid F^{s-\kappa}_{pp}(M)\|$ for some $\kappa>0$, and hence by $c'\|f \mid B^s_{pq}(M)\|^{\psi}$, see (7.3.1/2). The proof of (12) is complete. Now (2,12) prove the theorem if $s > \frac{2n}{p}+4$.

Step 3. The extension of the proof from $s>\frac{2n}{p}+4$ to arbitrary values of s will be done in precisely the same way as in 7.2.6. For this purpose we need the counterpart of the Proposition 7.2.5/2. Let $s\in\mathbb{R}$, $0<p\leqq\infty$, and $0<q\leqq\infty$. Let $G\in\mathbb{N}$ with $s+2G>\frac{2n}{p}+4$, $N\in\mathbb{N}$ with $N>s+2G+\sigma_p$, $\varepsilon>0$ sufficiently small and $\rho>0$ sufficiently large. Let k_0 and k be the same means as in (7.2.5/16) then

$$\|k_0(\varepsilon, f) \mid L_p(M)\| + \left(\int_0^r t^{-(s+2G)q} \|k(t,f) \mid L_p(M)\|^q \frac{dt}{t}\right)^{1/q} \tag{15}$$

$$\sim \|k_0(\varepsilon, (\rho\,\mathrm{id} - \Delta)^G f) \mid L_p(M)\| + \left(\int_0^r t^{-sq} \|k(t, (\rho\,\mathrm{id} - \Delta)^G f \mid L_p(M)\|^q \frac{dt}{t}\right)^{1/q}$$

(modification if $q = \infty$) in the sense of equivalent quasi-norms in $B^{s+2G}_{pq}(M)$. Recall that Δ stands for the Laplace–Beltrami operator, see (7.2.5/1). The proof of the equivalence (15) is the same as in Proposition 7.2.5/2, based on (7.2.5/19) where the remainder terms can be treated in the same way as above.

Step 4. Let again $s \in \mathbb{R}$, $0 < p \leqq \infty$, $0 < q \leqq \infty$, and let G and ρ be the same numbers as in the previous step. We proved in 7.2.6 that $(\rho\,\mathrm{id} - \Delta)^G$ yields an isomorphic mapping from $F^{s+2G}_{pp}(M)$ onto $F^s_{pp}(M)$. However by (7.3.1/2) and the interpolation property it follows that $(\rho\,\mathrm{id} - \Delta)^G$ yields also an isomorphic mapping from $B^{s+2G}_{pq}(M)$ onto $B^s_{pq}(M)$. Now Steps 1, 2 and (15) prove the theorem for all values of s.

7.4 Further properties of the spaces F^s_{pq} and B^s_{pq} on manifolds

7.4.1 The discrete version of Theorems 7.2.4 and 7.3.2

In Theorems 2.4.6 and 2.5.3 we dealt with equivalent quasi-norms based on local means in the spaces F^s_{pq} and B^s_{pq}, respectively, on $\mathbb{R}^n$. We studied two versions, a continuous version, see (2.4.6/7) and (2.5.3/3), and a discrete version, see (2.4.6/6) and (2.5.3/2). The continuous versions look more handsome, the discrete versions are often more useful. In Theorems 7.2.4 and 7.3.2 we extended the continuous versions from $\mathbb{R}^n$ to the above manifolds M. Besides a lot of technicalities and a heavy machinery the basic idea was simple: reduction of the invariant Riemannian means (7.2.4/6) to their euclidean counterparts (2.4.6/1) and reduction of (7.2.4/10) and (7.3.2/1) to (2.4.6/7) and (2.5.3/3), respectively. However there is no problem to replace the continuous means in the arguments in 7.2.4 – 7.2.6 and 7.3.2 by their discrete counterparts with a reference to (2.4.6/6) instead of (2.4.6/7) and to (2.5.3/2) instead of (2.5.3/3). In other words, by the same arguments as in 7.2.4 – 7.2.6 and 7.3.2 we arrive at the following result.

Theorem. Let M be the Riemannian manifold in the sense of Hypotheses 7.2.1 with the injectivity radius $r_0 > 0$. Let $k(t, f)$ be given by (7.2.4/6), similarly $k_0(\varepsilon, f)$. Let $J \in \mathbb{N}$ with $2^{-J} < r_0$ and let $s \in \mathbb{R}$.

(i) Let either $0 < p < \infty$, $0 < q \leqq \infty$ or $p = q = \infty$. Let $\varepsilon > 0$ be sufficiently small and let $N \in \mathbb{N}$ in (7.2.4/3,5,6) be large enough, e.g., $N > \max(s, \frac{2n}{p} + 6) + \sigma_p$ with $\sigma_p = n(\frac{1}{p} - 1)_+$ is sufficient, then

$$\|k_0(\varepsilon, f) \mid L_p(M)\| + \left\| \left(\sum_{j=J}^{\infty} 2^{jsq} |k(2^{-j}, f)(\cdot)|^q \right)^{1/q} \mid L_p(M) \right\| \tag{1}$$

(modification if $q = \infty$) is an equivalent quasi-norm in $F^s_{pq}(M)$.

(ii) Let $0 < p \leqq \infty$ and $0 < q \leqq \infty$. Let $\varepsilon > 0$ and $N \in \mathbb{N}$ be the same numbers as in (i) then

$$\|k_0(\varepsilon, f) \mid L_p(M)\| + \left(\sum_{j=J}^{\infty} 2^{jsq} \|k(2^{-j}, f) \mid L_p(M)\|^q \right)^{1/q} \tag{2}$$

(modification if $q = \infty$) is an equivalent quasi-norm in $B^s_{pq}(M)$.

Proof. We collected the necessary comments and references in front of the above theorem.

7.4.2 Embeddings and density

We collect few results which can be obtained easily on the basis of the preceding considerations and corresponding $\mathbb{R}^n$-assertions. Preference is given to those facts which will be of some use later on. Recall that $D(M)$ stands for the collection of all compactly supported C^∞ functions on M. Furthermore for our later purposes it will be convenient for us to abbreviate

$$\mathcal{C}^\sigma(M) = F^\sigma_{\infty\infty}(M), \qquad \sigma > 0. \tag{1}$$

These are the Hölder–Zygmund spaces on M which will be characterized later on in terms of differences of functions, see 7.5.3.

Theorem. Let M be the Riemannian manifold in the sense of Hypotheses 7.2.1.

(i) Let $s \in \mathbb{R}$, $0 < p < \infty$, $0 < q < \infty$, then $D(M)$ is dense both in $F^s_{pq}(M)$ and $B^s_{pq}(M)$.

(ii) Let $s \in \mathbb{R}$, then

$$B^s_{p\min(p,q)}(M) \subset F^s_{pq}(M) \subset B^s_{p\max(p,q)}(M), \quad 0 < p < \infty, \quad 0 < q \leqq \infty, \tag{2}$$

$$B^s_{pq_0}(M) \subset B^s_{pq_1}(M), \qquad 0 < p \leqq \infty, \quad 0 < q_0 \leqq q_1 \leqq \infty, \tag{3}$$

$$F^s_{pq_0}(M) \subset F^s_{pq_1}(M), \qquad 0 < p < \infty, \quad 0 < q_0 \leqq q_1 \leqq \infty. \tag{4}$$

(iii) Let $-\infty < s_1 < s_0 < \infty$, $0 < p \leqq \infty$ (with $p < \infty$ in the case of the F-spaces), $0 < q_0 \leqq \infty$, $0 < q_1 \leqq \infty$, then

$$B^{s_0}_{pq_0}(M) \subset B^{s_1}_{pq_1}(M), \qquad F^{s_0}_{pq_0}(M) \subset F^{s_1}_{pq_1}(M). \tag{5}$$

(iv) Let $0 < p \leqq \infty$, $0 < q \leqq \infty$, and $s - \frac{n}{p} \geqq \sigma > 0$, then

$$B^s_{pq}(M) \subset \mathcal{C}^\sigma(M). \tag{6}$$

Let $0 < p \leqq 1$, $0 < q \leqq \infty$ and $s > n(\frac{1}{p} - 1)$ then

$$B^s_{pq}(M) \subset L_1(M). \tag{7}$$

Proof. Step 1. We prove (ii) and (iii). First we note that (3,4) follow from the monotonicity of the l_q-spaces and Theorem 7.4.1. Then we obtain (5). Finally the triangle inequality yields (2), see [Triα: Step 5 of the proof of Theorem 2.3.2] for details.

Step 2. To prove (iv) we recall the $\mathbb{R}^n$ assertions

$$F^s_{pq} \subset B^s_{p\infty} \subset \mathcal{C}^\sigma, \qquad 0 < p < \infty, \quad 0 < q \leqq \infty, \quad s - \frac{n}{p} \geqq \sigma > 0, \tag{8}$$

where $\mathcal{C}^\sigma = F^\sigma_{\infty\infty} = B^\sigma_{\infty\infty}$ are the Hölder–Zygmund spaces on $\mathbb{R}^n$, see (1) and [Triß: (2.7.1/12)]. By (7.2.2/1) and (8) it follows

$$F^s_{pq}(M) \subset \mathcal{C}^\sigma(M), \quad 0 < p < \infty, \quad 0 < q \leqq \infty, \quad s - \frac{n}{p} \geqq \sigma > 0. \tag{9}$$

Now (9), (7.3.1/2), and the interpolation property yield (6). Next we note

$$F^s_{pq}(M) \subset L_1(M), \quad 0 < p \leqq 1, \quad 0 < q \leqq \infty, \quad s > n\left(\frac{1}{p} - 1\right). \tag{10}$$

The $\mathbb{R}^n$-counterpart of (10) follows from (2.3.2/22,25). Again (7.3.1/2) and the interpolation property lead to (7).

Step 3. We prove (i). As far as the spaces $F^s_{pq}(M)$ with $s \in \mathbb{R}$, $0 < p < \infty$, $0 < q < \infty$ are concerned we refer to Remark 7.2.6. The corresponding assertion for the spaces $B^s_{pq}(M)$ follows from (7.3.1/2) and well-known properties of the real interpolation method, see [Triα: 1.6.2] which holds also for quasi-Banach spaces.

Remark. See also 2.3.2 where we formulated some embeddings of the above type in $\mathbb{R}^n$. Further embedding theorems for spaces on $\mathbb{R}^n$ may be found in [Triß: 2.7.1]. There are no principal problems to extend these results from $\mathbb{R}^n$ to the above manifold M. If only spaces of type F^s_{pq} are involved then Definition 7.2.2 provides the possibility to switch over from $\mathbb{R}^n$ to M. Then spaces of type B^s_{pq} can be incorporated afterwards via interpolation on the basis of Definition 7.3.1.

7.4.3 Lifting properties

Recall that Δ always stands for the Laplace–Beltrami operator given in local coordinates by (7.2.5/1). In the euclidean case, i.e., $M = \mathbb{R}^n$, it is well-known that $\rho\,\mathrm{id} - \Delta$ with $\rho > 0$ yields an isomorphic mapping from F^{s+2}_{pq} onto F^s_{pq} and from B^{s+2}_{pq} onto B^s_{pq} for all $s \in \mathbb{R}$, $0 < p \leqq \infty$ (with $p < \infty$ in the case of the F-spaces), $0 < q \leqq \infty$. We refer to [Triß: 2.3.8] where one finds more general lifting properties for spaces in $\mathbb{R}^n$ which correspond to fractional powers of $\rho\,\mathrm{id} - \Delta$. Later on we return to fractional powers of the Laplace–Beltrami operator, but for the time being we restrict ourselves to $\rho\,\mathrm{id} - \Delta$, where ρ is sufficiently large such that Proposition 7.2.5/2 with $G = 1$ can be applied.

Theorem. Let M be the Riemannian manifold in the sense of Hypotheses 7.2.1.

(i) Let $s \in \mathbb{R}$ and either $0 < p < \infty$, $0 < q \leqq \infty$ or $p = q = \infty$. Let $\rho > 0$ be a sufficiently large number, then $\rho\,\mathrm{id} - \Delta$ yields an isomorphic map from $F^{s+2}_{pq}(M)$ onto $F^s_{pq}(M)$.

(ii) Let $s \in \mathbb{R}$, $0 < p \leqq \infty$ and $0 < q \leqq \infty$. Let $\rho > 0$ be a sufficiently large number, then $\rho\,\mathrm{id} - \Delta$ yields an isomorphic map from $B^{s+2}_{pq}(M)$ onto $B^s_{pq}(M)$.

Proof. Let p, q, s be given, then we choose $G \in \mathbb{N}$ and $\rho > 0$ such that Proposition 7.2.5/2 with s and $s+2$ and its counterpart with $G+1$ instead of G can be applied. Then the proofs in 7.2.6 and Step 4 of the proof of Theorem 7.3.2 show that $(\rho\,\mathrm{id} - \Delta)^G$ yields an isomorphic map from $F^{s+2G}_{pq}(M)$ onto $F^s_{pq}(M)$ and from $B^{s+2G}_{pq}(M)$ onto $B^s_{pq}(M)$. Similarly with $G+1$ instead of G. Let $f \in F^{s+2}_{pq}(M)$ and $f = (\rho\,\mathrm{id} - \Delta)^G g$ with $g \in F^{s+2+2G}_{pq}(M)$. Then we have

$$\|(\rho\,\mathrm{id} - \Delta)f \mid F^s_{pq}(M)\| \sim \|g \mid F^{s+2+2G}_{pq}(M)\| \sim \|f \mid F^{s+2}_{pq}(M)\|. \tag{1}$$

Now the claimed lifting property is an easy consequence of (1) and of the above mapping propeties.

7.4.4 Interpolation

In [Triα: 2.4] and [Triß: 2.4] we developed an interpolation theory for the spaces B^s_{pq} and F^s_{pq} on $\mathbb{R}^n$, see also 1.6.7 and the preceding subsections. It seems to be reasonable to expect that these formulas can be carried over from $\mathbb{R}^n$ to the above manifold M. As far as the spaces $F^s_{pq}(M)$ are concerned this hope is fully justified by Definition 7.2.2. But there is no similar procedure to reduce $B^s_{pq}(M)$ to the corresponding spaces on $\mathbb{R}^n$, we have to rely solely on Definition 7.3.1. However on this basis something can be done. We formulate few results which will be useful later on. Let again $(\cdot,\cdot)_{\theta,q}$ be the real interpolation method introduced in 1.6.2 and let $[\cdot,\cdot]_\theta$ be the classical complex method which may be found in 1.6.3.

Theorem. Let M be the Riemannian manifold in the sense of Hypotheses 7.2.1.

(i) Let $0 < p < \infty$, $0 < q_0 \leqq \infty$, $0 < q_1 \leqq \infty$, $-\infty < s_0 < s_1 < \infty$. Let $0 < q \leqq \infty$, $0 < \theta < 1$ and

$$s = (1-\theta)s_0 + \theta s_1, \tag{1}$$

then

$$(B^{s_0}_{pq_0}(M), B^{s_1}_{pq_1}(M))_{\theta,q} = (F^{s_0}_{pq_0}(M), F^{s_1}_{pq_1}(M))_{\theta,q} = B^s_{pq}(M). \tag{2}$$

(ii) Let $s_0 \in \mathbb{R}$, $s_1 \in \mathbb{R}$, $1 < p_0 < \infty$, $1 < p_1 < \infty$, $1 < q_0 < \infty$, $1 < q_1 < \infty$. Let $0 < \theta < 1$, let s be given by (1), and

$$\frac{1}{p} = \frac{1-\theta}{p_0} + \frac{\theta}{p_1}, \qquad \frac{1}{q} = \frac{1-\theta}{q_0} + \frac{\theta}{q_1} \tag{3}$$

then

$$[F^{s_0}_{p_0q_0}(M), F^{s_1}_{p_1q_1}(M)]_\theta = F^s_{pq}(M). \tag{4}$$

Proof. Step 1. We prove (i). By (7.3.1/2,3) we have

$$B^s_{pq}(M) = (B^{s_0}_{pp}(M), B^{s_1}_{pp}(M))_{\theta,q}. \tag{5}$$

Then (2) follows from the reiteration theorem of interpolation theory, see [Triα: 1.10] extended to quasi-Banach spaces, and (7.4.2/2).

Step 2. We prove (ii). We use the operators Λ and Ψ from (7.2.3/5–7). We arrive at a standard situation of interpolation theory, the method of retraction-coretraction, see [Triα: 2.4], which reduces (4) to the question whether

$$[l_{p_0}(F^{s_0}_{p_0q_0}), l_{p_1}(F^{s_1}_{p_1q_1})]_\theta = l_p(F^s_{pq}) \tag{6}$$

holds, where $F^{s_0}_{p_0q_0}$, $F^{s_1}_{p_1q_1}$, and F^s_{pq} are the corresponding spaces on $\mathbb{R}^n$. However (6) is known, see [Triα: (1.18.1/4)], which shows that the left-hand side of (6) coincides with

$$l_p([F^{s_0}_{p_0q_0}, F^{s_1}_{p_1q_1}]_\theta), \tag{7}$$

and (1.6.7/6) as far as the inner complex interpolation is concerned, see also [Triα: (2.4.2/7)].

Remark. The two steps show clearly how to extend interpolation results for the spaces F^s_{pq} and B^s_{pq} from $\mathbb{R}^n$ to M: the first one proves corresponding assertions for the spaces $F^s_{pq}(M)$ via reduction to $\mathbb{R}^n$ in the sense of Step 2, secondly, one uses the definition of $B^s_{pq}(M)$ and general properties of interpolation methods in order to incorporate $B^s_{pq}(M)$.

7.4.5 Fractional Sobolev spaces

On $\mathbb{R}^n$ we discussed some distinguished types of spaces, included in the two general scales B^s_{pq} and F^s_{pq}, which are of peculiar interest both for applications and from a historical point of view. We refer to the Hölder–Zygmund spaces $\mathcal{C}^s$, see Definition 1.2.2/2, Theorem 1.5.1(ii), and 2.6.5; the Sobolev spaces W^k_p, see 1.2.3 and Theorem 1.5.1(iii); and their fractional extensions H^s_p, see 1.3.2, 1.3.3, and 2.6.5. It is quite natural to ask for corresponding spaces on Riemannian manifolds which are introduced in a similar way as in $\mathbb{R}^n$. Up to now we formally defined $\mathcal{C}^s(M)$ in (7.4.2/1) and we postpone the task to characterize these spaces in terms of invariant differences to the next section. In this subsection we deal with Sobolev spaces on Riemannian manifolds M and their fractional generalization. Sobolev spaces on M have been introduced by T. Aubin in [Aub1,2] (1976/82) imitating the classical definition (1.2.3/1,2). Looking for a promising way to introduce the fractional Sobolev spaces H^s_p on $\mathbb{R}^n$ one may think of the special case of (1.3.2/5):

$$(\mathrm{id} - \Delta)^{-s/2} L_p = H^s_p. \tag{1}$$

The generalization of (1) from $\mathbb{R}^n$ to M has been studied in detail by R.S. Strichartz and we shall follow his paper [Str2] (1983). It is quite clear that such an approach to introduce spaces via liftings is well prepared by our above results, see in particular 7.4.3. Further comments may be found in 1.11.1 and 1.11.3.

We always assume that M is a Riemannian manifold satisfying the Hypotheses 7.2.1. In order to define the Sobolev spaces $W_p^k(M)$ we need some preliminaries. The covariant derivatives with respect to a given local chart are denoted by ∇_j. If f is a complex C^∞ function on M and $k \in \mathbb{N}_0$ then we put

$$|\nabla^k f|^2 = g^{\alpha_1\beta_1} \cdots g^{\alpha_k\beta_k} \nabla_{\alpha_1} \cdots \nabla_{\alpha_k} f \cdot \nabla_{\beta_1} \cdots \nabla_{\beta_k} \overline{f}. \tag{2}$$

Of course, (2) is invariant. In normal geodesic coordinates, see 7.2.1, we have

$$|\nabla^k f|^2(0) = \sum_{\alpha_l=1}^{n} |\partial_{\alpha_1} \cdots \partial_{\alpha_k} f|^2(0) \tag{3}$$

at the origin, where ∂_j stands for partial derivatives, see 7.2.1. Recall that Δ always stand for the Laplace–Beltrami operator given in local coordinates by (7.2.5/1). Furthermore $\rho\,\mathrm{id} - \Delta$ with $\rho > 0$ and $D(M)$ as its domain of definition is a positive-definite and essentially self-adjoint operator in $L_2(M)$, see 7.2.5 for some details. Then the heat semi-group $\{e^{t\Delta}\}_{t \geqq 0}$ and the Bessel-potentials $(\rho\,\mathrm{id} - \Delta)^{-s/2}$ with $s > 0$ can be defined in $L_2(M)$ via the spectral theorem. They can be extended from $L_2(M)$ to $L_p(M)$ with $1 < p < \infty$, see [Str2: Theorem 3.5 and Section 4] for details. We refer also to Step 3 of the proof of Proposition 7.2.5/1 where we listed some of the deeper properties of resolvent kernels known nowadays.

Definition. Let M be the Riemannian manifold in the sense of Hypotheses 7.2.1.

(i) Let $1 < p < \infty$, $k \in \mathbb{N}$, and

$$\|f \mid W_p^k(M)\| = \sum_{l=0}^{k} \|\nabla^l f \mid L_p(M)\|, \tag{4}$$

then $W_p^k(M)$ is the completion of $\{h\colon C^\infty$ function on M with $\|f \mid W_p^k(M)\| < \infty\}$ in the norm (4).

(ii) Let $1 < p < \infty$ and $\rho > 0$. Let $s > 0$, then $H_p^s(M)$ is the collection of all $f \in L_p(M)$ such that $f = (\rho\,\mathrm{id} - \Delta)^{-s/2} h$ for some $h \in L_p(M)$, with the norm $\|f \mid H_p^s(M)\| = \|h \mid L_p(M)\|$. Let $s < 0$, then $H_p^s(M)$ is the collection of all $f \in D'(M)$ having the form $f = (\rho\,\mathrm{id} - \Delta)^k h$ with $h \in H_p^{2k+s}(M)$ where $k \in \mathbb{N}$ such that $2k + s > 0$, and $\|f \mid H_p^s(M)\| = \|h \mid H_p^{2k+s}(M)\|$. Let $s = 0$ then $H_p^0(M) = L_p(M)$.

Remark 1. Part (i) of the definition coincides essentially with the definitions in [Aub1] and [Aub2: p. 32]. The completion takes place in $L_p(M)$ and hence in $D'(M)$, where the latter has the same meaning as in 7.2.2. Part (ii) goes back to [Str2: in particular Definition 4.1]. The spaces $H^s_p(M)$ with $s < 0$ are independent of k in the sense of equivalent norms. Besides technicalities part (ii) is the extension of (1) from $\mathbb{R}^n$ to M.

Theorem. Let M be the Riemannian manifold in the sense of Hypotheses 7.2.1.

(i) (Paley–Littlewood theorem). Let $1 < p < \infty$ and $s \in \mathbb{R}$, then

$$H^s_p(M) = F^s_{p2}(M). \tag{5}$$

(ii) Let $1 < p < \infty$, and $k \in \mathbb{N}_0$, then

$$W^k_p(M) = H^k_p(M) = F^k_{p2}(M). \tag{6}$$

Proof. Step 1. Let $\psi = \{\psi_j\}_{j=1}^\infty$ be the same resolution of unity as in Definition 7.2.2, then it follows easily

$$\|f \mid W^k_p(M)\| \sim \left(\sum_{j=1}^\infty \|\psi_j f \mid W^k_p(M)\|^p \right)^{1/p}, \qquad k \in \mathbb{N}_0. \tag{7}$$

Furthermore

$$\|\psi_j f \mid W^k_p(M)\| \sim \|\psi_j f \circ \exp_{P_j} \mid W^k_p\|, \tag{8}$$

where W^k_p are the usual Sobolev spaces on $\mathbb{R}^n$, see 1.2.3. Of course we assume that $\psi_j f \circ \exp_{P_j}$ is extended by zero outside of $B(r)$, see (7.2.1/6). Recall the Paley–Littlewood assertion $W^k_p = F^k_{p2}$ referred to in Theorem 1.5.1(iii), see also [Triα: 2.3.3; Triβ: 2.5.6] for proofs. Then (7,8) and (7.2.2/1) yield

$$\|f \mid W^k_p(M)\| \sim \|f \mid F^k_{p2}(M)\|. \tag{9}$$

Recall that $D(M)$ is dense in $F^k_{pq}(M)$, see Theorem 7.4.2(i). We have

$$D(M) \subset \{h: C^\infty \text{ function on } M \text{ with } \|h \mid W^k_p(M)\| < \infty\} \subset F^k_{p2}(M), \tag{10}$$

where the last inclusion comes again from (7.2.2/1) and the above arguments. Hence we arrive at $W^k_p(M) = F^k_{p2}(M)$ and, in particular, $L_p(M) = F^0_{p2}(M)$, which proves half of (6).

Step 2. Let $\rho > 0$ be sufficiently large then $(\rho\,\mathrm{id} - \Delta)^k$ yields an isomorphic map from $F^{2k}_{p2}(M) = W^{2k}_p(M)$ onto $F^0_{p2}(M) = L_p(M)$, see Theorem 7.4.3(i), where we used Step 1. Hence we have by the above definition

$$H^{2k}_p(M) = W^{2k}_p(M) = F^{2k}_{p2}(M), \qquad k \in \mathbb{N}_0, \tag{11}$$

which proves the second half of (6) if k is even.

Step 3. By (11) and (7.4.4/4) we obtain

$$[L_p(M), W_p^{2k}(M)]_\theta = F_{p2}^{2k\theta}(M), \qquad k \in \mathbb{N}, \quad 0 < \theta < 1. \tag{12}$$

Furthermore by (11) and [Str2: Corollary 4.6] we have

$$[L_p(M), W_p^{2k}(M)]_\theta = H_p^{2k\theta}(M), \qquad k \in \mathbb{N}, \quad 0 < \theta < 1. \tag{13}$$

However (12) and (13) prove (5) if $s > 0$. Furthermore by Step 1 and (12,13) with $\theta = \frac{1}{2}$ we arrive at (6) for any $k \in \mathbb{N}_0$.

Step 4. We have (5) with $s = 0$. Finally let $s < 0$, then (5) follows from Step 3 and the lifting property which holds both for $H_p^s(M)$ by definition and for $F_{p2}^s(M)$ by Theorem 7.4.3.

Remark 2. By (7.4.4/4) and (5) we have

$$[H_{p_0}^{s_0}(M), H_{p_1}^{s_1}(M)]_\theta = H_p^s(M), \tag{14}$$

where $s_0 \in \mathbb{R}$, $s_1 \in \mathbb{R}$, $0 < \theta < 1$, $s = (1-\theta)s_0 + \theta s_1$, $1 < p_0 < \infty$, $1 < p_1 < \infty$, $\frac{1}{p} = \frac{1-\theta}{p_0} + \frac{\theta}{p_1}$, which was proved in [Str2: Corollary 4.6] on the basis that $(\rho \,\mathrm{id} - \Delta)^{i\kappa}$, $\kappa \in \mathbb{R}$, is bounded in L_p. Of course, the above theorem extends well-known assertions from $\mathbb{R}^n$ to M, see (2.6.5/2) and Theorem 1.5.1.

7.4.6 Spaces with $s > \sigma_p$

Recall $\sigma_p = n(\frac{1}{p} - 1)_+$ with $0 < p \leqq \infty$, see (2.3.3/1). Spaces of type F_{pq}^s and B_{pq}^s with $s > \sigma_p$ are of peculiar interest because they consist of regular distributions, in contrast to spaces with $s < \sigma_p$. In the case of $\mathbb{R}^n$ we dealt several times with spaces B_{pq}^s and F_{pq}^s with $s > \sigma_p$, see 2.3.3 and some subsections of 2.6. One of the typical features is that $\|f \mid L_p\|$ is one summand in the corresponding equivalent quasi-norms. Now we ask the same question for spaces on manifolds. For this purpose we return to Theorems 7.2.4 and 7.3.2, where we try to replace $\|k_0(\varepsilon, f) \mid L_p(M)\|$ by $\|f \mid L_p(M)\|$.

Let M be the above manifold and let $k(t, f)(P)$ be the same means as in 7.2.4, see in particular (7.2.4/1–6). As in 7.2.4, 7.3.2 etc., we do not indicate the dependence of $k(t, f)(P)$ on the chosen number $N \in \mathbb{N}$.

Theorem. Let M be the Riemannian manifold in the sense of Hypotheses 7.2.1 with the injectivity radius $r_0 > 0$. Let $0 < r < r_0$.

(i) Let either $0 < p < \infty$, $0 < q \leqq \infty$ or $p = q = \infty$, let $s > \sigma_p$, and let $N \in \mathbb{N}$ in (7.2.4/3,5,6) be sufficiently large, then

$$\|f \mid L_p(M)\| + \left\| \left(\int_0^r t^{-sq} |k(t,f)(\cdot)|^q \frac{dt}{t} \right)^{1/q} \mid L_p(M) \right\| \tag{1}$$

(modification if $q = \infty$) is an equivalent quasi-norm in $F^s_{pq}(M)$.

(ii) Let $0 < p \leqq \infty$, $0 < q \leqq \infty$ and $s > \sigma_p$, let $N \in \mathbb{N}$ in (7.2.4/3,5,6) be sufficiently large, then

$$\|f \mid L_p(M)\| + \left(\int_0^r t^{-sq} \|k(t,f) \mid L_p(M)\|^q \frac{dt}{t} \right)^{1/q} \tag{2}$$

(modification if $q = \infty$) is an equivalent quasi-norm in $B^s_{pq}(M)$.

Proof. Part (i) with $s > \frac{2n}{p} + 4$ coincides with Corollary 7.2.4. By Theorem 7.4.3(i) and Remark 7.2.5/4 we can extend this assertion to $s > \sigma_p$ which completes the proof of (i). Now one can follow the arguments in the proof of Theorem 7.3.2 where one replaces (7.2.4/10) by (1), see also (7.2.5/26). We arrive at (2).

7.4.7 Spaces on analytic manifolds

If one compares the euclidean means (2.4.6/1) with their Riemannian counterparts (7.2.4/6) then we substituted quite naturally the euclidean geodesics $x + ty$ by the Riemannian geodesics $c(P, X, t)$. The question arises whether either in the euclidean case or in its Riemannian generalization the geodesics can be replaced by the more general line systems. This would give the means a higher flexibility which one can try to adapt to PDE's or ψDE's or to domains in $\mathbb{R}^n$. In [Tri27] we employed this idea in order to study weighted spaces on domains where the weight is implanted in a suitable line system. But in this book we consider exclusively unweighted spaces although the distinction between weighted and unweighted spaces is somewhat questionable if one deals with Riemannian metrics. The aim of this subsection is a little bit different but in the same spirit. We wish to prepare 7.6 where we consider function spaces of type F^s_{pq} and B^s_{pq} on Lie groups. The basic idea is to convert the Lie group in a Riemannian manifold furnished with a left-invariant Riemannian metric. Unfortunately the corresponding Riemannian geodesics do not coincide in general with the Lie geodesics. So it is quite natural to ask whether the line system of the Riemannian geodesics can be replaced by the line system of the Lie geodesics. We return to this problem later on which will be based on the present consideration.

Let M be the Riemannian manifold in the sense of Hypotheses 7.2.1. The desired replacement of the Riemannian geodesics $c(P, X, t)$ by an other line system causes some trouble which forces us to strengthen our assumptions about M. We suppose in addition that M is analytic. Then we assume that M carries a second structure, an analytic covariant derivation, which may be different from the Levi–Cività derivation. Now we denote by $c(P, X, t)$ the geodesics with respect to this additional analytic covariant derivation. In local charts they obey the differential equations (7.2.1/3) where now the Christoffel symbol Γ^j_{ik} have a new meaning, see [Kli: 1.5.1] for details. Let again $(\Omega_P(r), \exp_P^{-1})$ be the above normal geodesic coordinates (with respect to the Levi–Cività derivation), $0 < r < r_0$, see 7.2.1. Then we assume that the new Christoffel symbol Γ^j_{ik} (of the additional covariant derivation) is analytic in $(\Omega_P(r), \exp_P^{-1})$ and that there exist two positive numbers b and c with

$$|D^\alpha \Gamma^j_{ik}| \leqq c\, \alpha!\, b^{|\alpha|} \tag{1}$$

in the normal geodesic coordinates of every local chart $(\Omega_P(r), \exp_P^{-1})$. This condition complements now (7.2.1/8) where $D^\alpha g_{ij}$ satisfy now an estimate of type (1). It ensures a uniform analyticity of these new Christoffel symbols. Now we re-interpret the means $k(t, f)(P)$ in (7.2.4/6): All symbols have the previous meaning with exception of $c(P, X, t)$ and its image $C(\varphi(P), \varphi_* X, t)$ which are now identified with the geodesics generated by the additional analytic covariant derivation. Similarly $k_0(\varepsilon, f)(P)$ is re-interpreted and we always assume that the corresponding kernels k and k_0 satisfy (7.2.4/1–5), where we do not indicate explicitly $N \in \mathbb{N}$.

Theorem. Let M be an analytic Riemannian manifold in the sense of Hypotheses 7.2.1 equipped with a double analytic structure: the metric and an independent covariant derivation whose Christoffel symbols satisfy (1). Let $k(t, f)$ and $k_0(\varepsilon, f)$ be the above newly interpreted means.

(i) Let $s \in \mathbb{R}$ and let either $0 < p < \infty$, $0 < q \leqq \infty$ or $p = q = \infty$. Let $\varepsilon > 0$ and $r > 0$ be sufficiently small and $N \in \mathbb{N}$ in (7.2.4/3,5,6) be sufficiently large (in dependence on s, p, q), then (7.2.4/10) (modification if $q = \infty$) is an equivalent quasi-norm in $F^s_{pq}(M)$. If $s > \sigma_p = n(\frac{1}{p} - 1)_+$ then $\|k_0(\varepsilon, f) \mid L_p(M)\|$ can be replaced by $\|f \mid L_p(M)\|$.

(ii) Let $s \in \mathbb{R}$ and let $0 < p \leqq \infty$, $0 < q \leqq \infty$. Let $\varepsilon > 0$ and $r > 0$ be sufficiently small and $N \in \mathbb{N}$ in (7.2.4/3,5,6) be sufficiently large, then (7.3.2/1) (modification if $q = \infty$) is an equivalent quasi-norm in $B^s_{pq}(M)$. If $s > \sigma_p$ then $\|k_0(\varepsilon, f) \mid L_p(M)\|$ can be replaced by $\|f \mid L_p(M)\|$.

Proof. Step 1. We follow the proof in 7.2.4. There are no changes up to the crucial expansions (7.2.4/19,20). The replacement of $\sum_{l=2}^{2L-1}$ in (7.2.4/19) by $\sum_{2\leqq|\alpha|\leqq 2L-1}$ in (7.2.4/20) remains unchanged because it is based on the differential equations (7.2.1/3) and now on (1). However the estimate of the remainder terms in (7.2.4/19, 20) was based on (7.2.4/21) which is now not guaranteed. But this is precisely the point where we need our stronger assumption. Instead of the remainder term in (7.2.4/19) we develop $C^a(x,X,t)$ in its full Taylor series. We use again (7.2.1/3) and now (1) which show that this expansion converges if $0<t<r$ and r is sufficiently small. Then we have (7.2.4/20) where again $|R^a_{2L}(x,X,t)| \leqq c$ for all $x\in U$, $0<t<r$, $|X|\leqq 1$ with $r>0$ small. The rest is now the same as in 7.2.4 which proves (i) if s is large. The extension of the proof from large values of s to arbitrary values of s can be done in the same way as in 7.2.5 and 7.2.6. Remainder estimates must always be treated via full Taylor series, for instance in (7.2.5/19) where tX is now perturbed analytically in the above way. The rest in 7.2.5 and 7.2.6 remains unchanged. The corresponding proof for the spaces $B^s_{pq}(M)$ in 7.3.2 causes no problem.

Step 2. Let $s>\sigma_p$, then we can apply the same arguments as in the proof of Theorem 7.4.6 including the mentioned references.

Remark. In the light of the indicated proof we return briefly to the beginning of this subsection. As we said, for further applications it would be highly desirable to relax the rigid constructions in the means (2.4.6/1) and (7.2.4/6) replacing the euclidean or Riemannian geodesics by more general line systems, maybe governed by differential equations of type (7.2.1/3) or so. The above proof shows clearly the difficulties which may arise and which are mostly connected with remainder estimates. In the original approach in 7.2.4 we used for this purpose that (7.2.1/4) is constant along the geodesics which resulted in the estimate (7.2.4/21). In general such a simple argument is not available and we circumvented this difficulty in the above theorem by imposing analyticity. This is not really necessary but we wish to emphasize that we incorporated the above theorem not so much for its own sake but to prepare our investigations about spaces on Lie groups. In general one would like to return to C^∞ manifolds or C^∞ domains in $\mathbb{R}^n$ or C^∞ weights etc. and general line systems. The first question is to find a substitute of (7.2.4/21). Let us assume that the line system obeys differential equations of type (7.2.1/3). Let $u(t)=\sum_{k=1}^n |\frac{dC^k}{dt}(t)|^2$, then (7.2.1/3) yields an inequality of the type

$$\left|\frac{dC^j}{dt}(t)\right| \leqq c + \int_0^t g(s)u(s)ds \quad \text{with} \quad c=|X|^2\leqq 1;\ j=1,\ldots,n. \tag{2}$$

But (2) can be used to find a substitute of (7.2.4/21) for small values of t. We shall not follow this way, we wanted simply to make clear that analyticity is not the last word in this context.

7.5 Characterizations of spaces on manifolds via differences and derivatives

7.5.1 The basic theorem

In 2.6.1 – 2.6.3 we studied characterizations of B^s_{pq} and F^s_{pq} on $\mathbb{R}^n$ in terms of differences and derivatives of functions. One can ask for similar descriptions of the spaces $B^s_{pq}(M)$ and $F^s_{pq}(M)$ on the above Riemannian manifold M. However there is a significant difference between the general manifold in the sense of Hypotheses 7.2.1 and $\mathbb{R}^n$ which enables us to find a counterpart of (2.6.2/1) for $F^s_{pq}(M)$ but not a counterpart of (2.6.1/2) for $B^s_{pq}(M)$. We return to this point later on. Furthermore there are some new technical problems. Our way to circumvent these difficulties is to prove a basic theorem which is an extension of Theorem 2.4.3/2 from $\mathbb{R}^n$ to M. Somewhat in contrast to $\mathbb{R}^n$ we take this theorem as the starting point of the following considerations. Let

$$l \in S'(\mathbb{R}) \quad \text{with} \quad \operatorname{supp} l \subset [-1, 1], \tag{1}$$

which coincides with (2.4.3/10). Then we have

$$|l(g)| \leqq c \sup_{|\lambda| \leqq 2} \sum_{j=0}^{T} \left| \frac{d^j g}{d\lambda^j}(\lambda) \right|, \qquad g \in S(\mathbb{R}), \tag{2}$$

for some $c > 0$ and $T \in \mathbb{N}_0$. By (1) the one-dimensional Fourier transform $\psi = F_1 l$ of l is an analytic function on $\mathbb{R}$. As in (2.4.3/11) we assume

$$0 < |\psi(\lambda)| \leqq c\lambda^m \quad \text{if} \quad 0 < \lambda < \delta \tag{3}$$

for some $\delta > 0$ and $m \in \mathbb{N}$. Let M be the manifold satisfying the Hypotheses of 7.2.1 and let $c(P, X, t)$ be the same geodesics as in 7.2.1. Then the natural counterpart of (2.4.3/12) is given by

$$l(X, f)(P) = \int_{\mathbb{R}} l(\lambda) f(c(P, X, \lambda)) d\lambda, \qquad P \in M, \quad 0 \neq X \in T_P M. \tag{4}$$

This is an integration along geodesics which makes sense at least if f is smooth. We adopt the same point of view as in front of Theorem 2.4.3/2: first we apply (4) to, say, C^∞ functions f on M, then we derive inequalities and equivalence assertions, and the rest is a matter of limiting arguments. The latter point is based on Theorem 7.4.2(i) in the case of $F^s_{pq}(M)$ with $0 < p < \infty$, $0 < q < \infty$. If $0 < p < \infty$, $q = \infty$, or $p = q = \infty$ then limiting arguments can be based on Fatou's lemma, in analogy to Step 4 of the proof 7.2.6 extended to $p = q = \infty$. In other words, we always assume that f in (4) is smooth enough and that the results can be extended to all $f \in F^s_{pq}(M)$ under consideration. Let

$$B_P(r) = \{X \in T_P M,\ \|X\| < r\}, \tag{5}$$

see (7.2.1/6) where we now emphasize that $B_P(r)$ is a ball in $T_P M$ and $\|X\| = \sqrt{g(X,X)}$. Let $H(x) \in S$ be the same function as in Theorem 2.4.3, i.e., $H(x)$ satisfies (2.4.1/2,3). Finally if $s \in \mathbb{R}$ then $[s]$ stands as usual for the largest integer less than or equal to s, and σ_{pq} has been defined by (2.4.1/1).

Theorem. Let M be the Riemannian manifold in the sense of Hypotheses 7.2.1 with the injectivity radius $r_0 > 0$. Let $s \in \mathbb{R}$ and let either $0 < p < \infty$, $0 < q \leqq \infty$ or $p = q = \infty$. Let $l(X, f)(P)$ be given by (4) where the distribution $l \in S'(\mathbb{R})$, its one-dimensional Fourier transform $\psi = F_1 l$, and the numbers $m \in \mathbb{N}$ and $T \in \mathbb{N}_0$ have the above meaning. Let for some $s_0 \in \mathbb{R}$ and some $a \in \mathbb{R}$ with $a > \frac{n}{\min(p,q)}$

$$\sup |h|^{-s_0 + \frac{b}{2}} \int_{\mathbb{R}^n} |(\psi^{(b)}(hz)H(z))^\vee(y)\ |\ (1 + |y|^a)dy < \infty \tag{6}$$

where the supremum is taken over all $h \in \mathbb{R}^n$ with $|h| \geqq 1$ and $b = 0, \ldots, [s]$. Let

$$m > \frac{3}{2}s, \qquad s > 2\left(T + 2 + \frac{n}{p}\right), \qquad s > \sigma_{pq} + s_0, \tag{7}$$

and $0 < r < r_0$, then

$$\|f \mid L_p(M)\| + \left\| \left(\int_{B_P(r)} \|X\|^{-sq} |l(X,f)(\cdot)|^q \frac{dX}{\|X\|^n} \right)^{1/q} \mid L_p(M) \right\| \tag{8}$$

is an equivalent quasi-norm in $F^s_{pq}(M)$.

Proof. Step 1. In 7.2.4 we proved Theorem 7.2.4 for $s > \frac{2n}{p} + 4$ by reduction to Theorem 2.4.6. For this purpose we used the expansion (7.2.4/19) in order to prove the crucial equivalence (7.2.4/14). Now we adapt that proof to the situation under consideration and reduce the above assertion to Theorem 2.4.3/2. The crucial point in the proof in 7.2.4 transferred to our situation is the following. Let $\psi = \{\psi_j\}_{j=1}^\infty$ be the same resolution of unity as in 7.2.4 and let again $h = \psi_j f \circ \exp_{P_j}$ and $U = B(r')$ as in Step 3 in 7.2.4 and (7.2.4/16). Then the counterpart of the decisive equivalence (7.2.4/17) reads as follows,

$$\|h \mid F^s_{pq}\| \sim \|h \mid L_p(U)\| + \left\| \left(\int_{B(r)} |X|^{-sq} |l(X,h)(\cdot)|^q \frac{dX}{|X|^n} \right)^{1/q} \mid L_p(U) \right\| \tag{9}$$

with $h(X) = 0$ if $|X| > \delta$. The equivalence in (9) must be understood independently of j. All other arguments from 7.2.4 can be taken over with obvious modification. Similar as in 7.2.4 we may assume that $l(X, h)$ in the underlying local chart $(\Omega, \varphi) = (\Omega_{P_j}(r'), \exp_{P_j}^{-1})$ is given by

$$l(X,h)(x) = \int_{\mathbb{R}} l(\lambda) h(C(x, X, \lambda)) d\lambda, \qquad x \in U, \tag{10}$$

$|X| \leqq r$ and $C(x, X, \lambda) \subset U$.

Step 2. In order to prove (9) with (10) we follow Step 3 in 7.2.4. We use again the Taylor expansion (7.2.4/19). Let $(C^a)^{(j)}(x, X, \lambda) = \frac{d^j C^a}{d\lambda^j}(x, X, \lambda)$. We may assume $r = 1$ in (9). Then we have

$$\begin{aligned} C^a(x, X, \lambda) = x^a + \lambda X^a + \sum_{j=2}^{2L-1} \frac{\lambda^j}{j!} (C^a)^{(j)}(x, X, 0) \\ + \frac{\lambda^{2L}}{(2L-1)!} \int_0^1 (1-\tau)^{2L} (C^a)^{(2L)}(x, X, \tau\lambda) d\tau, \end{aligned} \tag{11}$$

see (3.4.4/4). By $C^a(x, X, \lambda) = C^a(x, \frac{X}{|X|}, |X|\lambda)$ we arrive at

$$\begin{aligned} C^a(x, X, \lambda) = x^a + \lambda X^a + \sum_{j=2}^{2L-1} \frac{\lambda^j |X|^j}{j!} (C^a)^{(j)} \left(x, \frac{X}{|X|}, 0 \right) \\ + \frac{\lambda^{2L} |X|^{2L}}{(2L-1)!} \int_0^1 (1-\tau)^{2L} (C^a)^{(2L)} \left(x, \frac{X}{|X|}, |X|\tau\lambda \right) d\tau \end{aligned} \tag{12}$$

and, similarly as in (7.2.4/20),

$$\begin{aligned} C^a(x, X, \lambda) = x^a + \lambda X^a + \sum_{2 \leqq |\alpha| \leqq 2L-1} \lambda^{|\alpha|} b^a_\alpha(x) X^\alpha \\ + \lambda^{2L} |X|^{2L} \tilde{R}^a_{2L}(x, X, \lambda), \end{aligned} \tag{13}$$

where the remainder term $\tilde{R}^a_{2L}(x, X, \lambda)$ is essentially the integral in (12). By the same arguments as in 7.2.4 the terms $\lambda^{2L} \tilde{R}^a_{2L}(x, X, \lambda)$ and all their derivatives with respect to λ up to order T can be estimated uniformly. The counterpart of (7.2.4/22) reads now as follows,

$$\begin{aligned} \left| D^\beta h(C(x, X, \lambda)) - D^\beta h \left(x + \lambda X + \sum_{2 \leqq |\alpha| \leqq 2L-1} \lambda^{|\alpha|} b_\alpha(x) X^\alpha \right) \right| \\ \leqq c |X|^{2L} \sum_{|\gamma| \leqq T+1} |D^\gamma h(\cdots)| \end{aligned} \tag{14}$$

with $0 \leqq |\beta| \leqq T$. Now we insert (13) in (10), then we use (2,14) and put the resulting remainder term in the second term on the right-hand side of (9) which, consequently, can be estimated from above by

$$c \left\| \sup_{|x-y| \leqq c} \left| \sum_{|\gamma| \leqq 1+T} D^\gamma h(y) \right| \mid L_p(U) \right\| \leqq c \| h \mid F^\sigma_{pq} \|, \tag{15}$$

provided that $2L > s$ and $1 + T + \frac{n}{p} < \sigma$, see (7.2.4/24). Next we use (7.2.4/25) and follow the arguments afterwards with the above modifications. Then we arrive at the counterpart of (7.2.4/26), now with $\sum_{|\beta| \leqq L+T}$ instead of $\sum_{|\beta| = L}$. Hence we have to assume $\kappa > L + T + \frac{n}{p}$. In other words, instead of (7.2.4/12) we need now

$$L + T + \frac{n}{p} < s - 1 < s < 2L. \tag{16}$$

We follow the arguments after (7.2.4/26). Let

$$l^{euc}(X, h)(x) = \int_{\mathbb{R}} l(\lambda) h(x + \lambda x) d\lambda, \qquad x \in \mathbb{R}^n, \ \ 0 \neq X \in \mathbb{R}^n, \tag{17}$$

see (2.4.3/12). The term with $|\beta| = 0$ in (7.2.4/25) inserted in (10) yields (17) which we put in the second term on the right-hand side of (9). Then we have (2.4.3/15) with $\|h \mid L_p(U)\|$ instead of $\|\varphi_0(D) h \mid L_p\|$. The latter replacement is justified by the proof of Corollary 7.2.4. In other words, the term with $|\beta| = 0$ in (7.2.4/25) generates also in the present case the desired equivalence (9). There remain the counterparts of (7.2.4/27), and as in 7.2.4 only the terms with $|\alpha| \leqq s$ must be considered in detail, which look like

$$X^\alpha \int_{\mathbb{R}} l(\lambda) \lambda^{|\alpha|} D^\beta h(x + \lambda x) d\lambda, \qquad 0 < |\alpha| \leqq [s], \ \ 2|\beta| \leqq |\alpha|. \tag{18}$$

We insert this expression in the second term on the right-hand side of (9) which can be estimated from above by

$$\left\| \left(\int_{B(1)} |X|^{-(s-|\alpha|)q} \left| \int l(\lambda) \lambda^{|\alpha|} D^\beta h(\cdot + \lambda X) d\lambda \right|^q \frac{dX}{|X|^n} \right)^{1/q} \mid L_p(U) \right\|. \tag{19}$$

In (19) we may replace $s - |\alpha|$ by $s - \frac{|\alpha|}{2} - \varepsilon$ with $0 < \varepsilon \leqq \frac{1}{2}$. Recall $F_1(\lambda^b l(\lambda)) = \psi^{(b)}$. By (6), $m - b > s - \frac{b}{2}$ for $b = 0, \ldots, [s]$ and (2.4.3/13, 14) the quasi-norm (19) can be estimated from above by

$$c \left\| D^\beta h \mid F_{pq}^{s - \frac{|\alpha|}{2} - \varepsilon} \right\| \leqq c' \left\| h \mid F_{pq}^{s + |\beta| - \frac{|\alpha|}{2} - \varepsilon} \right\| \leqq c'' \| h \mid F^{s-\varepsilon}_{pq} \| \tag{20}$$

for some $\varepsilon > 0$. But this is just what we want, see (7.2.4/28,29). Now the rest is the same as in 7.2.4. Especially by the same argument as in (7.2.4/12) we have (16) if (7) is satisfied.

Remark. The above theorem extends Theorem 2.4.3/2 from $\mathbb{R}^n$ to M. Unfortunately we had been forced in the course of the proof to replace the rather natural restrictions (2.4.3/13) by the a little awkward conditions (7). In Theorem 2.5.2/2 we proved the counterpart of Theorem 2.4.3/2 for B^s_{pq} spaces on $\mathbb{R}^n$. We applied this result also occasionally, see Step 2 of the proof of Theorem 2.6.1. However for a general Riemannian manifold M of the above type the situation is completely different. There is no chance to extend Theorem 2.5.2/2 from B^s_{pq} on $\mathbb{R}^n$ to $B^s_{pq}(M)$ in the sense of the above theorem. For such a purpose a natural one-to-one map between the two tangent spaces would be necessary. This can be done if M has a distant parallelism, but this is essentially the euclidean case. Hence we have another example of the priority of $F^s_{pq}(M)$ over $B^s_{pq}(M)$.

7.5.2 Characterizations via differences

In Remark 2.4.3/2 we indicated the possibility to apply Theorem 2.4.3/2 in order to get equivalent quasi-norms for F^s_{pq} on $\mathbb{R}^n$ in terms of differences of functions. However our considerations in 2.6.2 were mainly based on other techniques although we referred to Theorem 2.4.3/2 occasionally. Now the situation is somewhat different. All what will be done in the sequel is based on Theorem 7.5.1, the counterpart of Theorem 2.4.3/2. We use the same notations as in the preceding subsection. Let $F_1 l = \psi(\lambda) = (e^{i\nu\lambda} - 1)^m$ with $\nu = \frac{1}{m}$ and $m \in \mathbb{N}$. Then l is a linear combination of δ-distributions and it satisfies (7.5.1/1–3) with $T = 0$. Furthermore we have

$$l(X, f)(P) = \sum_{j=0}^{m} (-1)^{m-j} \binom{m}{j} f\left(c\left(P, X, \frac{j}{m}\right)\right) = \Delta^m_X f(P) \tag{1}$$

which stands for the mth differences along the geodesic $c(P, X, t)$. Let again $B_P(r)$ be given by (7.5.1/5).

Theorem. Let M be the Riemannian manifold in the sense of Hypotheses 7.2.1 with the injectivity radius $r_0 > 0$. Let either $0 < p < \infty$, $0 < q \leqq \infty$ or $p = q = \infty$. Let

$$s > \frac{2n}{\min(p,q)} + 2n \left(\frac{1}{\min(p,q)} - 1 \right)_+ + 4 \quad \text{and} \quad m > \frac{3}{2} s \tag{2}$$

with $m \in \mathbb{N}$. Let $0 < r < r_0$, then

$$\| f \mid L_p(M)\| + \left\| \left(\int_{B_P(r)} \|X\|^{-sq} |\Delta^m_X f(\cdot)|^q \frac{dX}{\|X\|^n} \right)^{1/q} \mid L_p(M) \right\| \tag{3}$$

(modification if $q = \infty$) is an equivalent quasi-norm in $F^s_{pq}(M)$.

Proof. We apply Theorem 7.5.1 to $F_1 l = \psi(\lambda) = (e^{i\nu\lambda} - 1)^m$ with $\nu = \frac{1}{m}$. As we remarked, (7.5.1/1–3) are satisfied and (7.5.1/8) coincides with (3), see (1). Furthermore (7.5.1/6) holds if $-s_0 + \frac{s}{2} + a \leqq 0$ where a is any number $a > \frac{n}{\min(p,q)}$; hence any $s_0 > \frac{s}{2} + \frac{n}{\min(p,q)}$ is sufficient. By (2) we have (7.5.1/7) if s_0 is chosen in an appropriate way, where we used $\sigma_{pq} = n(\frac{1}{\min(p,q)} - 1)_+$, see (2.4.1/1).

Remark. This theorem is the extension of (2.6.2/1) from $\mathbb{R}^n$ to M. The restrictions (2) are not natural, they depend on our method. It would be desirable to have similar restrictions as in Theorem 2.6.2. In particular it should be possible to replace $m > \frac{3}{2}s$ by $m > s$. In general this causes some trouble, but see the interesting case $p = q = \infty$ which will be treated in detail in the following subsection. There is no counterpart of the above theorem for the spaces $B^s_{pq}(M)$, see Remark 7.5.1.

7.5.3 The Hölder–Zygmund spaces

Mostly for sake of brevity and in analogy to $\mathbb{R}^n$ we introduced in (7.4.2/1) the spaces $\mathcal{C}^s(M) = F^s_{\infty\infty}(M)$ where $s > 0$ and called the Hölder–Zygmund spaces. In the case $M = \mathbb{R}^n$ this notation is justified by (2.6.5/1) (with $B^s_{\infty\infty} = F^s_{\infty\infty}$) and the given references in particular to 1.2.2. Now we look for corresponding assertions for general manifols M of the above type. Unfortunately the shortcomings of Theorem 7.5.2 force us to introduce a further class $\mathfrak{C}^s(M)$ directly defined via differences (7.5.2/1) on M.

Definition. Let M be the Riemannian manifold in the sense of Hypotheses 7.2.1. Let $s > 0$, then

$$\mathcal{C}^s(M) = F^s_{\infty\infty}(M) \tag{1}$$

and

$$\mathfrak{C}^s(M) = \Big\{ f \in L_\infty(M)\colon \|f \mid \mathfrak{C}^s(M)\| \tag{2}$$

$$= \sup_{P\in M} |f(P)| + \sup_{P\in M}\ \sup_{X\in T_PM} \|X\|^{-s} |\Delta_X^{1+[s]} f(P)| < \infty \Big\}.$$

Remark. Formula (1) coincides with (7.4.2/1). Let $M = \mathbb{R}^n$: It is obvious that $|h| \leqq 1$ in (2.6.1/2) with $p = q = \infty$ can be replaced by $\mathbb{R}^n$ in the sense of an equivalent norm in $B^s_{\infty\infty} = F^s_{\infty\infty}$. In other words, by Corollary 2.6.1/1 we have

$$\mathcal{C}^s(\mathbb{R}^n) = \mathfrak{C}^s(\mathbb{R}^n) = B^s_{\infty\infty}(\mathbb{R}^n), \qquad s > 0. \tag{3}$$

We refer also to (2.6.5/1) and 1.2.2. Hence there are good reasons to conjecture $\mathcal{C}^s(M) = \mathfrak{C}^s(M)$ for all $s > 0$, but our technique covers only $s > 4$.

Theorem. Let M be the Riemannian manifold in the sense of Hypotheses 7.2.1.

(i) Let $m > s > 0$ with $m \in \mathbb{N}$, then $\mathfrak{C}^s(M)$ is a Banach space and

$$\|f \mid \mathfrak{C}^s(M)\|^m = \sup_{P \in M} |f(P)| + \sup_{P \in M} \sup_{X \in T_P M} \|X\|^{-s} |\Delta^m_X f(P)| \tag{4}$$

is an equivalent norm.

(ii) Let $s > 4$ then

$$\mathcal{C}^s(M) = \mathfrak{C}^s(M). \tag{5}$$

Proof. Step 1. We prove (i). We obtain by standard arguments that $\mathfrak{C}^s(M)$ is linear and complete. Let $m_1 > m_2 > s > 0$ with $m_1 \in \mathbb{N}$ and $m_2 \in \mathbb{N}$. Let $0 \neq X \in T_P M$, then $X(t)$ stands for the parallel displacement of X along the geodesic $c(P, X, t)$, $t \in \mathbb{R}$. We have $\|X(t)\| = \|X\|$,

$$\|X\|^{-s} |\Delta^{m_1}_X f(P)| \leqq c \sup_{t \in \mathbb{R}} \|X(t)\|^{-s} |\Delta^{m_2}_{X(t)} f(c(P, X, t))|, \tag{6}$$

and consequently

$$\|f \mid \mathfrak{C}^s(M)\|^{m_1} \leqq c \|f \mid \mathfrak{C}^s(M)\|^{m_2}. \tag{7}$$

In order to prove the converse assertion we begin with the identity

$$(z-1)^m = 2^{-m}(z^2-1)^m + (z-1)^{m+1} P(z), \qquad z \in \mathbb{C}, \quad m \in \mathbb{N}, \tag{8}$$

where $P(z)$ is an appropriate polynomial, $m \in \mathbb{N}$. Let $\xi \in \mathbb{R}^n$ and $h \in \mathbb{R}^n$ then we have

$$(e^{ih\xi} - 1)^m = 2^{-m}(e^{2ih\xi} - 1)^m + (e^{ih\xi} - 1)^{m+1} P(e^{ih\xi}). \tag{9}$$

We multiply (9) with $\hat{f}(\xi)$, apply the inverse Fourier transform to the result, and arrive at

$$(\Delta_h^m f)(x) = 2^{-m}(\Delta_{2h}^m f)(x) + \Delta_h^{m+1} \sum a_l f(x+lh), \tag{10}$$

where $\sum$ with $a_l \in \mathbb{R}$ is a finite sum generated by the above polynomial and Δ_h^m are the usual differences on $\mathbb{R}^n$, see (7.5.2/1) with $M = \mathbb{R}^n$ or (2.6.1/1,6). Of course there is an immediate analog of (10) along the geodesic $c(P, X, t)$ in terms of $\Delta_X^m f(P)$. Using the abbreviation (4) and a similar argument as in (6) we arrive at

$$\|X\|^{-s}|\Delta_X^m f(P)| \leqq 2^{-m+s}\|2X\|^{-s}|\Delta_{2X}^m f(P)| + c\|f \mid \mathfrak{C}^s(M)\|^{m+1}. \tag{11}$$

Let $m > s$ and $f \in \mathfrak{C}^s(M)$, then, by (7), any $\|f \mid \mathfrak{C}^s(M)\|^m$ is finite. Now $m > s$ in (11) and interation yield

$$\|f \mid \mathfrak{C}^s(M)\|^{m_2} \leqq c\|f \mid \mathfrak{C}^s(M)\|^{m_1}, \qquad m_1 > m_2 > s > 0, \tag{12}$$

which is the converse of (7). Hence, all the quasi-norms (4) are equivalent to each other.

Step 2. We prove (ii). Let $r_0 > 0$ be the injectivity radius of M and let $B_P(r)$ be given by (7.5.1/5) then it follows immediately that

$$\sup_{P\in M} |f(P)| + \sup_{P\in M} \sup_{X\in B_P(r)} \|X\|^{-s}|\Delta_X^m f(P)| \tag{13}$$

with $m > s$, $m \in \mathbb{N}$, and $0 < r < r_0$ is also an equivalent norm on $\mathfrak{C}^s(M)$. By Theorem 7.5.2 it follows that $\|f \mid F^s_{\infty\infty}(M)\|$ and $\|f \mid \mathfrak{C}^s(M)\|$ for $s > 4$ are equivalent to each other, at least for smooth functions. However any $f \in F^s_{\infty\infty}(M)$ or $f \in \mathfrak{C}^s(M)$ can be approximated by smooth functions, not necessarily in the corresponding norm, but at least pointwise. But this is sufficient in order to prove (5) via a limiting argument.

7.5.4 Characterizations via means of differences

Our next aim is to extend Theorem 2.6.3 from $\mathbb{R}^n$ to the above manifold M. For this purpose we adapt the function K from 2.6.3 to the present situation in the same way as in (7.2.4/1,2,4). Let B be the unit ball in $\mathbb{R}^n$. Let K be a non-negative C^∞ function in $\mathbb{R}^n$ with

$$\operatorname{supp} K \subset B \quad \text{and} \quad \hat{K}(0) > 0, \tag{1}$$

see (7.2.4/1,2). Of course $\hat{K}(0) > 0$ simply means that K does not vanish identically. As in (7.2.4/4) we assume in addition that K is rotational-invariant, i.e.,

$$K(y) = k(|y|), \qquad y \in \mathbb{R}^n. \tag{2}$$

Let again M be the above manifold and let $\Delta_X^m f(P)$ be the differences along geodesics introduced in (7.5.2/1). Let r_0 be the injectivity radius of M and $0 < r < r_0$, then the extension of the means (2.6.3/1) from $\mathbb{R}^n$ to M is given by

$$K_m(t,f)(P) = \int_{T_P M} k(\|X\|)\Delta_{tX}^m f(P)dX, \qquad P \in M, \quad 0 < t < r, \tag{3}$$

$m \in \mathbb{N}$, which must be interpreted in the same way as in (7.2.4/6). Recall $\sigma_{pq} = n(\frac{1}{\min(p,q)} - 1)_+$, see (2.4.1/1).

Theorem. Let M be the Riemannian manifold in the sense of Hypotheses 7.2.1 with the injectivity radius $r_0 > 0$. Let $0 < r < r_0$.

(i) Let either $0 < p < \infty$, $0 < q \leqq \infty$, or $p = q = \infty$, let

$$s > 2\max\left(\frac{n}{p} + 2, \sigma_{pq}\right), \qquad m > \frac{3}{2}s \text{ with } m \in \mathbb{N}, \tag{4}$$

then

$$\|f \mid L_p(M)\| + \left\| \left(\int_0^r t^{-sq} |K_m(t,f)(\cdot)|^q \frac{dt}{t} \right)^{1/q} \mid L_p(M) \right\| \tag{5}$$

(modification if $q = \infty$) is an equivalent quasi-norm in $F_{pq}^s(M)$.

(ii) Let $0 < p \leqq \infty$, $0 < q \leqq \infty$, and

$$s > 2\left(\frac{n}{p} + 2\right), \qquad m > \frac{3}{2}s \text{ with } m \in \mathbb{N}, \tag{6}$$

then

$$\|f \mid L_p(M)\| + \left(\int_0^r t^{-sq} \|K_m(t,f) \mid L_p(M)\|^q \frac{dt}{t} \right)^{1/q} \tag{7}$$

(modification if $q = \infty$) is an equivalent quasi-norm in $B_{pq}^s(M)$.

Proof. Step 1. The proof of part (i) follows precisely the same scheme as in the proof of Theorem 7.5.1: reduction of the above theorem to its euclidean counterpart in Theorem 2.6.3 by imitating the proof in 7.2.4. The counterpart of (7.5.1/9) reads as follows,

$$\|h \mid F_{pq}^s\| \sim \|h \mid L_p(U)\| + \left\| \left(\int_0^r t^{-sq} |K_m(t,h)(\cdot)|^q \frac{dt}{t} \right)^{1/q} \mid L_p(U) \right\|, \tag{8}$$

where we used the same notations as in Step 1 in the proof of 7.5.1. We have (7.5.1/11,16) with $T = 0$. The remainder estimate does not cause any problem. Then one has to cope with the counterparts of (7.5.1/18–20), where l has the same meaning as in 7.5.2 and (7.5.1/18) must be modified in the sense of (3), i.e., we have to multiply with $k(\|X\|)$ and to integrate over $T_P M$. Otherwise the arguments remain unchanged and the needed restrictions for s are now given by

$$\sigma_{pq} < s - \frac{k}{2} < m - k \quad \text{with} \quad k = 0, \ldots, [s], \tag{9}$$

where the first inequality comes from Theorem 2.6.3(i). Together with $s > \frac{2n}{p} + 4$ we arrive at (4).

Step 2. On the basis of (i) one proves (ii) in the same way as in the Steps 1 and 2 of the proof of Theorem 7.3.2.

Remark. The above theorem is the extension of Theorem 2.6.3 from $\mathbb{R}^n$ to M. The conditions (4) are not natural somewhat in contrast to the corresponding restrictions in Theorem 2.6.3. The main advantage of the above theorem compared with Theorem 7.5.2 is that the spaces $B^s_{pq}(M)$ are now included.

7.5.5 Characterizations via derivatives and differences

Recall that Theorem 7.5.1 is always our starting point. In 7.5.2 we specified l by $F_1 l = (e^{i\nu\lambda} - 1)^m$. Another distinguished candidate is given by $F_1 l = \lambda^k (e^{i\nu\lambda} - 1)^{m-k}$ with $k \in \mathbb{N}$, $m \in \mathbb{N}$ and $m > k$. We insert this l in (7.5.1/4), then $(e^{i\nu\lambda} - 1)^{m-k}$ generates differences in the sense of (7.5.2/1), whereas λ^k refers to derivatives. Hence first we have to find an invariant description of derivatives of functions on the above manifold M. Let $P \in M$, $X \in T_P M$, and $\alpha = (\alpha_1, \ldots, \alpha_k)$ where $\alpha_j \in \mathbb{N}$ and $1 \leqq \alpha_j \leqq n$. Let again r_0 be the injectivity radius of M and $0 < r < r_0$. Then we put in a suitable local chart, say, $(\Omega_P(r), \exp_P^{-1})$,

$$\nabla_\alpha = \nabla_{\alpha_1} \cdots \nabla_{\alpha_k} \quad \text{and} \quad X^\alpha = X^{\alpha_1} \cdots X^{\alpha_k}, \tag{1}$$

where the ∇_j's are the usual covariant derivatives and X^j stands for the components of X. Then

$$\nabla^k_X f(P) = \nabla_\alpha f(P) \frac{X^\alpha}{\|X\|^k}, \qquad P \in M, \quad 0 \neq X \in T_P M, \tag{2}$$

summation over all $\alpha = (\alpha_1, \ldots, \alpha_k)$, $\alpha_j \in \mathbb{N}$, $1 \leqq \alpha_j \leqq n$, has an invariant meaning. Let $c(P, X, t)$ be the above geodesic where $X \in T_P M$ and let $Q = c(P, X, t_Q)$ with $t_Q \|X\| = \text{dist}(P, Q)$ and $0 < \text{dist}(P, Q) < r < r_0$, then we abbreviate

$$\nabla_X^k f(Q) = \nabla_{X(Q)}^k f(Q) \tag{3}$$

where $X(Q) \in T_Q M$ is the parallel displacement of $X \in T_P M$ along $c(P, X, t)$. Recall that Δ_X^m has been defined in (7.5.2/1). Let again $B_P(r)$ be given by (7.5.1/5).

Theorem. Let M be the Riemannian manifold in the sense of Hypotheses 7.2.1 with the injectivity radius $r_0 > 0$. Let either $0 < p < \infty$, $0 < q \leqq \infty$, or $p = q = \infty$. Let

$$s > \frac{2n}{\min(p,q)} + 2n\left(\frac{1}{\min(p,q)} - 1\right)_+ + 2k + 4 \quad \text{and} \quad m > \frac{3}{2}s \tag{4}$$

with $k \in \mathbb{N}_0$ and $m \in \mathbb{N}$. Let $0 < r < r_0$, then

$$\|f \mid L_p(M)\| + \left\| \left(\int_{B_P(r)} \|X\|^{-(s-k)q} |\Delta_X^{m-k} \nabla_X^k f(\cdot)|^q \frac{dX}{\|X\|^n} \right)^{1/q} \mid L_p(M) \right\| \tag{5}$$

(modification if $q = \infty$) is an equivalent quasi-norm in $F_{pq}^s(M)$.

Proof. The theorem is an extension of Theorem 7.5.2 and its proof. We apply Theorem 7.5.1 to $F_1 l = \psi(\lambda) = \lambda^k (e^{i\nu\lambda} - 1)^{m-k}$ with $\nu = \frac{1}{m-k}$. Then we have (7.5.1/1–3) with $T = k$. Furthermore (7.5.1/6) is satisfied if

$$-s_0 + \frac{s}{2} + k + \frac{n}{\min(p,q)} < 0. \tag{6}$$

By (4) there exists an s_0 with (6) such that (7.5.1/7) with $T = k$ is satisfied. Then (7.5.1/8) with the above l is an equivalent quasi-norm in $F_{pq}^s(M)$. By (7.5.1/8) and (5) we have to prove that $l(X, f)$ coincides with $\|X\|^k \Delta_X^{m-k} \nabla_X^k f$, where $l(X, f)$ with $F_1 l = \lambda^k (e^{i\nu\lambda} - 1)^{m-k}$ is given by (7.5.1/4). By (7.5.2/1) this question can be reduced to

$$\frac{d^k f}{d\lambda^k}(c(P, X, \lambda)) = (\nabla_X^k f)(c(P, X, \lambda)) \|X\|^k, \qquad X \in T_P M, \tag{7}$$

where we used the abbreviation (3). Both sides of (7) have an invariant meaning. Hence we may use normal geodesic coordinates, see 7.2.1, in particular $c(P, X, \lambda) = \lambda X$. Then we have invariantly written

$$\frac{d^k f}{d\lambda^k}(c(P, X, \lambda)) = (\nabla_{\alpha_1} \cdots \nabla_{\alpha_k} f)(c(P, X, \lambda)) X^{\alpha_1} \cdots X^{\alpha_k}(c(P, X, \lambda)), \tag{8}$$

summation over α_j form 1 to n. However by (2) we arrive at (7) and the proof is complete.

Remark. The above theorem generalizes Theorem 7.5.2. In Corollary 2.6.2/2 we proved the $\mathbb{R}^n$-counterpart under more natural restrictions for the involved parameters. As in Theorems 7.5.1 and 7.5.2 there is no chance to prove a corresponding assertion for the spaces $B^s_{pq}(M)$, see Remark 7.5.1. On the other hand there is no problem to replace some differences in (7.5.3/4) by some derivatives in the above sense.

7.5.6 Sobolev spaces

Let $W^k_p(M)$ be the Sobolev spaces introduced in Definition 7.4.5(i) and let $\nabla^k_X f(P)$ be given by (7.5.5/2). We introduce the means

$$\nabla^k_r f(P) = \int_{B_P(r)} |\nabla^k_X f(P)| dX, \qquad 0 < r < r_0, \tag{1}$$

where $B_P(r)$ has the same meaning as in (7.5.1/5), $r_0 > 0$ is the injectivity radius of M and (1) must be interpreted in the same way as in (7.2.4/6).

Theorem. Let M be the Riemannian manifold in the sense of Hypotheses 7.2.1 with the injectivity radius $r_0 > 0$. Let $0 < r < r_0$, $1 < p < \infty$, and $k \in \mathbb{N}$, then

$$\sum_{j=0}^{k} \|\nabla^j_r f \mid L_p(M)\| \tag{2}$$

is an equivalent norm in $W^k_p(M)$.

Proof. We have (7.4.5/7,8). As in Step 3 of the proof of the theorem in 7.2.4 we write $h = \psi_j f \circ \exp_{P_j}$, and let

$$\nabla^j_r h(x) = \int_{B(r)} |\nabla^j_X h(x)| dX, \qquad |x| < r, \tag{3}$$

where $B(r)$ stands for a ball in $\mathbb{R}^n$ of radius r, centered at the origin. By (7.5.5/8) it is not hard to see that the proof of the above theorem can be reduced to

$$\sum_{j=0}^{k} \|\nabla_r^j h \mid L_p\| \sim \|h \mid W_p^k\|, \qquad \operatorname{supp} h \subset B(r). \tag{4}$$

We claim that there exist two positive numbers c_1 and c_2 with

$$c_1 \sum_{|\alpha| \leqq k} |D^\alpha h(x)| \leqq \sum_{j=0}^{k} |\nabla_r^j h(x)| \leqq c_2 \sum_{|\alpha| \leqq k} |D^\alpha h(x)|. \tag{5}$$

Because x in (3,5) is fixed and $\nabla_X^j h(x)$ is invariant under transformations of coordinates we may choose normal geodesic coordinates with respect to x in order to prove (5). Furthermore by (7.5.5/2) the integration in (3) may be restricted to, say, $\frac{r}{2} < |X| < r$. Then (5) reduces to

$$c_1 \sum_{|\alpha| \leqq k} |D^\alpha h(x)| \leqq \sum_{j=0}^{k} \int_{\frac{r}{2} < |X| < r} \left| \sum_{|\alpha| = j} D^\alpha h(x) X^\alpha \right| dX \leqq c_2 \sum_{|\alpha| \leqq k} |D^\alpha h(x)|. \tag{6}$$

With $a_\alpha = D^\alpha h(x)$ all expressions in (6) are norms in the finite-dimensional space of polynomials of degree k. Hence they are equivalent. We arrive at (6) and then at (5). But (5) proves (4).

7.6 Spaces on Lie groups

7.6.1 Lie groups as analytic Riemannian manifolds

The main subject of this chapter is the study of function spaces on Riemannian manifolds. This section may be considered as an appendix where we reduce corrsponding problems for Lie groups to those ones for Riemannian manifolds. Function spaces on Lie groups attracted some attention in recent years, see the references and comments given in 1.11.1 which we shall not repeat here. In 1.11.5 we gave a first description of the main results. Now we return to this subject in detail.

We always assume that G stands for an n-dimensional connected Lie group. (But we wish to emphasize that it would be sufficient to assume that G consists of a finite number of connected components.) Let e be the identity and let $\hat{G} = T_e G$ be the Lie algebra with respect to G, identified with the tangent spaces $T_e G$. We equip $\hat{G}$ with a scalar product g_e, a positive-definite symmetric bilinear form, that is,

$$g_e(X, Y) = g_e(Y, X) \ \text{ if } \ X \in \hat{G}, \ \ Y \in \hat{G}, \tag{1}$$

and $g_e(X, X) > 0$ if $X \neq 0$. For each $a \in G$ we define the left translation L_a by $L_a x = ax$, $x \in G$. We transfer the metric on $\hat{G}$ to $T_a G$ by the pullback operation

$$g_a = (L_{a^{-1}})^*_a g_e, \qquad a \in G, \tag{2}$$

which means in more explicit terms

$$g_a(X,Y) = g_e(L_{a^{-1}*a}X, L_{a^{-1}*a}Y), \quad X \in T_aG, \;\; Y \in T_aG. \tag{3}$$

Furnished with this Riemannian metric g the Lie group G becomes a connected n-dimensional analytic Riemannian manifold. Furthermore, g is left-invariant: Let $a \in G$, $b \in G$, $X \in T_aG$, and $Y \in T_aG$, then by (3)

$$\begin{aligned} g_{ba}(L_{b^*a}X, L_{b^*a}Y) &= g_e(L_{a^{-1}b^{-1}*ba}L_{b^*a}X, \ldots) \\ &= g_e(L_{a^{-1}*a}X, \ldots) = g_a(X,Y). \end{aligned} \tag{4}$$

In particular, G, furnished with g, is a complete Riemannian manifold with positive injectivity radius and bounded geometry. Hence it satisfies the Hypotheses 7.2.1 and we can apply the theory developed in this chapter. However at the end one wishes to have an intrinsic description of the corresponding spaces $F^s_{pq}(G)$ and $B^s_{pq}(G)$ completely in terms of the Lie group and not in terms of the introduced Riemannian metric.

Recall that the exponential map exp yields a diffeomorphic map from the ball

$$B(r) = \{X \in \hat{G}\colon |X| < r\} \;\text{ onto }\; \Omega(r) = \exp B(r), \tag{5}$$

provided that the positive number r is sufficiently small. Identifying $\hat{G}$ with $\mathbb{R}^n$ we have the distinguished local chart $(\Omega(r), \exp^{-1})$ around the identity e. In (5) we may choose any norm $|X|$, for instance $|X| = g_e(X,X)^{1/2}$ with respect to the above metric g_e. Recall that $\exp tX$ with $0 \neq X \in \hat{G}$ and $t \in \mathbb{R}$ are the one-dimensional subgroups of G and $x \exp tX$ are their left translates, called Lie geodesics. Let $a \in G$, then $(L_a\Omega(r), \exp^{-1} \circ L_{a^{-1}})$ are local charts around a.

Remark 1. The above notations and sketched results cover the general background we need in the sequel. Further informations and details may be found in [Kli, Hel]. Furthermore we refer in particular to [Pri: Chapter 4], where one finds the above-described connection between Lie groups and Riemannian manifolds. For more specific results we shall give more detailed references.

Definition. Let G be an n-dimensional connected Lie group equipped in the above way with a left-invariant Riemannian metric. Let $\psi = \{\psi_j\}_{j=1}^{\infty}$ be a related resolution of unity in the sense of Definition 7.2.2.

(i) Let $s \in \mathbb{R}$ and let either $0 < p < \infty$, $0 < q \leqq \infty$, or $p = q = \infty$. Then

$$F^s_{pq}(G) = \Big\{ f \in D'(G) \colon \|f \mid F^s_{pq}(G)\|^\psi \tag{6}$$

$$= \Big(\sum_{j=1}^\infty \|\psi_j f \circ L_{a_j} \circ \exp \mid F^s_{pq}\|^p \Big)^{1/p} < \infty \Big\}$$

(modification if $p = \infty$).

(ii) Let $0 < p \leqq \infty$, $0 < q \leqq \infty$, and $-\infty < s_0 < s < s_1 < \infty$. Then

$$B^s_{pq}(G) = (F^{s_0}_{pp}(G), F^{s_1}_{pp}(G))_{\theta,q} \tag{7}$$

with $s = (1-\theta)s_0 + \theta s_1$.

Remark 2. As we said, G equipped with the left-invariant Riemannian metric g satisfies the Hypotheses 7.2.1. Then (i) coincides with Definition 7.2.2, specified to the above local charts, and (ii) coincides with Definition 7.3.1. There one finds also the necessary explanations. We have Theorems 7.2.3 and 7.3.1, in particular (6) is independent of the chosen local charts $(L_{a_j}\Omega(r), \exp^{-1} \circ L_{a_j^{-1}})$ and the resolution of unity ψ. Furthermore, $F^s_{pq}(G)$, and hence also $B^s_{pq}(G)$, is also independent of the left-invariant Riemannian metric g. This justifies the notation (6,7). In the sense of Remark 7.2.1/2 we always deal with the case $\psi = \{\psi_j\}_{j=1}^\infty$, but it is clear that our approach covers also finite resolutions of unity $\psi = \{\psi_j\}_{j=1}^J$ with $J \in \mathbb{N}$ which refer to compact Lie groups.

Remark 3. We prefer to equip the connected Lie group G with a left-invariant Riemannian metric (2,3). But in the same way one can introduce a right-invariant metric and corresponding spaces of type F^s_{pq} and B^s_{pq}. In [Tri15: Remark 9] we compared these two types of metrics and spaces. In general they are different but this has not yet been studied in detail.

Remark 4. One can ask under what conditions a Lie group G has an invariant Riemannian metric, that is, a metric which is both left-invariant and right-invariant. A detailed study of this problem may be found in [Pri: Theorem 4.2.3]. In this case the Riemannian geodesics (with respect to this invariant metric) coincide with the Lie geodesics $x \exp tX$, see [Pri: Theorem 4.3.3 and 4.3.6]. In general this is not the case. It may even happen that $\exp\colon \hat{G} \to G$ is not onto, see [Kah: p. 272] for an example, cf. also [Pri: 4.3.5]. On the other hand if G is compact (or abelian) then G has an invariant Riemannian metric, see [Pri: 4.2.5].

7.6.2 The main theorem

Let again G be the above n-dimensional connected Lie group equipped with a left-invariant Riemannian metric. First we remark that the corresponding left-invariant Riemannian volume element generates a left-invariant Haar measure. Recall that left-invariant Haar measures on G are uniquely determined besides a positive constant. Hence the spaces $L_p(G)$, $0 < p \leqq \infty$, with respect to a left-invariant Haar measure are naturally defined and they coincide with the corresponding spaces when G is interpreted as a Riemannian manifold in the above sense, see the explanations in front of Theorem 7.2.4.

Definition 7.6.1 shows that the spaces $F^s_{pq}(G)$ and $B^s_{pq}(G)$ can be intrinsically described in terms of local means in the sense of Theorems 7.2.4 and 7.3.2, where the means $k(t,f)$ are given by (7.2.4/6) with the Riemannian geodesics $c(P,X,t)$ generated by the above-constructed left-invariant Riemannian metric. In the context of Lie group theory this is not satisfactory. One would strongly prefer to substitute the Riemannian geodesics $c(P,X,t)$ in the means (7.2.4/6) by the Lie geodesics $x \exp tX$. This is also supported by their joint euclidean ancestors (2.4.6/1): If $\mathbb{R}^n$ is interpreted as an (abelian) Lie group then $x \exp tX$ is the immediate generalization of $x + ty$.

We use the same basic functions as 7.2.4. In other words, K_0, K^0, K and k_0, k^0, k have the same meaning as in (7.2.4/1–5). As in 7.2.4 we do not indicate the values of $N \in \mathbb{N}$ in (7.2.4/3,5). Then the Lie counterpart of (7.2.4/6) is given by

$$k(t,f)(x) = \int_{\hat{G}} k(|X|) f(x \exp tX) dX, \qquad x \in G, \ 0 < t < r, \tag{1}$$

where dX stands for the Lebesgue measure on $\hat{G}$, and r has the same meaning as in (7.6.1/5). In the same way one defines $k_0(t,f)(x)$.

Theorem. Let G be an n-dimensional connected Lie group and let $F^s_{pq}(G)$ and $B^s_{pq}(G)$ be the spaces introduced in Definition 7.6.1.

(i) Let either $0 < p < \infty$, $0 < q \leqq \infty$, or $p = q = \infty$, and let $s \in \mathbb{R}$. Let both $r > 0$ and $\varepsilon > 0$ be sufficiently small and let $N \in \mathbb{N}$ in (7.2.4/5), and hence also in (1), be sufficiently large (in dependence on s, p, q), then

$$\|k_0(\varepsilon, f) \mid L_p(G)\| + \left\| \left(\int_0^r t^{-sq} |k(t,f)(\cdot)|^q \frac{dt}{t} \right)^{1/q} \mid L_p(G) \right\| \tag{2}$$

(with the usual modification if $q = \infty$) is an equivalent quasi-norm in $F^s_{pq}(G)$.

(ii) Let $0 < p \leqq \infty$, $0 < q \leqq \infty$, and $s \in \mathbb{R}$. Let both $r > 0$ and $\varepsilon > 0$ be sufficiently small and let $N \in \mathbb{N}$ in (7.2.4/5), and hence also in (1), be sufficiently large (in dependence on s, p, q), then

$$\|k_0(\varepsilon, f) \mid L_p(G)\| + \left(\int_0^r t^{-sq} \|k(t,f) \mid L_p(G)\|^q \frac{dt}{t} \right)^{1/q} \tag{3}$$

(with the usual modification if $q = \infty$) is an equivalent quasi-norm in $B^s_{pq}(G)$.

Proof. We reduce the theorem to Theorem 7.4.7. By 7.6.1 we know that G equipped with a left-invariant Riemannian metric is an analytic Riemannian manifold satisfying Hypotheses 7.2.1. As an additional structure we introduce the left-invariant covariant derivation in the sense of [Kli: 1.7.7], see also [Hel: p. 102]. It comes out that the corresponding geodesics coincide with the left translates of the 1-dimensional subgroups of G, that is, with the Lie geodesics $x \exp tX$, see [Kli: 1.7.10 and 1.7.13] or [Hel: pp. 102, 104]. Furthermore (7.4.7/1) is satisfied. Then the above theorem is a special case of Theorem 7.4.7.

Remark 1. Let $s > \sigma_p = n(\frac{1}{p} - 1)_+$, then $\|k_0(\varepsilon, f) \mid L_p(G)\|$ in (2,3) can be replaced by $\|f \mid L_p(G)\|$. This assertion is also covered by Theorem 7.4.7.

Remark 2. We reduced the proof of the above theorem to Theorem 7.2.4, which in turn was reduced to the arguments in 7.2.4 – 7.2.6. Hence by (7.2.5/30) it is sufficient to choose an $N \in \mathbb{N}$ with

$$N > \max\left(s, \frac{2n}{p} + 6\right) + n\left(\frac{1}{p} - 1\right)_+ . \tag{4}$$

Remark 3. In [Tri15, 17, 20] we gave three different proofs of the above theorem. The presented one coincides with the proof in [Tri17]. The original proof in [Tri15] was based on completely different arguments: Instead of expansions (7.2.4/19,20) based on the differential equations (7.2.1/3) we used the Campbell–Baker–Hausdorff formula

$$\exp Y \cdot \exp tX = \exp Z(t) \tag{5}$$

with

$$Z(t) = Y + tX + \frac{t}{2}[Y,X] + \frac{t}{12}[[Y,X],Y] + \frac{t^2}{12}[[Y,X],X] + \cdots \tag{6}$$

where [,] is the Lie product on the Lie algebra $\hat{G}$. The higher terms $+\cdots$ are the so-called Dynkin polynomials, see [Tri15: p. 335] for details and further references. This way has also its advantages: One does not need that the line system in question, in our case the Lie geodesics, satisfies differential equations of type (7.2.1/3). Maybe this can be used to study function spaces generated by vector fields which are not necessarily connected with Lie groups. Again Campbell–Baker–Hausdorff formulas should be of great service, we refer in this context to the appendix in [NSW]. In [Tri20] we gave a third proof of the above theorem which is based on a more detailed study of means in Lie groups and which combines ideas of the two preceding proof.

7.6.3 Characterizations via differences

In 7.5 we described equivalent quasi-norms for some spaces $F^s_{pq}(M)$ and $B^s_{pq}(M)$ where M is a manifold in the sense of Hypotheses 7.2.1. There is no doubt that we have corresponding assertions for spaces $F^s_{pq}(G)$ and $B^s_{pq}(G)$ on the above Lie group. Since the situation is more special one can even hope to get additional results. We describe a typical example: the counterpart of Theorem 7.5.2 and its extension to $B^s_{pq}(G)$.

Let again G be the above n-dimensional connected Lie group with the Lie geodesics $x \exp tX$, $x \in G$, $X \in \hat{G}$, $t \in \mathbb{R}$, then

$$\Delta^m_X f(x) = \sum_{j=0}^{m} (-1)^{m-j} \binom{m}{j} f\left(x \exp \frac{j}{m} X\right). \tag{1}$$

This is the natural counterpart of (7.5.2/1). Let $B(r)$ be given by (7.6.1/5).

Theorem. Let G be an n-dimensional connected Lie group and let $F^s_{pq}(G)$ and $B^s_{pq}(G)$ be the spaces introduced in Definition 7.6.1.

(i) Let either $0 < p < \infty$, $0 < q \leqq \infty$ or $p = q = \infty$. Let

$$s > \frac{2n}{\min(p,q)} + 2n\left(\frac{1}{\min(p,q)} - 1\right)_+ + 4 \quad \text{and} \quad m > \frac{3}{2}s \tag{2}$$

with $m \in \mathbb{N}$. Let $r > 0$ be sufficiently small, then

$$\|f \mid L_p(G)\| + \left\| \left(\int_{B(r)} |X|^{-sq} |\Delta^m_X f(\cdot)|^q \frac{dX}{|X|^n} \right)^{1/q} \mid L_p(G) \right\| \tag{3}$$

(with the usual modification if $q = \infty$) is an equivalent quasi-norm in $F^s_{pq}(G)$.

(ii) Let $0 < p \leqq \infty$ and $0 < q \leqq \infty$. Let s and $m \in \mathbb{N}$ be given by (2). Let $r > 0$ be sufficiently small, then

$$\|f \mid L_p(G)\| + \left(\int_{B(r)} |X|^{-sq} \|\Delta_X^m f \mid L_p(G)\|^q \frac{dX}{|X|^n} \right)^{1/q} \tag{4}$$

(with the usual modification if $q = \infty$) is an equivalent quasi-norm in $B^s_{pq}(G)$.

Proof. We apply Theorem 7.5.1 in the same way as in the proof of Theorem 7.5.2 with $x \exp tX$ in (7.5.2/1) instead of $c(P, X, \frac{j}{m})$. This replacement is justified by the technique developed in 7.4.7 and in the proof of Theorem 7.6.2. This proves part (i) where (2) coincides with (7.5.2/2). The proof of part (ii) is based on part (i) and the arguments in the Steps 1 and 2 of the proof of Theorem 7.3.2.

Remark. Theorems 7.5.1 and 7.5.2 have no counterparts for the spaces $B^s_{pq}(M)$, see Remark 7.5.1 for explanations. This comes from the observation that the interchange of the integrations of $X \in B_P(r)$ and $x \in M$ in (7.5.2/3) is not possible. In the case of the above Lie group G we are in a better position: we have the standard ball $B(r)$ which is the same for all $x \in G$. In the euclidean case, that is $G = \mathbb{R}^n$, we have (2.6.1/2) and (2.6.2/1) as counterparts of (4,3), respectively. In this sense the Lie group G stands somewhat between $\mathbb{R}^n$ and the general Riemannian manifold M satisfying Hypotheses 7.2.1. In the same way one has corresponding counterparts of the Theorems 7.5.4 and 7.5.5, see also the following subsection as far as derivatives are concerned.

7.6.4 Sobolev and Hölder–Zygmund spaces

There are no problems to adapt Definition 7.4.5(i) and 7.5.3 of Sobolev and Hölder–Zygmund spaces to the present situation. Let Δ_X^m be given by (7.6.3/1).

Definition. Let G be an n-dimensional connected Lie group.

(i) (Sobolev spaces). Let $m \in \mathbb{N}$, $1 < p < \infty$, and let $X_1(x), \ldots, X_n(x)$ be n linearly independent left-invariant vector fields on G, then

$$W_p^m(G) = \Big\{ f \in L_p(G) \colon \|f \mid W_p^m(G)\| \tag{1}$$

$$= \sum_{k=0}^{m} \sum_{1 \leqq n_i \leqq n} \|X_{n_1} \cdots X_{n_k} f \mid L_p(G)\| < \infty \Big\}$$

where the derivatives must be understood in the sense of distributions.

(ii) (Hölder–Zygmund spaces). Let $s > 0$ then

$$\mathcal{C}^s(G) = F^s_{\infty\infty}(G) \tag{2}$$

and

$$\mathfrak{C}^s(G) = \Big\{ f \in L_\infty(G) \colon \| f \mid \mathfrak{C}^s(G) \| \tag{3}$$

$$= \sup_{x \in G} |f(x)| + \sup_{x\in G} \sup_{X \in \hat{G}} |X|^{-s} |\Delta_X^{1+[s]} f(x)| < \infty \Big\}.$$

Remark 1. Of course, (1) and (2,3) are the obvious counterparts of (7.4.5/4) and (7.5.3/1,2), respectively. Both the Sobolev spaces $W^m_p(G)$ and the Hölder–Zygmund spaces $\mathfrak{C}^s(G)$ are Banach spaces. Both are "left" spaces. Similarly one may introduce "right" spaces based on the right translates $(\exp tX)x$ of the one-dimensional subgroups of G in (7.6.3/1) and on right-invariant vector fields. In general, the "left" spaces and the "right" ones are different. We put $L_p(G) = W^0_p(G)$.

Theorem. Let G be an n-dimensional connected Lie group.

(i) (Paley–Littlewood theorem). Let $m \in \mathbb{N}_0$ and $1 < p < \infty$, then

$$W^m_p(G) = F^m_{p2}(G). \tag{4}$$

(ii) Let $m > s > 0$ with $m \in \mathbb{N}$, then

$$\sup_{x \in G} |f(x)| + \sup_{x\in G} \sup_{X \in \hat{G}} |X|^{-s} |\Delta_X^m f(x)| \tag{5}$$

is an equivalent quasi-norm in the Banach space $\mathfrak{C}^s(G)$.

(iii) Let $s > 4$ then

$$\mathcal{C}^s(G) = \mathfrak{C}^s(G). \tag{6}$$

Proof. All spaces, $W^m_p(G)$, $F^m_{p2}(G)$, $\mathcal{C}^s(G)$, and $\mathfrak{C}^s(G)$ are left-invariant. Decompositions in the sense of Step 1 of the proof of Theorem 7.4.5 and Definition 7.6.1 reduce (4) to the standard local chart $(\Omega(r), \exp^{-1})$. Now one can argue in the same way as in Step 1 of the proof of Theorem 7.4.5 in order to prove (4). The proof of (ii) and (iii) is the same as in the proof of Theorem 7.5.3, now based on Theorem 7.6.3.

Remark 2. Recall that (i) is the counterpart of the corresponding assertions in Theorem 7.4.5, whereas (ii) and (iii) are the counterpart of Theorem 7.5.3. Again it would be desirable to prove (6) for any $s > 0$.

7.6.5 A Fourier-analytical remark

In Chapters 2–4 we developed the theory of the spaces B^s_{pq} and F^s_{pq} on $\mathbb{R}^n$. One of the most distinguished features was the reduction of this theory to the spectral theory of the Laplacian Δ on $\mathbb{R}^n$, see Definition 2.3.1 and 1.3 for further explanations. On the other hand, the theory of the spaces $F^s_{pq}(M)$ and $B^s_{pq}(M)$ in the sense of Definitions 7.2.2 and 7.3.1 was mostly local and the Laplace–Beltrami operator (7.2.5/1) lost its dominant position, however its reputation was at least partly restored in 7.4.3 and 7.2.5. It is hard to believe that the spectral theoretical approach to spaces of type F^s_{pq} and B^s_{pq} which is so typical for spaces on $\mathbb{R}^n$ can be extended to general manifolds M in the sense of Hypotheses 7.2.1. The question is what happens when we specify M such that a substantial Fourier analysis on M is available. Let $M = T^n$ be the n-torus, then it is by no means a surprise that one has a full counterpart of the theory presented in Chapters 2–4, which may be found in [ScT: Chapter 3] (with exception of some more recent developments). From the Fourier-analytical point of view globally symmetric spaces M and in particular hyperbolic spaces M are of peculiar interest. These manifolds M have similar features as the above treated Lie groups G, in particular the Lie group of all isometries on M acts transitively on M which is a substitute for the left translations for the above Lie group. There exists an elaborated Fourier analysis, convolutions, etc. In [Tri20] we studied these questions in some detail, there one can also find the necessary explanations and references. We shall not go into detail here. We mention only one aspect. Let M be a manifold in the sense of Hypotheses 7.2.1. Then we modify the means (7.2.4/6) by

$$k(t,f)(P) = \int_{T_P M} k(\|X\|) f(c(P,X,t)) \sqrt{|\det g_{c(P,X,t)}|}\, dX \tag{1}$$

where all symbols have the same meaning as in (7.2.4/6). By the same arguments as in 7.2.4 – 7.2.6, Theorem 7.2.4 is also valid for these modified means. Let M be a Riemannian globally symmetric space then (1) can be expressed as a convolution, and let M be a hyperbolic space, for instance Poincaré's model of the non-euclidean plane, then we find C^∞ functions φ^t such that

$$k(t,f)(P) = \varphi^t(-\Delta) f(P), \tag{2}$$

where the right-hand side of (2) stands for the spectral means of the corresponding Laplace–Beltrami operator. We refer again for details and necessary explanations to [Tri20]. Hence Theorem 7.2.4 with the modified means (1), interpreted by (2), reminds of the original Fourier-analytical Definition 2.3.1 of the spaces B^s_{pq} and F^s_{pq} on $\mathbb{R}^n$. But in contrast to $\mathbb{R}^n$ the connection (2) is more complicated and it is not so clear how to take advantage of it. But nevertheless under severe restrictions for M there is a link between the spectral theory of the Laplace–Beltrami operator on the one hand and the spaces $F^s_{pq}(M)$ and $B^s_{pq}(M)$ on the other hand. See also [Skr1,2] where the author studies spaces of Sobolev type on symmetric manifolds in a somewhat different approach.

References

[Ada] Adams, R.A., Sobolev Spaces. New York, Academic Press, 1975.

[ADN] Agmon, S., Douglis, A., and Nirenberg, L., Estimates near the boundary for solutions of elliptic partial differential equations satisfying general boundary conditions, I. Comm. Pure Appl. Math. *12* (1959), 623–727.

[AlH] Alvarez, J. and Hounie, J., Estimates for the kernel and continuity properties of pseudo-differential operators. Ark. Mat. *28* (1990), 1–22.

[Arn] Arnold, V.I., Mathematical Methods of Classical Mechanics. New York, Springer 1978 (translation from Russian).

[Aro] Aronszajn, N., Boundary values of functions with finite Dirichlet integral. Techn. Report *14*, Univ. of Kansas, 1955, 77–94.

[ArS] Aronszajn, N. and Smith, K.T., Theory of Bessel potentials, I. Ann. Inst. Fourier (Grenoble) *11* (1961), 385–476.

[Aub1] Aubin, T., Espaces de Sobolev sur les variétés Riemanniennes. Bull. Sci. Math. *100* (1976), 149–173.

[Aub2] Aubin, T., Nonlinear Analysis on Manifolds. Monge-Ampère Equations, New York, Springer, 1982.

[Bea1] Beals, R., L^p and Hölder estimates for pseudodifferential operators: necessary conditions. In: Harmonic analysis in Euclidean spaces, Proc. Symp. Pure Math. *35*, II. Providence, Amer. Math. Soc., 1979, 153–157.

[Bea2] Beals, R., L^p and Hölder estimates for pseudodifferential operators: sufficient conditions. Ann. Inst. Fourier (Grenoble) *29* (1979), 239–260.

[Bea3] Beals, R.M., L^p boundedness of Fourier integral operators. Mem. Amer. Math. Soc. *38*, 264 (1982).

[BeL] Bergh, J. and Löfström, J., Interpolation Spaces, an Introduction. Berlin, Springer, 1976.

[Bes1] Besov, O.V., On a family of function spaces. Embedding theorems and extensions (Russian). Dokl. Akad. Nauk SSSR *126* (1959), 1163–1165.

[Bes2] Besov, O.V., On a family of function spaces in connection with embeddings and extensions (Russian). Trudy Mat. Inst. Steklov *60* (1961), 42–81.

H. Triebel, *Theory of Function Spaces II*, Modern Birkhäuser Classics,
DOI 10.1007/978-3-0346-0419-2, © Birkhäuser Verlag 1992

[Bes3] Besov, O.V., On Sobolev–Liouville and Lizorkin–Triebel spaces in domains (Russian). Trudy Mat. Inst. *192* (1990), 20–34.

[BiG] Bishop, R.L. and Goldberg, S.I., Tensor Analysis on Manifolds, New York, Dover Publ., 1980.

[BIN] Besov, O.V., Il'in, V.P. and Nikol'skij, S.M., Integral Representations of Functions and Embedding Theorems (Russian). Moskva, Nauka 1975, (there are two English translations).

[BKLN] Besov, O.V., Kudrjavcev, L.D., Lizorkin, P.I., and Nikol'skij, S.M., Studies on the theory of spaces of differentiable functions of several variables (Russian). Trudy Mat. Inst. Steklov *182* (1988), 68–127. (English translation, Proc. Steklov Inst. Math. *182* (1990), 73–140).

[BKS] Brudnyj, Ju.A., Krejn, S.G., and Semenov, E.M., Interpolation of linear operators (Russian). In: Itogi nauki i techniki, Ser. mat. analiz *24*, Moskva, Akademija nauk SSSR, 1986, 3–163.

[Boa] Boas, R.P., Entire Functions. New York, Adademic Press, 1954.

[Boj] Bojarski, B., Sharp maximal operators of fractional order and Sobolev imbedding inequalities. Bull. Polish Acad. Sci., Ser. Math. *33* (1985), 7–16.

[Bom] Boman, J., Equivalence of generalized moduli of continuity, Ark. Mat. *18* (1980), 73–100.

[Bou1] Bourdaud, G., L^p estimates for certain non-regular pseudodifferential operators. Comm. Part. Diff. Equa. *7* (1982), 1023–1033.

[Bou2] Bourdaud, G., Localisation et multiplicateurs des espaces de Sobolev homogenes. Manuscripta Math. *60* (1988), 93–130.

[Bou3] Bourdaud, G., Une algebre maximale d'operateurs pseudodifferentiels. Comm. Part. Diff. Equations *13* (1988), 1059–1083.

[Bou4] Bourdaud, G., Analyse fonctionnelle dans l'espaces euclidien. Paris, Publ. Math. Univ. Paris VII, 1988.

[Bru] Brudnyj, Ju.A., Spaces defined by means of local approximation (Russian). Trudy Moskov. Mat. Obsc. *24* (1971), 69–132.

[BuB] Butzer, P.L. and Berens, H., Semi-Groups of Operators and Approximation. Berlin, Springer, 1967.

[BuG] Burenkov, V.I. and Gol'dman, M.L., On the extension of functions of L_p (Russian). Trudy Mat. Inst. Steklov *150* (1979), 31–51.

[Bui1] Bui, H.Q., Some aspects of weighted and non-weighted Hardy spaces. Kôkyûroku Res. Inst. Math. Sci. *383* (1980), 38–56.

[Bui2] Bui, H.Q., On Besov, Hardy and Triebel spaces for $0 < p < 1$. Ark. Mat. *21* (1983), 159–184.

[Bui3] Bui, H.Q., Characterizations of weighted Besov and Triebel–Lizorkin spaces via temperatures. J. Funct. Anal. *55* (1984), 39–62.

[Bui4] Bui, H.Q., Representation theorems and atomic decomposition of Besov spaces. Math. Nachr. *132* (1987), 301–311.

[BuN] Butzer, P.L. and Nessel, R.J., Fourier Analysis and Approximation. Basel, Birkhäuser, 1971.

[Cal1] Calderón, A.P., Lebesgue spaces of differential functions. Conf. Part. Diff. Equations, Univ. of California, 1960.

[Cal2] Calderón, A.P., Lebesgue spaces of differentiable functions and distributions. In: "Part. Diff. Equations," Proc. Symp. Math. *4*, Amer. Math. Soc. (1961), 33–49.

[Cal3] Calderón, A.P., Intermediate spaces and interpolation. Studia Math., Seria specjalna, *1* (1963), 31–34.

[Cal4] Calderón, A.P., Intermediate spaces and interpolation, the complex method. Studia Math. *24* (1964), 113–190.

[Cal5] Calderón, A.P., Estimates for singular integral operators in terms of maximal functions. Studia Math. *44* (1972), 563–582.

[Cam1] Campanato, S., Proprietà di inclusione per spaci di Morrey. Ricerche Mat. *12* (1963), 67–86.

[Cam2] Campanato, S., Proprietà di hölderianità di alcune classi di funzioni. Ann. Scuola Norm. Sup. Pisa *17* (1963), 175–188.

[Cam3] Campanato, S., Proprietà di una famiglia di spazi funzionali. Ann. Scuola Norm. Sup. Pisa *18* (1964), 137–160.

[Cam4] Campanato, S., Teoremi di interpolazione per transformazioni che applicano L^p in $C^{k,\alpha}$. Ann. Scuola Norm. Sup. Pisa *18* (1964), 345–360.

[CaS] Calderón, A.P. and Scott, R., Sobolev type inequalities for $p > 0$. Studia Math. *62* (1978), 75–92.

[CaT] Calderón, A.P. and Torchinsky, A., Parabolic maximal functions associated with a distribution, I,II. Advances in Math. *16* (1975), 1–64, and *24* (1977), 101–171.

[CGT] Cheeger, J., Gromov, M. and Taylor, M., Finite propagation speed, kernel estimates for functions of the Laplace operator, and the geometry of complete Riemannian manifolds. J. Differential Geometry *17* (1982), 15–53.

[Chi] Ching, C.-H., Pseudodifferential operators with non-regular symbols. J. Differential Equations *11* (1972), 436–447.

[Chr] Christ, M., The extension problem for certain function spaces involving fractional orders of differentiability. Ark. Mat. *22* (1984), 63–81.

[CMS1] Coifman, R.R., Meyer, Y. and Stein, E.M., Some new function spaces and their applications to harmonic analysis. J. Funct. Anal. *62* (1985), 304–335.

[CMS2] Cwikel, M., Milman, M. and Sagher, Y. Complex interpolation of some quasi-Banach spaces. J. Funct. Anal. *65* (1986), 339–347.

[Coi] Coifman, R.R., A real variable characterization of H^p. Studia Math. *51* (1974), 269–274.

[CoR] Coifman, R.R. and Rochberg, R. Representation theorems for holomorphic and harmonic functions in L^p. Astérisque *77* (1980), 11–66.

[CoS1] Coulhon, T. and Saloff-Coste, L., Theorie de Hardy–Littlewood–Sobolev pour les semi-groupes d'operateurs et application aux groupes de Lie unimodulaires. Sem. d'Analyse, Univ. de Clermont II, 1987–1988, exp. 21.

[CoS2] Coulhon, T. and Saloff-Coste, L., Theoremes de Sobolev pour les semigroupes d'operateurs et application aux groupes de Lie unimodulaires.

[CoW] Coifman, R.R. and Weiss, G., Extensions of Hardy spaces and their use in analysis, Bull. Amer. Math. Soc. *83* (1977), 569–645.

[CwS] Cwikel, M. and Sagher, Y., Analytic families of operators on some quasi-Banach spaces. Proc. Amer. Math. Soc. *102* (1988), 979–984.

[Dau1] Daubechies, I. Orthonormal bases of compact supported wavelets. Comm. Pure Appl. Math. *41* (1988), 909–996.

[Dau2] Daubechies, I., The wavelet transform, time-frequency localization and signal analysis. IEEE Trans. Inform. Theory *36* (1990), 5, 961–1005.

[Dav] Davies, E.B., Heat Kernels and Spectral Theory. Cambridge, Univ. Press, 1989.

[DeS] DeVore, R.A. and Sharpley, R.C., Maximal functions measuring smoothness. Mem. Amer. Math. Soc. *47*, 293 (1984), 1–115.

[Dor1] Dorronsoro, J.R., A characterization of potential spaces. Proc. Amer. Math. Soc. *95* (1985), 21–31.

[Dor2] Dorronsoro, J.R., Poisson integrals of regular functions. Trans. Amer. Math. Soc. *297* (1986), 669–685.

[EdE] Edmunds, D.E. and Evans, W.D., Spectral Theory and Differential Operators. Oxford, Clarendon Press, 1987.

[EdG] Edwards, R.E. and Gaudry, G.I. Littlewood–Payley and Multiplier Theory. Berlin, Springer, 1977.

[Eic1] Eichhorn, J., Elliptic differential operators on noncompact manifolds. In "Seminar Analysis Karl–Weierstrass–Institute 1986/87," Teubner-Texte Math. *106*. Teubner, Leipzig 1988, 4–169.

[Eic2] Eichhorn, J., The invariance of Sobolev spaces over noncompact manifolds. In: "Symp. Partial Differential Equations, Holzhau 1988," Teubner-Texte Math. *112* (1989), 73–107.

[Fef] Fefferman, C., Characterizations of bounded mean oscillation. Bull. Amer. Math. Soc. *77* (1971), 587–588.

[FeG1] Feichtinger, H.G. and Gröchenig, K., A unified approach to atomic decompositions via integral group representations. In "Function Spaces and Applications," Proc. Conf. Lund 1986, Lect. Notes Math. *1302*. Berlin, Springer 1988, 52–73.

[FeG2] Feichtinger, H.G. and Gröchenig, K., Banach spaces related to integrable group representations and their atomic decompositions, I. J. Funct. Anal. *86* (1989), 307–340. II. Monatsh. Math. *108* (1989), 129–148.

[FeS1] Fefferman, C. and Stein, E.M., Some maximal inequalities. Amer. J. Math. *93* (1971), 107–115.

[FeS2] Fefferman, C. and Stein, E.M., H^p spaces of several variables. Acta Math. *129* (1972), 137–193.

[FHJW] Frazier, M., Han, Y.-S., Jawerth, B., and Weiss, G., The T1 theorem for Triebel–Lizorkin spaces. In "Harmonic Analysis and Partial Differential Equations (El Escorial, 1987)," Lect. Notes Math. *1384*, Springer, Berlin 1989, 168–181.

[FJW] Frazier, M., Jawerth, B., and Weiss, G., Littlewood–Paley theory and the study of function spaces. CBMS-AMS Regional Conf. Ser. *79*, 1991.

[Fle1] Flett, T.M., Temperatures, Bessel potentials and Lipschitz spaces. Proc. London Math. Soc. *32* (1971), 385–451.

[Fle2] Flett, T.M., Lipschitz spaces of functions on the circle and the disc. J. Math. Anal. Appl. *39* (1972), 125–158.

[Fol1] Folland, G.B., Subelliptic estimates and function spaces on nilpotent Lie groups. Ark. Mat. *13* (1975), 161–207.

[Fol2] Folland, G.B., Lipschitz classes and Poisson integrals on stratified groups. Studia Math. *66* (1979), 37–55.

[FoS] Folland, G.B. and Stein, E.M., Hardy Spaces on Homogeneous Groups. Princeton, Univ. Press, 1982.

[Fra] Franke, J., On the spaces $F^s_{p,q}$ of Triebel–Lizorkin type: Pointwise multipliers and spaces on domains. Math. Nachr. *125* (1986), 29–68.

[Fra2] Franke, J., Besov–Triebel–Lizorkin spaces and boundary value problems. In "Seminar Analysis 1984/85," Berlin, Akad. Wissenschaften DDR, 1985, 89–104.

[Fra3] Franke, J., Elliptische Randwertprobleme in Besov–Triebel–Lizorkin–Räumen. Thesis, Univ. Jena, 1985.

[FrJ1] Frazier, M. and Jawerth, B., Decomposition of Besov spaces. Indiana Univ. Math. J. *34* (1985), 777–799.

[FrJ2] Frazier, M. and Jawerth, B., The φ-transform and applications to distribution spaces. In "Function Spaces and Applications," Proc. Conf. Lund 1986, Lect. Notes Math. *1302*. Berlin, Springer, 1988, 223–246.

[FrJ3] Frazier, M. and Jawerth, B., A discrete transform and decomposition of distribution spaces. J. Funct. Anal. *93* (1990), 34–170.

[Gag] Gagliardo, E., Proprietà di alcune classi di funzioni in più variabili. Ricerche Mat. *7* (1958), 102–137.

[Gei1] Geisler, M., Besov spaces on compact Lie groups, Math. Nachr. *139* (1988), 193–205.

[Gei2] Geisler, M., Morrey–Campanato spaces on manifolds. Comment. Math. Univ. Carolinae *29* (1988), 309–318.

[Gib] Gibbons, G., Opérateurs pseudo-différentiels et espace de Besov. C.R. Acad. Sci. Paris (Sér. A-B) *286* (1978), 895–897.

[Gol1] Goldberg, D., A local version of real Hardy spaces, Duke Math. J. *46* (1979), 27–42.

[Gol2] Goldberg, D., Local Hardy spaces. In "Harmonic Analysis in Euclidean Spaces," Proc. Symp. Pure Math. *35*, I. Amer. Math. Soc. 1979, 245–248.

[Gol3] Gol'dman, M.L., On the extension of functions of $L_p(R_n)$ in spaces with a larger number of dimensions (Russian). Mat. Zametki *25* (1979), 513–520.

[Gre] Grevholm, B., On the structure of the spaces $\mathcal{L}_k^{p,\lambda}$. Math. Scand. *26* (1970), 241–254.

[Gri] Grisvard, P., Commutativité de deux foncteurs d'interpolation et applications. J. Math. Pures Appl. *45* (1966), 143–290.

[Grö1] Gröchenig, K., Unconditional bases in translation and dilation invariant functions spaces on $\mathbb{R}^n$. In "Constructive theory of functions (Varna 1987)," 174–183. Sofia, Bulgar. Acad. Sci., 1988.

[Grö2] Gröchenig, K., Describing functions: Atomic decompositions versus frames.

[Gru1] Grubb, G., Pseudo-differential boundary problems in L_p spaces. Comm. Part. Diff. Equations *15* (1990), 289–340.

[Gru2] Grubb, G., Parabolic pseudo-differential boundary problems and applications. Preprint, Univ. Copenhagen, 1989.

[GrW] Greene, R.E. and Wu, H. Function Theory on Manifolds which Possess a Pole. Lect. Notes Math. *699*. Berlin, Springer, 1979.

[HaL1] Hardy, G.H. and Littlewood, J.E., A maximal theorem with function-theoretical applications. Acta Math. *54* (1930), 81–116.

[HaL2] Hardy, G.H. and Littlewood, J.E., Some properties of fractional integrals, I, II. Math. Z. *27* (1927), 565–606; *34* (1932), 403–439.

[Har] Hardy, G.H., The mean value of the modulus of an analytic function. Proc. L.M.S. *14* (1914), 269–277.

[Hel] Helgason, S., Differential Geometry, Lie Groups and Symmetric Spaces. New York, Academic Press, 1978.

[HiK] Hinz, A.M. and Kalf, H., Subsolution estimates and Harnack's inequality for Schrödinger operators. J. Reine Angew. Math. *404* (1990), 118–134.

[HLP] Hardy, G.H., Littlewood, J.E., and Pólya, G., Inequalities. Cambridge, Univ. Press, 1964, sec. ed. (first ed. 1934).

[Hör1] Hörmander, L., Estimates for translation invariant operators in L_p spaces. Acta Math. *104* (1960), 93–140.

[Hör2] Hörmander, L., Linear Partial Differential Operators. Berlin, Springer, 1963.

[Hör3] Hörmander, L., The Analysis of Linear Partial Differential Operators, 1–4. Berlin, Springer, 1983/85.

[Hör4] Hörmander, L., Pseudo-differential operators of type 1,1. Comm. Part. Diff. Equations *13* (1988), 1085–1111.

[Hör5] Hörmander, L., Continuity of pseudo-differential operators of type 1,1. Comm. Part. Diff. Equations *14* (1989), 231–243.

[Ill] Illner, R.A., A class of L^p-bounded pseudo-differential operators. Proc. Amer. Math. Soc. *51* (1975), 347–355.

[Jaw1] Jawerth, B., Some observations on Besov and Lizorkin–Triebel spaces, Math. Scand. *40* (1977), 94–104.

[Jaw2] Jawerth, B., The trace of Sobolev and Besov spaces if $0 < p < 1$. Studia Math. *62* (1978), 65–71.

[JoN] John, F. and Nirenberg, L., On functions of bounded mean oscillation. Comm. Pure Appl. Math. *14* (1961), 415–426.

[JoW] Jonsson, A. and Wallin, H., Function spaces on subsets of $\mathbb{R}^n$. Math. Reports 2,1 (1984), 1–221.

[Kah] Kahn, D.W., Introduction to Global Analysis, New York, Academic Press, 1980.

[Kal1] Kaljabin, G.A., Characterization of functions of spaces of Besov–Lizorkin–Triebel type (Russian). Dokl. Akad. Nauk SSSR *236* (1977), 1056–1059.

[Kal2] Kaljabin, G.A., The description of functions of classes of Besov–Lizorkin–Triebel type (Russian). Trudy Mat. Inst. Steklov *156* (1980), 82–109.

[Kal3] Kaljabin, G.A., Functional classes of Lizorkin–Triebel type in domains with Lipschitz boundary (Russian). Dokl. Akad. Nauk SSSR *271* (1983), 795–798.

[Kal4] Kaljabin, G.A., Theorems on extensions, multipliers and diffeomorphisms for generalized Sobolev–Liouville classes in domains with Lipschitz boundary (Russian). Trudy Mat. Inst. Steklov *172* (1985), 173–186.

[Kal5] Kaljabin, G.A., A characterization of certain function spaces by means of generalized differences (Russian). Dokl. Akad. Nauk SSSR *284* (1985), 1305–1308.

[Kal6] Kaljabin, G.A., Characterizations of spaces of Besov–Lizorkin–Triebel type by generalized differences (Russian). Trudy Mat. Inst. Steklov *181* (1988), 95–116.

[KaL] Kaljabin, G.A. and Lizorkin, P.I., Spaces of functions of generalized smoothness. Math. Nachr. *133* (1987), 7–32.

[KJF] Kufner, A., John, O., and Fučik, S., Function Spaces. Prague, Academia, 1977.

[Kli] Klingenberg, W., Riemannian Geometry. Berlin, W. de Gruyter, 1982.

[Kom] Komatsu, H., Fractional powers of operators, II. Interpolation spaces. Pacif. J. Math. *21* (1967), 89–111.

[KPS] Krejn, S.G., Petunin, Ju.I., and Semenov, E.M., Interpolation of Linear Operators. Providence, Amer. Math. Soc. 1982 (Translation from Russian, Moskva, Nauka, 1978).

[Kra] Krantz, S.G., Lipschitz spaces on stratified groups. Trans. Amer. Math. Soc. *269* (1982), 39–66.

[KuN1] Kudrjavcev, L.D. and Nikol'skij, S.M. Spaces of differentiable functions of several variables and embedding theorems (Russian). In: Itogi nauki i techniki, Ser. Sovremennye problemj mat. *26.* Moskva: akad. nauk SSSR 1988, 5–157. (English translation: In: Analysis III, Spaces of differentiable functions, Encyclopaedia of Math. Sciences 26. Heidelberg, Springer 1990, 4–140).

[KuN2] Kumano-go, H. and Nagase, M., Pseudodifferential operators with nonregular symbols and applications. Funkcial. Ekvac. *21* (1978), 151–192.

[Lat] Latter, R., A characterization of $H^p(\mathbb{R}^n)$ in terms of atoms. Studia Math. *62* (1978), 93–101.

[LiP] Lions, J.L. and Peetre, J., Sur une classe d'espaces d'interpolation. Inst. Hautes Études Sci. Publ. Math. *19* (1964), 5–68.

[LiY] Li, P. and Yau, S.T., On the parabolic kernel of the Schrödinger operator. Acta Math. *156* (1986), 153–201.

[Liz1] Lizorkin, P.I., A Littlewood–Paley theorem for multiple Fourier integrals (Russian). Trudy Mat. Inst. Steklov *89* (1967), 214–230.

[Liz2] Lizorkin, P.I., On Fourier multipliers in the spaces $L_{p,\theta}$ (Russian). Trudy Mat. Inst. Steklov *89* (1967), 231–248.

[Liz3] Lizorkin, P.I., Operators connected with fractional derivatives and classes of differentiable functions (Russian). Trudy Mat. Inst. Steklov *117* (1972), 212–243.

[Liz4] Lizorkin, P.I., Generalized Hölder classes of functions in connection with fractional derivatives (Russian). Trudy Mat. Inst. Steklov *128* (1972), 172–177.

[Liz5] Lizorkin, P.I., Properties of functions of the spaces $\Lambda^r_{p\theta}$ (Russian). Trudy Mat. Inst. Steklov *131* (1974), 158–181.

[Liz6] Lizorkin, P.I., Classes of functions constructed on the basis of spherical means (the case of spaces of Sobolev type) (Russian). Trudy Mat. Inst. Steklov *192* (1990), 122–139.

[Loh] Lohoué, N., Comparaison des champs de vecteurs et des puissances du Laplacien sur une variété Riemannienne à courbure non positive. J. Funct. Anal. *61* (1985), 164–201.

[Lor] Lorentz, G.G., Approximation of Functions. New York: Holt, Rinehart & Winston, 1966.

[Mal] Maligranda, L., A bibliography on "Interpolation of operators and applications" (1926–1988). Lulea, Dep. Math. Univ. 1988.

[Mar1] Marschall, J., Pseudo-differential operators with coefficients in Sobolev spaces. Trans. Amer. Math. Soc. *307* (1988), 335–361.

[Mar2] Marschall, J., Pseudo-differential operators with non-regular symbols in the class $S^m_{\rho\delta}$. Comm. Part. Diff. Equations *12* (1987), 921–965. Correction, ibidem *13* (1988), 129–130.

[Mar3] Marschall, J., Weighted parabolic Triebel spaces of product type. Fourier Multipliers and pseudo-differential operators. Forum Math. *3* (1991), 479–511.

[MaR] Madych, W.R. and Rivière, N.M., Multipliers of the Hölder classes. J. Funct. Anal. *21* (1976), 369–379.

[MaS] Maz'ya, V.G. and Shaposhnikova, T.O., Theory of Multipliers in Spaces of Differentiable Functions. Boston, Pitman 1985.

[Maz] Maz'ja, V.G., Sobolev Spaces. Berlin, Springer 1985 (Translation from Russian, Leningrad, Univ. 1985).

[Mey1] Meyers, G.N., Mean oscillation over cubes and Hölder continuity. Proc. Amer. Math. Soc. *15* (1964), 717–721.

[Mey2] Meyer, Y., Régularité des solutions des équations aux dérivées partielles non linéaires (d' après J.M. Bony). Sém. Bourbaki, 32e année, 1979/80, no. 560.

[Mey3] Meyer, Y., Remarques sur un théorème de J.M. Bony. Suppl. Rend. Circ. Mat. Palermo, Atti Sem. Analisi Armonica, Pisa 1980, Sér. II, n. 1, 1981.

[Mic1] Michlin, S.G., On multipliers of Fourier integrals (Russian). Dokl. Akad. Nauk SSSR *109* (1956), 701–703.

[Mic2] Michlin, S.G., Fourier integrals and multiple singular integrals (Russian). Vestnik Leningrad. Univ. Mat. Meh. Astronom. *7* (1957), 143–155.

[Mic3] Michel, D., Estimées des coefficients du Laplacien d'une variété Riemannienne. Bull. Sci. Math. *102* (1978), 15–41.

[Mir] Miranda, C., Partial Differential Equations of Elliptic Type. Sec. ed., Berlin, Springer, 1970.

[Miy1] Miyachi, A., Estimates of pseudo-differential operators with exotic symbols. J. Fac. Sci. Univ. Tokyo Sect. IA Math. *34* (1987), 81–100.

[Miy2] Miyachi, A., Estimates of pseudo-differential operators of class $S^m_{\rho,\delta}$ in L^p, h^p and *bmo*. In: Proc. Analysis Conf. Singapore 1986, 177–187. Amsterdam, North-Holland, Math. Stud. *150*, 1988.

[Miy3] Miyachi, A., Estimates for pseudo-differential operators of class $S_{0,0}$. Math. Nachr. *133* (1987), 135–154.

[MiY1] Miyachi, A. and Yabuta, K., L^p-boundedness of pseudo-differential operators with non-regular symbols. Bull. Fac. Sci. Ibaraki Univ. Ser. A *17* (1985), 1–20.

[MiY2] Miyachi, A. and Yabuta, K., Sharp function estimates for pseudo-differential operators of class $S^m_{\rho,\delta}$. Bull. Fac. Sci. Ibaraki Univ. *19* (1987), 17–30.

[Mor] Morrey, C.B., On the solutions of quasi linear elliptic partial differential equations. Trans. Amer. Math. Soc. *43* (1938), 126–166.

[MuN] Muramuta, T. and Nagase, M., On sufficient conditions for the boundedness of pseudo-differential operators. Proc. Japan Acad. Ser. A, Math. Sci. *55* (1979), 293–296.

[Mur] Muramuta, T., On imbedding theorems for Besov spaces of functions defined in general regions. Publ. Res. Inst. Math. Sci. Kyoto Univ. *7* (1971/72), 261–285.

[Nag1] Nagase, M., The L^p-boundedness of pseudo-differential operators with non-regular symbols. Comm. Part. Diff. Equ. *2* (1977), 1045–1061.

[Nag2] Nagase, M., On the boundedness of pseudo-differential operators in L^p-spaces. Sci. Rep. College Ed. Osaka Univ. *32* (1983), 9–19.

[Nag3] Nagase, M., On a class of L^p-bounded pseudo-differential operators. Sci. Rep. College Ed. Osaka Univ. *33* (1984), 1–7.

[Nag4] Nagase, M., On some classes of L^p-bounded pseudo-differential operators. Osaka J. Math. *23* (1986), 425–440.

[NaS1] Nagel, A. and Stein, E.M., A new class of pseudo-differential operators. Proc. Nat. Acad. Sci. U.S.A. *75* (1978), 582–585.

[NaS2] Nagel, A. and Stein, E.M., Lectures on Pseudo-Differential Operators (Math. Notes 24). Princeton, Univ. Press 1979.

[Net1] Netrusov, Ju.V., Embedding theorems of Besov spaces into ideal spaces (Russian). Zap. Naučn. Sem. Leningrad. Otdel. Mat. Inst. Steklov (LOMI) *159* (1987), 69–82.

[Net2] Netrusov, Ju.V., Embedding theorems for Lizorkin–Triebel spaces (Russian). Zap. Naučn. Sem. Leningrad. Otdel. Mat. Inst. Steklov (LOMI) *159* (1987), 103–112.

[Net3] Netrusov, Ju.V., Sets of singularities of functions of spaces of Besov and Lizorkin–Triebel type (Russian). Trudy Mat. Inst. Steklov *187* (1989), 162–177.

[Net4] Netrusov, Ju.V., Metric estimates of the capacities of sets in Besov spaces (Russian). Trudy Mat. Inst. Steklov *190* (1989), 159–185.

[Nik1] Nikol'skij, S.M., Inequalities for entire functions of finite order and their application in the theory of differentiable functions of several variables (Russian). Trudy Mat. Inst. Steklov *38* (1951), 244–278.

[Nik2] Nikol'skij, S.M., Approximation of Functions of Several Variables and Imbedding Theorems (Russian). Sec. ed., Moskva, Nauka 1977. English translation of the first ed., Berlin, Springer 1975.

[NLL] Nikolsky, S.M., Lions, J.L., and Lisorkin, P.I., Integral representation and isomorphism properties of some classes of functions, Ann. Scuola Norm. Sup. Pisa *19* (1965), 127–178.

[NSW] Nagel, A., Stein, E.M., and Wainger, S., Balls and metrics defined by vector fields, I: Basic properties. Acta Math. *155* (1985), 103–147.

[Ovc] Ovchinnikov, V.I., The method of orbits in interpolation theory. Math. Reports 1,2 (1984), 349–516.

[Päi1] Päivärinta, L., On the spaces $L_p^A(l_q)$: Maximal inequalities and complex interpolation. Ann. Acad. Sci. Fenn. Ser. AI Math. Dissertationes *25*, Helsinki 1980.

[Päi2] Päivärinta, L., Equivalent quasi-norms and Fourier multipliers in the Triebel spaces $F^s_{p,q}$. Math. Nachr. *106* (1982), 101–108.

[Päi3] Päivärinta, L., Pseudo differential operators in Hardy–Triebel spaces. Z. Analysis Anwendungen *2* (1983), 235–242.

[PäS] Päivärinta, L. and Somersalo, E., A generalization of Calderón–Vaillancourt theorem to L^p and h^p. Math. Nachr. *138* (1988), 145–156.

[Pee1] Peetre, J., Sur les espaces de Besov. C.R. Acad. Sci. Paris, Sér. A-B *264* (1967), 281–283.

[Pee2] Peetre, J., On the theory of $\mathcal{L}_{p,\lambda}$ spaces. J. Funct. Anal. *4* (1969), 71–87.

[Pee3] Peetre, J., Remarques sur les espaces de Besov. Le cas $0 < p < 1$. C.R. Acad. Sci. Paris, Sér. A-B *277* (1973), 947–950.

[Pee4] Peetre, J., H_p spaces. Lect. Notes, Lund 1974.

[Pee5] Peetre, J., On spaces of Triebel–Lizorkin type. Ark. Mat. *13* (1975), 123–130.

[Pee6] Peetre, J., New Thoughts on Besov Spaces. Duke Univ. Math. Series. Durham, Univ. 1976.

[Pee7] Peetre, J., The trace of Besov space — a limiting case. Technical Report, Lund 1975.

[Per] Peral, J.C., L^p estimates for the wave equation. In: Proc. Symp. "Harmonic Analysis in Euclidean Spaces," Providence, Amer. Math. Soc. 1979, 171–174.

[Pes1] Pesenson, J.Z., On interpolation spaces on Lie groups (Russian). Dokl. Akad. Nauk SSSR *246* (1979), 1298–1303.

[Pes2] Pesenson, J.Z., Nikol'skij–Besov spaces connected with representations of Lie groups (Russian). Dokl. Akad. Nauk SSSR *273* (1983), 45–49.

[Pes3] Pesenson, J.Z., On the abstract theory of Nikol'skij–Besov spaces (Russian). Izv. Vysš. Učebn. Zaved. Matematika *6* (1988), 59–68.

[Pes4] Pesenson, J.Z., On best approximations in spaces of Lie group representations (Russian). Dokl. Akad. Nauk SSSR *302* (1988), 1055–1058.

[Pet] Petrow, A.S., Einstein–Räume. Berlin, Akademie-Verl. 1964.

[PlP] Plancherel, M. and Pólya, G., Fonctions entières et intégrales de Fourier multiples. Math. Helv. *9* (1937), 224–248.

[Pri] Price, J.F., Lie Groups and Compact Groups. Cambridge, Univ. Press, 1979.

[Qin1] Qing-jiu, Q., The Besov space boundedness for certain Fourier integral operators. Acta Math. Scientia *5* (1985), 167–174.

[Qin2] Qing-jiu, Q., On L_p-estimates for certain Fourier integral operators. Scientia Sinica, Ser. A, *29* (1986), 350–362.

[ReS] Rempel, S. and Schulze, B.-W., Index Theory of Elliptic Boundary Problems. Berlin, Akademie-Verl. 1982.

[Rie] Riesz, F., Über die Randwerte einer analytischen Funktion. Math. Z. *18* (1923), 87–95.

[RiT] Ricci, F. and Taibleson, M., Boundary values of harmonic functions in mixed norm spaces and their atomic structure. Ann. Scuola Norm. Sup. Pisa Cl. Sci. (4) *10* (1983), 1–54.

[Riv] Rivière, N.M., Classes of smoothness, the Fourier method. Unpublished Lect. Notes.

[Rob1] Robinson, D.W., Lie groups and Lipschitz spaces. Duke Math. J. *57* (1988), 357–395.

[Rob2] Robinson, D.W., Lie algebras, the heat semigroups, and Lipschitz spaces. Res. Report *38*, 1987, Dep. Math., Australian Nat. Univ.

[RoC] Rouhuai, W. and Chengzhang, L., On the L^p-boundedness of several classes of pseudo-differential operators. Chin. Ann. of Math. *5* (1984), 193–213.

[Run1] Runst, T., Pseudo-differential operators of the "exotic" class $L^0_{1,1}$ in spaces of Besov and Triebel–Lizorkin type. Annals Global Analysis Geometry *3* (1985), 13–28.

[Run2] Runst, T., Solvability of semilinear elliptic boundary value problems in Besov–Triebel–Lizorkin spaces. In: "Surveys on Analysis, Geometry and Mathematical Physics," Teubner-Texte Math. *117*. Leipzig, Teubner 1990, 198–291.

[RVVW] Rochberg, R., Tabacco Vignati, A., Vignati, M. and Weiss, G., Interpolation of quasinormed spaces by the complex method. In: "Function Spaces and Applications," Proc. Conf. Lund 1986, Lect. Notes Math. 1302. Berlin, Springer 1988, 91–98.

[Sak1] Saka, K., Besov spaces and Sobolev spaces on a nilpotent Lie group. Tôhoku Math. J. *31* (1979), 383–437.

[Sak2] Saka, K., Besov spaces on Riemannian manifolds and its application to L_p-L_q-estimates for wave equations. Mem. Coll. Education Akita Univ., Nat. Sc. *39* (1988), 81–86.

[San] Sands, S.E., A representation theorem for temperatures on mixed norm spaces. PhD. Thesis, Univ. of Maryland, 1981.

[ScT] Schmeisser, H.-J. and Triebel, H., Topics in Fourier Analysis and Function Spaces. Leipzig, Geest & Portig 1987, Chichester, Wiley 1987.

[See1] Seeger, A., A note on Triebel–Lizorkin spaces. In: "Approximation and Function Spaces," Banach Centre Publ. *22*, Warsaw, PWN Polish Sci. Publ. 1989, 391–400.

[See2] Seeger, A., Remarks on singular convolution operators. Studia Math. *97* (1990), 91–114.

[SeS] Seeger, S. and Sogge, C.D., On the boundedness of functions of (pseudo-) differential operators on compact manifolds. Duke Math. J. *59* (1989), 709–736.

[Sha1] Shapiro, H.S., A Tauberian theorem related to approximation theory. Acta Math. *120* (1968), 279–292.

[Sha2] Shapiro, H.S., Topics in Approximation Theory. Lect. Notes Math. *187*, Berlin, Springer 1971.

[Sic1] Sickel, W., On pointwise multipliers in Besov–Triebel–Lizorkin spaces. In: "Seminar Analysis Karl–Weierstrass Institute 1985/86," Teubner-Texte Math. *96*. Leipzig, Teubner 1987, 45–103.

[Sic2] Sickel, W., Spline representations of functions in Besov–Triebel–Lizorkin spaces on R^n. Forum Math. *2* (1990), 451–475.

[Sjö] Sjölin, P., An H^p inequality for strongly singular integrals. Math. Z. *165* (1979), 231–238.

[Skr1] Skrzypczak, L., Function spaces of Sobolev type on Riemannian symmetric manifolds. Forum Math. *3* (1991), 339–353.

[Skr2] Skrzypczak, L., Anisotropic Sobolev spaces on Riemannian symmetric manifolds. I. Function spaces, Proc. Conf. Poznán 1989, Teubner-Texte Math. *120*. Leipzig: Teubner 1991, 252–264. II. Comment. Math. Prace Mat.

[Slo] Slobodeckij, L.N., Generalized Sobolev spaces and their applications to boundary value problems of partial differential equations (Russian). Leningrad. Gos. Ped. Inst. Učep. Zap. *197* (1958), 54–112.

[Sob1] Sobolev, S.L., The Cauchy problem in a function space (Russian). Dokl. Akad. Nauk SSSR *3* (1935), 291–294.

[Sob2] Sobolev, S.L., Méthode nouvelle à resoudre le problème de Cauchy pour les équations linéaires hyperboliques normales. Mat. Sb. *1* (1936), 39–72.

[Sob3] Sobolev, S.L., On a theorem of functional analysis (Russian). Mat. Sb. *4* (1938), 471–497.

[Sob4] Sobolev, S.L., Einige Anwendungen der Funktionalanalysis auf Gleichungen der mathematischen Physik. Berlin, Akademie Verlag 1964. (Translation from Russian, Leningrad 1950).

[Spa] Spanne, S., Sur l'interpolation entre les espaces $\mathcal{L}_k^{p\Phi}$. Ann. Scuola Norm. Pisa *20* (1966), 625–648.

[Sta1] Stampacchia, G., $\mathcal{L}^{(p,\lambda)}$-spaces and interpolation. Comm. Pure Appl. Math. *17* (1964), 293–306.

[Sta2] Stampacchia, G., The spaces $\mathcal{L}^{(p,\lambda)}$, $N^{(p,\lambda)}$ and interpolation. Ann. Scuola Norm. Sup. Pisa *19* (1965), 443–462.

[Ste1] Stein, E.M., Singular Integrals and Differentiability Properties of Functions. Princeton, Univ. Press 1970.

[Ste2] Stein, E.M., Topics in Harmonic Analysis, Related to the Littlewood–Paley Theory. Princeton, Univ. Press 1970.

[Str1] Strichartz, R.S., Multipliers on fractional Sobolev spaces, J. Math. Mechanics *16* (1967), 1031–1060.

[Str2] Strichartz, R.S., Analysis of the Laplacian on the complete Riemannian manifold. J. Funct. Anal. *52* (1983), 48–79.

[StW1] Stein, E.M. and Weiss, G., On the theory of harmonic functions of several variables, I. The theory of H^p spaces. Acta Math. *103* (1960), 25–62.

[StW2] Stein, E.M. and Weiss, G., Introduction to Fourier Analysis on Euclidean Spaces. Princeton, Univ. Press, 1971.

[Sug1] Sugimoto, M., L^p-boundedness of pseudo-differential operators satisfying Besov estimates. I. J. Math. Soc. Japan *40* (1988), 105–122. II. J. Fac. Sci. Univ. Tokyo Sect. IA Math. *35* (1988), 149–162.

[Sug2] Sugimoto, M., Pseudo-differential operators on Besov spaces. Tsukuba J. Math. *12* (1988), 43–63.

[Tai] Taibleson, M.H., On the theory of Lipschitz spaces of distributions on Euclidean n-space, I,II. J. Math. Mech. *13* (1964), 407–479, *14* (1965), 821–839.

[TaW] Taibleson, M.H. and Weiss, G., The molecular characterization of certain Hardy spaces. Astérisque *77* (1980), 67–149.

[Tay] Taylor, M.E., Pseudodifferential Operators. Princeton, Univ. Press 1981.

[Tor] Torres, R.H., Continuity properties of pseudodifferential operators of type 1,1. Comm. Partial Differential Equations *15* (1990), 1313–1328.

[Tre] Treves, F., Introduction to Pseudodifferential and Fourier Integral Operators 1,2. New York, Plenum Press 1980.

[Triα] Triebel, H., Interpolation Theory, Function Spaces, Differential Operators. Berlin, VEB Deutsch. Verl. Wissenschaften 1978, Amsterdam, North-Holland 1978.

[Triß] Triebel, H., Theory of Function Spaces. Leipzig, Geest & Portig 1983, Basel, Birkhäuser 1983.

[Tri1] Triebel, H., Über die Existenz von Schauderbasen in Sobolev–Besov–Räumen. Isomorphiebeziehungen, Studia Math. *46* (1973), 83–100.

[Tri2] Triebel, H., Spaces of distributions of Besov type on Euclidean n-space. Duality, interpolation. Ark. Mat. *11* (1973), 13–64.

[Tri3] Triebel, H., A remark on embedding theorems for Banach spaces of distributions. Ark. Mat. *11* (1973), 65–74.

[Tri4] Triebel, H., Fourier Analysis and Function Spaces. Teubner-Texte Math. *7* Leipzig, Teubner 1977.

[Tri5] Triebel, H., Spaces of Besov–Hardy–Sobolev Type. Teubner-Texte Math. *15*, Leipzig, Teubner 1978.

[Tri6] Triebel, H., On spaces of $B^s_{\infty,q}$ type and C^s type. Math. Nachr. *85* (1978), 75–90.

[Tri7] Triebel, H., Strongly degenerate elliptic differential operators in Besov spaces: the case $0 < p < \infty$. Beiträge Anal. *13* (1979), 27–47.

[Tri8] Triebel, H., Theorems of Littlewood–Paley type for BMO and for anisotropic Hardy spaces. In: "Constructive Function Theory '77," Proc. Sofia, Bulg. Acad. Publ. House 1980, 525–532.

[Tri9] Triebel, H., On the spaces $F^s_{p,q}$ of Hardy–Sobolev type: Equivalent quasinorms, multipliers, spaces on domains, regular elliptic boundary value problems. Comm. Part. Diff. Equations *5* (1980), 245–291.

[Tri10] Triebel, H., On $L_p(l_q)$-spaces of entire analytic functions of exponential type: Complex interpolation and Fourier multipliers. The case $0 < p < \infty$, $0 < q < \infty$. J. Approximation Theory *28* (1980), 317–328.

[Tri11] Triebel, H., Complex interpolation and Fourier multipliers for the spaces $B^s_{p,q}$ and $F^s_{p,q}$ of Besov–Hardy–Sobolev type: The case $0 < p \leqq \infty$, $0 < q \leqq \infty$. Math. Z. *176* (1981), 495–510.

[Tri12] Triebel, H., Characterizations of Besov–Hardy–Sobolev spaces via harmonic functions, temperatures and related means. J. Approximation Theory *35* (1982), 275–297.

[Tri13] Triebel, H., Spaces of Besov–Hardy–Sobolev type on complete Riemannian manifolds. Ark. Mat. *24* (1986), 299–337.

[Tri14] Triebel, H., Characterizations of function spaces on a complete Riemannian manifold with bounded geometry. Math. Nach. *130* (1987), 312–346.

[Tri15] Triebel, H., Function spaces on Lie groups, the Riemannian approach. J. London Math. Soc. *35* (1987), 327–338.

[Tri16] Triebel, H., Characterizations of Besov–Hardy–Sobolev spaces: a unified approach. J. Approximation Theory *52* (1988), 162–203.

[Tri17] Triebel, H., Function spaces on Lie groups and on analytic manifolds. In: "Function Spaces and Applications," Proc. Conf. Lund 1986, Lect. Notes Math. 1302. Berlin, Springer 1988, 384–396.

[Tri18] Triebel, H., Diffeomorphism properties and pointwise multipliers for function spaces. In "Function Spaces," Proc. Conf. Poznań 1986. Teubner-Texte Math. *103*. Leipzig, Teubner 1988, 75–84.

[Tri19] Triebel, H., Atomic representations of F^s_{pq} spaces and Fourier integral operators. In: "Seminar Analysis Karl–Weierstrass Institute 1986/87," Teubner-Texte Math. *106*. Leipzig, Teubner 1988, 297–305.

[Tri20] Triebel, H., How to measure smoothness of distributions on Riemannian symmetric manifolds and Lie groups? Z. Analysis Anwendungen *7* (1988), 471–480.

[Tri21] Triebel, H., Inequalities in the theory of function spaces: A tribute to Hardy, Littlewood and Polya. In: "Inequalities," Proc. Conf. Birmingham (U.K.). Lecture Notes Pure Appl. Math. *119*. New York, Marcel Dekker 1991, 231–248.

[Tri22] Triebel, H., Atomic decompositions of F^s_{pq} spaces. Applications to exotic pseudodifferential and Fourier integral operators. Math. Nachr. *144* (1989), 189–222.

[Tri23] Triebel, H., Local approximation spaces. Z. Analysis Anwendungen *8* (1989), 261–288.

[Tri24] Triebel, H., Characterizations of F^s_{pq} spaces via local means, the extension problem (Russian). Trudy Mat. Inst. Steklov *192* (1990), 207–220.

[Tri25] Triebel, H., On Besov–Hardy–Sobolev spaces in domains and regular elliptic boundary value problems. The case $0 < p \leqq \infty$. Comm. Partial Differential Equations *3* (1978), 1083–1164.

[Tri26] Triebel, H., Pseudo-differential operators in F^s_{pq}-spaces. Z. Analysis Anwendungen *6* (1987), 143–150.

[Tri27] Triebel, H., On a class of weighted function spaces and related pseudodifferential operators. Studia Math. *90* (1988), 37–68.

[Vig] Vignati, A., Tabacco. Complex interpolation for families of quasi-Banach spaces. Indiana Univ. Math. J. *37* (1988), 1–21.

[Wal] Wallin, H., New and old function spaces. In: "Function Spaces and Applications," Proc. Conf. Lund 1986. Lect. Notes Math. 1302. Berlin, Springer 1988, 97–114.

[Wie] Wiener, N., The ergodic theorem. Duke Math. J. *5* (1939), 1–18.

[Wil] Wilson, J.M., On the atomic decomposition for Hardy spaces. Pacific J. Math. *116* (1985), 201–207.

[Yab1] Yabuta, K., Singular integral operators on Besov spaces. Math. Nachr. *136* (1988), 163–175.

[Yab2] Yabuta, K., Singular integral operators on Triebel–Lizorkin spaces. Bull. Fac. Sci. Ibaraki Univ. Ser. A *20* (1988), 9–17.

[Yam1] Yamazaki, M., The L^p-boundedness of pseudo-differential operators satisfying estimates of parabolic type and product type. Proc. Japan Acad. Ser. A, Math. Sci. *60* (1984), 279–282, II. Ibidem *61* (1985), 95–98.

[Yam2] Yamazaki, M., Continuité des opérateurs pseudo-différentiels et para-différentiels dans les espaces de Besov et les espaces de Triebel–Lizorkin non-isotropes. C.R. Acad. Paris (Sér. I) *296* (1983), 533–536.

[Yam3] Yamazaki, M., A quasi-homogeneous version of paradifferential operators, I: Boundedness on spaces of Besov type. J. Fac. Sci. Univ. Tokyo, Sect. IA Math. *33* (1986), 131–174. II: A symbolic calculus. Ibidem *33* (1986), 311–345.

[Yam4] Yamazaki, M., The L^p-boundedness of pseudodifferential operators with estimates of parabolic type and product type. J. Math. Soc. Japan *38* (1986), 199–225.

[Yos] Yosida, K., Functional Analysis. Berlin, Springer 1965.

[You1] Youssfi, A., Localisation des espaces de Besov et de Lizorkin–Triebel. Thesis, Univ. Paris VII, 1988.

[You2] Youssfi, A., Localisation des espaces de Lizorkin–Triebel homogènes. Math. Nachr. *147* (1990), 107–121.

[Zyg1] Zygmund, A., Smooth functions. Duke Math. J. *12* (1945), 47–76.

[Zyg2] Zygmund, A., Trigonometric Series, I,II. Cambridge Univ. Press 1977 (First ed., Warsaw 1935).

Symbols

Index

GPSR Compliance

The European Union's (EU) General Product Safety Regulation (GPSR) is a set of rules that requires consumer products to be safe and our obligations to ensure this.

If you have any concerns about our products, you can contact us on ProductSafety@springernature.com

In case Publisher is established outside the EU, the EU authorized representative is:

Springer Nature Customer Service Center GmbH
Europaplatz 3
69115 Heidelberg, Germany

Batch number: 09640028

Printed by Printforce, the Netherlands